AF803132

Principles of Naval Architecture Second Revision

Volume II • Resistance, Propulsion and Vibration

Edward V. Lewis, Editor

Published by
The Society of Naval Architects and Marine Engineers
601 Pavonia Avenue
Jersey City, NJ

Copyright © 1988 by The Society of Naval Architects and Marine Engineers.

It is understood and agreed that nothing expressed herein is intended or shall be construed to give any person, firm, or corporation any right, remedy, or claim against SNAME or any of its officers or members.

Library of Congress Catalog Card No. 88-60829
ISBN No. 0-939773-01-5
Printed in the United States of America
First Printing, November, 1988

Preface

The aim of this second revision (third edition) of the Society's successful *Principles of Naval Architecture* was to bring the subject matter up-to-date through revising or rewriting areas of greatest recent technical advances, which meant that some chapters would require many more changes than others. The basic objective of the book, however, remained unchanged: to provide a timely survey of the basic principles in the field of naval architecture for the use of both students and active professionals, making clear that research and engineering are continuing in almost all branches of the subject. References are to be included to available sources of additional details and to ongoing work to be followed in the future.

The preparation of this third edition was simplified by an earlier decision to incorporate a number of sections into the companion SNAME publication, *Ship Design and Construction,* which was revised in 1980. The topics of Load Lines, Tonnage Admeasurement and Launching seemed to be more appropriate for the latter book, and so Chapters V, VI, and XI became IV, V and XVII respectively, in *Ship Design and Construction.* This left eight chapters, instead of 11, for the revised *Principles of Naval Architecture,* which has since become nine in three volumes.

At the outset of work on the revision, the Control Committee decided that the increasing importance of high-speed computers demanded that their use be discussed in the individual chapters instead of in a separate appendix as before. It was also decided that throughout the book more attention should be given to the rapidly developing advanced marine vehicles.

In regard to units of measure, it was decided that the basic policy would be to use the International System of Units (S.I.). Since this is a transition period, conventional U.S. (or "English") units would be given in parentheses, where practical, throughout the book. This follows the practice adopted for the Society's companion volume, *Ship Design and Construction.* The U.S. Metric Conversion Act of 1975 (P.L. 94-168) declared a national policy of increasing the use of metric systems of measurement and established the U.S. Metric Board to coordinate voluntary conversion to S.I. The Maritime Administration, assisted by a SNAME ad hoc task group, developed a *Metric Practice Guide* to "help obtain uniform metric practice in the marine industry," and this guide was used here as a basic reference. Following this guide, ship displacement in metric tons (1000 kg) represents mass rather than weight, (In this book the familiar symbol, Δ, is reserved for the displacement mass). When forces are considered, the corresponding unit is the kilonewton (kN), which applies, for example, to resistance and to displacement weight (symbol W, where $W = \rho\Delta g$) or to buoyancy forces. When conventional or English units are used, displacement weight is in the familiar long ton unit (2240

(Continued)

lb), which numerically is 1.015 × metric ton. Power is usually in kilowatts (1 kW = 1.34 hp). A conversion table also is included in the Nomenclature at the end of each volume

The first volume of the third edition of *Principles of Naval Architecture*, comprising Chapters I through IV, covers almost the same subject matter as the first four chapters of the preceding edition. Thus, it deals with the essentially static principles of naval architecture, leaving dynamic aspects to the remaining volumes. Chapter I deals with the graphical and numerical description of hull forms and the calculations needed to deal with problems of flotation and stability that follow. Chapter II considers stability in normal intact conditions, while Chapter III discusses flotation and stability in damaged conditions. Finally, Chapter IV deals with principles of hull structural design, first under static calm water conditions, and then introducing the effect of waves which also are covered more fully in Volume III Chapter VIII, Motions in Waves.

For Volume II it seemed desirable, on the basis of subject matter and space requirements, to include Chapter V, Resistance, Chapter VI, Propulsion and Chapter VII, Vibration. The first two of these were covered in a single chapter in the preceding edition. The new chapters have been extensively revised, with considerable new material, particularly dealing with high performance craft and new propulsion devices. Chapter VII, Vibration, which is the third in Volume II, has been almost completely rewritten to take advantage of new developments in the field.

May 1988

EDWARD V. LEWIS
Editor

Table of Contents
Volume II

Acknowledgments

The authors of Chapters V and VI, J.D. van Manen and P. van Oossanen, wish to acknowledge their indebtedness to the author of Chapter V in the preceding edition, Frederick H. Todd. Extensive use has been made of the original text and figures. The authors also wish to recognize the assistance provided by U. Nienhuis of the Maritime Institute Netherlands in working through the entire text a second time, making additions and corrections whenever necessary. And valuable ideas and suggestions regarding high-speed displacement and planing hulls in Section 9 of Chapter V were provided by Daniel Savitsky, Director of the Davidson Laboratory and are acknowledged with thanks.

The author of Chapter VII, William S. Vorus, expresses his appreciation of the pioneering work of Frank M. Lewis, as distilled in Chapter X of the preceding edition of this book, which provided a foundation for the new chapter. He appreciates the reveiw and comments on early drafts by Edward F. Noonan, of NFK Engineering Associates, Inc., and John P. Breslin of Stevens Institute of Technology.

The Control Committee provided essential guidance, as well as valuable assistance in the early stages. Members are:

John J. Nachtsheim, Chairman
Thomas M. Buermann
William A. Cleary, Jr.
Richard B. Couch
Jerome L. Goldman
Jacques B. Hadler
Ronald K. Kiss
Donald P. Roseman
Stanley G. Stiansen
Charles Zeien

Finally, the Editor wishes to thank all of the authors for their fine work and for their full cooperation in making suggested revisions. He acknowledges the indispensible efforts of Trevor Lewis-Jones in doing detailed editing and preparing text and figures in proper format for publication.

May 1988

E. V. Lewis
Editor

CHAPTER V

J. D. van Manen
P. van Oossanen

Resistance

Section 1
Introduction

1.1 The Problem. A ship differs from any other large engineering structure in that—in addition to all its other functions—it must be designed to move efficiently through the water with a minimun of external assistance. In Chapters I-III of Vol. I it has been shown how the naval architect can ensure adequate buoyancy and stability for a ship, even if damaged by collision, grounding, or other cause. In Chapter IV the problem of providing adequate structure for the support of the ship and its contents, both in calm water and rough seas, was discussed.

In this chapter we are concerned with how to make it possible for a structure displacing up to 500,000 tonnes or more to move efficiently across any of the world's oceans in both good and bad weather. The problem of moving the ship involves the proportions and shape—or form—of the hull, the size and type of propulsion plant to provide motive power, and the device or system to transform the power into effective thrust. The design of power plants is beyond the scope of this[1] book (see *Marine Engineering,* by R.L. Harrington, Ed., SNAME 1971). The nine sections of this chapter will deal in some detail with the relationship between hull form and resistance to forward motion (or drag). Chapter VI discusses propulsion devices and their interaction with flow around the hull.

The task of the naval architect is to ensure that, within the limits of other design requirements, the hull form and propulsion arrangement will be the most efficient in the hydrodynamic sense. The ultimate test is that the ship shall perform at the required speed with the minimum of shaft power, and the problem is to attain the best combination of low resistance and high propulsive efficiency. In general this can only be attained by a proper matching of hull and propeller.

Another factor that influences the hydrodynamic design of a ship is the need to ensure not only good smooth-water performance but also that under average service conditions at sea the ship shall not suffer from excessive motions, wetness of decks, or lose more speed than necessary in bad weather. The assumption that a hull form that is optimum in calm water will also be optimum in rough seas is not necessarily valid. Recent research progress in oceanography and the seakeeping qualities of ships has made it possible to predict the relative performance of designs of varying hull proportions and form under different realistic sea conditions, using both model test and computing techniques. The problem of ship motions, attainable speed and added power requirements in waves are discussed in Chapter VIII, Vol. III. This chapter is concerned essentially with designing for good smooth-water performance.

Another consideration in powering is the effect of deterioration in hull surface condition in service as the result of fouling and corrosion and of propeller roughness on resistance and propulsion. This subject is discussed in this chapter.

As in the case of stability, subdivision, and structure, criteria are needed in design for determining acceptable levels of powering. In general, the basic contractual obligation laid on the shipbuilder is that the ship shall achieve a certain speed with a specified power in good weather on trial, and for this reason smooth-water performance is of great importance. As previously noted, good sea performance, particularly the maintenance of sea speed, is often a more important requirement, but one that is much more difficult to define. The effect of sea condition is customarily allowed for by the provision of a service power margin above the power required in smooth water, an allowance which depends on the type of ship and the average weather on the sea routes on which the ship is designed to operate. The determination of this service allowance depends on the accumulation of sea-performance data on similar ships in similar trades. Powering criteria in the form of conventional service allowances for both

[1] Complete references are listed at end of chapter.

sea conditions and surface deterioration are considered in this chapter.

The problem of controlling and maneuvering the ship will be covered in Chapter IX, Vol. III.

1.2 Types of Resistance. The resistance of a ship at a given speed is the force required to tow the ship at that speed in smooth water, assuming no interference from the towing ship. If the hull has no appendages, this is called the bare-hull resistance. The power necessary to overcome this resistance is called the towrope or effective power and is given by

$$P_E = R_T V \tag{1a}$$

where P_E = effective power in kWatt (kW)
R_T = total resistance in kNewton (kN)
V = speed in m/sec

$$\text{or ehp} = R_T V_k / 326 \tag{1b}$$

where ehp = effective power in English horsepower
R_T = total resistance in lb
V_k = speed in knots

To convert from horsepower to S.I. units there is only a slight difference between English and metric horsepower:

hp (English) $\times$ 0.746 = kW
hp (metric) $\times$ 0.735 = kW
Speed in knots $\times$ 0.5144 = m/sec

This total resistance is made up of a number of different components, which are caused by a variety of factors and which interact one with the other in an extremely complicated way. In order to deal with the question more simply, it is usual to consider the total calm water resistance as being made up of four main components.

(a) The frictional resistance, due to the motion of the hull through a viscous fluid.

(b) The wave-making resistance, due to the energy that must be supplied continuously by the ship to the wave system created on the surface of the water.

(c) Eddy resistance, due to the energy carried away by eddies shed from the hull or appendages. Local eddying will occur behind appendages such as bossings, shafts and shaft struts, and from stern frames and rudders if these items are not properly streamlined and aligned with the flow. Also, if the after end of the ship is too blunt, the water may be unable to follow the curvature and will break away from the hull, again giving rise to eddies and separation resistance.

(d) Air resistance experienced by the above-water part of the main hull and the superstructures due to the motion of the ship through the air.

The resistances under *(b)* and *(c)* are commonly taken together under the name residuary resistance. Further analysis of the resistance has led to the identification of other sub-components, as discussed subsequently.

The importance of the different components depends upon the particular conditions of a design, and much of the skill of naval architects lies in their ability to choose the shape and proportions of hull which will result in a combination leading to the minimum total power, compatible with other design constraints.

In this task, knowledge derived from resistance and propulsion tests on small-scale models in a model basin or towing tank will be used. The details of such tests, and the way the results are applied to the ship will be described in a later section. Much of our knowledge of ship resistance has been learned from such tests, and it is virtually impossible to discuss the various types of ship resistance without reference to model work.

1.3 Submerged Bodies. A streamlined body moving in a straight horizontal line at constant speed, deeply immersed in an unlimited ocean, presents the simplest case of resistance. Since there is no free surface, there is no wave formation and therefore no wave-making resistance. If in addition the fluid is assumed to be without viscosity (a "perfect" fluid), there will be no frictional or eddymaking resistance. The pressure distribution around such a body can be determined theoretically from considerations of the potential flow and has the general characteristics shown in Fig. 1(*a*).

Near the nose, the pressure is increased above the hydrostatic pressure, along the middle of the body the pressure is decreased below it and at the stern it is again increased. The velocity distribution past the hull, by Bernoulli's Law, will be the inverse of the pressure distribution—along the midportion it will be greater than the speed of advance V and in the region of bow and stern it will be less.

Since the fluid has been assumed to be without viscosity, the pressure forces will everywhere be normal to the hull (Fig. 1(*b*)). Over the forward part of the hull, these will have components acting towards the stern and therefore resisting the motion. Over the after part, the reverse is the case, and these components are assisting the motion. It can be shown that the resultant total forces on the fore and after bodies are equal, and the body therefore experiences no resistance.[2]

In a real fluid the boundary layer alters the virtual shape and length of the stern, the pressure distribution there is changed and its forward component is reduced. The pressure distribution over the forward portion is but little changed from that in a perfect fluid. There is therefore a net force on the body acting against the motion, giving rise to a resistance which is variously referred to as form drag or viscous pressure drag.

In a real fluid, too, the body experiences frictional resistance and perhaps eddy resistance also. The fluid immediately in contact with the surface of the body is

[2] This was first noted by the French mathematician d'Alembert in 1744, and is known as d'Alembert's paradox.

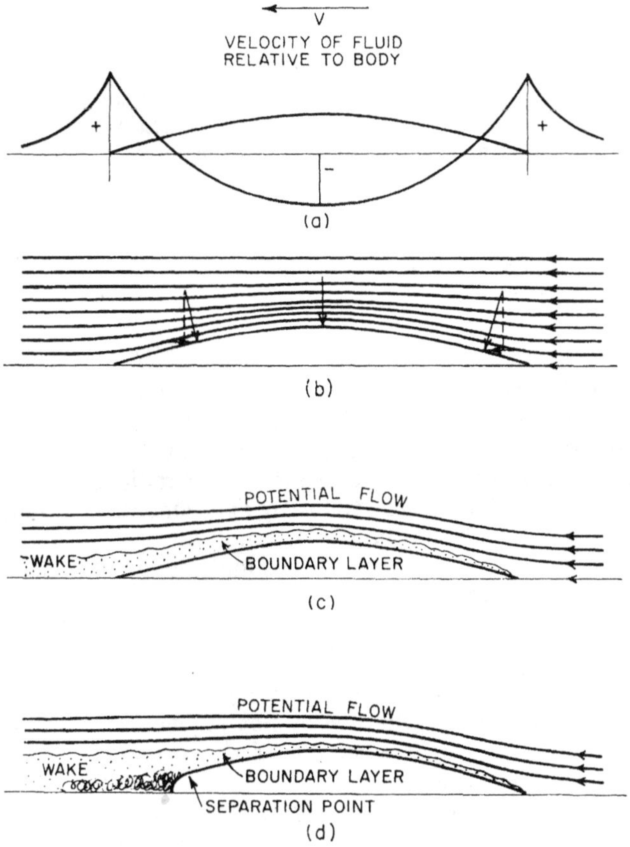

Fig. 1 Examples of flow about a submerged body

carried along with the surface, and that in the close vicinity is set in motion in the same direction as that in which the body is moving. This results in a layer of water, which gets gradually thicker from the bow to the stern, and in which the velocity varies from that of the body at its surface to that appropriate to the potential flow pattern (almost zero for a slender body) at the outer edge of the layer (Fig. 1(*c*)). This layer is called the boundary layer, and the momentum supplied to the water in it by the hull is a measure of the frictional resistance. Since the body leaves behind it a frictional wake moving in the same direction as the body (which can be detected far astern) and is continually entering undisturbed water and accelerating it to maintain the boundary layer, this represents a continual drain of energy. Indeed, in wind-tunnel work the measurement of the velocities of the fluid behind a streamlined model is a common means of measuring the frictional drag.

If the body is rather blunt at the after end, the flow may leave the form at some point—called a *separation point*—thus reducing the total pressure on the afterbody and adding to the resistance. This *separation resistance* is evidenced by a pattern of eddies which is a drain of energy (Fig. 1(*d*)).

1.4 Surface Ships. A ship moving on the surface of the sea experiences frictional resistance and eddy-making, separation, and viscous pressure drag in the same way as does the submerged body. However, the presence of the free surface adds a further component. The movement of the hull through the water creates a pressure distribution similar to that around the submerged body; i.e., areas of increased pressure at bow and stern and of decreased pressure over the middle part of the length.

But there are important differences in the pressure distribution over the hull of a surface ship because of the surface wave disturbance created by the ship's forward motion. There is a greater pressure acting over the bow, as indicated by the usually prominent bow wave build-up, and the pressure increase at the stern, in and just below the free surface, is always less than around a submerged body. The resulting added resistance corresponds to the drain of energy into the wave system, which spreads out astern of the ship and has to be continuously recreated. (See Section 4.3). Hence, it has been called wave-making resistance. The result of the interference of the wave systems originating at bow, shoulders (if any) and stern is to produce a series of divergent waves spreading outwards from the ship at a relatively sharp angle to the centerline and a series of transverse waves along the hull on each side and behind in the wake (Fig. 7).

The presence of the wave systems modifies the skin friction and other resistances, and there is a very complicated interaction among all the different components.

Section 2
Dimensional Analysis

2.1 General. Dimensional analysis is essentially a means of utilizing a partial knowledge of a problem when the details are too obscure to permit an exact analysis. See Taylor, E. S. (1974). It has the enormous advantage of requiring for its application a knowledge only of the variables which govern the result. To apply it to the flow around ships and the corresponding resistance, it is necessary to know only upon what variables the latter depends. This makes it a powerful tool, because the correctness of a dimensional solution does not depend upon the soundness of detailed analyses, but only upon the choice of the basic variables. Dimensional solutions do not yield numerical answers, but they provide the form of the answer so that every experiment can be used to the fullest advantage in determining a general empirical solution.

2.2 Dimensional Homogeneity. Dimensional analysis rests on the basic principle that every equation which expresses a physical relationship must be dimensionally homogeneous. There are three basic quantities in mechanics—mass, length and time—which are represented by the symbols M, L, and T. Other quantities, such as force, density, and pressure, have dimensions made up from these three basic ones.

Velocity is found by dividing a length or distance by a time, and so has the dimensions L/T. Acceleration, which is the change in velocity in a certain time, thus has dimensions of $(L/T)/T$, or L/T^2.

Force, which is the product of mass and acceleration, has dimensions of $M \times L/T^2$ or ML/T^2.

As a simple case to illustrate the principle of dimensional analysis, suppose we wish to determine an expression for the time of swing of a simple pendulum.

If T is the period of such a pendulum in vacuo (so that there is no frictional damping), it could depend upon certain physical quantities such as the mass of the pendulum bob, m, the length of the cord, l, (supposed to be weightless) and the arc of swing, s. The force which operates to restore the pendulum to its original position when it is disturbed is its weight, mg, and so the acceleration due to gravity, g, must be involved in the problem.

We can write this in symbols as

$$T = f(m, l, s, g)$$

where f is a symbol meaning "is some function of."

If we assume that this function takes the form of a power law, then

$$T = m^a\, l^b\, s^c\, g^d$$

If this equation is to fulfill the principle of dimensional homogeneity, then the dimensions on each side must be the same. Since the left-hand side has the dimension of time only, so must the right-hand side.

Writing the variables in terms of the fundamental units, we have

$$T^1 = M^a L^b L^c\, (L/T^2)^d$$

Equating the exponents of each unit from each side of the equation, we have

$$\begin{aligned} a &= 0 \\ b + c + d &= 0 \\ -2d &= 1 \end{aligned}$$

Hence

$$\begin{aligned} d &= -1/2 \\ a &= 0 \\ b + c &= 1/2 \end{aligned}$$

The expression for the period of oscillation T seconds is therefore

$$T = \text{constant} \times l^{1/2-c} \times s^c \times g^{-1/2}$$

$$= \text{constant} \times \sqrt{l/g} \times (s/l)^c$$

The solution indicates that the period does not depend on the mass of the bob, but only on the length, the acceleration due to gravity, and the ratio of length of arc to length of pendulum. The principle of dimensions does not supply the constant of proportionality, which must be determined experimentally.

The term (s/l) is a mere number, each quantity being of dimension L, and dimensionally there is no restriction on the value of c. We can therefore write

$$\text{T} = \text{constant} \times \sqrt{l/g} \times f(s/l) \tag{2}$$

Although the form of the function f is undetermined, it is explicitly indicated by this equation that it is not the arc s itself which is important, but its ratio to l: i.e., the maximum angle of swing, s/l radians.

The function f can be found by experiment, and must approach the value unity for small swings, so as to lead to the usual formula for a simple pendulum under such conditions:

$$T = \text{constant} \times \sqrt{l/g}$$

The most important question regarding any dimensional solution is whether or not physical reasoning has led to a proper selection of the variables which govern the result.

Applying dimensional analysis to the ship resistance problem, the resistance R could depend upon the following:

(*a*) Speed, V.

(*b*) Size of body, which may be represented by the linear dimension, L.

(*c*) Mass density of fluid, ρ (mass per unit volume)

(*d*) Viscosity of fluid, μ

(*e*) Acceleration due to gravity, g

(*f*) Pressure per unit area in fluid, p

It is assumed that the resistance R can now be written in terms of unknown powers of these variables:

$$R \propto \rho^a V^b L^c \mu^d g^e p^f \tag{3}$$

Since R is a force, or a product of mass times acceleration, its dimensions are ML/T^2.

The density ρ is expressed as mass per unit volume, or M/L^3.

In a viscous fluid in motion the force between adjacent layers depends upon the area A in contact, the coefficient of viscosity of the liquid and upon the rate at which one layer of fluid is moving relative to the next one. If u is the velocity at a distance y from the boundary of the fluid, this rate or velocity gradient is given by the expression du/dy.

The total force is thus

$$F = \mu A du/dy$$

du/dy being a velocity divided by a distance has dimensions of $(L/T)/L$, *or* $1/T$, and the dimensional equation becomes

$$ML/T^2 = \mu L^2 \times 1/T$$

or

$$\mu = M/LT$$

p is a force per unit area, and its dimensions are $(ML/T^2)/L^2$, or M/LT^2.

The ratio μ/ρ is called the kinematic viscosity of the liquid, ν, and has dimensions given by

$$\nu = \mu/\rho = (M/LT)\cdot(L^3/M) = L^2/T$$

Introducing these dimensional quantities into Equation (3), we have

$$ML/T^2 = (M/L^3)^a\,(L/T)^b\,(L)^c\,(M/LT)^d \times (L/T^2)^e\,(M/LT^2)^f \quad (4)$$

whence

$$\left.\begin{aligned} a + d + f &= 1 \\ -3a + b + c - d + e - f &= 1 \\ b + d + 2e + 2f &= 2 \end{aligned}\right\}$$

or

$$\left.\begin{aligned} a &= 1 - d - f \\ b &= 2 - d - 2e - 2f \end{aligned}\right\}$$

and

$$\begin{aligned} c &= 1 + 3a - b + d - e + f \\ &= 1 + 3 - 3d - 3f - 2 + d \\ &\quad + 2e + 2f + d - e + f \\ &= 2 - d + e \end{aligned}$$

Then from Equation (3)

$$R \propto \rho V^2 L^2 f\left[\left(\frac{\rho V L}{\mu}\right)^{-d}\left(\frac{gL}{V^2}\right)^{e}\left(\frac{p}{\rho V^2}\right)^{f}\right] \quad (5)$$

All three expressions within the brackets are nondimensional, and are similar in this respect to the s/l term in Equation (2). There is therefore no restriction dimensionally on the exponents d, e, and f. The form of the function f must be found by experiment, and may be different for each of the three terms.

Writing ν for μ/ρ and remembering that for similar shapes the wetted surface S is proportional to L^2, Equation (5) may be written

$$\frac{R}{\frac{1}{2}\rho S V^2} = f\left[\frac{VL}{\nu}, \frac{gL}{V^2}, \frac{p}{\rho V^2}\right] \quad (6)$$

where the left-hand side of the equation is a nondimensional resistance coefficient. Generally in this chapter R will be given in kN and ρ in kg/L (or t/m^3), although N and kg/m^3 are often used (as here) in the cases of model resistance and ship air/wind resistance.

[3] A term first suggested by Dr. E.V. Telfer.

Equation (6) states in effect that if all the parameters on the right-hand side have the same values for two geometrically similar but different sized bodies, the flow patterns will be similar and the value of $R/\frac{1}{2}\rho S V^2$ will be the same for each.

2.3 Corresponding Speeds. Equation (6) showed how the total resistance of a ship depends upon the various physical quantities involved, and that these are associated in three groups, VL/ν, gL/V^2 and $p/\rho V^2$.

Considering first the case of a nonviscous liquid in which there is no frictional or other viscous drag, and neglecting for the moment the last group, there is left the parameter gL/V^2 controlling the surface wave system, which depends on gravity. Writing the wave-making or residuary resistance as R_R and the corresponding coefficient as C_R, C_R can be expressed as

$$C_R = \frac{R_R}{\frac{1}{2}\rho S V^2} = f_1(V^2/gL) \quad (7)$$

This means that *geosims*[3] (geometrically similar bodies) of different sizes will have the same specific residuary resistance coefficient C_R if they are moving at the same value of the parameter V^2/gL.

According to Froude's *Law of Comparison*[4]: "The (residuary) resistance of geometrically similar ships is in the ratio of the cube of their linear dimensions if their speeds are in the ratio of the square roots of their linear dimensions." Such speeds he called *corresponding speeds.*[5] It will be noted that these corresponding speeds require $V/\sqrt{L}$ to be the same for model and ship, which is the same condition as expressed in Equation (7). The ratio $V_K/\sqrt{L}$, commonly with V_k in knots and L in feet, is called the speed-length ratio. This ratio is often used in presenting resistance data because of the ease of evaluating it arithmetically, but it has the drawback of not being nondimensional. The value of $V/\sqrt{gL}$, on the other hand, is nondimensional and has the same numerical value in any consistent system of units. Because of Froude's close association with the concept of speed-length ratio, the parameter $V/\sqrt{gL}$ is called the Froude number, with the symbol Fn.

When V_k is expressed in knots, L in feet, and g in ft/sec^2, the relation between $V/\sqrt{L}$ and Froude number is

$$\text{Fn} = 0.298\, V_k/\sqrt{L}$$

or

$$V_k/\sqrt{L} = 3.355\text{Fn}$$

[4] Stated in 1868 by William Froude (1955) who first recognized the practical necessity of separating the total resistance into components, based on the general law of mechanical similitude, from observations of the wave patterns of models of the same form but of different sizes.

The residuary resistances of ship (R_{RS}) and of model (R_{RM}) from Equation (7) will be in the ratio

$$\frac{R_{RS}}{R_{RM}} = \frac{\frac{1}{2}\rho S_S V_S^2 C_{RS}}{\frac{1}{2}\rho S_M V_M^2 C_{RN}}$$

where subscripts S and M refer to ship and model, respectively.

If both model and ship are run in water of the same density and at the same value of V^2/gL, as required by Equation (7), i.e.

$$(V_S)^2/gL_S = (V_M)^2/gL_M$$

then C_R will be the same for each, and

$$R_{RS}/R_{RM} = S_S\,(V_S)^2/S_M\,(V_M)^2 = (L_S)^2/(L_M)^2\ L_S/L_M$$
$$= (L_S)^3/(L_M)^3 = \Delta_S/\Delta_M \quad (8)$$

where Δ_S and Δ_M are the displacements of ship and model, respectively.

This is in agreement with Froude's law of comparison.

It should be noted from Equation (8) that at corresponding speeds, i.e., at the same value of $V/\sqrt{L}$

$$R_{RS}/\Delta_S = R_{RM}/\Delta_M \quad (9)$$

i.e., the residuary resistance per unit of displacement is the same for model and ship. Taylor made use of this in presenting his contours of residuary resistance in terms of pounds resistance per long ton of displacement (Section 8.6).

If the linear scale ratio of ship to model is λ, then the following relations hold:

$$\begin{aligned} L_S/L_M &= \lambda \\ V_S/V_M &= \sqrt{L_S}/\sqrt{L_M} = \sqrt{\lambda} = \lambda^{1/2} \\ R_{RS}/R_{RM} &= (L_S)^3/(L_M)^3 = \Delta_S/\Delta_M = \lambda^3 \end{aligned} \quad (10)$$

The "corresponding speed" for a small model is much lower than that of the parent ship. In the case of a 5 m model of a 125 m ship (linear scale ratio $\lambda = 25$), the model speed corresponding to 25 knots for the ship is $25/\lambda^{1/2}$, or $25/\sqrt{25}$, or 5 knots. This is a singularly fortunate circumstance, since it enables ship models to be built to reasonable scales and run at speeds which are easily attainable in the basin.

Returning to Equation (6), consider the last term, $p/\rho V^2$. If the atmospheric pressure above the water surface is ignored and p refers only to the water head, then for corresponding points in model and ship p will vary directly with the linear scale ratio λ. At corresponding speeds V^2 varies with λ in the same way so that $p/\rho V^2$ will be the same for model and ship. Since the atmospheric pressure is usually the same in model and ship, when it is included in p, so that the latter is the *total* pressure at a given point, the value of $p/\rho V^2$ will be much greater for model than for ship. Fortunately, most of the hydrodynamic forces arise from differences in *local* pressures, and these are proportional to V^2, so that the forces are not affected by the atmospheric pressure so long as the fluid remains in contact with the model and ship surfaces. When the pressure approaches very low values, however, the water is unable to follow surfaces where there is some curvature and cavities form in the water, giving rise to *cavitation*. The similarity conditions are then no longer fulfilled. Since the absolute or total pressure is greater in the model than in the ship, the former gives no warning of such behavior. For tests in which this danger is known to be present, special facilities have been devised, such as variable-pressure water tunnels, channels or towing basins, where the correctly scaled-down total pressure can be attained at the same time that the Froude condition is met.

In the case of a deeply submerged body, where there is no wavemaking, the first term in Equation (6) governs the frictional resistance, R_F. The frictional resistance coefficient is then

$$C_F = \frac{R_F}{\frac{1}{2}\rho S V^2} = f_2(VL/\nu) \quad (11)$$

and C_F will be the same for model and ship provided that the parameter VL/ν is the same. This follows essentially from the work of Osborne Reynolds (1883), for which reason the product VL/ν is known as Reynolds number, with the symbol Rn.

If both model and ship are run in water at the same density and temperature, so that ν has the same value, it follows from (11) that $V_S L_S = V_M L_M$. This condition is quite different from the requirement for wave-making resistance similarity. As the model is made smaller, the speed of test must increase. In the case already used as an illustration, the 5-m model of a 125-m, 25-knot ship would have to be run at a speed of 625 knots.

The conditions of mechanical similitude for both friction and wave-making cannot be satisfied in a single test. It might be possible to overcome this difficulty by running the model in some other fluid than water, so that the change in value of ν would take account of the differences in the VL product. In the foregoing example, in order to run the model at the correct wave-making corresponding speed, and yet keep the value of VL/ν the same for both model and ship, a fluid would have to be found for use with the model which had a kinematic viscosity coefficient only 1/125 that of water. No such fluid is known. In wind-tunnel work, similitude can be attained by using compressed air in the model tests, so decreasing ν and increasing VL/ν to the required value.

The practical method of overcoming this fundamental difficulty in the use of ship models is to deal with

[5] This same law had previously been put forward by the French Naval Constructor Reech in 1832, but he had not pursued it or demonstrated how it could be applied to the practical problem of predicting ship resistance (Reech, 1852).

the frictional and the wave-making resistances separately, by writing

$$C_T = C_R + C_F \tag{12}$$

This is equivalent to expressing Equation (6) in the form

$$C_T = \frac{R_T}{\frac{1}{2}\rho S V^2} = f_1(V^2/gL) + f_2(VL/\nu) \tag{13}$$

Froude recognized this necessity, and so made ship-model testing a practical tool. He realized that the frictional and residuary resistances do not obey the same law, although he was unaware of the relationship expressed by Equation (11).

2.4 Extension of Model Results to Ship. To extend the model results to the ship, Froude proposed the following method, which is based on Equation (12). Since the method is fundamental to the use of models for predicting ship resistance, it must be stated at length:

Froude noted:

(*a*) The model is made to a linear scale ratio of λ and run over a range of "corresponding" speeds such that $V_S/\sqrt{L_S} = V_M/\sqrt{L_M}$.

(*b*) The total model resistance is measured, equal to R_{TM}.

(*c*) The frictional resistance of the model R_{FM} is calculated, assuming the resistance to be the same as that of a smooth flat plank of the same length and surface as the model.

(*d*) The residuary resistance of the model R_{RM} is found by subtraction:

$$R_{RM} = R_{TM} - R_{FM}$$

(*e*) The residuary resistance of the ship R_{RS}, is calculated by the law of comparison, Equation (10):

$$R_{RS} = R_{RM} \times \lambda^3$$

This applies to the ship at the corresponding speed given by the expression

$$V_S = V_M \times \lambda^{1/2}$$

(*f*) The frictional resistance of the ship R_{FS} is calculated on the same assumption as in footnote (4), using a frictional coefficient appropriate to the ship length.

(*g*) The total ship resistance (smooth hull) R_{TS} is then given by

$$R_{TS} = R_{FS} + R_{RS}$$

This principle of extrapolation from model to ship is still used in all towing tanks, with certain refinements to be discussed subsequently.

Each component of resistance will now be dealt with in greater detail.

Section 3
Frictional Resistance

3.1 General One has only to look down from the deck of a ship at sea and observe the turbulent motion in the water near the hull, increasing in extent from bow to stern, to realize that energy is being absorbed in frictional resistance. Experiments have shown that even in smooth, new ships it accounts for 80 to 85 percent of the total resistance in slow-speed ships and as much as 50 percent in high-speed ships. Any roughness of the surface will increase the frictional resistance appreciably over that of a smooth surface, and with subsequent corrosion and fouling still greater increases will occur. Not only does the nature of the surface affect the drag, but the wake and propulsive performance are also changed. Frictional resistance is thus the largest single component of the total resistance of a ship, and this accounts for the theoretical and experimental research that has been devoted to it over the years. The calculation of wetted surface area which is required for the calculation of the frictional resistance, Equation (11), is discussed in Chapter I.

3.2 Froude's Experiments on Friction. Froude, knowing the law governing residuary resistance and having concluded that the model-ship extrapolation problem could only be solved by dividing the resistance into two components, undertook a basic investigation into the frictional resistance of smooth planks in his tank at Torquay, England, the results of which he gave to the British Association (Froude, W., 1872, 1874).

The planks varied in lengths from 0.61 m (2 ft) to 15.2 m (50 ft) and the speed range covered was from 0.5 m/sec (1.67 fps) to 4.1 m/sec (13.3 fps), the maximum for the 15.2 m plank being 3.3 m/sec (10.8 fps). Froude found that at any given speed the specific resistance per unit of surface area was less for a long plank than for a shorter one, which he attributed to the fact that towards the after end of the long plank the water had acquired a forward motion and so had a lower relative velocity.

He gave an empirical formula for the resistance in the form

$$R = fSV^n \tag{14}$$

where

R = resistance, kN or lb
S = total area of surface, m^2 or ft^2
V = speed, m/sec or ft/sec

Table 1—Froude's Skin-Friction Coefficients[a]

$R = f \cdot S \cdot V^n$ R = resistance, lb S = area of plank, sq ft V = speed, fps

Nature of surface	2			8			20			50		
	f	n	k	f	n	k	f	n	k	f	n	k
Varnish	0.00410	2.00	0.00390	0.00460	1.88	0.00374	0.00390	1.85	0.00337	0.00370	1.83	0.00335
Paraffin	0.00425	1.95	0.00414	0.00360	1.94	0.00300	0.00318	1.93	0.00280	—	—	—
Calico	0.01000	1.93	0.00830	0.00750	1.92	0.00600	0.00680	1.89	0.00570	0.00640	1.87	0.00570
Fine sand	0.00800	2.00	0.00690	0.00580	2.00	0.00450	0.00480	2.00	0.00384	0.00400	2.06	0.00330
Medium sand	0.00900	2.00	0.00730	0.00630	2.00	0.00490	0.00530	2.00	0.00460	0.00490	2.00	0.00460
Coarse sand	0.01000	2.00	0.00880	0.00710	2.00	0.00520	0.00590	2.00	0.00490	—	—	—

(Length of surface, or distance from cutwater, ft)

[a] W. Froude's results for planks in fresh water at Torquay (British Association 1872 and 1874).
NOTE: The values of k represent the f-values for the last square foot of a surface whose length is equal to that given at the head of the column.

f and n depended upon length and nature of surface, and are given in Table 1.

For the smooth varnished surface, the value of the exponent n decreased from 2.0 for the short plank to 1.83 for the 15.2 m (50 ft) plank. For the planks roughened by sand, the exponent had a constant value of 2.0.

For a given type of surface, the f-value decreased with increasing length, and for a given length it increased with surface roughness.

In order to apply the results to ships, the derived skin-friction coefficients had to be extrapolated to much greater lengths and speeds. W. Froude did not give these extrapolated figures in his reports, but suggested two methods which might be used for their derivation. In his own words, "it is at once seen that, at a length of 50 feet, the decrease, with increasing length, of the friction per square foot of every additional length is so small that it will make no very great difference in our estimate of the total resistance of a surface 300 ft long whether we assume such decrease to continue at the same rate throughout the last 250 feet of the surface, or to cease entirely after 50 feet; while it is perfectly certain that the truth must lie somewhere between these assumptions." Payne, (1936) has reproduced the curve Froude used at Torquay in 1876 for ships up to 152.4 m (500 ft) in length. This curve is almost an arithmetic mean between those which would be obtained by the two methods suggested. W. Froude (1874) also obtained some full-scale information in an attempt to confirm his law of comparison and to assist in the extrapolation of the frictional coefficients to ship lengths by carrying out towing tests on the sloop HMS *Greyhound,* a wooden ship 52.58 m (172 ft 6 in.) in length, with copper sheathing over the bottom. The results of the towing tests and the predictions made from the model are given in Table 2.

Table 2—Results of Towing Trials on HMS *Greyhound*

Speed V, fpm	600	800	1000	1200
Resistance R_S, lb, from ship	3100	5400	9900	19100
R_M, lb, predicted from model	2300	4500	8750	17500
Percent difference	35	20	13	9
$(R_S/V^2) \times 10^2$ for ship	0.86	0.84	0.99	1.33
$(R_M/V^2) \times 10^2$, model prediction	0.64	0.70	0.87	1.22
Difference in $(R/V^2) \times 10^2$	0.22	0.14	0.12	0.11

The actual ship resistance was everywhere higher than that predicted from the model, the percentage increase becoming less with increasing speed. The difference in R/V^2, however, is almost the same at all speeds, except the lowest, and decreases only slowly with increasing speed, as might occur if this additional resistance were of viscous type and varying at some power less than the second. Froude pointed out that the additional resistance could be accounted for by assuming that the copper-sheathed hull was equivalent to smooth varnish over 2/3 of the wetted surface and to calico over the rest. This he considered reasonable, and the two resistance curves were then almost identical, which he took as a visible demonstration of the correctness of his law of comparison.

In his paper on the *Greyhound* trials, Froude states quite clearly how he applied his idea of the "equivalent plank" resistance: "For this calculation the immersed skin was carefully measured, and the resistance due to it determined upon the hypothesis that it is equivalent to that of a rectangular surface of equal area, and of length (in the line of motion) equal to that of the model, moving at the same speed."

The 1876 values of frictional coefficients were stated to apply to new, clean, freshly painted steel surfaces, but they lie considerably above those now generally accepted for smooth surfaces. The original curves have been modified and extended from time to time by R.E. Froude, up to a length of 366 m (1200 ft), but these extended curves had no experimental basis beyond the 15.2 m (50 ft) plank tests made in 1872, (Froude, R. E. 1888). Nevertheless, they are still used today in some towing tanks.

3.3 Two-dimensional Frictional Resistance Formulations. In the experiments referred to in Section 2.3, Osborne Reynolds made water flow through a glass tube, introducing a thin stream of dye on the centerline at the entrance to the tube. When the velocity was small, the dye remained as a straight filament parallel to the axis of the tube. At a certain velocity, which Reynolds called the critical velocity V_c, the filament began to waver, became sinuous and finally lost all definiteness of outline, the dye filling the whole tube. The resistance experienced by the fluid over a given length of pipe was measured by finding the loss of pressure head. Various diameters of the tube, D, were used, and the kinematic viscosity was varied by heating the water. Reynolds found that the laws of resistance exactly corresponded for velocities in the ratio ν/D, and when the results were plotted logarithmically

$$V_c = 2000\nu/D$$

Below the critical velocity the resistance to flow in the pipe varied directly as the speed, while for higher velocities it varied at a power of the speed somewhat less than 2.

When the foregoing relationship is written in the form

$$V_c D/\nu = 2000$$

the resemblance to Equation (11) is obvious.

Stanton, et al. (1952) showed that Reynolds' findings applied to both water and air flowing in pipes, and also that the resistance coefficients for models of an airship on different scales were sensibly the same at the same value of VL/ν. Baker (1915) plotted the results of much of the available data on planks in the form of the resistance coefficient

$$C_F = \frac{Rf}{\frac{1}{2}\rho S V^2}$$

to a base of VL/ν, and found that a mean curve could be drawn passing closely through Froude's results except at low values of VL/ν.

Experiments such as those performed by Reynolds suggested that there were two separate flow regimes possible, each associated with a different resistance law. At low values of VD/ν, when the dye filament retained its own identity, the fluid was evidently flowing in layers which did not mix transversely but slid over one another at relative speeds which varied across the pipe section. Such flow was called laminar and was associated with a relatively low resistance. When the Reynolds number VD/ν increased, either by increasing VD or by decreasing ν, the laminar flow broke down, the fluid mixed transversely in eddying motion, and the resistance increased. This flow is called turbulent.

In modern skin-friction formulations the specific frictional resistance coefficient C_F is assumed to be a function of the Reynolds number Rn or VL/ν. As early as 1904 Blasius had noted that at low Reynolds numbers the flow pattern in the boundary layer of a plank was laminar (Blasius, 1908). He succeeded in calculating

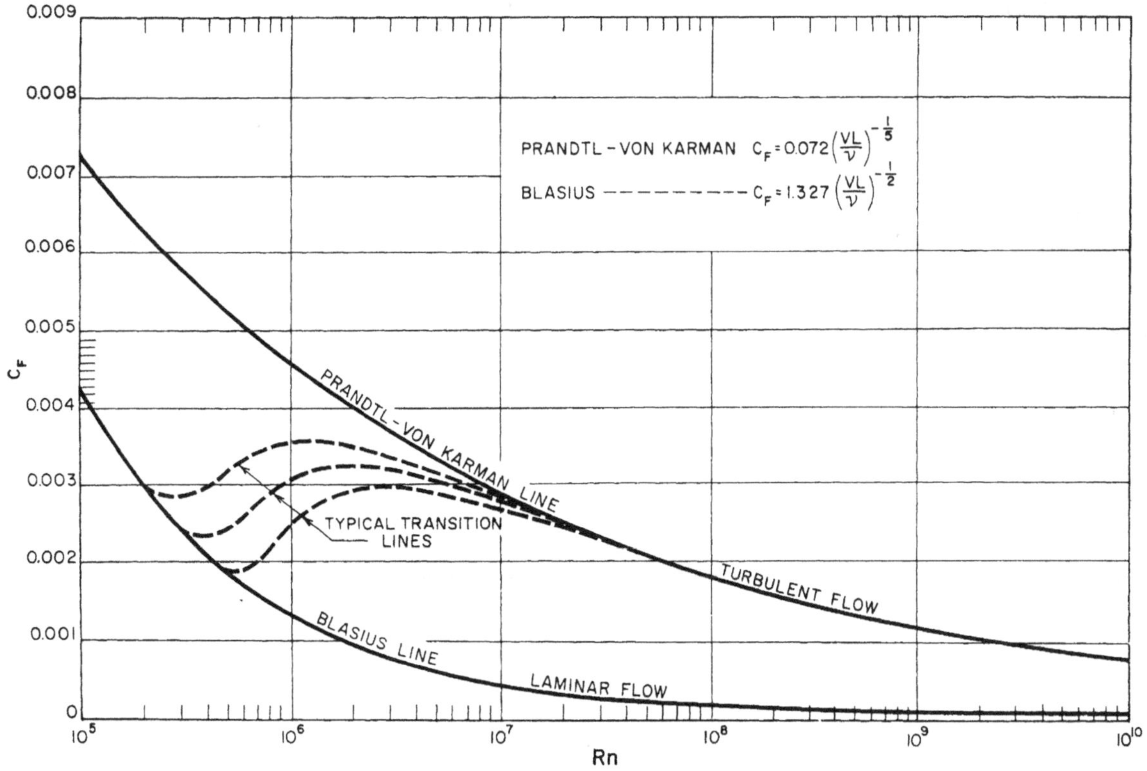

Fig. 2 Skin friction lines, turbulent and laminar flow

the total resistance of a plank in laminar flow by integrating across the boundary layer to find the momentum transferred to the water, and gave the formula for C_F in laminar flow in terms of Rn:

$$C_F = \frac{R_F}{\frac{1}{2}\rho S V^2} = 1.327\,(VL/\nu)^{-\frac{1}{2}} \tag{15}$$

This line is plotted in Fig. 2. Blasius found good agreement between his calculated resistances and direct experiment, but found that the laminar flow became unstable at Reynolds numbers of the order of 4.5×10^5, beyond which the resistance coefficients increased rapidly above those calculated from his equation.

Prandtl and von Karman (1921) separately published the equation

$$C_F = \frac{R_F}{\frac{1}{2}\rho S V^2} = 0.072(VL/\nu)^{-\frac{1}{5}} \tag{16}$$

for turbulent flow, which is also shown in Fig. 2. This equation was based on an analytical and experimental investigation of the characteristics of the boundary layer, as well as on the available measurements of overall plank resistance, principally those of Froude and further experiments run by Gebers in the Vienna tank (Gebers, 1919).

At low values of Reynolds number, and with quiet water, the resistance of a smooth plank closely follows the Blasius line, the flow being laminar, and from Equation (15) it is seen that the resistance R varies as $V^{1.5}$.

For turbulent flow, the value of the resistance coefficient is considerably higher than for laminar flow, and varies as a higher power of the speed; according to Equation (16) as $V^{1.8}$.

The transition from laminar flow to turbulent flow does not occur simultaneously over the whole plank. Transition begins when the Reynolds number reaches a critical value R_c. As the velocity V increases beyond this value, the transition point moves forward so that the local value of the Reynolds number, Vx/ν, remains equal to the critical value, x being the distance of the transition point from the leading edge of the plank. This is called the "local Reynolds number," and for the constant value of this local Rn at which transition takes place, x will decrease as V increases, and more and more of the plank surface will be in turbulent flow and so experience a higher resistance. The value of C_F will thus increase along a transition line of the type shown in Fig. 2, and finally approach the turbulent line asymptotically. It should be noted that there is no unique transition line, the actual one followed in a given case depending upon the initial state of turbulence in the fluid, the character of the plank surface, the shape of the leading edge, and the aspect ratio.

These transition lines for smooth planks occur at values of Reynolds number within the range over which most plank-friction tests have been run, and if such plank results are to be used to predict the values of C_F at Reynolds numbers appropriate to a ship—100 times or so larger than the highest plank values—only those results for fully turbulent flow can properly be used.

3.4 Development of Frictional Resistance Formulations in the United States. With the completion of the Experimental Model Basin (EMB) in Washington in 1900, new experiments were made on planks and new model coefficients were derived from these tests. For the ship coefficients, those published by Tideman (1876) were adopted. These did not represent any new experiments, being simply a re-analysis of Froude's results by a Dutch naval constructor. This combination of friction coefficients—EMB plank results for model, Tideman's coefficients for ship—was in use at EMB from 1901 to 1923 (Taylor, D. W., 1943).

By this time the dependence of frictional resistance on Reynolds number was well established, and a formulation was desired which was in accord with known physical laws. In 1923, therefore, EMB changed to the use of frictional coefficients given by Gebers for both the model and ship range of Reynolds number (Gebers, 1919). This practice continued at that establishment and at the new David Taylor Model Basin (DTMB) until 1947 (now DTRC, David Taylor Research Center).

Schoenherr (1932) collected most of the results of plank tests then available, and plotted them as ordinates of C_F to a base of Rn as is shown in Fig. 3. He included the results of experiments on 6.1 m (20 ft) and 9.1 m (30 ft) planks towed at Washington, and at the lower Reynolds numbers some original work on 1.8 m (6 ft) catamarans with artificially-induced turbulent flow. At the higher Reynolds numbers he was guided largely by the results given by Kempf (1929) for smooth varnished plates. Kempf's measurements were made on small plates inserted at intervals along a 76.8 m (252 ft) pontoon, towed in the Hamburg tank. The local specific resistances so measured were integrated by Schoenherr to obtain the total resistance for surfaces of different lengths. In order to present these data in conformity with rational physical principles, Schoenherr examined his results in the light of the theoretical formula of Prandtl and von Karman, which was of the form

$$A/\sqrt{C_F} = \log_{10}(\text{Rn}\,C_F) + M$$

He found he could get a good fit to the experimental data by making M zero and A equal to 0.242, so arriving at the well-known Schoenherr formulation

$$0.242/\sqrt{C_F} = \log_{10}(\text{Rn}\,C_F) \tag{17}$$

The Schoenherr coefficients as extended by this formula to the ship range of Reynolds numbers apply to a perfectly smooth hull surface. For actual ship hulls with structural roughnesses such as plate seams, welds or rivets, and paint roughness, some allowance,

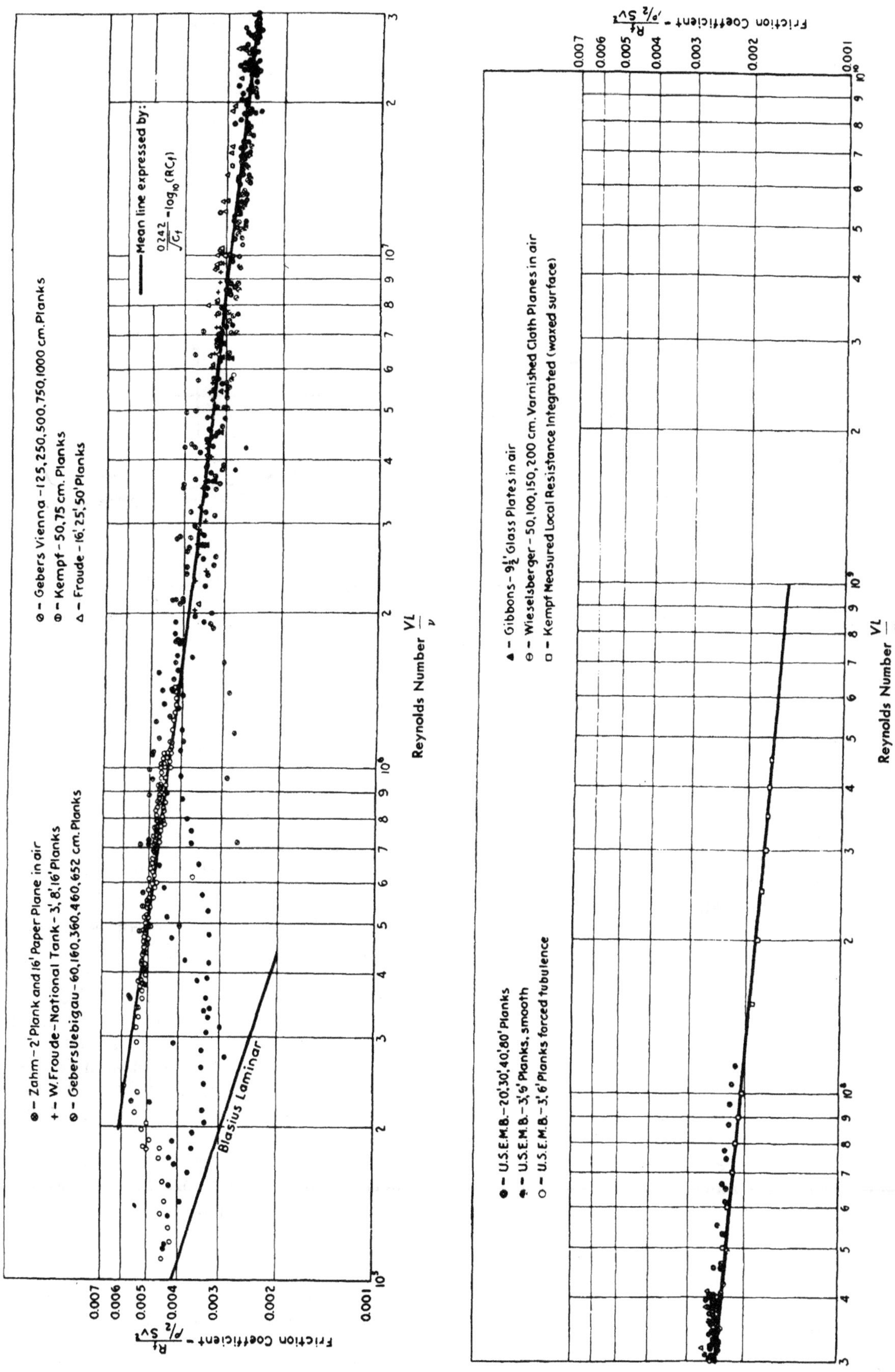

Fig. 3 Schoenherr's log-log chart for friction formulation

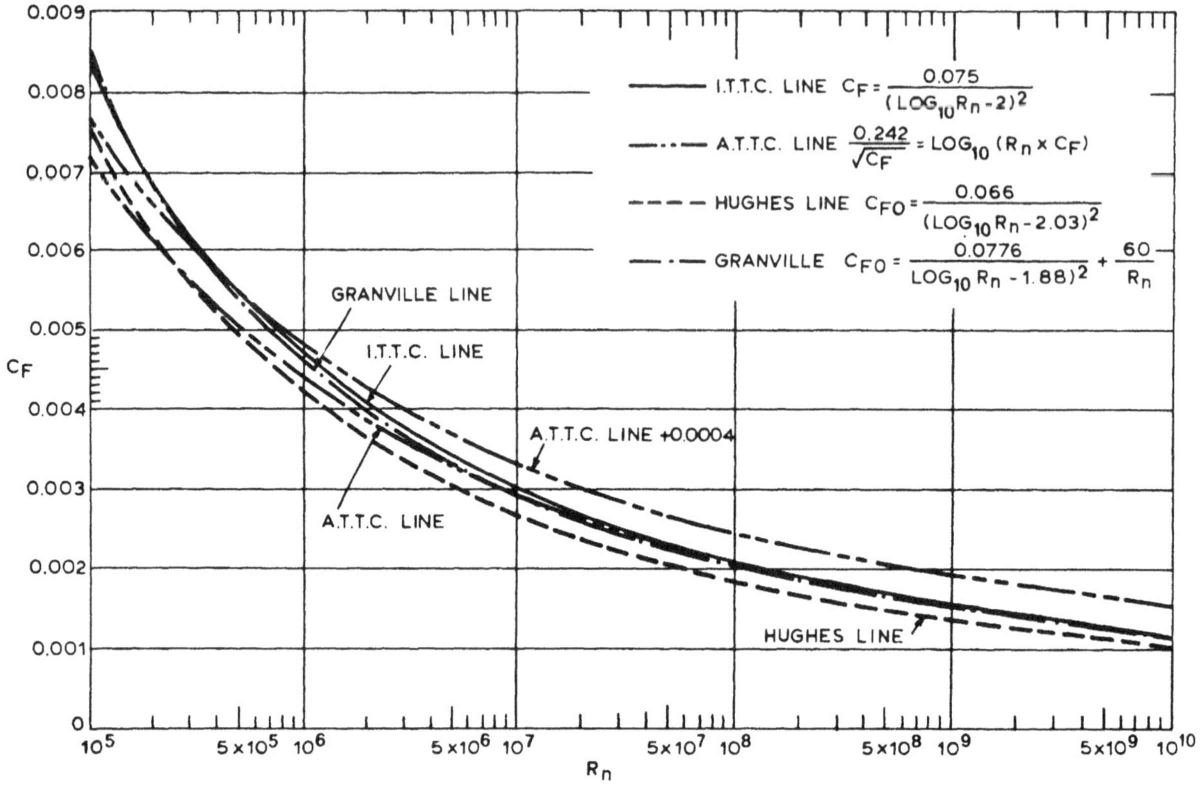

Fig. 4 Skin friction lines

the magnitude of which is discussed later, is necessary to give a realistic prediction.

3.5 The Work of the Towing Tank Conferences. The International Conference of Ship Tank Superintendents (ICSTS) was a European organization founded in 1932 to provide a meeting place for towing-tank staffs to discuss problems peculiar to their field. In 1935, the ICSTS agreed to adopt the Froude method of model extrapolation, among the decisions recorded being the following:

"*V*—on the determination of length and wetted surface:

(a) For every kind of vessel, the length on the water line should be used.

(b) The mean girth multiplied by the length is adopted as the wetted surface.[6]

VI—Froude's method of calculation:[7]

(a) The Committee adheres to the skin friction deduced from Froude's 0 values,[7] and takes these to be represented by the formula below, since this gives the same values of friction for model and ship within the limits of experimental errors:

$$R_F = \left[\frac{0.000418 + 0.00254}{8.8 + 3.281L}\right] S V_K^{1.825} \qquad (18)^{8}$$

where

R_F = resistance in kNewton;
L = length in meters;
S = wetted surface in square meters;
V_K = speed in knots.

(b) All model results should be corrected to a standard temperature of 15 deg C (= 59 deg F) by a correction of −0.43 percent of the frictional resistance per + 1 deg C or −0.24 percent per + 1 deg F."

In 1946 the American Towing Tank Conference (ATTC) began considering the establishment of a uniform practice for the calculation of skin friction and the expansion of model data to full size. In 1947 the following two resolutions were adopted (SNAME, 1948):

"1. Analysis of model tests will be based on the Schoenherr mean line. Any correction allowances applied to the Schoenherr mean line are to be clearly stated in the report."

[6] That is, no "obliquity" correction.

[7] These were the Froude frictional coefficients presented in a particular notation—see Froude (1888).

[8] As pointed out by Nordstrom (ITTC Proceedings, Washington, 1951) this formula applies to salt water. For fresh water the corresponding formula is

$$R_F = [0.000407 + 0.00248/(8.8 + 3.281L)] S . V_K^{1.825}$$

"2. Ship effective power calculations will be based on the Schoenherr mean line with an allowance that is ordinarily to be +0.0004 for clean, new vessels, to be modified as desired for special cases and in any event to be clearly stated in the report."

No decision was made as regards a standard temperature for ship predictions, but this has subsequently been taken as 15 deg C (59 deg F) in conformity with the ICSTS figure (ATTC, 1953). It was also agreed that the Schoenherr line shall be known as the "1947 ATTC line" (ATTC, 1956). This line, both with and without the 0.0004 allowance, is shown in Fig. 4. The method of applying the coefficients has been described in detail by Gertler (1947). He also gave tables of their values for a wide range of Reynolds numbers, together with values of ρ and ν for fresh and salt water.

New values of ν were adopted by the ITTC[9] (1963) at the 10th Conference in London in 1963. These are also reproduced together with the C_F coefficients in a SNAME Bulletin (1976).

The allowance referred to in the second resolution of the ATTC was originally considered necessary because of the effect of hull roughness upon resistance. However, the difference between the ship resistance as deduced from full-scale trials and that predicted from the model depends upon other factors also, as is discussed in Section 6.4 and at the ITTC meeting in 1963 it was agreed to refer to it as a "model-ship correlation allowance" and to give it the symbol C_A (ITTC, 1963).

The 5th Conference of the ICSTS was held in London in 1948, and was attended for the first time by delegates from the United States and Canada. There was much discussion on the model-extrapolation problem, and unanimous agreement was reached "in favor of departing from Froude's coefficients and selecting a substitute in line with modern concepts of skin friction." However, the delegates were unable to agree upon any such alternative, largely because it was felt that the progress in knowledge might in the near future demand a further change. The Conference therefore agreed that in published work either the Froude or Schoenherr coefficients could be used, and at the same time set up a Skin Friction Committee to recommend further research to establish a minimum turbulent-friction line for both model and ship use.

The Committee was instructed that any proposed friction formulation should be in keeping with modern concepts of physics, and the coefficient C_F should be a function of Reynolds number Rn. The Schoenherr (ATTC) line already fulfilled this requirement, but the slope was not considered sufficiently steep at the low Reynolds numbers appropriate to small models, so that it did not give good correlation between the results of small and large models. With the introduction of welding, ships' hulls had become much smoother and for long, all-welded ships the correlation allowance C_A necessary to reconcile the ship resistance with the prediction from the model using the ATTC line was sometimes zero or negative. Also, Schoenherr had used data from many sources, and the planks were in no sense geosims, so that the experimental figures included aspect ratio or edge effects (the same applied to Froude's results). Telfer (1927, 1950, 1951, 1952) suggested methods for taking edge effects into account and developed an "extrapolator" for predicting ship resistance from model results which was an inverse function of Reynolds number. Hughes (1952), (1954) carried out many resistance experiments on planks and pontoons, in the latter case up to 77.7 m (255 ft) in length, and so attained Reynolds numbers as high as 3×10^8. These plane surfaces covered a wide range of aspect ratios, and Hughes extrapolated the resistance coefficients to infinite aspect ratio, obtaining what he considered to be a curve of minimum turbulent resistance for plane, smooth surfaces in two-dimensional flow. This curve had the equation

$$C_{FO} = 0.066/(\log_{10}\text{Rn} - 2.03)^2 \qquad (19)$$

and is shown in Fig. 4. C_{FO} denotes the frictional resistance coefficient in two-dimensional flow.[10]

The ITTC Friction Committee, with the knowledge of so much new work in progress, did not feel able in 1957 to recommend a final solution to the problem of predicting ship resistance from model results. Instead, it proposed two alternative single-line, interim engineering solutions. One was to use the ATTC line for values of Rn above 10^7, and below this to use a new line which was steeper than the ATTC line. The latter would, in the Committee's opinion, help to reconcile the results between large and small models, while using the ATTC line above Rn $= 10^7$ would make no difference in ship predictions from large models. The second proposal was to use an entirely new line, crossing the ATTC line at about Rn $= 10^7$, and being slightly steeper throughout. This would result in lower ship predictions, and so would tend to increase the correlation allowance C_A and avoid negative allowances for long ships.

The Conference in Madrid in 1957 adopted a slight variation of the second proposal, and agreed to

$$C_F = 0.075/(\log_{10}\text{Rn} - 2)^2 \qquad (20)$$

This line is also shown in Fig. 4.

The Conference adopted this as the "ITTC 1957 model-ship correlation line," and was careful to label

---

[9] The International Conference of Ship Tank Superintendents (ICSTS) became the International Towing Tank Conference (ITTC) in 1957.

[10] ITTC Presentation Committee Report, Ottawa 1975. Also published by the British Ship Research Association, now British Maritime Technology (BMT), as Technical Memorandum No. 500.

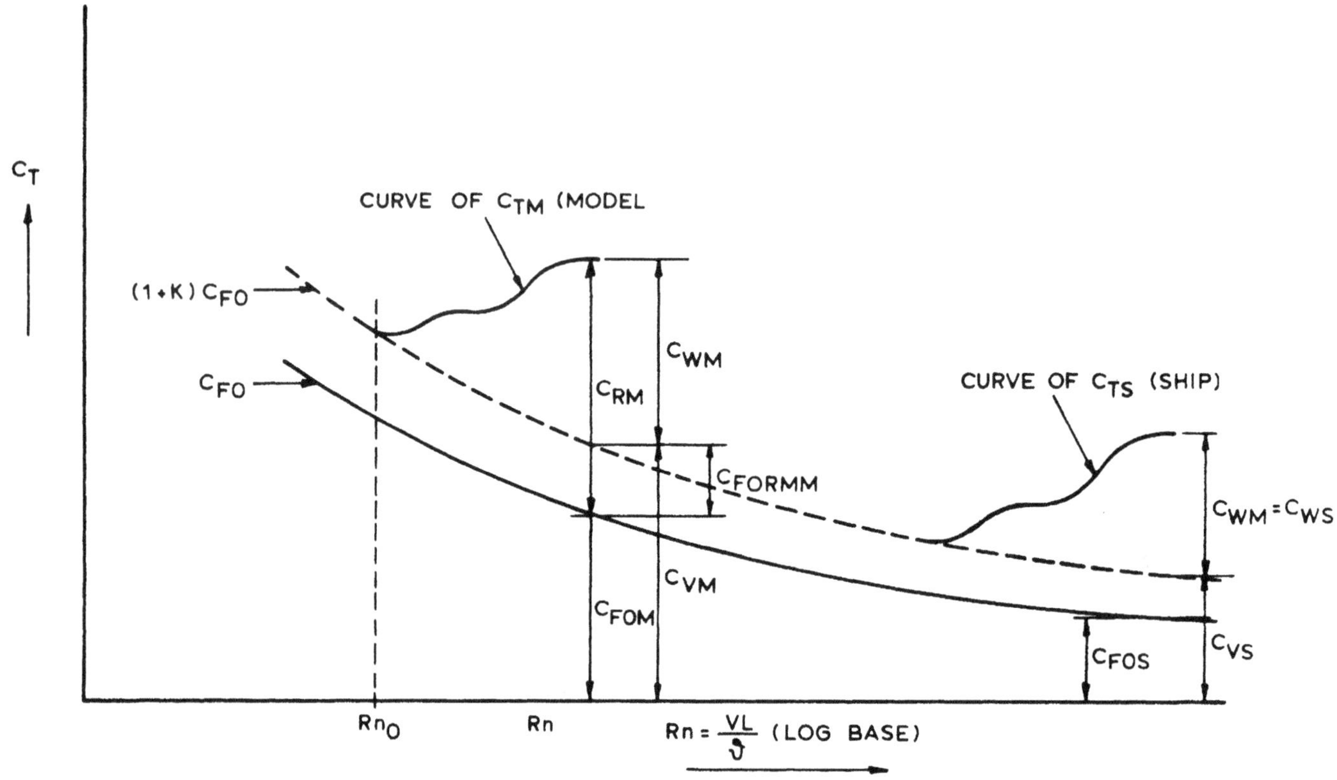

Fig. 5 Extrapolation of model results to ship using the form factor method

it as "only an interim solution to this problem for practical engineering purposes," (ITTC 1957). Equation (20) was called a model-ship correlation line, and not a frictional resistance line; it was not meant to represent the frictional resistance of plane or curved surfaces, nor was it intended to be used for such a purpose.

The Hughes proposal in Equation (19) is of the same general type as the ITTC line but gives much lower values of C_F than either the ITTC 1957 formulation or the ATTC 1947 line. On the other hand, the Hughes line does claim to be a true friction line for smooth plates in fully turbulent, two-dimensional flow, but its low values have been criticized by many other workers in this field. The 1957 ITTC line, in fact, gives numerical values of C_F which are almost the same as those of the Hughes line with a constant addition of 12 percent.

Granville (1977) showed that the 1957 ITTC model-ship correlation line can also be considered as a turbulent flat plate (two-dimensional) frictional resistance line. From fundamental considerations involving the velocity distribution in the boundary layer, he derived the general formula

$$C_{FO} = a/(\log_{10}\mathrm{Rn} - b)^2 + c/\mathrm{Rn} \qquad (21)$$

with $a = 0.0776$, $b = 1.88$ and $c = 60$. This formula is a generalization of the form of the 1957 ITTC line as given by Equation (20), with a = 0.075, b = 2 and c = 0. Good agreement of Equation (21) with the 1957 ITTC line is obtained for values of Rn less than 5×10^5. At values of Rn above 1×10^8, the 1957 ITTC, the 1947 ATTC, and the Granville lines are all in good agreement, as shown in Fig. 4.

3.6 Three-Dimensional Viscous Resistance Formulations. In association with his two-dimensional line, Hughes proposed a new method of extrapolation from model to ship. He assumed that the total model resistance coefficient C_{TM} could be divided into two parts, C_{VM} and C_{WM}, representing the viscous and wavemaking resistance, respectively. At low Froude numbers, C_{WM} will become very small, and at a point where wavemaking can be neglected, the curve of C_{TM} will become approximately parallel to the two-dimensional friction line. Hughes called this point the *run-in* point. The value of C_{TM} at this point can then be identified with the total viscous resistance coefficient C_{VM} at the same point Rn_o.

The form resistance coefficient, due at least in part to the curvature of the hull (see Fig. 5), is defined by

$$1 + k = \frac{C_{TM}(\mathrm{Rn}_o)}{C_{FO}(\mathrm{Rn}_o)}$$

The three-dimensional model viscous resistance for arbitrary Rn can now be written as $C_{VM} = (1 + k)\ C_{FO}$ (Rn) where C_{FO} is the equivalent flat-plate resistance coefficient. The factor k accounts for the three-dimensional form, and is appropriately termed the form factor. The form factor $(1 + k)$ is assumed to be invariant with Rn and the line $(1 + k)\ C_{FO}$ is now taken as the extrapolator for the hull form concerned, and the ship curve of C_{TS} can be drawn above the $(1 + k)C_{FO}$ curve at the appropriate values of the Reynolds number. In the Froude method the whole of the model residuary-resistance coefficient C_R is transferred to the ship unchanged, while in the form factor method only that part of C_R attributed to viscous effects (C_{FORMM} in Fig. 5) is reduced in the transfer. Accordingly, the three-dimensional method gives substantially lower ship predictions and so calls for larger values of the correlation allowance C_A. This procedure avoids the negative allowances sometimes found when using the Froude method. It should also be noted that in the case of the Froude method only the slope of the two-dimensional friction line matters while in the case of the form factor approach the vertical position of the line also affects the ship prediction. The choice of the basic line becomes an essential factor in the case of the three-dimensional approach.

The study carried out by the ITTC Performance Committee has shown that the introduction of the form factor philosophy has led to significant improvements in model-ship correlation (ITTC, 1978). The ITTC has recommended that for all practical purposes, for conventional ship forms, a form factor determined on an experimental basis, similar to Prohaska's method, is advisable; i.e.,

$$C_{TM}/C_{FO} = (1 + k) + c\ \mathrm{Fn}^n/C_{FO} \qquad (23)$$

where n is some power of Fn, $4 \leq n \leq 6$, and c and k are coefficients, chosen so as to fit the measured C_{TM}, Fn data points as well as possible (Prohaska, 1966). (A numerical example of how Prohaska's method is used is given in Section 6.4). This requires that the resistance of the model be measured at very low speeds, generally at Fn $\leq$ 0.1. This is a drawback because unwanted Reynolds scale effects are then often introduced. For this reason sometimes empirically-derived form factors values are adopted. However, no satisfactory method to derive appropriate values of such form factors has as yet been found. The ITTC Performance Committee, which reviews, collates and tests the various proposed methods, states in its 1978 report: "With regard to the influence of form on the various components of the viscous resistance no clear conclusion can be drawn. Results reported by Tagano (1973) and Wieghardt (1976) show that the form mainly influences the viscous pressure drag, while Dyne (1977) stated that the pressure drag is low and its influence on k is practically negligible. Furthermore, the interaction between different resistance components is hindering the isolation of a single significant factor."

Section 4
Wave-Making Resistance

4.1 General. The wave-making resistance of a ship is the net fore-and-aft force upon the ship due to the fluid pressures acting normally on all parts of the hull, just as the frictional resistance is the result of the tangential fluid forces. In the case of a deeply submerged body, travelling horizontally at a steady speed far below the surface, no waves are formed, but the normal pressures will vary along the length. In a non-viscous fluid the net fore-and-aft force due to this variation would be zero, as previously noted.

If the body is travelling on or near the surface, however, this variation in pressure causes waves which alter the distribution of pressure over the hull, and the resultant net fore-and-aft force is the wave-making resistance. Over some parts of the hull the changes in pressure will increase the net sternward force, in others decrease it, but the overall effect must be a resistance of such magnitude that the energy expended in moving the body against it is equal to the energy necessary to maintain the wave system. The wave-making resistance depends in large measure on the shapes adopted for the area curve, waterlines and transverse sections, and its determination and the methods by which it can be reduced are among the main goals of the study of ships' resistance. Two paths have been followed in this study—experiments with models in towing tanks and theoretical research into wave-making phenomena. Neither has yet led to a complete solution, but both have contributed greatly to a better understanding of what is a very complicated problem. At present, model tests remain the most important tool available for reducing the resistance of specific ship designs, but theory lends invaluable help in interpreting model results and in guiding model research.

4.2 Ship Wave Systems. The earliest account of the way in which ship waves are formed is believed to be that due to Lord Kelvin (1887, 1904). He considered a single pressure point travelling in a straight line over the surface of the water, sending out waves which combine to form a characteristic pattern. This consists of a system of transverse waves following behind the point, together with a series of divergent waves radiating from the point, the whole pattern being contained within two straight lines starting from the pressure point and making angles of 19 deg 28 min

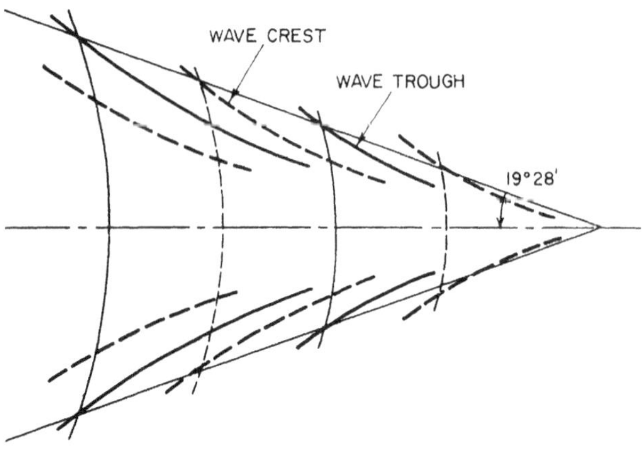

Fig. 6 Kelvin wave pattern

on each side of the line of motion, Fig. 6. The heights of successive transverse-wave crests along the middle line behind the pressure point diminish going aft. The waves are curved back some distance out from the centerline and meet the diverging waves in cusps, which are the highest points in the system. The heights of these cusps decrease less rapidly with distance from the point than do those of the transverse waves, so that eventually well astern of the point the divergent waves become the more prominent (see Fig. 7).

The Kelvin wave pattern illustrates and explains many of the features of the ship-wave system. Near the bow of a ship the most noticeable waves are a series of divergent waves, starting with a large wave at the bow, followed by others arranged on each side along a diagonal line in such a way that each wave is stepped back behind the one in front in echelon (Fig. 8) and is of quite short length along its crest line. Between the divergent waves on each side of the ship, transverse waves are formed having their crest lines normal to the direction of motion near the hull, bending back as they approach the divergent-system waves and finally coalescing with them. These transverse waves are most easily seen along the middle portion of a ship or model with parallel body or just behind a ship running at high speed. It is easy to see the general Kelvin pattern in such a bow system.

Similar wave systems are formed at the shoulders, if any, and at the stern, with separate divergent and transverse patterns, but these are not always so clearly distinguishable because of the general disturbance already present from the bow system.

Since the wave pattern as a whole moves with the ship, the transverse waves are moving in the same direction as the ship at the same speed V, and might

Fig. 7(a) Pattern of diverging waves

Fig. 7(b) Typical ship wave pattern

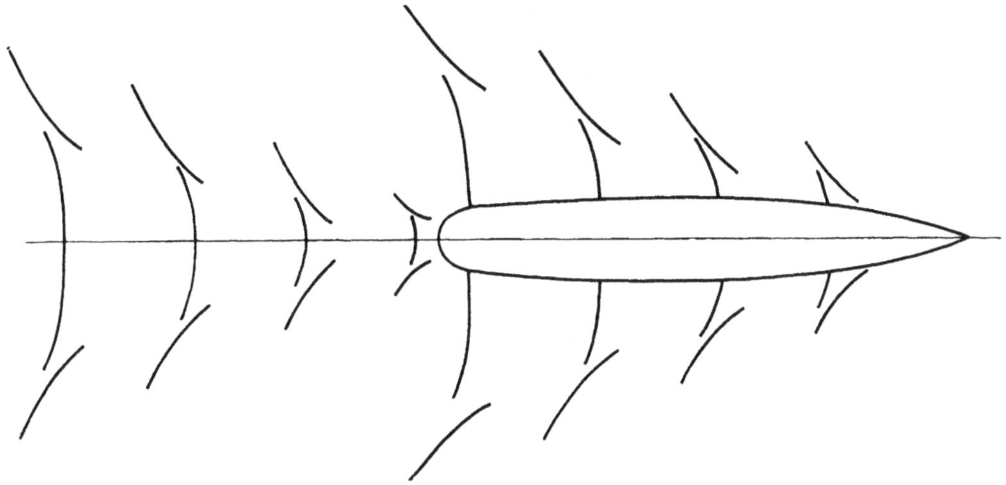

Fig. 8 Schematic diagram of bow and stern wave systems

be expected to have the length appropriate to free waves running on surface at that speed

$$L_W = 2\pi V^2/g$$

Actually, the waves in the immediate vicinity of a model are found to be a little shorter, but they attain the length L_W about two wave lengths astern.

The divergent waves will have a different speed along the line normal to their crests (Fig. 9). In this case, the component of speed parallel to the line of the ship's motion must be equal to the ship's speed in order to retain the fixed pattern relative to the ship. If the line normal to the crest of a divergent wave makes an angle θ with the ship's course, the speed in that direction will be $V\cos\theta$, and the corresponding wave length

$$L'_W = (2\pi V^2/g)\cos^2\theta$$

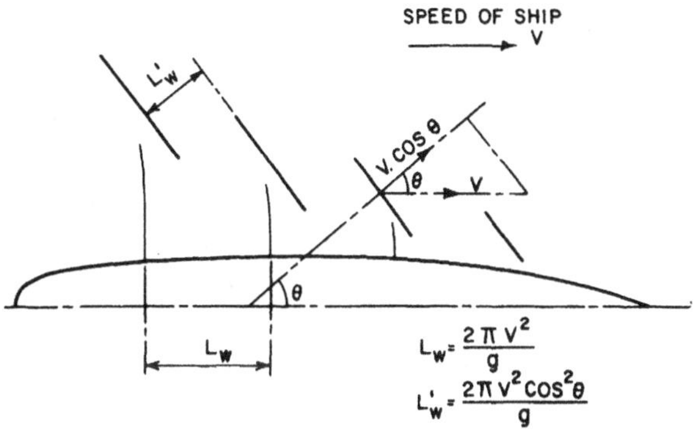

Fig. 9 Speed and length of divergent waves

4.3 Wave-Making Resistance of Surface Ships. At low speeds, the waves made by the ship are very small, and the resistance is almost wholly viscous in character. Since the frictional resistance varies at a power of the speed a little less than the square, when the coefficient of total resistance $C_T = R_T/\frac{1}{2}\rho SV^2$ is plotted to a base of Froude number Fn (or of $V_K/\sqrt{L}$), at first the value of C_T decreases with increase of speed (Fig. 10). With further increase in speed, the value of C_T begins to increase more and more rapidly, and at Froude numbers approaching 0.45 ($V_K/\sqrt{L} = 1.5$) the resistance may vary at a power of V of 6 or more. However, this general increase in C_T is usually accompanied by a number of humps and hollows in the resistance curve. As the speed of the ship increases, the wave pattern must change, for the length of the waves will increase and the relative positions of their crests and troughs will alter. In this process there will be a succession of speeds when the crests of the two systems reinforce one another, separated by other speeds at which crests and troughs tend to cancel one another. The former condition leads to higher wave heights, the latter to lower ones, and as the energy of the systems

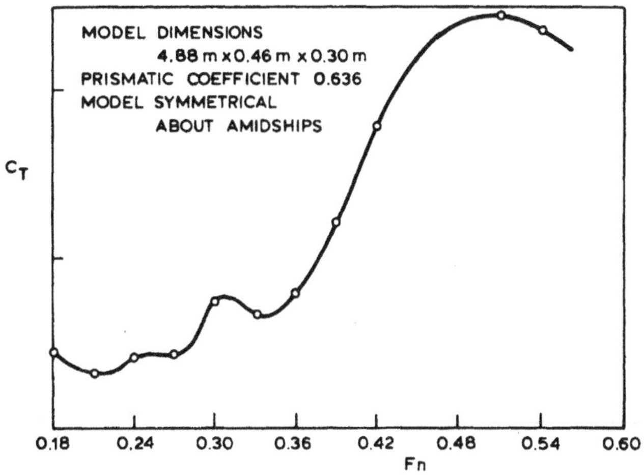

Fig. 10 Typical resistance curve, showing interference effects

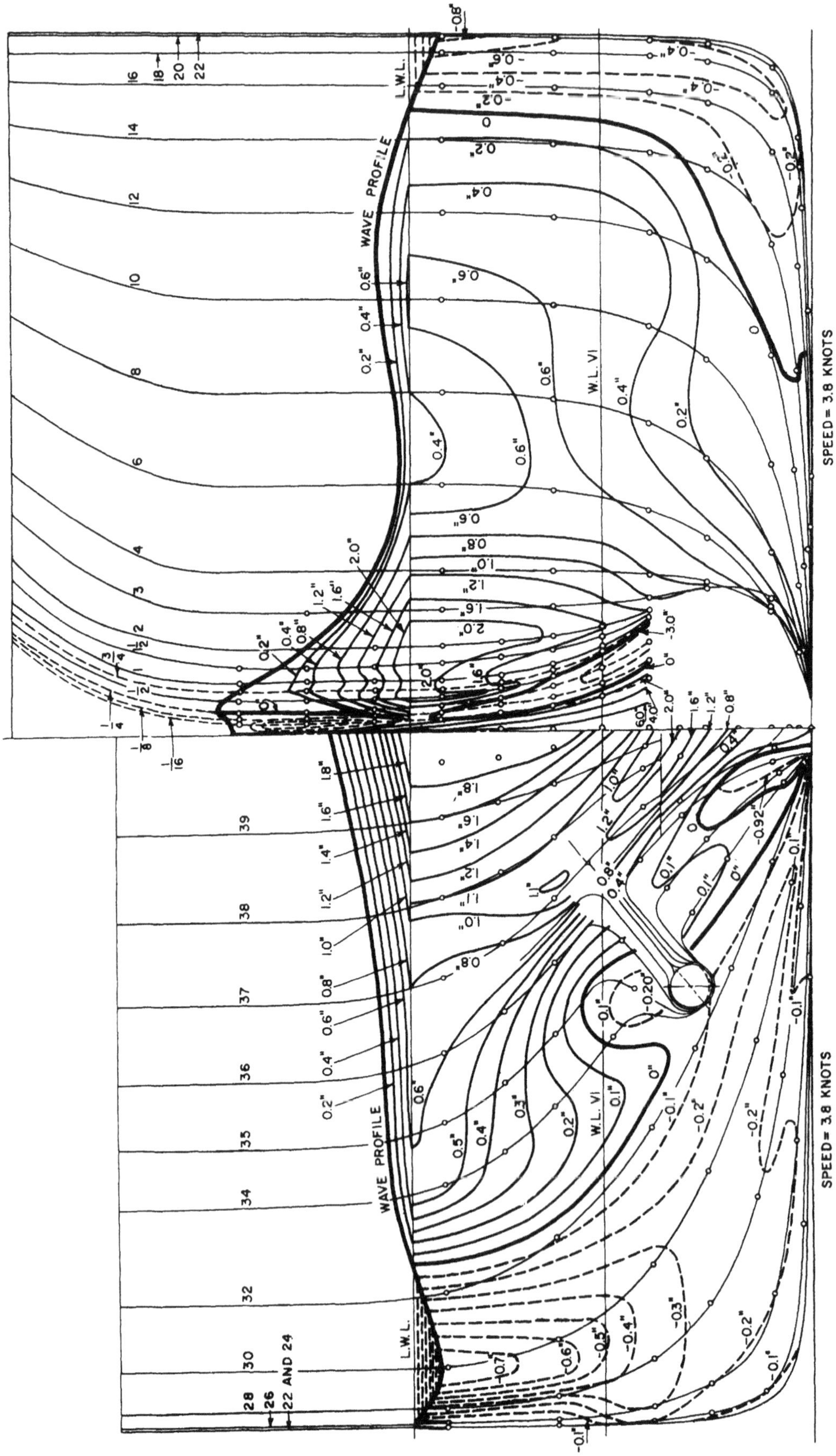

Fig. 11 Contours of change of pressure on model hull

Positive pressures indicate an increase in pressure above that existing in the static condition. Pressures are expressed in inches of water.

$\mathrm{Fn} = \frac{V}{\sqrt{gL}} = 0.253.$ 1 in. = 25.4 mm

depends upon the square of the wave heights, this means alternating speeds of higher and lower than average resistance. The humps and hollows in the C_T-curve are due to these interference effects between the wave systems, and it is obviously good design practice to ensure whenever possible that the ship will be running under service conditions at a *favorable speed.* As will be seen later, it is the dependence of these humps and hollows on the Froude number that accounts for the close relationship between economic speeds and ship lengths.

The mechanism by which wave-making resistance is generated is well illustrated by experiments made by Eggert (1939). He measured the normal pressure distribution over the ends of a model and plotted resulting pressure contours on a body plan (Fig. 11). By integrating the longitudinal components of these pressure forces over the length, he showed that the resulting resistance agreed fairly well with that measured on the model after the estimated frictional resistance had been subtracted. Fig. 12a shows curves of longitudinal force per meter length; Fig. 12b shows form resistance derived from pressure experiments and residuary resistance model tests. One important point brought out by these experiments is that a large proportion of the wave-making resistance is generated by the upper part of the hull near the still waterline.

4.4 Theoretical Calculation of Wave-Making Resistance. Much research has been devoted to theoretical methods of calculating wave-making resistance and to their experimental verification (Lunde, 1957).

One method is to determine the flow around the hull and hence the normal pressure distribution, and then to integrate the fore-and-aft components of these pressures over the hull surface. This method was developed by Michell (1898) for a slender hull moving over the surface of a nonviscous fluid. It corresponds to the experimental technique employed by Eggert. The pioneer work of Michell was unfortunately overlooked and neglected for many years until rescued from obscurity by Havelock (1951).

A second method is to calculate the wave pattern generated by the ship at a great distance astern, as done by Havelock, the wave-making resistance then being measured by the flow of energy necessary to maintain the wave system. This method has been used experimentally by Gadd, et al (1962), Eggers (1962), Ward (1962) and many others.

Both methods lead to the same final mathematical expression, and in each case the solution is for a nonviscous and incompressible fluid, so that the ship will experience only wave-making resistance (Timman, et al, 1955).

Michell obtained the mathematical expression for the flow around a "slender" ship of narrow beam when placed in a uniform stream. From the resultant velocity potential the velocity and pressure distribution over the hull can be obtained, and by integrating the fore-and-aft components of the pressure an expression can be derived for the total wave-making resistance.[11]

The theory as developed by Michell is valid only for certain restrictive conditions:

(a) The fluid is assumed to be nonviscous and the flow irrotational. Under these circumstances the motion can be specified by a velocity potential ϕ, which in addition must satisfy the necessary boundary conditions.

(b) The hull is narrow compared with its length, so that the slope of the surface relative to the centerline plane is small.

(c) The waves generated by the ship have heights small compared with their lengths, so that the squares of the particle velocities can be neglected compared with the ship speed.

(d) The ship does not experience any sinkage or trim.

The boundary conditions to be satisfied by the velocity potential ϕ are:

(a) At all points on the surface of the hull, the normal velocity relative to the hull must be zero.

(b) The pressure everywhere on the free surface of the water must be constant and equal to the atmospheric pressure.

To make the problem amenable to existing mathematical methods, Michell assumed that the first boundary condition could be applied to the centerline plane rather than to the actual hull surface, so that the results applied strictly to a vanishingly thin ship, and that the condition of constant pressure could be applied to the original flat, free surface of the water, the distortion of the surface due to the wave pattern being neglected.

The alternative method developed by Havelock, in which the wave-making resistance is measured by the energy in the wave system, makes use of the idea of sources and sinks.[12] This is a powerful tool with which to simulate the flow around different body shapes and so to find the wave pattern, pressure distribution, and resistance. A "thin" ship, for example, can be simulated by a distribution of sources on the centerline plane of the forebody and of sinks in the afterbody, the sum of their total strength being zero. The re-

[11] The velocity potential ϕ has the property that the velocity of the flow in any given direction is the partial derivative of ϕ with respect to that direction. Thus for a uniform stream of velocity U m the negative x-direction, the velocity potential will be defined by the expression

$$\frac{\partial \phi}{\partial x} = -U, \text{ or } \phi = -Ux$$

and

$$\frac{\partial \phi}{\partial y} = \frac{\partial \phi}{\partial z} = 0$$

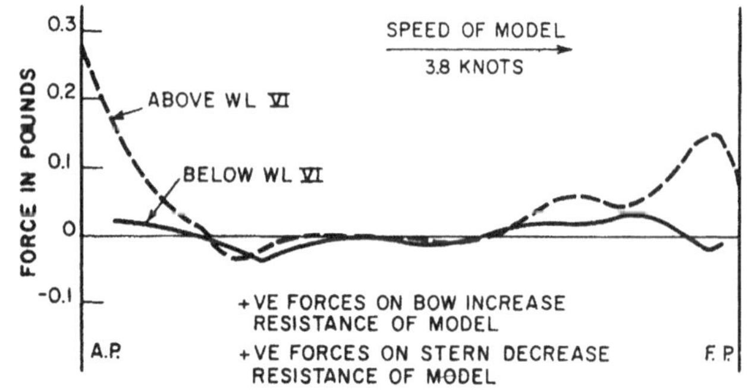

Fig. 12(a) Pressure forces on model

Curves show longitudinal force per inch length

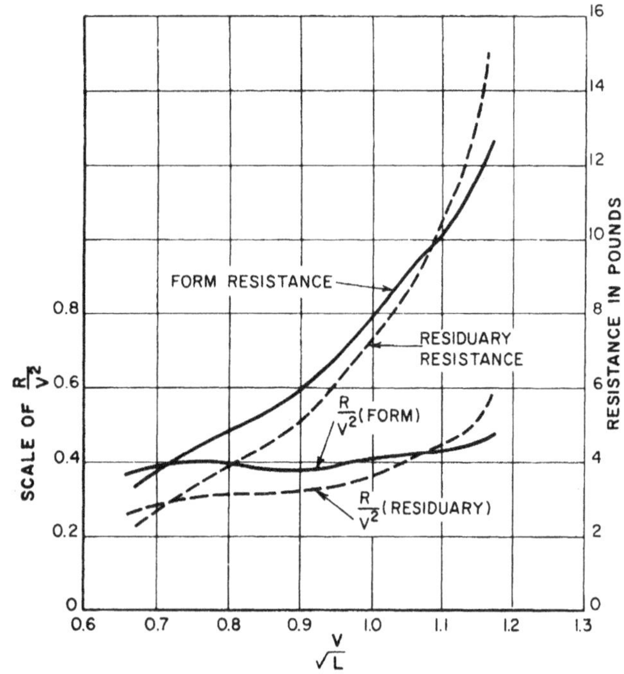

Fig. 12(b) Curves of form resistance and residuary resistance

Form resistance derived from pressure experiments. Residuary resistance derived from towing experiments

striction to a "thin" ship can be removed if the sources and sinks are distributed over the hull surface itself. If the strengths of the sources and sinks are assumed to be everywhere proportional to the slope of the hull surface, this will result in a total strength of zero, and the total velocity potential will be the sum of those due to the individual sources and sinks and the uniform flow.

Each source and sink when in motion in a fluid, on or near the surface, gives rise to a wave system, and by summing these up the total system for the ship can be obtained. Havelock by this method found the wave pattern far astern, and from considerations of energy obtained the wave-making resistance.

Much of the research into wave-making resistance has been done on models of mathematical form, having sections and waterlines defined by sine, cosine, or parabolic functions. When the calculations are applied to actual ship forms, the shape of the latter must be expressed approximately by the use of polynomials (Weinblum, 1950, Wehausen, 1973); or by considering the hull as being made up of a number of elementary wedges (Guilloton, 1951).

In recent years, a great deal of work on the calculation of wave-making resistance has been carried out in Japan by Professor Inui and his colleagues (Inui, 1980). They used a combination of mathematical and experimental work and stressed the importance of observing the wave pattern in detail as well as simply measuring the resistance. Instead of starting with a given hull geometry, Professor Inui began with an assumed source-sink distribution, with a view to obtaining better agreement between the measured and calculated wave systems, both of which would refer to the same hull shape. The wave pattern and the wave-making resistance were then calculated from the amplitudes of the elementary waves by using Havelock's concept.

Professor Inui tried various distributions of sources and sinks (singularities) by volume over the curved surface, in a horizontal plane and over the vertical middle-line plane. For displacement ships at Froude numbers from 0.1 to 0.35, he found the geometry of the ends to be most important, and these could be represented quite accurately by singularities on the middle-line plane. For higher Froude numbers, the dis-

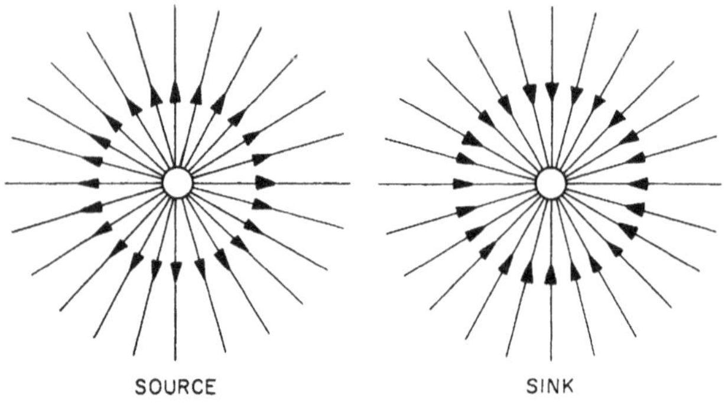

Fig. 13 Flow patterns for source and sink

[12] A source may be looked upon as a point in a fluid at which new fluid is being continuously introduced, and a sink is the reverse, a point where fluid is being continuously abstracted. The flow out of a source or into a sink will consist of radial straight stream lines, Fig. 13. If a source and an equal sink be imagined in a uniform stream flow, the axis of the source and sink being parallel to the flow, the streamlines can be combined as shown in Fig. 14, and there will be one completely closed streamline *ABCD*. Since the source and sink are of equal strength, all fluid entering at $+s$ will be removed at $-s$, and no fluid will flow across *ABCD*, and the space inside this line could be replaced by a solid body.

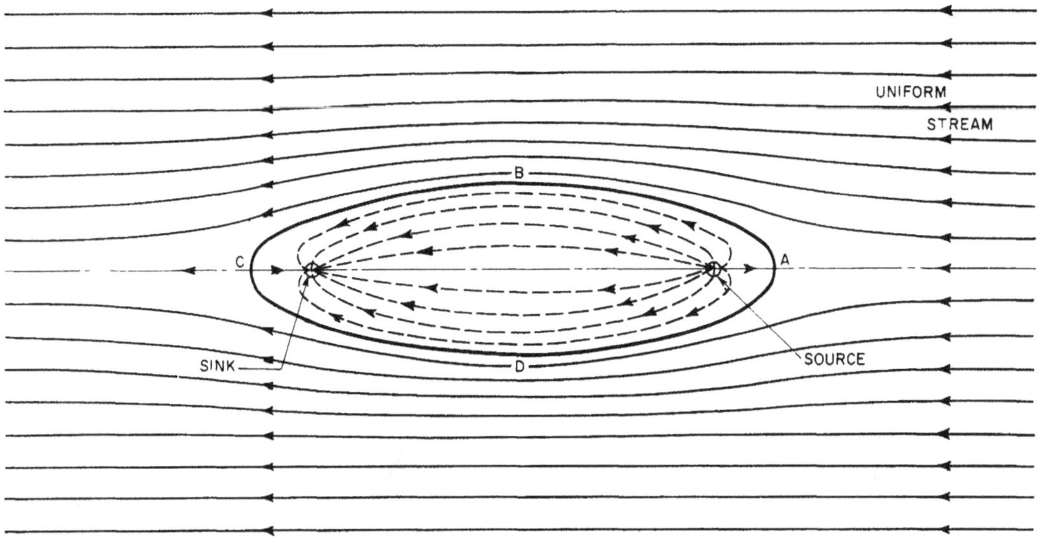

Fig. 14 Flow patterns for a source and sink in a uniform stream

tribution of sources along the whole length becomes important. In summary, the method is to choose a singularity distribution which will give good resistance qualities, obtain the corresponding hull geometry, carry out resistance and wave-observation tests and modify the hull to give a more ship-shape form amidships. In this way Inui has been able to obtain forms with considerably reduced wave-making resistance, usually associated with a bulb at the stem and sometimes at the stern also.

Recent developments in wave-making resistance theory can be divided into four main categories. The first concerns applications of linearized potential flow theory, either with empirical corrections to make it more accurate, or uncorrected for special cases where the errors due to linearization are not serious. The second concerns attempts to improve on linearized potential flow theory, by analysis of non-linear effects on the free-surface condition, or by an assessment of the effects of viscosity. Thirdly, attempts have been made to apply wave resistance theory to hull form design. Fourthly there has been an increase in the number of primarily numerical approaches to ship wave resistance estimation. In the second category (non-linear calculations) the work of Daube (1980), (1981) must be mentioned. He uses an iterative procedure where at each step a linear problem is solved. To this end an initial guess of the location of the free surface is made which is subsequently changed to fulfill a free surface condition. In the computation of the free surface elevation the assumption is made that the projection of the free surface streamlines on the horizontal plane, $z = 0$, always agree with the double model streamlines. This is in fact a low-speed assumption. The non-linear calculation method has been applied to a Wigley hull and a Series 60 ship. Comparison with measurements show a qualitatively satisfactory agreement, and quantitatively the calculations are better than with linear theory for these cases. Part of the discrepancies between measurements and calculations (at least for the higher speed range) can be ascribed to trim and sinkage effects which have not been properly included in Daube's method.

An interesting development has been the determination of pure wave-making resistance from measurements of model wave patterns. The attempts to improve hull forms using the data of wave pattern measurement combined with linearized theory are particularly interesting. For example, Baba (1972) measured the difference in wave pattern when a given hull was modified according to the insight gained from wave resistance theory and thereby gained an improvement.

To a certain degree, the hull forms of relatively high-speed merchant ships have improved because of the application of wave resistance theory. Pien et al (1972) proposed a hull-form design procedure for high-speed displacement ships aided by wave resistance theory. Inui and co-workers have applied the streamline tracing method to practical hull forms with flat bottoms and a design method for high-speed ships with the aid of minimum wave resistance theory has been proposed by Maruo et al (1977). The development of special types of hull forms for drastically reduced wave making have also been guided to a certain extent by wave resistance theory. One of these is the small waterplane area twin-hull (SWATH) ship, discussed in Section 9. The accuracy and usefulness of wave resistance theory was recently demonstrated at a workshop organized by the DTRC (Bai, et al, 1979).

The results of theoretical work would therefore seem at present to be most useful in giving guidance in the choice of the secondary features of hull shape for given proportions and fullness, such as the detail

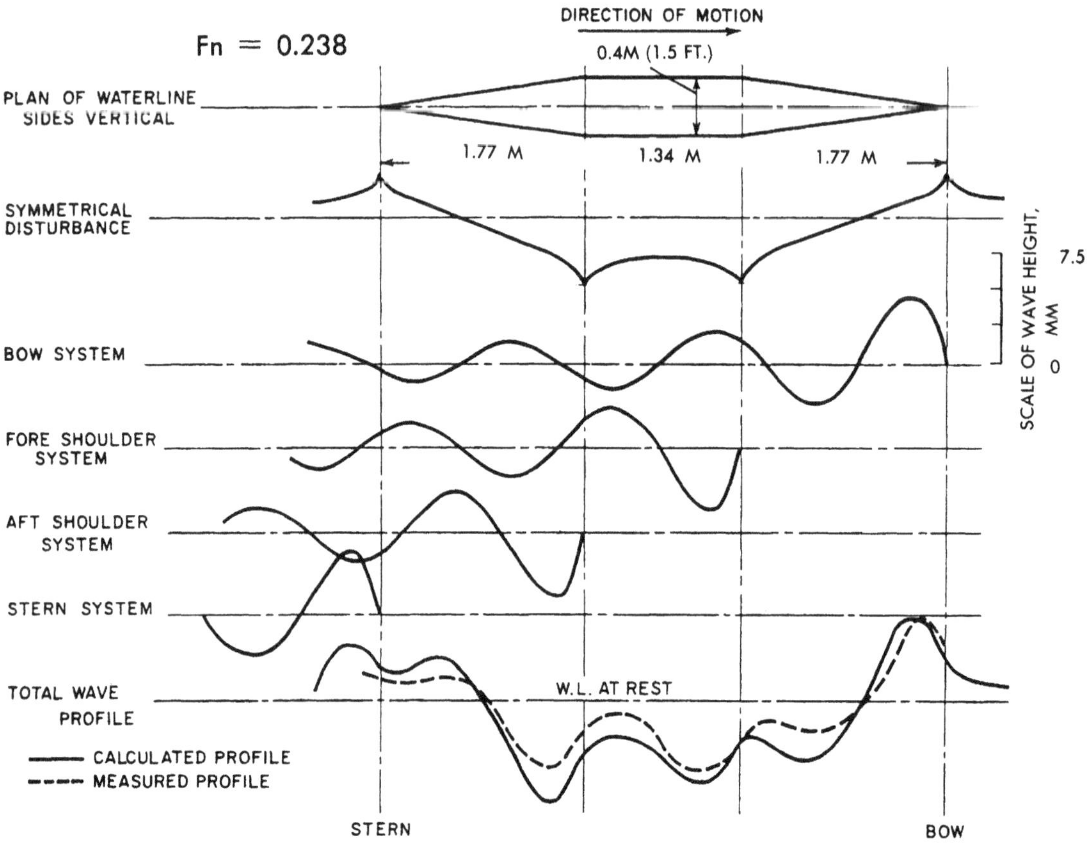

Fig. 15 Wave systems for a simple wedge-shaped form

shapes of waterlines and sections, and choice of size and location of bulbs. The calculation of resistance cannot yet be done with sufficient accuracy to replace model experiments, but it is a most valuable guide and interpreter of model work. The advent of the computer has placed new power in the hands of the naval architect, however, and has brought much closer the time when theory can overcome its present limitations and begin to give meaningful numerical answers to the resistance problem.

4.5 Interference Effects. The results of mathematical research have been most valuable in providing an insight into the effects of mutual wave interference upon wave-making resistance. A most interesting example is that of a double-wedge-shaped body with parallel inserted in the middle, investigated by Wigley (1931). The form of the hull and the calculated and measured wave profiles are shown in Fig. 15. He showed that the expression for the wave profile along the hull contained five terms:

(a) A symmetrical disturbance of the surface, which has a peak at bow and stern and a trough along the center, dying out quickly ahead and astern of the hull. It travels with the hull and because of its symmetry does not absorb any energy at constant speed, and four wave systems, generated at

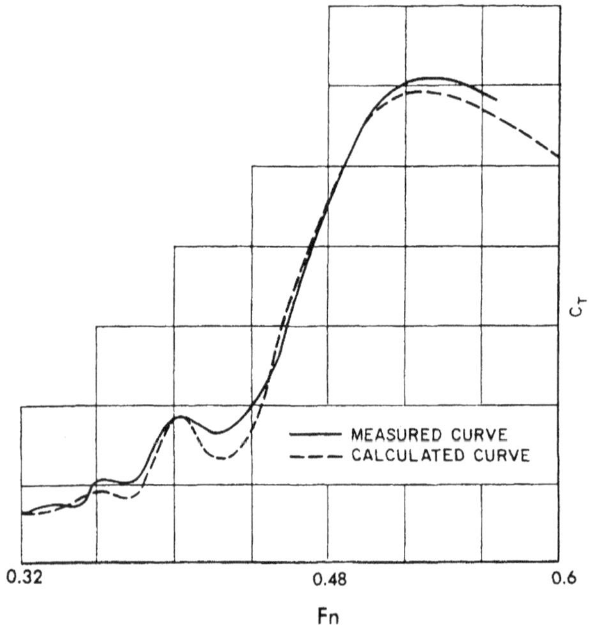

Fig. 16 Resistance curves for wedge-shaped model

Dimensions: 4.8 × 0.46 × 0.30 m, prismatic coefficient 0.636.

(b) the bow, beginning with a crest;

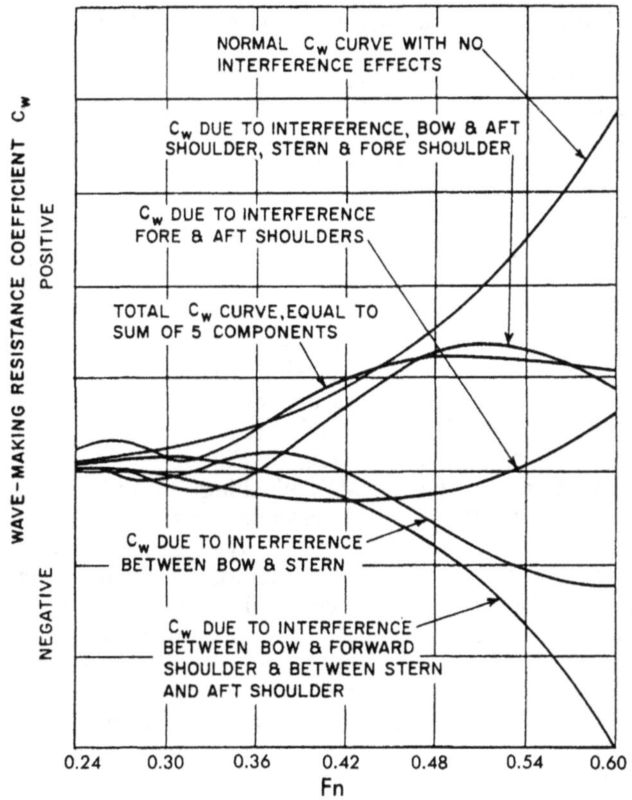

Fig. 17 Analysis of wave-making resistance into components for wedge-shaped model shown above

(*c*) the forward shoulder, starting with a trough;
(*d*) the after shoulder also starting with a trough;
(*e*) the stern, beginning with a crest.

These five systems are shown in Fig. 15. Considerably aft of the form, all four systems become sine curves of continuously diminishing amplitude, of a length appropriate to a free wave travelling at the speed of the model, this length being reached after about two waves.

The calculated profile along the model is the sum of these five systems, and the measured profile was in general agreement with it so far as shape and positions of the crests and troughs were concerned, but the heights of the actual waves towards the stern were considerably less than those calculated (Fig. 15).

This simple wedge-shaped body illustrates clearly the mechanism of wave interference and its effects upon wave-making resistance. Because of the definite sharp corners at bow, stern, and shoulders, the four free-wave systems have their origins fixed at points along the hull. As speed increases, the wave lengths of each of the four systems increase. Since the primary crests and troughs are fixed in position, the total wave profile will continuously change in shape with speed as the crests and troughs of the different systems pass through one another. At those speeds where the interference is such that high waves result, the wave-making resistance will be high, and vice-versa.

In this simple wedge-shaped form the two principal types of interference are between two systems of the same sign, e.g., bow and stern, or the shoulder systems, and between systems of opposite sign, e.g., bow and forward shoulder. The second type is the most important in this particular case, because the primary hollow of the first shoulder system can coincide with the first trough of the bow system before the latter has been materially reduced by viscous effects.

Wigley calculated the values of $V/\sqrt{gL}$ for minima and maxima of the wave-making resistance coefficient C_W for this form, and found them to occur at the following points:

	Values of Fn							
Minima C_W		—	0.187	—	0.231	—	0.345	—
Maxima C_W	0.173	—	0.205	—	0.269	—	0.476	

The mathematical expression for the wave-making resistance R_W is of the form

$$R_W \propto V^6 \text{ (constant term + 4 oscillating terms)}$$

so that the wave-making resistance coefficient C_W is

$$C_W = R_W/\tfrac{1}{2}\rho S V^2 = V^4 \text{ (constant term + 4 oscillating terms)} \quad (23)$$

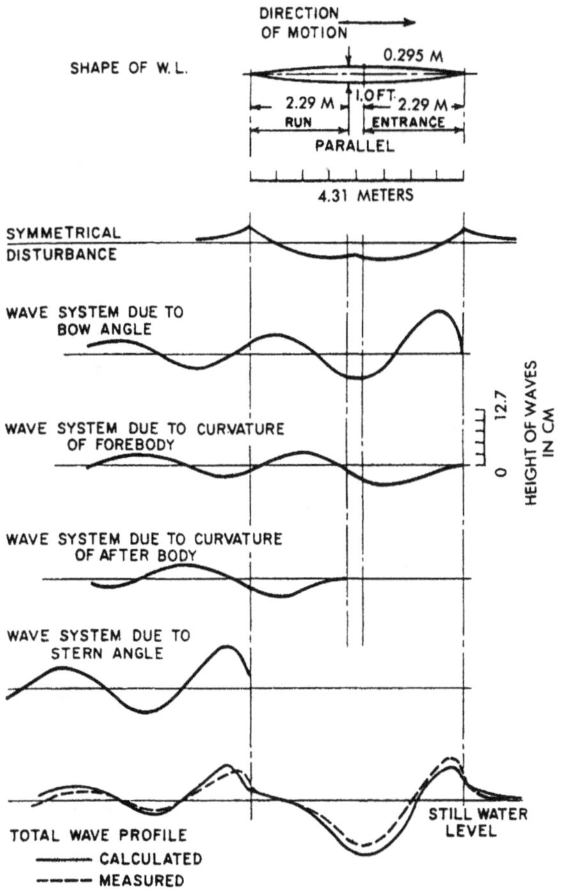

Fig. 18 Wave profile for model with parabolic waterline

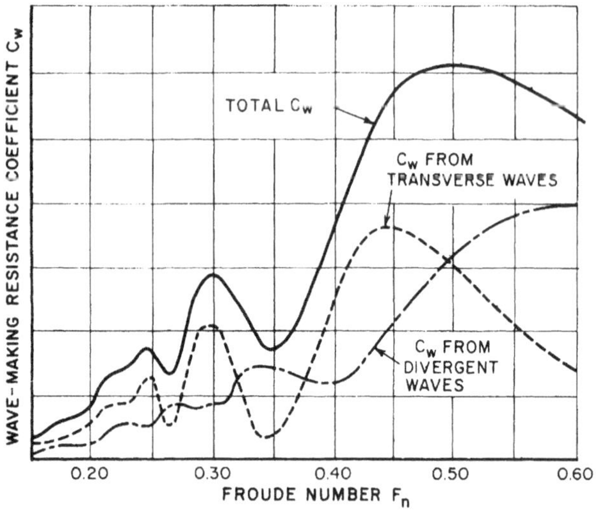

Fig. 19 Contributions made by transverse and divergent wave systems to wave resistance

The C_W curve is thus made up of a steady increase varying as V^4 due to the constant term and four oscillating curves due to the interference between the different free-wave systems (Fig. 17). These latter ultimately, at very high speeds, cancel both each other and the steady increase in C_W, and there is no further hump beyond that occurring at a $V \sqrt{gL}$ value of about 0.45 after which the value of R_W continuously decreases with further increase in speed. However, at these high speeds the hull will sink bodily and change trim so much that entirely new phenomena arise.

For more ship-shaped forms, where the waterlines are curved and have no sharp discontinuities, the wave pattern still consists of five components—a symmetrical disturbance and four free-wave systems (Wigley, 1934). Two systems begin with crests, one at the bow and one at the stern, and are due to the change in the angle of the flow at these points. The other two systems, like the shoulder systems in the straight line form, begin with hollows, but are no longer tied to definite points, since the change of slope is now gradual and spread over the whole entrance and run. They commence at the bow and after shoulder, respectively, as shown in Fig. 18, much more gradually than in the case of the wedge-shaped form. The one due to entrance curvature, for example, may be looked upon as a progressive reduction of that due to the bow angle as the slope of the waterline gradually becomes less in going aft.

Wigley also made calculations to show the separate contributions to the wave-making resistance of the transverse and divergent systems (Wigley, 1942). Up to a Froude number of 0.4 the transverse waves are mainly responsible for the positions of the humps and hollows, Fig. 19. Above this speed the contribution from the divergent waves becomes more and more important, and the interference of the transverse waves alone will not correctly determine the position of the higher humps, particularly the last one at Fn = 0.5.

The existence of interference effects of this kind was known to naval architects long before such mathematical analysis was developed. The Froudes demonstrated them in a striking way by testing a number of models consisting of the same bow and stern separated by different lengths of parallel body (Froude, W., 1877 and Froude, R.E., 1881). W. Froude's sketch of the bow wave system is shown in Fig. 20. As the ship advances but the water does not, much of the energy given to the water by the bow is carried out laterally and away from the ship. This outward spreading of the energy results in a decrease in the height of each succeeding wave of each system with no appreciable change in wave length. Fig. 21 shows a series of tests made at the EMB, Washington, and the corresponding curve of model residuary resistance plotted against length of parallel body (Taylor, 1943). The tests were not extended to such a length of parallel body that the bow system ceased to affect that at the stern. It is clear, however, that its effect is decreasing and would eventually died out, as suggested by the dotted extension of the resistance curve.

Fig. 22 shows a series of curves for the same form at various speeds. In this chart the change of parallel middle-body length which results in successive humps on any one curve is very nearly equal to the wave length for the speed in question, as shown for speeds of 2.6 and 3.2 knots. This indicates that ship waves do have substantially the lengths of deep-sea waves of the same speed.

If all the curves in Fig. 22 are extended in the direction of greater parallel-body length until the bow system ceases to affect the stern system, as was done in Fig. 21, the mean residuary resistances for this form, shown by the dashed lines at the left of the chart, are found to increase approximately as the sixth power of the speed. They are, in fact, the actual resistances stripped of interference effects and represent the true residuary resistances of the two ends. This rate of variation with speed is the same as that given by theory for the basic wave-making resistance before taking into account the interference effects (Fig. 17).

The mathematical theory indicates that the wave resistance is generated largely by those parts of the hull near the surface, which is in agreement with the experimental results obtained by Eggert. This suggests that from the point of view of reducing wave-making resistance the displacement should be kept as low down as possible. The relatively small effect of the lower part of the hull on the wave systems also means that the wave-making resistance is not unduly sensitive to the midship section shape (Wigley, et al, 1948).

4.6 Effects of Viscosity on Wave-Making Resistance. Calculations of wave-making resistance have so far been unable to take into account the effects of viscosity, the role of which has been investigated by Havelock (1923), (1935) and Wigley (1938). One of these

Fig. 20 W. Froude's sketch of characteristic bow wave train

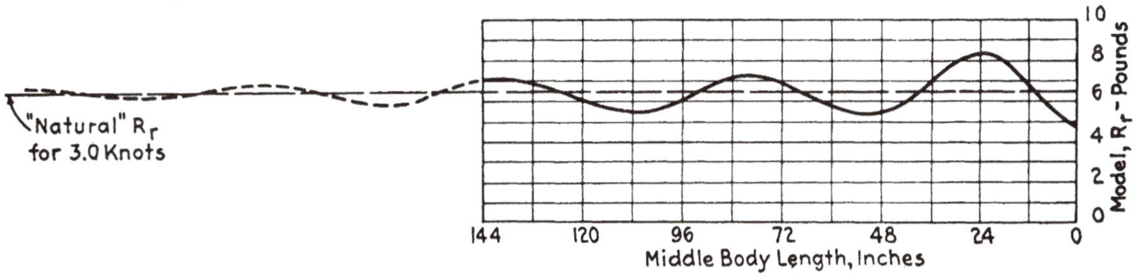

Fig. 21 Quantitative effects of altering length of parallel middle body (English units)

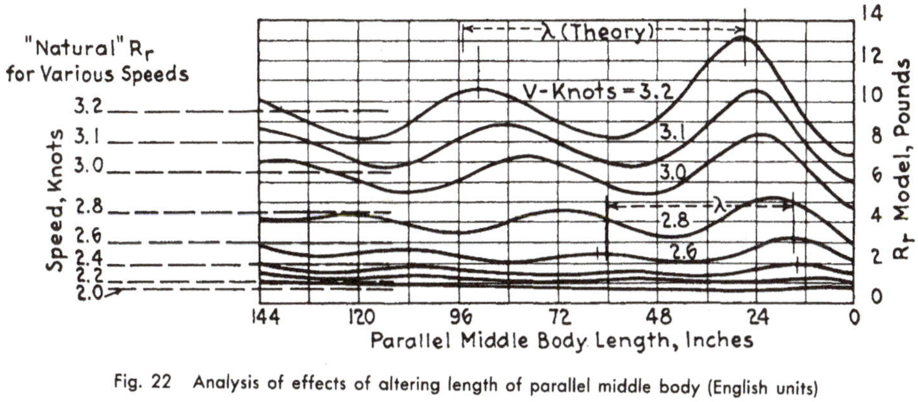

Fig. 22 Analysis of effects of altering length of parallel middle body (English units)

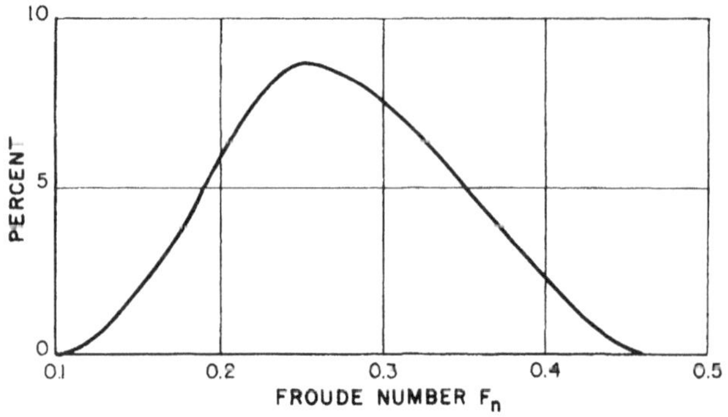

Fig. 23 Scale effect on wave-making resistance

effects is to create a boundary layer close to the hull, which separates the latter from the potential-flow pattern with which the theory deals. This layer grows thicker from stem to stern, but outside of it the fluid behaves very much in accordance with the potential flow theory. Havelock (1926) stated that the direct influence of viscosity on the wave motion is comparatively small, and the "indirect effect might possibly be allowed for later by some adjustment of the effective form of the ship." He proposed to do this by assuming that the after body was virtually lengthened and the aft end waterlines thereby reduced in slope, so reducing the after-body wavemaking. Wigley (1962) followed up this suggestion by comparing calculated and measured wave-making resistance for 14 models of mathematical forms, and deriving empirical correction factors. He found that the remaining differences in resistance were usually within 4 percent, and that the virtual lengthening of the hull due to viscosity varied between 2 and 8 percent.

The inclusion of a viscosity correction of this nature also explains another feature of calculated wave-making resistance. For a ship model which is unsymmetrical fore and aft, the theoretical wave-making resistance in a nonviscous fluid is the same for both directions of motion, while the measured resistances are different. With the viscosity correction included, the calculated resistance will also be different.

Professor Inui (1980) in his wave-making resistance work also allows for viscosity by means of two empirical coefficients, one to take care of the virtual lengthening of the form, the other to allow for the effect of viscosity on wave height.

4.7 Scale Effect on Wave-Making Resistance. Wigley (1962) has investigated the scale effect on C_W due to viscosity, pointing out that the calculated curves of C_W are usually higher than those measured in experiments and also show greater oscillations. These differences he assigned to three major causes:

(a) Errors due to simplifications introduced to make the mathematical work possible.

(b) Errors due to neglect of the effects of viscosity on R_W.

(c) Errors due to the effects of wave motion on R_F.

Errors under *(a)* will decrease with increasing speed, since they depend on the assumption that the velocities due to the wave motion are small compared with the speed of the model, which is more nearly fulfilled at high speed.

Errors under *(b)* will depend on Reynolds number, and therefore on the size of the model, decreasing as size increases. From experiments on unsymmetrical models tested moving in both directions, these errors cease to be important for Fn greater than 0.45.

At low speeds errors under *(c)* are negligible, but become important when Fn exceeds 0.35, ($V_k/\sqrt{L}$ = 1.15) as evidenced by the sinkage and trim, which increase very rapidly above this speed.

A practical conclusion from this work is the effect on the prediction of ship resistance from a model. In a typical model the actual wave resistance is less than that calculated in a perfect fluid for Froude numbers less than about 0.35. This difference is partly due to viscosity, the effect of which will decrease with increased size, and C_W will increase with scale instead of being constant as assumed in extrapolation work. Wigley made estimates of the difference involved in calculating the resistance of a 121.9 m ship from that of a 4.88 m model at a Froude number of 0.245 and found that the resistance of the ship would be underestimated, using the usual calculations, by about 9 percent, the variation with speed being approximately as shown in Fig. 23. The effect disappears at low speeds and for values of Fn above 0.45.

4.8 Comparison Between Calculated and Observed Wave-Making Resistance. Many comparisons have been made between the calculated and measured wave-making resistances of models. Such a comparison is difficult to make, however. All that can be measured on the model is the total resistance R_T, and the value

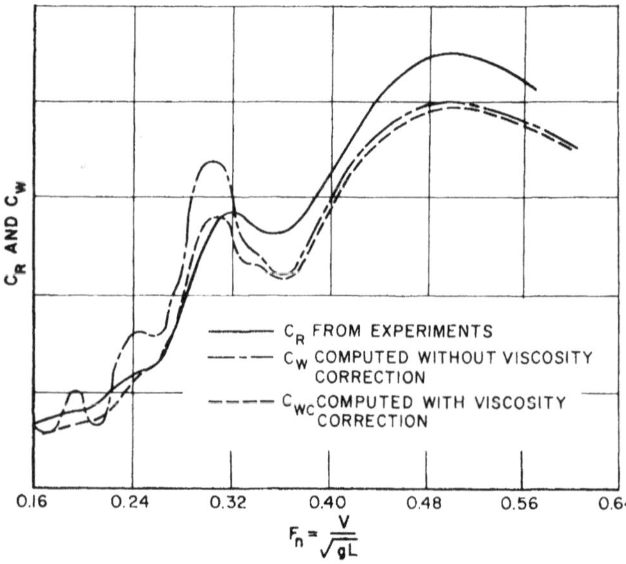

Fig. 24 Comparison of measured C_R and calculated C_W

of R_W can only be obtained by making assumptions as to the amount of frictional resistance, viscous pressure drag and eddy-making resistance, all quantities subject to considerable doubt. The wave-making resistance has been measured directly by observing the shape of the wave system astern of the model and computing its energy and the total viscous drag has been measured by a pitot tube survey behind the model (Wehausen, 1973). Both of these methods are relatively new, and there are problems in interpreting the results. In the meantime it is perhaps best to be content with comparing the differences between the calculated and measured resistances for pairs of models of the same overall proportions and coefficients but differing in those features which are likely to affect wave-making resistance.

A comparison of much of the available data has been made by Lunde, the measured C_R being derived from C_T on Froude's assumption and using his skin-friction coefficients, the calculated C_W being empirically corrected for viscosity (Lunde, 1957).

At low Froude numbers, less than 0.18, it is difficult to determine C_R with any accuracy. At higher speeds, the humps at Fn of 0.25 and 0.32 and the intervening hollow are much exaggerated in the calculated curves, and any advantage expected from designing a ship to run at the "hollow" speed would not be fully achieved in practice (Fig. 24). The general agreement in level of the curves over this range depends to some extent upon the form of the model, theory overestimating the resistance for full ships with large angles of entrance. Just above a speed of Fn = 0.32 the model becomes subject to increasing sinkage and stern trim, effects which are not taken into account in the calculations.

The last hump in the C_W-curve occurs at a Froude number of about 0.5, and here the calculated value of C_W is less than the measured C_R, again possibly due to the neglect of sinkage and trim.

In all cases the humps and hollows on the measured curves occur at higher values of Fn than those given by theory, by amounts varying from 2 to 8 percent. In other words, the model behaves as though it were longer than its actual length, and this is undoubtedly due mostly to the virtual lenghening of the form due to the viscous boundary layer. At very low speeds, Fn = 0.1, the wave-making resistance varies approximately as the square of the tangent of the half-angle of entrance, but its total value in terms of R_T is very small. At high speeds, with Fn greater than 1.0, the wave-making resistance varies approximately as the square of the displacement, illustrating the well-known fact that at very high speeds shape is relatively unimportant, the chief consideration being the displacement carried on a given length.

Section 5
Other Components of Resistance

5.1 Eddy Resistance, Viscous Pressure Drag, Separation Resistance and Wave-breaking Resistance. The turbulent frictional belt around a ship consists of eddies, so that all forms of frictional resistance are really due to eddy-making. However, the term is usually applied to the resistance due to eddy formation or disturbed streamline flow caused by abrupt changes of form, appendages or other projections, and excludes tangential skin friction.

When the total model resistance R_{TM} is measured over a range of speeds and plotted as the coefficient $C_{TM} = R_{TM}/\frac{1}{2}\rho SV^2$ against log Rn, the curve will be of the general shape shown in Fig. 25. Also shown is a curve of the coefficient of frictional resistance C_{FOM} for a smooth plank in fully turbulent flow. The intercept C_{RM} between the curves of C_{FOM} for the plank and C_{TM} for the total model resistance is the so-called residuary-resistance coefficient.

In a typical case the C_{TM}-curve at the very low values of Rn is almost parallel to the C_{FOM}-curve but some distance above it. Since the primary component of the coefficient C_{WM} varies as the fourth power of the speed, (Equation 23) the wave-making resistance at very low values of Rn must be vanishingly small, and so the intercept C_{RM} (BC in Fig. 25) cannot be attributed to this cause. If a line CH is drawn parallel to the curve of C_{FOM}, running into the lowest part of the total resistance curve C_{TM}, it would be a nearer approach to the truth to say that the intercept FG represents the wave-making resistance coefficient,[13] $C_{WM} = (R_{WM}/\frac{1}{2}\rho SV^2)$.

On this assumption the intercept FE (= BC) must be due to some other cause. Models of submarines and airships run fully submerged where wave-making is absent have resistances higher than the C_{FOM}-values for the equivalent plank, and this increase has been called the form resistance or form drag. Moreover, the form-resistance coefficient for such shapes is higher the smaller the slenderness ratio (the length-diameter ratio, L/D); i.e., the "stumpier" the models. A similar

[13] The exact formulation used for C_{FOM} and the question whether the line CH should be exactly parallel to the curve C_{FOM} are not important for the present argument. These questions are discussed in Section 6.4.

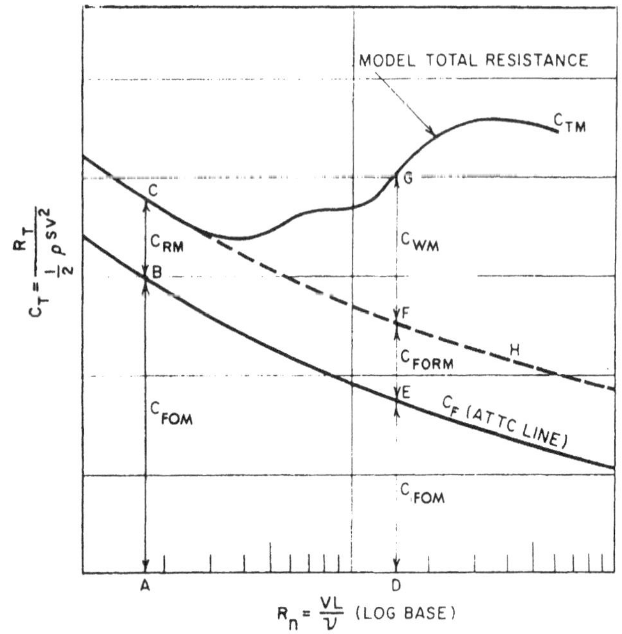

Fig. 25 Elements of model resistance

effect is found in ship models, the intercept BC being greater the fuller the model and the smaller the ratio $L/\nabla^{1/3}$.

There are three main causes of this form resistance. The ordinate of the curve C_{FOM} applies to a flat surface having the same length and wetted area as the model and so neglects any effects due to curvature of the hull. This curvature affects the pressure distribution along the length, causing the velocity to increase along most of the middle part and to decrease at the ends. The former effect outweighs the latter. Also, since the path along a streamline from bow to stern is longer on a shaped body than on a plank, the average velocity must be higher. Thus the real skin friction of a ship must be greater than that of the "equivalent plank." Since the pressure and velocity changes and the extra path length are greater the fuller and stumpier the form, such shapes would be expected to have greater form drag. This is borne out by experiments on bodies of revolution in air, where, for example, Young (1939) found form drag percentages varying with length-diameter ratio as follows:

L/D	10	5	3.33
Form drag, percent	5	17	30

Similar figures have been found for bodies of revolution run deeply submerged in water. For a given volume of displacement, increases of L/D ratio beyond a certain point, while it may still reduce the form drag, will increase the frictional resistance because of the greater surface area and so in terms of total resistance there will be some optimum value of the L/D ratio. The value depends upon the exact shape and upon the amount of appendages necessary to give directional stability, and varies between 5 and 7. For surface ships the intercept C_{RM} has been found to vary from 5 to 15 percent of C_{FOM} in warships and up to 40 percent or more in full cargo ships. These increases, however, cannot be attributed wholly to curvature effects, which leads to the other causes of form effect.

In discussing wave-making resistance, it was pointed out that the existence of the boundary layer had the virtual effect of lengthening the form and reducing the slopes of the after waterlines. This is a region where the normal pressure on the hull is higher than the static pressure, owing to the closing in of the streamlines, and the forward components of these excess pressures will exert a forward thrust overcoming some of the ship's resistance. The presence of the boundary layer reduces these forward components, resulting in an increase in resistance as compared with that which would be experienced in a nonviscous fluid, and so is called the *viscous pressure drag*.

If the curvature near the stern becomes too abrupt, the bilge radius too hard, the after end sections too U-shaped or there are other discontinuities in the hull shape, the water may no longer be able to follow the hull and breaks away, and the intervening space between the hull and the smooth-flowing water is filled with eddies, as illustrated in Fig. 1(d). A point at which this happens is called a separation point, and the resulting resistance is the third element of form drag, called separation resistance. Separation of this kind can also affect the pressure distribution on the hull, and so modify the viscous-pressure drag.

To explain the failure of the streamlines to follow the hull it is necessary to consider the variation of pressure and velocity along the length.

The water particles immediately in contact with the hull are assumed to be carried along with it. Due to viscosity, these particles draw the next layer of particles with them, and this effect spreads outwards from the hull. The spread of the boundary layer continues until the velocity of the outer particles at any point is just equal to the potential flow velocity at this point (Fig. 26). The boundary layer gets thicker from bow to stern due to the continual entrainment of more water. Within this layer the velocity gradients are very much greater than those existing in the potential flow, and most of the fluid shear responsible for the skin friction must occur within this boundary layer. This leads naturally to the idea of a boundary layer of finite thickness, within which the influence of viscosity is important and beyond which viscosity may be neglected, a concept which has proved useful in analyses of various problems in aerodynamics (Prandtl, 1904). Since the velocity in the boundary layer approaches the potential flow velocity asymptotically, its thickness is usually taken to the point where the velocity is 0.99 of that of the undisturbed fluid. The body shape defined by the outer limits of the layer may be considered to move without friction and normal pressures appear to be transmitted across the boundary layer without sensible distortion. Fluid particles moving aft from midships, relative to the body, have their velocities diminished both by the shearing stresses and by the increasing pressures. Some of them may have insufficient kinetic energy to overcome the adverse pressure gradient and so come to rest before reaching the stern or even start to move forward. Following particles are then forced outward away from the body, setting up pressures which tend to move them back towards the hull, thus causing large scale vortices in the boundary layer. From this point the flow is separated from the hull and a widening band of eddying water intervenes between the hull and the smooth flow outside it. These eddies carry away the kinetic energy which has been expended in forming them and so give rise to separation drag. Sufficient is not yet known to divide the total viscous drag into its separate components, and this fact has an important bearing on the extrapolation of model results to the ship.

In addition to form and separation resistance, eddy-making resistance is also caused by struts, shafts, bossings and other appendages, as is discussed in Section 5.3.

Especially in the case of bluff hull forms the phenomenon of *wave-breaking* and wave-breaking resistance have to be considered as well. For this type of hull the flow ahead of the bow becomes irregular and complex, usually leading to a breaking wave.

At very low Froude numbers, below approximately 0.10, wave-making hardly occurs and the free surface at the stern rises to a height approximately equal to $V^2/2g$, where V is the speed of the ship and g the acceleration due to gravity, in accordance with the Bernoulli equation. As the ship speed increases however, this rise of the wave at the stern no longer occurs and instead the bow-wave breaks.

The resistance associated with wave-breaking at the stern has recently been the subject of extensive investigations. Important studies on this topic were made by Taneda, et al (1969), who presented the results of flow observations around a fishery training ship and its corresponding scaled model. Baba (1969) was the first to term the observed flow around the bow "wave-breaking" and presented measurements of momentum loss due to wave-breaking by means of a wake survey far behind a ship model. Baba proposed a hydraulic jump model as a means of calculating the momentum loss due to breaking waves. He showed that wave-breaking resistance may contribute an appreciable part of the total resistance of full forms. Dagan, et al (1969) carried out a theoretical study of the two-dimensional flow past a blunt body. The drag associated with wave-breaking was obtained by calculating the loss of momentum of the flow. Ogilvie (1973) obtained analytical results for the case of a fine wedge-shaped bow and obtained a universal curve for the shape of the bow wave on the hull. Experimental results generally confirm their predictions.

The most recent work in this area was carried out by Kayo et al (1981). They carried out systematic experiments on the effect of shear on the free surface. They concluded that bow wave-breaking can be considered to be due to flow separation at the free surface.

Bow wave-breaking can generally be avoided by requiring that the tangent to the curve of sectional areas at the forward perpendicular be not too steep. Taniguchi, et al (1966) derived a criterion for full form, low-speed ships that can be applied in this connection. The work carried out by Inui, et al (1979), Miyata, et al (1980) and Kayo, et al (1981) has resulted in a criterion based on the half-entrance angle of the waterline. Finally, the work carried out by Taylor, G.I. (1950) reveals that at a certain speed the free surface becomes unstable and breaks when the radius of curvature of the curved streamlines result in a value of the centrifugal acceleration V^2/R greater than a critical value. This so-called Taylor instability criterion, when applied to the case of the flow around the stem of a ship with radius R, results in the approximate expression that $R \geq V^2/50$, with R in meters and V in m per sec, to avoid wave-breaking.

5.2 Air and Wind Resistance. A ship sailing on a smooth sea and in still air experiences a resistance due to the movement of the above-water hull through the

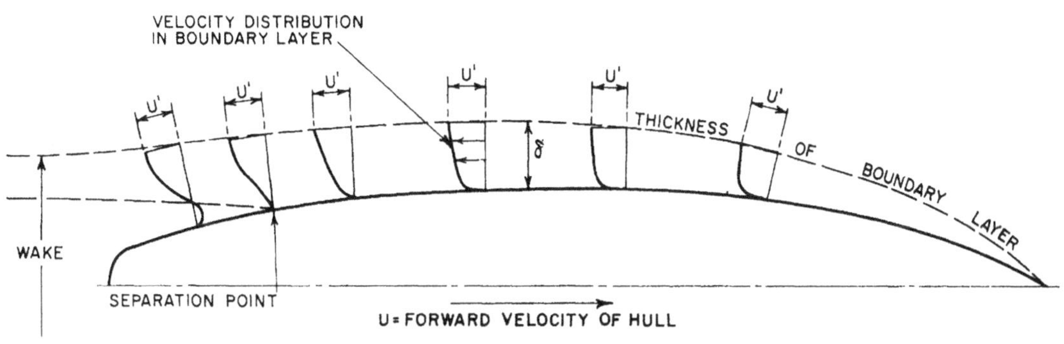

Fig. 26 Schematic diagram of boundary-layer flow

U' = Velocity at any point on hull in potential flow without viscosity
The velocity in boundary layer approaches U' asymptotically, and the thickness of layer, δ, is usually measured to the point where the velocity is 0.99 U'

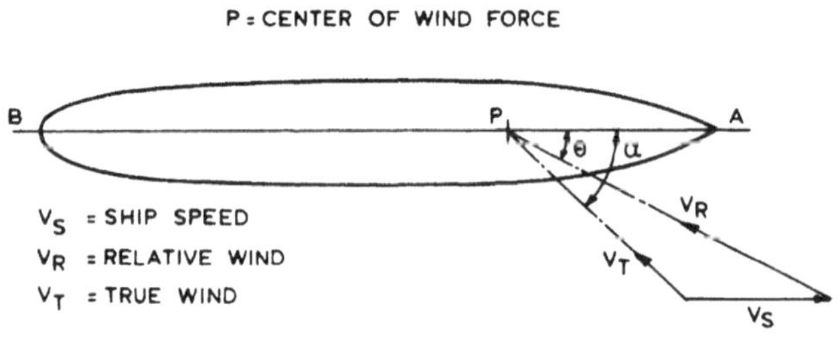

Fig. 27 Diagram of wind force

air. This resistance depends upon the ship's speed and upon the area and shape of the upper works.

When a wind is blowing, the resistance depends also upon the wind speed and its relative direction. In addition, the wind raises waves which may cause a further increase in resistance. The effects of waves are dealt with in Chapter VIII, Vol. III.

The "true" wind is termed to be the wind which is due to natural causes and exists at a point above the sea whether or not the ship is there. Zero true wind is still air. The "relative" or "apparent" wind is the vectorial summation of the velocities and directions of the ship and the true wind (see Fig. 27).

Because of the many functions which superstructures have to fulfill, they cannot be adequately streamlined, and in any case this would be effective only in winds from nearly dead ahead. The reduction in total ship resistance which can be realized by such means is therefore relatively small.

Most of the resistance of superstructures is due to eddy-making, and therefore varies with the square of the speed, and the effects of Reynolds number changes can be neglected. For a ship moving in still air the air resistance can therefore be written as

$$R_{AA} = \text{coefficient} \times \tfrac{1}{2}\rho A_T V^2 \qquad (24)$$

where

A_T = transverse projected area of above-water hull
V = ship speed

and the coefficient will have a value depending on the shape of the hull and erections.

D.W. Taylor (1943) suggested that the air resistance of ordinary ships in a head wind could be assumed equal to that of a flat plate set normal to the direction of motion and having a width B equal to the beam of the ship and a height equal to $B/2$. From experiments in air, he derived a resistance coefficient of 1.28, so that

$$\begin{aligned} R_{AA} &= 1.28 \times \tfrac{1}{2}\rho A_T (V_R)^2 \\ &= 1.28 \times \tfrac{1}{2} \times 1.223 \times \tfrac{1}{2}B^2 \times (V_R)^2 \qquad (25) \\ &= 0.783 \times \tfrac{1}{2}B^2 \times (V_R)^2 \end{aligned}$$

where V_R is the "apparent" wind velocity, or wind velocity relative to the ship, in m per sec, B is in m, ρ is in kg/m^3 and R_{AA} is in newtons. In still air, $V_R = V$.

An extensive study of the resistance of ships' superstructures has been made by Hughes (1930). Models were made of the above-water hull and erections, and were towed upside down in water at different speeds and at different angles to simulate various relative wind strengths and directions. Three models were used, representing a typical tanker, cargo ship and Atlantic liner, small structures such as railings and rigging being omitted.

The simulated relative wind velocity V_R was determined, Fig. 27, and the total force F acting on the model was measured, together with its direction and point of application. For a given arrangement at a constant angle θ of the relative wind off the bow, the value of $F/(V_R)^2$ was found to be constant for all speeds up to those at which wave-making began to be important.

A typical plot of $F/(V_R)^2$ and α is shown in Fig. 28, where α is the angle between the centerline of ship and the resultant wind force. The value of $F/(V_R)^2$ is a maximum when the relative wind is on the beam, with the maximum area presented to the wind. It does not correspond with maximum wind resistance to ahead motion, since it is acting approximately at right angles to the direction of motion. As shown later, maximum wind resistance occurs when the relative wind is about 30 deg off the bow.

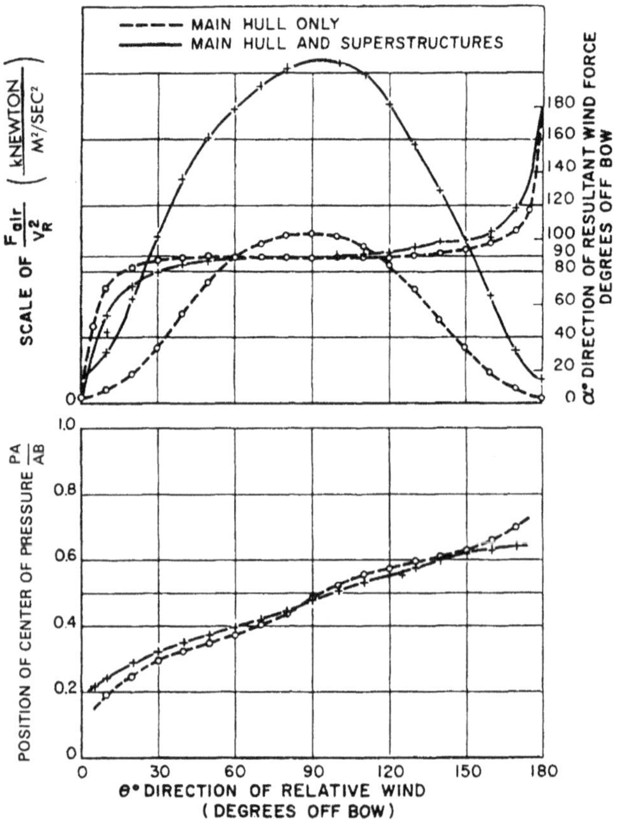

Fig. 28 Resultant wind force and center of pressure

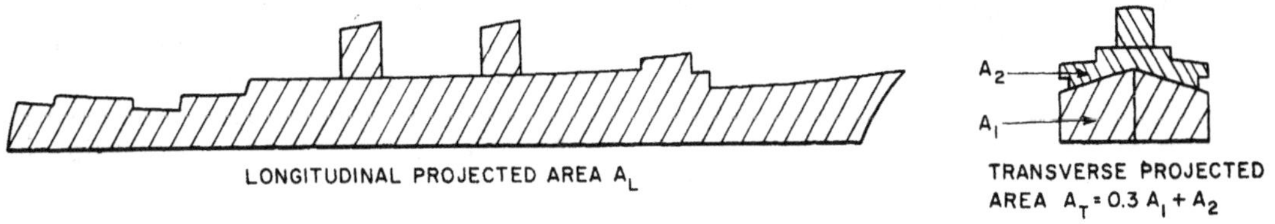

Fig. 29 Areas for air resistance

The center of wind force is close to the bow for winds nearly head on, moving aft with increasing values of θ to a point near the stern when the wind is nearly astern.

For a beam wind, most of the area, both main hull and superstructure, is normal to the wind, and has the same resistance value, so that the effective area is approximately equal to the longitudinal projected area A_L, Fig. 29. For a head wind, the main hull below the weather deck has a much lower specific resistance than the frontal area of the superstructures. Hughes found this ratio to be 0.31, 0.27 and 0.26 for the tanker, cargo ship and liner, respectively. For practical purposes the "equivalent area" can thus be found by adding 0.3 of the projected main hull area to the projected superstructure area, giving the transverse projected area A_T (Fig. 29).

Hughes developed an equation

$$F = K\rho(V_R)^2(A_L \sin^2\theta + A_T \cos^2\theta)/\cos(\alpha - \theta) \quad (26)$$

where again F is in newtons, V_R is in m/sec, ρ is the mass density of air $= 1.223$ kg/m^3, and A_L and A_T are in square meters. He found that K had a value of approximately 0.6 for all values of θ, varying between 0.5 and 0.65.

For a head wind, $\theta = \alpha = 0$, and the wind resistance will be

$$R_{AA} = F = K\rho A_T(V_R)^2, \text{ with } V_R \text{ in meter}$$

per second, so that

$$K = \frac{F}{\rho A_T(V_R)^2}$$

which is nondimensional in consistent units. Putting $K = 0.6$

$$R_{AA} = 0.734 A_T(V_R)^2 \quad (27)$$

which is practically the same as Taylor's expression (25) although the area A_T is somewhat different.

For small angles of wind off the bow or stern, the wind force in the line of the ship's motion will be approximately $F \cos\alpha$. Values of $F\cos\alpha/\rho A_T(V_R)^2$, or $K\cos\alpha$, for varying values of θ are shown in Fig. 30 for a tanker, a cargo ship and a liner with normal superstructures. These curves show that, although cos α decreases with an increasing angle of the apparent wind off the bow or stern, the experimentally determined values of F increase so rapidly, because of the rapid increase of area presented to the wind as θ departs from 0 or 180 deg, that the product $F\cos\alpha$ increases with θ and the maximum resistance to ahead motion on all three types of ship occurs when the wind is about 30 deg off the bow. This has been confirmed by full-scale data obtained on the German ship *Hamburg* by Kempf, et al (1928). The ahead resistance is given by

$$R_{AA} = F\cos\alpha = K\rho A_T(V_R)^2\cos\alpha \quad (28a)$$

With R_{AA} in newtons, V_R in m/sec, and $\rho = 1.223$

$$R_{AA} = 1.223\,A_T(V_R)^2(K\cos\alpha) \quad (28b)$$

where $(K\cos\alpha)$ is the ordinate from Fig. 30 at the desired value of θ.

For a head wind, $\alpha = 0$, and K from Fig. 30 is about 0.6, so that

$$R_{AA} = 1.223 \times 0.6 A_T(V_R)^2$$
$$= 0.734\,A_T(V_R)^2$$

in agreement with Equation (25).

For flat plates of area A_T normal to the wind, it was found the value of $R/A_T(V_R)^2$ to be 0.710 in agreement with Hughes' values derived from tests in water.

In a second paper, Hughes (1932) investigated the effects of changes in shape, type and arrangement of superstructures, measuring their single and combined resistances by attaching them to the underside of a raft rather than to a specific hull. The resistance of superstructures can be reduced either by a reduction

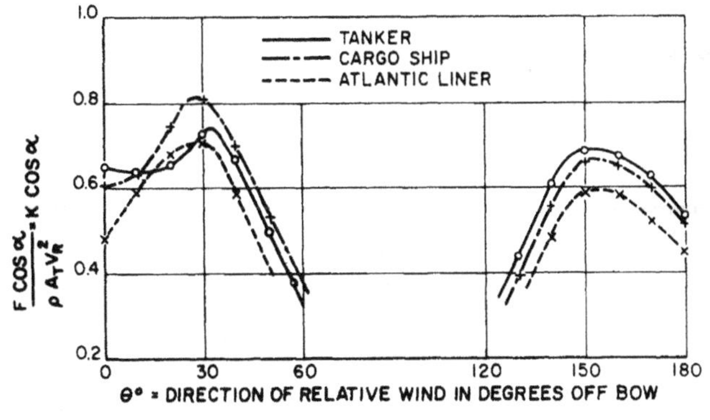

Fig. 30 Resistance coefficients for relative wind ahead or astern. All models with normal superstructures

of the projected area, a change in arrangement or by changes in shape and proportions. For ahead winds, Hughes found the principal effects to be as follows:

(a) The total resistance of a number of separate units is in general less than the sum of their separate resistances, because of shielding effects, but these effects decrease with streamlining.

(b) Rounding, tapering or stepping back the fore ends of the main erections reduce the ahead resistance, but changes of this kind at the aft ends have little effect.

(c) The shear on the main hull forward has a considerable shielding effect.

(d) There is probably some scale effect on the resistance of large rounded structures such as masts and smokestacks, but this will offset to some extent the omission of small fittings and rigging on the model.

The savings in speed and power for the three designs tested by Hughes when the superstructures were changed in the way indicated by the experiments for ships in still air, i.e., relative wind is ahead and equal to ship's speed, and against head winds of 20 and 40 knots are summarized in Table 3.

These estimates apply only to a head wind and give the additional wind resistance only, ignoring any effects of the seas which would accompany high winds. For winds up to 30 deg off the bow, the additional resistance to ahead motion may be up to 30 percent greater than the values given in Table 3. The effects of streamlining will be much less for winds other than from ahead. One of the points brought out by the foregoing figures is the much greater relative effect of wind resistance on the slowest ship. In a 40-knot head wind the normal tanker loses 3.27 knots, or 32.7 percent while the Atlantic liner only loses 1.73 knots, or 6.9 percent. For similar ships with streamlined superstructures, the figures are, respectively, 2.2 knots (22 percent) and 1.2 knots (4.8 percent). Thus on a percentage basis much larger allowances are required for wind resistance in slow-speed ships than in fast ones, and streamlining should by no means be confined to the latter. In all three types of ship, the reduction in ahead-wind resistance in calm air by streamlining the erections amounts to about 30 percent.

Air-resistance information is valuable in the analysis of measured mile and voyage data and the estimation of mooring and towing forces, and the British Ship Research Association (BSRA) now British Maritime Technology (BMT), has carried out tests on models of modern counterparts of the same three types of ships as used by Hughes (1932). The superstructures were typical of present-day ships, and in no case was the streamlining excessive. The experiments were carried out in a large wind tunnel, not by towing in water. Near to the sea surface the wind is slowed down, the velocity gradient very close to the water being very steep. The tests made by Hughes did not simulate this gradient and are strictly applicable only to estimates of air resistance when the ship is moving in still air. In the BMT experiments some tests were made in a velocity gradient which compared very closely with that measured over the sea and over very smooth snow (Fig. 31).

The passenger liner had an average *K*-value of 0.59 with no velocity gradient, which compares with Hughes' value of 0.63 for a similar type of ship. In the tests with a velocity gradient the value of *K* was reduced by some 28 percent of 0.425 using the value of V_R corresponding to no gradient in each case. The average values for the tanker and cargo-ship models with the velocity gradient were 0.32 and 0.34 as compared with Hughes' figures of 0.58 and 0.6 without gradient, a reduction of about 45 percent. The wind gradient reduces the resultant wind force by appre-

Table 3—Effects of Superstructure Changes on Speed and Power in Head Winds

Condition			Tanker	Cargo Liner	Atlantic Liner
Still Air	Dispt in ton		16,256	15,037	38,608
	Normal speed, knots		10	14	25
	P_E Bare hull, kW		888	2,072	25,760
	Ratio as percent (still air resist/hull water resist)	With original erections	2.5	2.15	2.1
		With streamlined erections	1.7	1.55	1.45
With 20 knots head wind	Normal erections	Ship speed, knots	8.89	13.13	24.17
		Rel. wind in knots	28.9	33.1	44.2
		R_{air}/R_{water}, percent	26.5	13.5	7.0
	Streamlined erections	Ship speed, knots	9.25	13.37	24.42
		Rel. wind in knots	29.25	33.37	44.42
		R_{air}/R_{water}, percent	17.0	10.0	5.0
		Gain in speed, knots	0.36	0.24	0.25
With 40 knots head wind	Normal erections	Ship speed, knots	6.73	11.76	23.27
		Rel. wind in knots	46.7	51.76	63.3
		R_{air}/R_{water}, percent	122.0	42.0	15.5
	Streamlined erections	Ship speed, knots	7.81	12.39	23.80
		Rel. wind in knots	47.8	52.4	63.8
		R_{air}/R_{water}, percent	64.0	27.5	10.4
		Gain in speed, knots	1.08	0.63	0.53

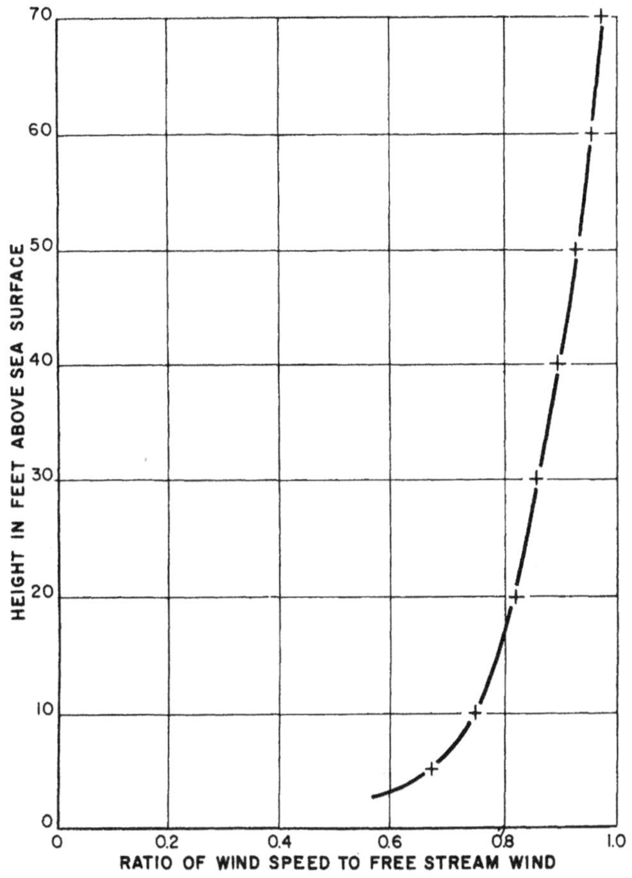

Fig. 31 Wind gradient above sea surface

ciable amounts, but less for the passenger ship with high erections which extend further into the higher wind-speed range.

When a ship is moving against a natural wind, the resultant relative wind is made up partly of the natural wind, which has a velocity gradient, and partly of the wind created by the ship's speed, which has no gradient. The effective gradient in any case therefore depends upon the relative values of wind speed and ship speed. Shearer, et al (1960) estimated that in applying their model results, obtained in a graded wind, to the case of a ship advancing into still air, the air resistance would be underestimated by some 40 and 25 percent for the tanker and cargo ships in loaded and light condition, respectively, and by about 21 percent for the passenger liner. In a head wind equal to the ship's speed, these differences would be reduced by about one half, and still more for higher relative winds.

Wilson, et al (1970) analyzed the available wind data and defined a wind drag coefficient C_{AA_0} for a head wind (defined as the zero relative wind heading)

$$C_{AA_0} = \frac{R_{AA_0}}{\frac{1}{2}\rho A_T V_R^2} \tag{29}$$

An average value of 0.45 was found for aircraft carriers and 0.70 for other combatant ships. A value of 0.75 was found for naval auxiliaries at the full-load condition which is also in agreement with Equation (25). They also defined a heading coefficient C_γ according to

$$C_\gamma = \frac{C_{AA_\gamma}}{C_{AA_0}}$$

where R_{AA_γ} is wind drag at non-zero relative wind heading in newtons and C_{AA_γ} is the corresponding wind wind drag coefficient. Their analysis indicated that the behavior of C_γ as a function of the relative wind heading γ is essentially the same for all combatant ships. Plots of the heading coefficient C_γ versus the relative wind heading γ are presented in Figs. 32 and 33, for naval combatants and auxiliaries, respectively. These plots also show that the maximum value of C_γ occurs near relative wind headings of 30 degrees and 150 degrees. Also, there is typically a rather flat spot in the curve at about 80 degrees.

The wind drag and the change in effective power or speed due to that drag can be determined from

$$P_{E_{\text{wind}}} = \frac{R_{AA_\gamma} V}{1000} \text{ in kW} \tag{30}$$

where $R_{AA_\gamma} = C_{AA_0} A_T V_R^2 C_\gamma$ in newtons

V = ship speed in m/sec.

The change in ship speed δV_{wind} is calculated from the slope of the relevant speed-power curve at the speed V of interest, viz.

$$\delta V_{\text{wind}} = \frac{P_{E_{\text{wind}}}}{\delta P_E / \delta V} \tag{31}$$

where $\dfrac{\delta P_E}{\delta V}$ = slope of the speed-power curve.

If there is a strong wind on the beam, the ship will make leeway, which leads to an important increase in hydrodynamic resistance. Wagner (1967) proposed a method for calculating the resistance of a ship for the effects of beam winds. He computes an effective wind resistance, including the hydrodynamic component due to leeway when subjected to winds at angles of attack other than zero. This resistance component can become quite sizeable for hull forms that are unable to produce an appreciable hydrodynamic sideforce at small drift angles or for those with large superstructures. Wieghardt (1973) presented results of measurements for the ferry *Kronprins Harald* and the research vessel *Meteor*. The results for the ferry display an important increase, while the results for the research vessel display a negligible increase in the hydrodynamic resistance due to beam winds. Jørgensen, et al (1966) reported that in some cases the existence of leeway can influence the wake and hence the propulsive efficiency.

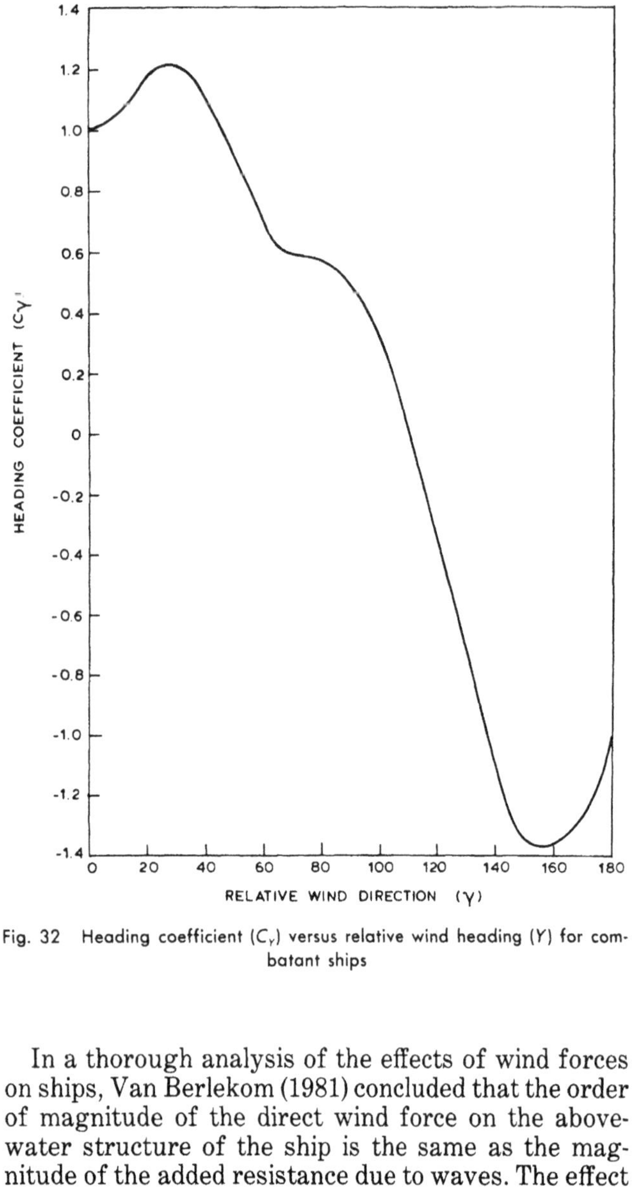

Fig. 32 Heading coefficient (C_Y) versus relative wind heading (Y) for combatant ships

In a thorough analysis of the effects of wind forces on ships, Van Berlekom (1981) concluded that the order of magnitude of the direct wind force on the above-water structure of the ship is the same as the magnitude of the added resistance due to waves. The effect of leeway caused by wind forces is less important according to this study. It was also concluded that the wind coefficients are in general very much dependent on the frontal and lateral areas exposed to the wind. The yawing moment on the vessel due to wind depends on the position of the main superstructure. Variations in deck house configurations are relatively of less importance with respect to the value of the wind coefficients.

5.3 Added Resistance due to Waves. Winds are seldom encountered at sea without wind-generated waves, sometimes from distant storms. Such waves approaching a ship from forward can cause appreciable added resistance, in part from the diffraction effect of the moving hull on the encountered waves and partly from the indirect effect of the pitching and heaving motions caused by the waves. In beam and quartering seas, there may be heavy rolling and some yawing, both of which will add to the resistance. Required rudder action may also make a significant addition. The evaluation of these effects by both model tests and theoretical calculations is discussed in Chapter VIII, Vol. III.

5.4 Appendage Resistance. In single-screw ships the principal appendages are the bilge keels and rudder, while in multiscrew ships there are bossings or open shafts and struts and there may be two rudders. All these items give rise to additional resistance, which is best determined by model experiments.

The lines of flow along the bilge are measured by the use of dye or of small flags, and the bilge keels arranged to conform with them. By these means the additional resistance can be kept to little more than that corresponding to the additional wetted surface, which amounts to 1 to 3 percent of that of the main hull, depending upon the extent and depth of the keels.

The resistance of rudders can be measured by model tests or calculated from a knowledge of their shape, using drag coefficients for airfoils of similar characteristics and Reynolds numbers appropriate to their length and speed. When rudders are not in the pro-

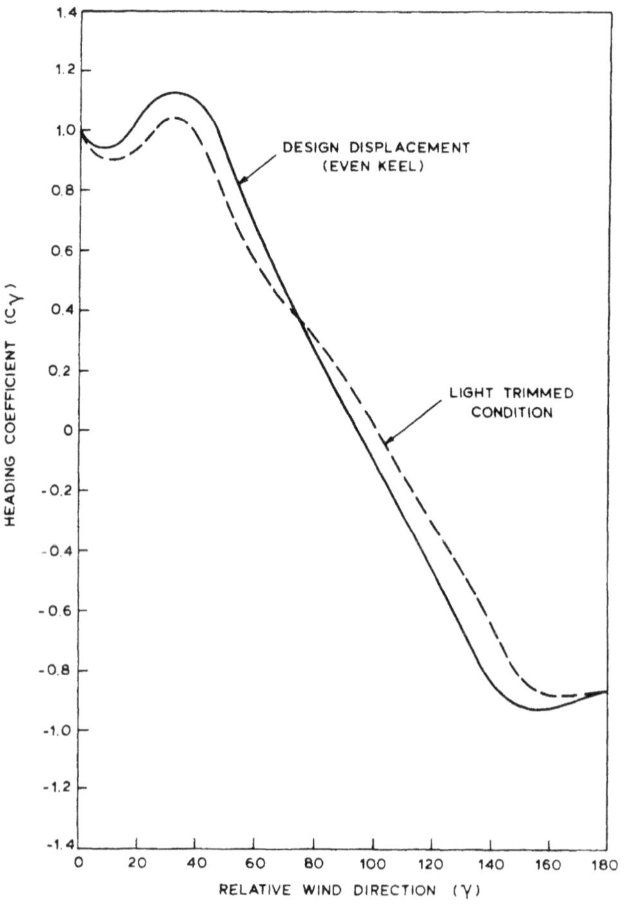

Fig. 33 Heading coefficient (C_Y) versus relative wind heading (Y) for naval auxiliary ships

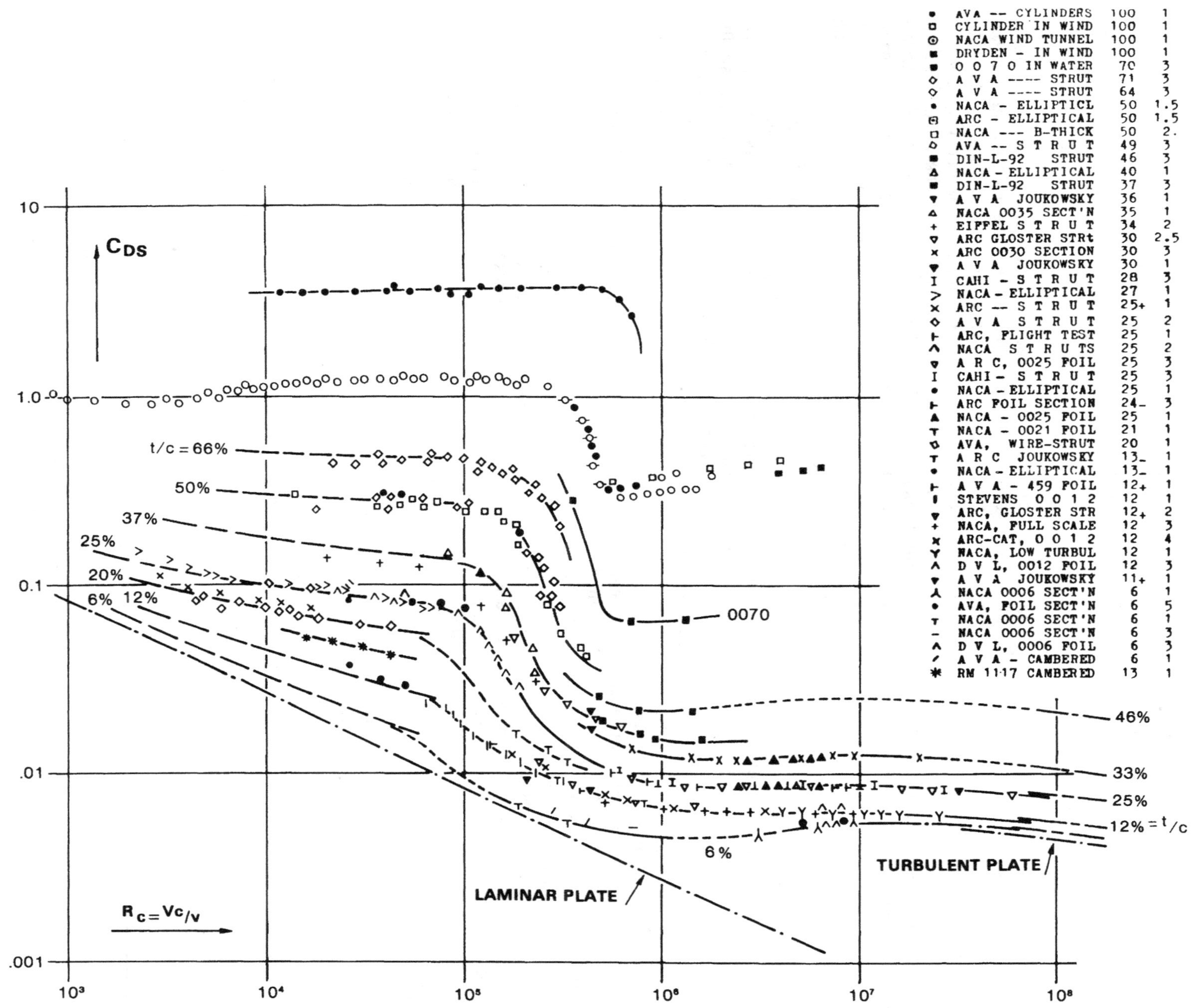

Fig. 34 Data on sectional drag (at near-zero lift) of *streamline foil and strut* sections. Many of the experimental results are obtained by wake-survey technique; in others, drag of blunt wing tips has been subtracted from the original values.

peller race, the velocity of water past them is somewhat less than the ship speed due to wake effects, but when in the race the latter as a rule more than offsets the wake, and the velocity will usually be more than the ship speed. In model self-propulsion tests it is usual to ignore the resistance of rudders in the race, the effect being absorbed in the propulsive efficiency. For ships with twin rudders, model tests are advisable to determine the optimum zero setting for the rudders, because with the closing in of the streamlines at the stern, this setting may not be parallel to the ship's centerline. In certain unfavorable resonant conditions, however, such zero settings may give rise to rudder flutter and hull vibration, and it may be necessary to choose different rudder settings and accept the additional resistance.

To design bossings and struts which will give minimum resistance, it is necessary to explore the flow around the stern, and this can be done by dye, flags or pitot tubes, either in a towing tank or in a circulating water channel.

With bossings correctly aligned with the flow, the minimum possible increase in resistance to be expected would be that due to the wetted surface, the amount of which depends upon the propeller diameter, which governs the spread of the shafts, and fineness of the afterbody, which determines the length of bossing outside the hull. MARIN has investigated this problem on a number of twin-screw 6 m models (Lap, 1956).

The increase in resistance due to the added wetted surface alone amounted to some 1 to 5 percent, but due to the greater curvature of the bossing surface the specific frictional resistance was greater than that of the main hull, so that the total increase in resistance was some 5 to 9 percent of the frictional resistance of the hull. For stern arrangements with open shafts and struts, the increase in wetted surface, including shafts, stern tubes, struts and barrels, was 0.9 to 4.0 percent of the main hull wetted surface and for well-designed strut arrangements the resistance increase was from 6 to 9 percent of the frictional resistance, much the same as for bossings. In making any such estimates, allowance must be made for any unusual features, such as additional intermediate struts on very long open shafts, and in some cases the total shaft and strut resistance on a model may amount to as much as 16 or 18 percent of the bare model resistance.

The expansion of such estimates to the ship is a very difficult question, which is not yet satisfactorily solved. The model appendages themselves are very small, so that the Reynolds numbers based upon their speed and dimensions are also small, and scale effect is likely to be important. This is especially so with struts and open shafts.

When planning experiments with models it is often impossible to select model sizes and test conditions so that the flow on model appendages satisfies scaling requirements. The prediction of ship performance from models where appendages yield significant contributions to the measured values of drag is therefore in error if standard procedures are adopted, such as the performance prediction method adopted by the 15th ITTC in 1978 (see Section 5.2, Chapter VI), in which no separate Reynolds scaling of the appendages is included. Use of the Froude hypothesis for extrapolating model resistance in those cases leads to significant inaccuracies, and improved testing techniques and separate scaling relations have to be devised and adopted.

Model sizes are limited because of limited towing tank dimensions, limited speeds of the towing carriage, etc. Also, the cost of model construction and the cost of testing increases with model size. These are usually the reasons for adopting model sizes that are insufficiently large to allow the development of fully turbulent flow over the appendages. Even large models sometimes have small appendages, which experience incorrect flow conditions because of low Reynolds numbers.

The dependence of drag on the value of the Reynolds number has been reasonably documented for most types of two-dimensional foil sections. Fig. 34, taken from Hoerner (1965), shows the value of the drag coefficient C_D as a function of Reynolds number for various types of airfoil sections of various thicknesses. Below about 1×10^5 the boundary layer is completely laminar. At these Reynolds number values, sections with moderate to high thickness-chord ratios show high drag coefficients because of laminar boundary layer separation.

Between Reynolds numbers of about 1×10^5 and 1×10^6 the sections show a rapid decrease of their drag coefficient, caused by transition of laminar to turbulent flow which, for moderate to thick sections, separates further along the body (closer to the trailing edge), than in the case of laminar flow.

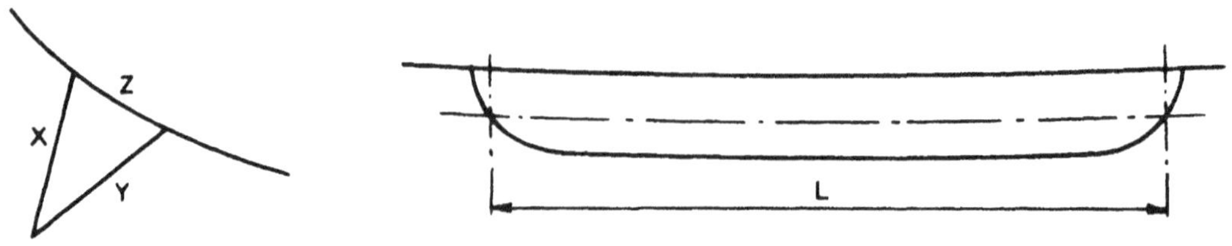

Fig. 35 Definition of bilge keel geometry for use in Peck's equation for bilge keel resistance.

For very thin sections, in absence of boundary layer separation, the transition from laminar to turbulent flow is accompanied by a relative increase in the value of the drag coefficient. The Reynolds number at which transition of the boundary layer occurs is termed the critical Reynolds number.

After transition of the boundary layer, the transition point remains comparatively fixed for a range of Reynolds number until, at about 1×10^6, the transition point moves steadily ahead, causing the drag coefficient to increase slightly. In excess of Reynolds numbers around 1×10^7 the drag coefficient decreases slightly with increasing Reynolds number.

For the sections commonly used in the design of appendages such as control surfaces (NACA 4-digit type sections for example), complete turbulent flow is usually obtained at a Reynolds number around 1×10^6 in high-turbulent flow and at about 5×10^6 in low-turbulent flow.

Various workers in the field have devised empirical relations for the calculation of the drag of appendages. Some of these relations are independent of Reynolds number and cannot be used for the estimation of scale effects. The more important relations are as follows:

(a) Bilge keels. The resistance of bilge keels, according to Peck (1976), can be divided into two parts:

1. Skin friction due to additional wetted surface.
2. Interference drag at junction between bilge keel and hull. Interference drag reduces as the angle between the hull and bilge keel plating increases; i.e., as dimension z increases (see Fig. 35). Thus when $z = x + y$, i.e., no bilge keel, additional drag = 0 and when $z = 0$, i.e., a plate keel, interference drag is taken as equal to skin friction drag $= \frac{1}{2}\rho SV^2 C_F$.

In any intermediate situation

$$D = \tfrac{1}{2}\rho SV^2 C_F\left(2 - \frac{2z}{x+y}\right) \qquad (32)$$

in which S = wetted surface of the bilge keel
and L = average length of the bilge keel to be used in calculating C_F
V = ship speed.

If the shape of the bilge keels is such that $\dfrac{z}{x+y}$ varies greatly, a mean value may be found by taking:

$$\frac{\text{area of hull covered by bilge keel}}{\text{wetted surface of bilge keel}}$$

(b) Control surfaces (rudders, shaft brackets, stabilizer fins, etc). Peck's equation for control surfaces is as follows:

$$D = \tfrac{1}{2}\rho SV^2 C_F\left[1.25\frac{c}{c_f} + \frac{S}{A} + 40\left(\frac{t}{c_a}\right)^3\right] \qquad (33)$$

where c = mean chord length $= c_f + c_a$ (see Fig. 36)
S = wetted area
A = frontal area of section of max thickness

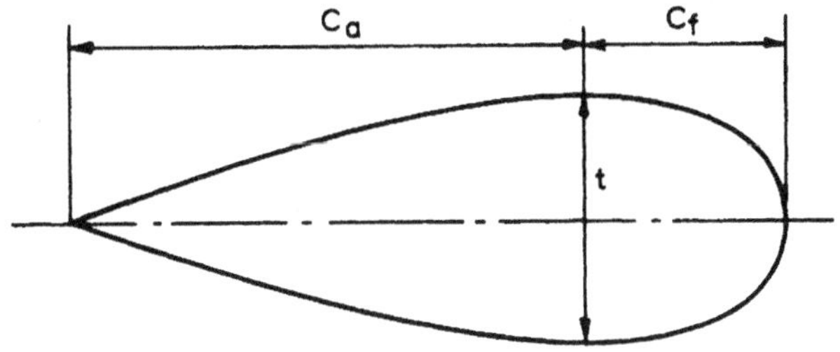

Fig. 36 Definition of control surface section geometry for use in Peck's equation for the drag of control surfaces.

t = maximum thickness
V = ship speed.

Peck suggests substituting $1.1V$ for V for rudders because of the increased speed over the rudder due to the propeller slipstream.

If the rudder or strut penetrates the water surface a spray drag component is to be added which, according to Hoerner (1965) is:

$$D_{\text{spray}} = 0.24\,(\tfrac{1}{2}\rho V^2 t_w^{\,2}) \qquad (34)$$

where t_w = maximum control surface thickness at the water surface.

For the case propeller struts have palms mounted on the hull rather than mounted flush with the hull surface, Hoerner provides a formula for the additional drag of the palm,

$$D_{\text{palm}} = 0.75\, C_{D_{\text{palm}}}(h_p/\delta)^{1/3} w\, h_p\, \tfrac{1}{2}\rho V^2$$

where h_p = height of palm above the surface
w = frontal width of palm
δ = boundary layer thickness (approximately equal to $0.016\,x_{b1}$, where x_{b1} = distance from leading edge to the point where the plate is attached).
$C_{D_{\text{palm}}}$ = 0.65 if the palm is rectangular in shape with rounded edges
V = ship speed

Hoerner also gives a formula for the drag component due to the interference of control surface-type appendages with the flow along the hull:

$$D_{\text{int}} = \tfrac{1}{2}\rho V^2 t^2\left[0.75(t/c) - \frac{0.0003}{(t/c)^2}\right] \qquad (35)$$

where t = maximum thickness of control surface at the hull
c = chord length of control surface at the hull.

According to Hoerner the drag coefficient of two-dimensional foil sections, with a location of maximum thickness at about 30 percent of the chord length from the leading edge, is:

$$C_D = C_F[1 + 2\,(t/c) + 60\,(t/c)^4] \qquad (36)$$

At Reynolds numbers below 1×10^7, this equation is not accurate enough, since the skin friction coefficient to be used in this formula is that for a turbulent boundary layer and the formula is therefore only valid for Reynolds numbers in excess of about 5×10^6. Kirkman, et al (1980) developed the following formulas for the drag of foil-type appendages:

For $Rn \leq 5 \times 10^4$

$$C_D = 1.46\,(Rn)^{-0.507}\ (\text{for } t/c = 0)$$
$$C_D = 0.466\,(Rn)^{-0.259}\ (\text{for } t/c = 0.2)$$

For $5 \times 10^4 \leq Rn \leq 5 \times 10^5$

$$C_D = 0.172\,(Rn)^{-0.310}\ (\text{for } t/c = 0)$$
$$C_D = 181\,(Rn)^{-0.810}\ (\text{for } t/c = 0.2) \qquad (37)$$

For intermediate t/c-values, the appropriate C_D values can be determined by linear interpolation.

For $5 \times 10^5 \leq Rn \leq 1 \times 10^7$

$$C_D = 0.00293\,[1 + 2\,(t/c) + 60(t/c)^4]$$

For $Rn \geq 1 \times 10^7$

$$C_D = 0.03\,Rn^{-0.143}\,[1 + 2\,(t/c) + 60(t/c)^4] \qquad (38)$$

These formulas can be used to derive an estimate of control surface drag at low Reynolds numbers.

The reference area to be used in the calculation of the drag according to these formulas is equal to twice the planform area. In the formulas for the Reynolds number an average value for the chord length is to be used. Foil interference drag is to be added which can be calculated from Equation (31). According to Kirkman et al, tip parasitic drag can be neglected.

(c) Shafts and bossings. Usually propeller shafts are inclined at some angle to the flow. This leads to appreciable lift and drag forces on the shaft and strut bossing. According to Hoerner (1965) for a Reynolds number value (based on the diameter of the shaft) smaller than 5×10^5.

$$D_{\text{Shaft}} = \tfrac{1}{2}\,\rho L_{SH} d V^2 (1.1 \sin^3 \epsilon + \pi C_F) \qquad (39)$$

where L_{SH} = total length of shaft and bossing
d = diameter of shaft or shaft and bossing
ϵ = angle of flow relative to shaft axis

Kirkman, et al (1980) have derived approximate formulas for the calculation of the drag coefficient of the cylindrical portions of ship appendages as follows. For the pressure drag of a cylinder:

For $Rn \leq 1 \times 10^5$.
$$C_{DP} = 1.1 \sin^3 \alpha$$
For $1 \times 10^5 \leq Rn \leq 5 \times 10^5$, and $\alpha \leq \beta$:
$$C_{DP} = -0.7154 \log_{10} Rn\ 4.677$$
For $1 \times 10^5 \leq Rn \leq 5 \times 10^5$, and $\alpha \leq \beta$:
$$C_{DP} = (-0.7154 \log_{10} Rn + 4.677)\,[\sin^3 (1.7883) \log_{10} Rn - 7.9415)\,\alpha] \qquad (40)$$
For $Rn \geq 5 \times 10^5$, and $0 \leq \alpha \leq 40°$:
$$C_{DP} = 0.60 \sin^3 (2.25\alpha)$$
For $Rn \geq 5 \times 10^5$, and $40 \leq \alpha \leq 90°$:
$$C_{DP} = 0.60$$

where $Rn = \dfrac{Vd}{\nu}$

α = angle between flow lines and cylinder axis (see Fig. 37)
$\beta = -71.54 \log_{10} Rn + 447.7$ (see Fig. 38).

The reference area used in the pressure drag calculation is equal to the projected area (i.e., length $\times$ diameter) of the cylinder.

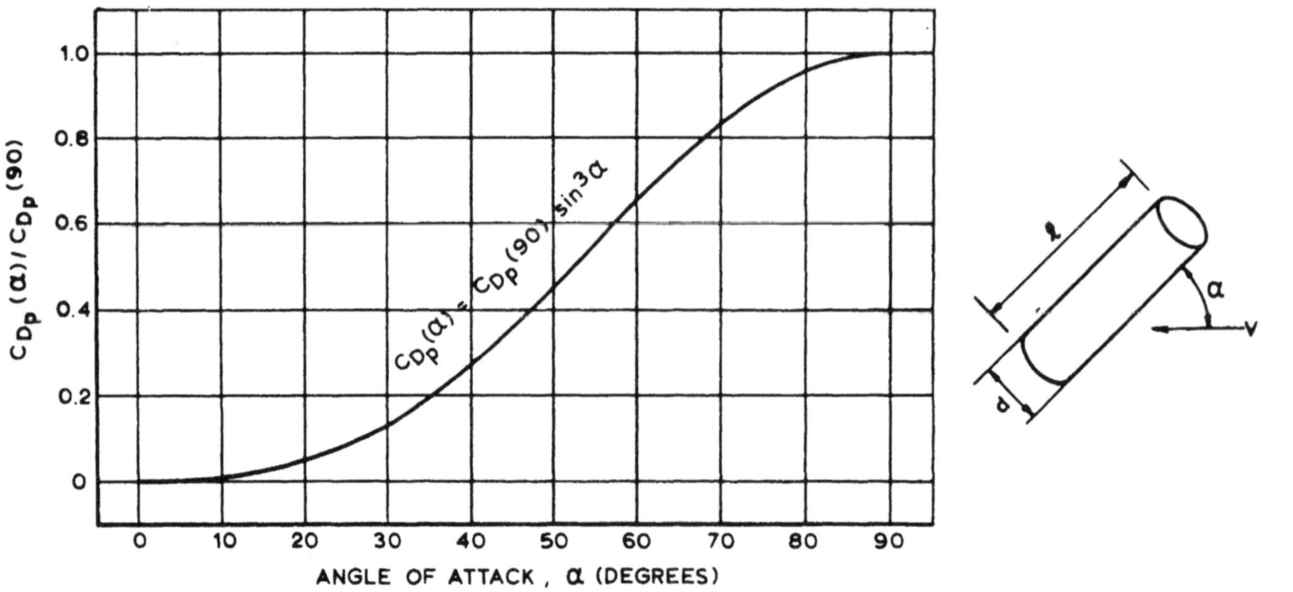

Fig. 37 Graphic representation of the cross-flow drag principle used in the subcritical area

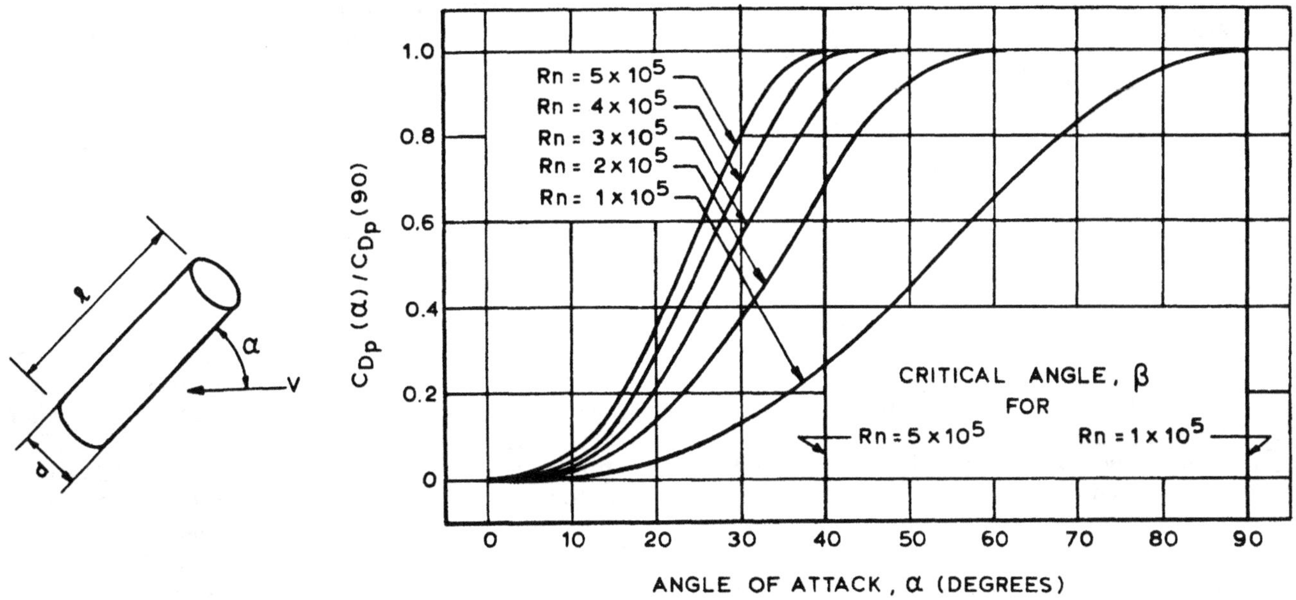

Fig. 38 Graphic representation of the sliding cross-flow treatment used in the transitional regime (between Rn $\approx 1 \times 10^5$ and 5×10^5)

The frictional drag of a cylinder in a cross-flow, according to Kirkman, et al is:

For Rn $\leq 5 \times 10^5$.

$$C_F = 1.327\,(\text{Rn})^{-0.5} \tag{41}$$

For Rn $\geq 5 \times 10^5$.

$$C_F = \frac{1}{(3.46 \log_{10} \text{Rn} - 5.6)^2} - \frac{1700}{\text{Rn}}$$

where $\text{Rn} = \dfrac{V\ell c}{\nu}$

and $\ell c = \dfrac{\ell}{\tan \alpha}$; $\ell c \geq \ell$

ℓ is length of cylinder, α is angle between flow lines and cylinder axis, see Fig. 37.

The reference area used in the frictional drag calculations is the wetted surface of the cylinder ($\pi \times$ length $\times$ diameter).

To the drag as calculated using the above formulas is to be added the drag of the forward and aft end of the cylinder (if applicable), viz.:

$C_{Db} = 0.01 \cos^3 \alpha$ (forward end if faired; see Fig. 39).

$C_{Db} = 0.90 \cos^3 \alpha$ (forward end with sharp edges, see Fig. 39).

$$C_{Db} = \frac{0.029}{\sqrt{C_F S_{\text{wet}} / S_{\text{base}}}} \cos^3 \alpha \text{ (aft end)} \tag{42}$$

where C_F is frictional drag coefficient for a Reynolds number based on the characteristic length of the cylinder, S_{wet} is $2\pi R\ell$ = wetted area of cylinder, S_{base} is πR^2 = base area of cylinder.

The reference area to be used in calculating cylinder base drag is the base area (πR^2), or the net base area [$\pi(R^2 - r^2)$] for the case where a cylinder of smaller diameter is located immediately downstream of the cylinder for which the drag is being calculated.

(d) Skegs. The drag of a skeg can be calculated from

$$D_{SK} = \tfrac{1}{2} \rho V_m^2 C_F A_{SK} \tag{43}$$

where V_m = average velocity over hull at the location of the skeg
A_{SK} = lateral area of skeg
C_F = frictional drag coefficient based on wetted length of skeg

Appendage drag formulations such as given above can be used in deriving full-scale resistance predictions as follows:

(a) The full-scale resistance of the bare hull is determined from model tests to which is added the resistance of appendages, as calculated, at the full-scale Reynolds number.

(b) The resistance and the propulsion characteristics of the appended model is determined experimentally. The resistance of the appendages is then determined at the model Reynolds numbers from the formulations and subtracted from the measured total resistance. After scaling-up the resulting resistance values to full-scale, the calculated appendage resistance for the full-scale Reynolds number is added.

The method described under *(b)* is considered more accurate than that described under *(a)* because the running trim of the model in the appended case, which is usually greater than the running trim in the unappended case, is then taken into account more accurately.

Notable recent attempts to determine the degree of correlation between the appendage drag as calculated from the formulae given above and experimental data were carried out by Kirkman, et al (1980) and Von Kerczek, et al (1983). Kirkman, et al found for 11 test cases that the used formulas over-predicted the appendage resistance by between 30 to 40 percent. Von Kerczek showed for 2 cases that the formulas lead to quite acceptable results if the appropriate boundary layer velocities are used to calculate the value of Rn and the dynamic lead $\frac{1}{2}\rho V^2$. This had not been done before, the speed of advance $V_A = V_S(1 - w)$ being used for the drag contributions due to struts, shaft bossings and shafts. In the calculations carried out by Von Kerczek et al, the fluid velocity in way of each appendage is determined using the results from boundary layer calculations along the hull. They found that particularly the shaded areas in Fig. 40 are affected by the boundary layer.

The work carried out by Kirkman, et al and Von Kerczek et al mainly concern ship speeds up to Fn = 0.5. For higher ship speeds the application of the formulations normally lead to serious under-prediction of the appendage drag. Gregory, et al (1979) reported a discrepancy of 100 percent, while Clement (1957) reported discrepancies of 45 percent for a $\frac{1}{10}$-scale model to virtually no error for a $\frac{1}{2}$-scale model. Results of this nature have often been reported. Reasons for discrepancies can be sought in the adopted value of the inflow velocity, in lack of alignment of shaft struts, rudders, etc. to the flow, the occurrence of cavitation and ventilation and interference effects between appendages, appendages and the hull, and between appendages and the propulsor.

The lack of alignment of rudders, shaft struts, stabilizer fins and similar appendages to the flow will cause the drag to increase slowly for small angles and sharply for larger angles. The appropriate position and alignment of rudders and stabilizer fins is often determined through carrying out model tests for various angles of attack. If alignment tests are not carried out the resulting experimentally-determined appendage drag will be higher than the values following from theoretical formulations which do not properly account for the drag-due-to-lift of foil-type appendages. This is particularly the case with rudders situated in the slipstream of a propeller. In that case a correction must also be carried out for the greater flow speed along the rudder.

Effects of propeller loading, cavitation and ventilation on appendage drag are also largely unknown. Rutgersson (1982) reported a decreasing value of rudder drag with increasing propeller loading and an increase with increasing amounts of cavitation. The decrease of rudder drag with increasing propeller loading is due to the forward force component on the rudder induced by the slipstream (the rudder acts as a stator, recovering rotational energy from the propeller slipstream). With the occurrence of cavitation and ventilation on various appendages the appendage drag will increase. No account thereof is provided by the available drag formulations.

Rutgersson (1982) also reported on the correlation between calculated strut and shaft bossing drag, as calculated by Hoerner's formula, with experimental values. At zero propeller loading the measured values were 20 to 30 percent higher. With increasing propeller loading the drag increases significantly, which effect is probably due to the increase of the induced flow (i.e., due to the increased velocity) over the appendages as propeller loading increased. Again the occurrence of cavitation was found to increase the drag.

The ultimate test of any model geosim series designed to explore appendage scale effect would be to measure the appendage resistance on the actual ship, and this was done by the British Ship Research Association on the *Lucy Ashton*. These experiments espoused by Lackenby (1955), were part of a larger program in which the old Clyde paddle steamer, 58 m long, with paddles removed, was propelled by aircraft jet engines fixed on deck, the measured thrust of which gave the ship resistance, unaffected by any interference from a towing ship. The ship results were compared with those from a series of six geosim models.

Dummy appendages were fitted aft, and measured mile trials carried out with bare hull, with open shafts and struts and with full bossings. Streamline-flow experiments on a model indicated that the best angle for the bossings would be 40 deg to the horizontal. These showed an increase of only 3 to 4 percent in resistance on the model, which was considered too small for accuracy in the type of experiments involved. The bossings were therefore fitted at 20 deg, being across the flow, and the increase in resistance was thereby doubled. The wetted surface of the bossings was only 2 percent of that of the bare hull. The percentage increase in resistance measured on the ship was fairly constant at about 5 percent up to 12 knots, after which it declined to almost nothing at 15 knots. The increase due to the A-brackets and open shafts was reasonably constant at about $3\frac{1}{2}$ percent.

The geosim models were tested at NPL and had lengths of 2.74, 3.66, 4.88, 6.10, 7.32 and 9.14 m. The increase in resistance due to appendages never exceeded 7 percent of the barehull resistance.

The measured increase in C_T on the ship is compared with that on the different models in Table 4.

Over the speed range from 8 to 12 knots and for models 3.66 to 6.10 m in length, which cover the sizes in common use in most tanks, the ratio is between 0.5

and 0.6 for both bossings and open shafts with struts.

These results are not in agreement with the geosim tests made by Allan (1950), in which there was no apparent scale effect on the bossings. There is an important difference between the two cases, however, in that the bossings used by Allan were designed to be in the flow, while those on the *Lucy Ashton* were across the flow. The resistance of the former was therefore likely to be mostly frictional, and the scale effect would be small in terms of the total resistance being measured. If the *Lucy Ashton* bossings had been aligned with the flow, their resistance might have been less than that for the shafts and struts. This difference in the bossings in the two cases makes any conclusions rather doubtful, and further research is needed to clarify the situation.

Tests carried out by the Bureau of Ships on models for four different ships showed very little difference in required power between well-designed bossings and exposed shafts and struts.

As a means of making approximate estimates of appendage resistance for design purposes, Mandel (1953) quotes overall figures derived from model tests, no reduction being made for scale effect, Table 5.

The whole question of appendage resistance is in an unsatisfactory state, both as regards making estimates of its magnitude in a given case and the application of model results to the ship. There is scope here for a great deal more research both with models and full-scale trials of ships to clarify the problem of scale effect.

5.5 Trim Effects. Owing to the change in pressure distribution around a ship at different speeds, it will rise or sink bodily and also trim. At low speeds there is a general sinkage and a slight trim by the bow as compared with the at-rest condition (Fig. 41.) As speed increases the movement of the bow is reversed and at Fn = 0.30 or thereabouts the bow begins to rise appreciably, the stern sinks still further and the ship takes on a decided trim by the stern (Fig. 42).

As D.W. Taylor (1943) pointed out, large trim changes or sinkage of the center of gravity are symptoms rather than causes of high resistance. Nevertheless they may indicate the desirability of altering the at-rest trim by shifting the center of gravity longitudinally. The reductions of resistance which can be effected by such changes of trim as are practicable in large displacement craft are very small, but in high-speed planing craft the position of the center of gravity and the resultant still-water trim have a most important influence on performance. In both cases the possible effects can be investigated on model scale.

In the average merchant-ship form, additional trim by the stern in the at-rest condition usually results in an increase in resistance at low speeds and a decrease at high speeds. At low speeds the increased draft aft makes the stern virtually fuller, with a consequent increase in form and separation resistance, whereas

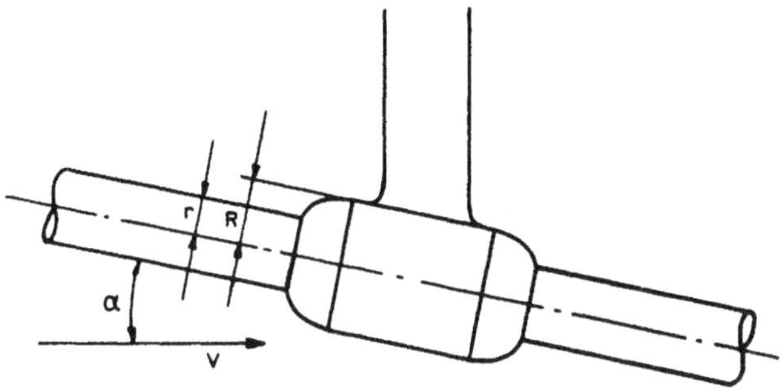

STRUT BARREL WITH FAIRED ENDS

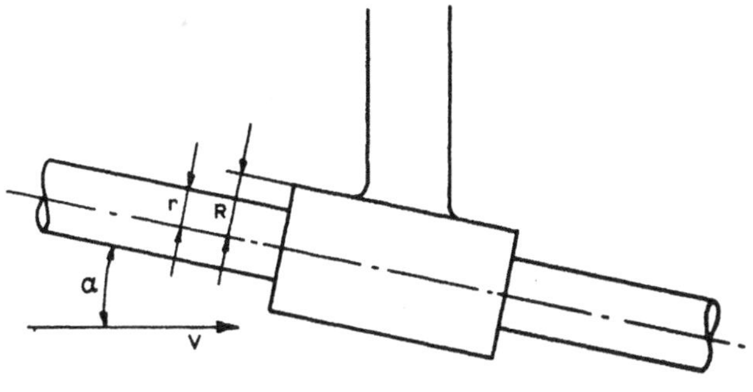

STRUT BARREL ENDS WITH SHARP EDGES

Fig. 39 Typical strut barrel ends

Table 4—Appendage Resistance on LUCY ASHTON

Ship speed in knots	Model length in m: 2.74	3.66	4.88	6.10	7.32	9.14
	Ratio $\frac{\text{increment in ship } C_r}{\text{increment in model } C_r}$ with bossings					
8	0.44	0.48	0.52	0.56	0.58	0.61
12	0.52	0.57	0.60	0.62	0.65	0.68
14½	0.10	0.12	0.14	0.16	0.17	0.20
	Ratio $\frac{\text{increment in ship } C_T}{\text{increment in model } C_T}$ with A brackets and open shafts					
8	0.48	0.52	0.56	0.58	0.61	0.67
12	0.43	0.47	0.52	0.54	0.57	0.61
14½	0.33	0.37	0.41	—	0.46	0.51

Table 5—Approximate Resistance of Appendages

Resistance expressed as percent of bare hull resistance.

Type of ship	Value of Fn: 0.21	0.30	0.48
Large, fast, 4 screws	10–16	10–16	—
Small, fast, 2 screws	20–30	17–25	10–15
Small, medium speed, 2 screws	12–30	10–23	—
Large, medium speed, 2 screws	8–14	8–14	—
All single-screw ships	2–5	2–5	—

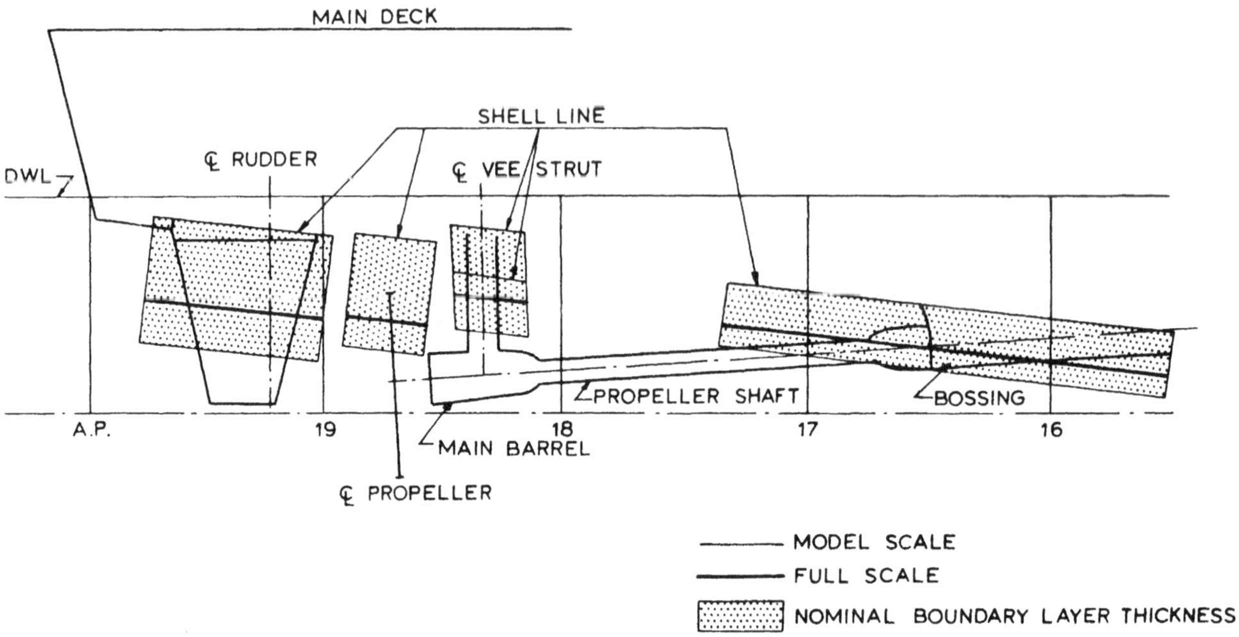

Fig. 40 Nominal boundary layer thickness in way of typical appendages (Von Kerczek, et al 1983)

at high speeds this is more than offset by the reduction in wave-making due to the finer entrance in the trimmed condition.

In ballast condition, at level trim, the wetted surface per unit of displacement is much increased, so that the frictional resistance is increased also, but because of the finer form at the reduced draft, the residuary resistance is decreased. In general, except in high-speed ships, the total resistance per unit of displacement will be greater, but because of the lower displacement the total resistance and power will be reduced, and the ship in ballast will make a higher speed at the same power.

In ballast condition it is usually necessary to carry considerable trim by the stern in order to ensure adequate immersion of the propeller, and this will have similar effects to those stated in the foregoing—higher resistance at low speeds, less at high speeds. For any ship which is likely to spend an appreciable part of her time at sea in ballast condition, model experiments are usually made to investigate these effects.

5.6 Shallow-Water Effects. The resistance of a ship is quite sensitive to the effects of shallow water.

In the first place there is an appreciable change in potential flow around the hull. If the ship is considered as being at rest in a flowing stream of restricted depth, but unrestricted width, the water passing below it must speed up more than in deep water, with a consequent greater reduction in pressure and increased sinkage, trim and resistance. If in addition the water is restricted laterally, as in a river or canal, these effects are further exaggerated. The sinkage and trim in very shallow water may set an upper limit to the speed at which ships can operate without touching bottom.

A second effect is the changes in the wave pattern which occur in passing from deep to shallow water. These changes have been studied by Havelock (1908) for a point pressure impulse travelling over a free water surface.

When the water is very deep, the wave pattern consists of the transverse and diverging waves shown in Fig. 6, the pattern being contained between the straight lines making an angle α of 19 deg 28 min on each side of the line of motion of the point.

As is discussed more fully in Chapter VIII, Vol. III, in water of depth h the velocity of surface waves is given by the expression

$$(V_c) = (gL_W/2\pi) \tanh 2\pi h/L_W \qquad (44)$$

where L_W is the length of wave from crest to crest.

As h/L_W increases, tanh $2\pi h/L_W$ approaches a value of unity, and for deep water this leads to the usual expression

$$(V_c)^2 = gL_W/2\pi \qquad (45)$$

As the depth h decreases, and the ratio h/L_W becomes small, tanh $2\pi h/L_W$ approaches the value $2\pi h/L_W$, and for shallow water the wave velocity is approximately given by the equation

$$(V_c)^2 = gh \qquad (46)$$

The wave pattern for the pressure point goes through a critical change when $V = \sqrt{gh}$ (see Fig. 43).

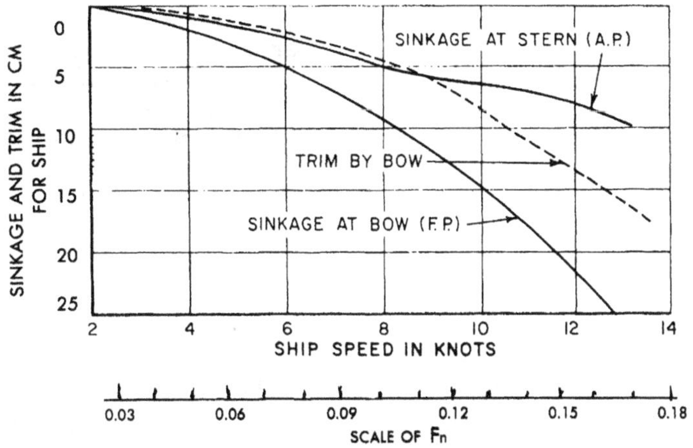

Fig. 41 Changes in sinkage and trim with speed for T.2 Tanker model. Ship dimensions: 155.4 × 20.7 × 9.2 m according to Norley (1948)

For speeds less than $V = \sqrt{gh}$, the system consists of a double set of waves, transverse and diverging as in deep water, advancing with the pressure point at velocity V. For values of V less than about $0.4\sqrt{gh}$, the pattern is enclosed between the straight lines having an angle α = 19 deg 28 min to the centerline, as for deep water. As V increases above this value, the angle α increases and approaches 90 deg as V approaches $\sqrt{gh}$, Fig. 43.

The pressure point is now generating a disturbance which is travelling at the same speed as itself, and all the wave-making effect is concentrated in a single crest through the point and at right angles to its direction of motion. This pattern agrees with observations on models and ships when running at the critical velocity in shallow water. The whole of the energy is transmitted with the wave, and the wave is called a *wave of translation*.

When V exceeds $\sqrt{gh}$, α begins to decrease again, the wave system being contained between the lines given by $\sin^2\alpha = gh/(V)^2$, Fig. 43. It now consists only of diverging waves, there being no transverse waves or cusps. The two straight lines themselves are the front crests of the diverging system, and the inner crests are concave to the line of advance instead of convex as in deep water.

The effect upon resistance due to these changes in wave pattern in shallow water has been investigated by Havelock (1908) for a pressure disturbance of linear dimension l travelling over water of depth h. The resistance curves are reproduced in Fig. 44. Each curve is marked with the value of the ratio of depth of water h to the characteristic length of the disturbance l, that marked ∞ being for deep water. When the ratio h/l is 0.75, there is a marked peak at a speed corresponding to a value of $V/\sqrt{gl} = 0.86$. Since $\sqrt{h/l} = 0.866$, this corresponds to a value of unity for $V/\sqrt{gh}$, so that the peak corresponds to the speed of the wave of translation for that particular depth of water, or the critical speed. At this speed the resistance is very much greater than in deep water, but ultimately at a sufficiently high speed it becomes less than in deep water. This depth effect has an important bearing on full-scale ship trials, and can cause misleading results on

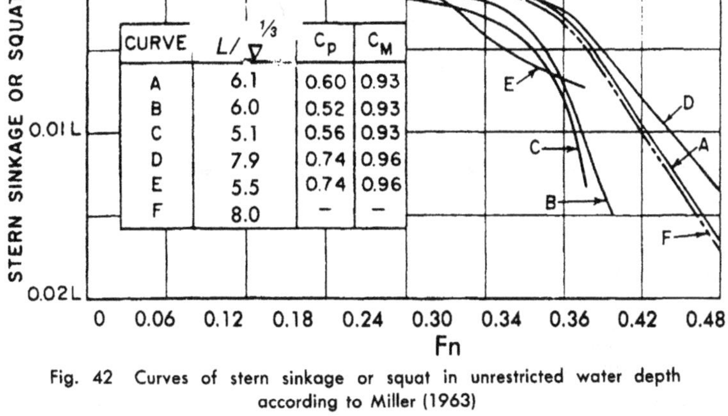

Fig. 42 Curves of stern sinkage or squat in unrestricted water depth according to Miller (1963)

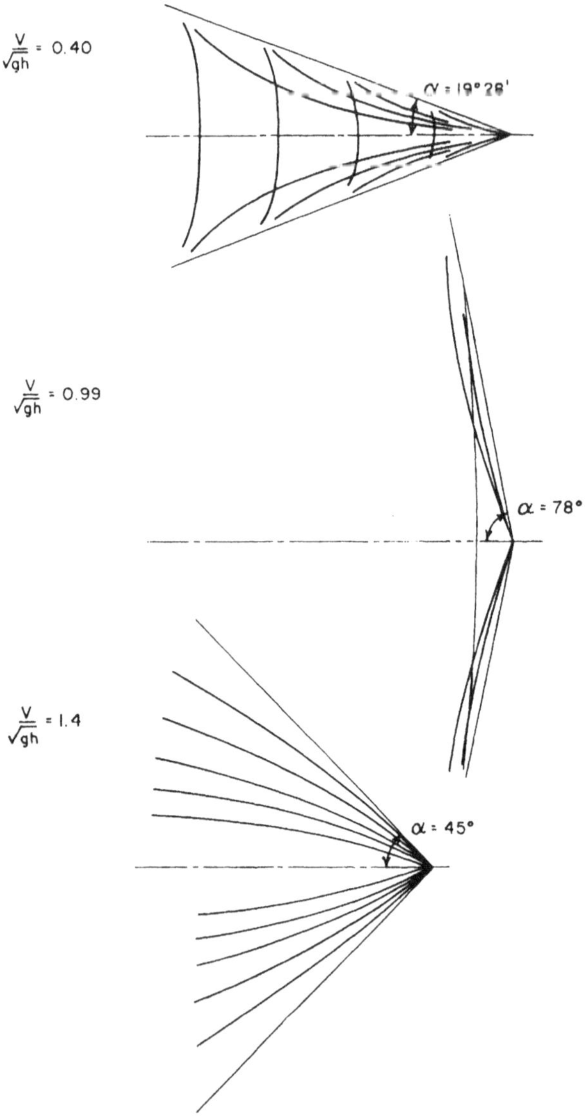

Fig. 43 Effect of shallow water on wave pattern

the relation between power and speed.

Speeds below and above $V = \sqrt{gh}$ are referred to as subcritical and supercritical, respectively. Nearly all displacement ships operate in the subcritical zone, the exceptions being destroyers, cross-channel ships and similar types. It is seen from Equation (32) that as the depth of water decreases the speed of a wave of given length decreases also. Thus to maintain the same wave pattern a ship moving in shallow water will travel at a lower speed than in deep water, and the humps and hollows in the resistance curve occur at lower speeds the shallower the water.

An analysis of shallow-water effects was made by Schlichting (1934). It covered the increase in resistance in shallow water at subcritical speeds, not the decrease at supercritical speeds, and was for shallow water of unlimited lateral extent. The analysis was based on theoretical considerations and on model experiments carried out in the Hamburg and Vienna tanks.

Typical frictional and total resistance curves for deep water are shown in Fig. 45 to a base of speed. At any particular speed V_∞ in deep water they are R_F and R_T, respectively. At this speed the wave pattern generated by the ship will have a wave length L_W given by

$$V_\infty^{\ 2} = gL_W/2\pi \tag{47}$$

In water of depth h the same wave length L_W would be generated at some lower or intermediate speed V_I, where

$$V_I^2 = (gL_W/2\pi) \tanh 2\pi h/L_W \tag{44}$$

and the ratio of the two speeds is

$$\begin{aligned} V_I/V_\infty &= (\tanh 2\pi h/L_W)^{1/2} \\ &= (\tanh gh/V_\infty^{\ 2})^{1/2} \end{aligned} \tag{48}$$

A curve of V_I/V_∞ is shown to a base of $V_\infty/\sqrt{gh}$ in Fig. 46. The reduction in speed on this account is

$$V_\infty - V_I = \delta C$$

in Fig. 45, and Schlichting assumed that the wave-making resistance in shallow water at speed V_I would be the same as that at speed V_∞ in deep water. The total resistance at speed V_I would then be found at point B by adding the wave-making resistance R_{W_∞} to the appropriate frictional resistance at this speed, R_{Fh}. The line AB is in fact parallel to EF.

There is a further loss in speed δV_P because of the increase in potential or displacement flow around the hull due to the restriction of area by the proximity of the bottom, giving as the final speed $V_h = V_I - \delta V_P$. Schlichting investigated this reduction in speed by

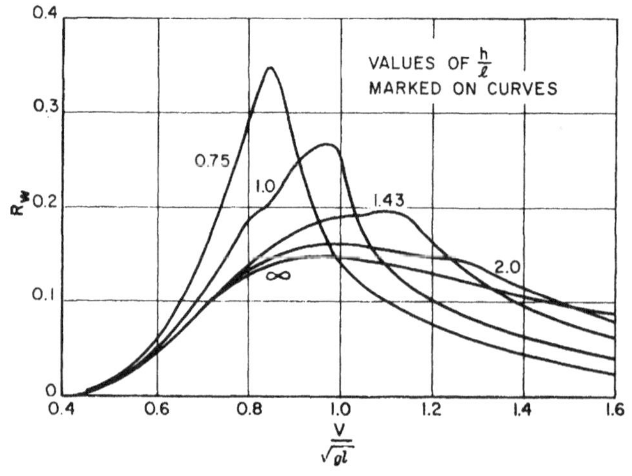

Fig. 44 Effect of shallow water on wave resistance

R_W = wave resistance
l = characteristic length
h = depth of water

model tests in deep and shallow water, using geosim models to detect any laminar flow on the one hand and tank wall interference on the other. He found that the principal factor controlling δV_P was the ratio

$$\sqrt{A_X}/h$$

where A_X = maximum cross-sectional area of the hull and h = depth of water. Fig. 46 shows the curve of V_h/V_I against $\sqrt{A_X}/h$ derived by Schlichting from his model tests and also the relation between V_I and V_∞ for different depths of water h. It should be noted that the ratio V_I/V_∞ is sensibly unity for values of $V_\infty/\sqrt{gh}$ less than 0.4, so that in this region the effect of shallow water on the wave-making part of the resistance is unimportant. If in Fig. 45 the distance δV_P is now set out horizontally from B to give BC, C will be a point on the curve of total resistance in shallow water. The corresponding speed is V_h. This construction can be made from a number of points to obtain the whole curve.

It should be noted that at point C the total resistance in shallow water at speed V_h is less than that in deep water at speed V_∞—point A. If it is desired to find the speed in shallow water for the same total resistance, this will be given approximately by the point H.

The total speed loss is

$$\delta V = \delta C + \delta V_P \tag{49}$$

which can be expressed in percentage terms as

$$\delta V/V_\infty \times 100 = (V_\infty - V_h)/V_\infty \times 100$$

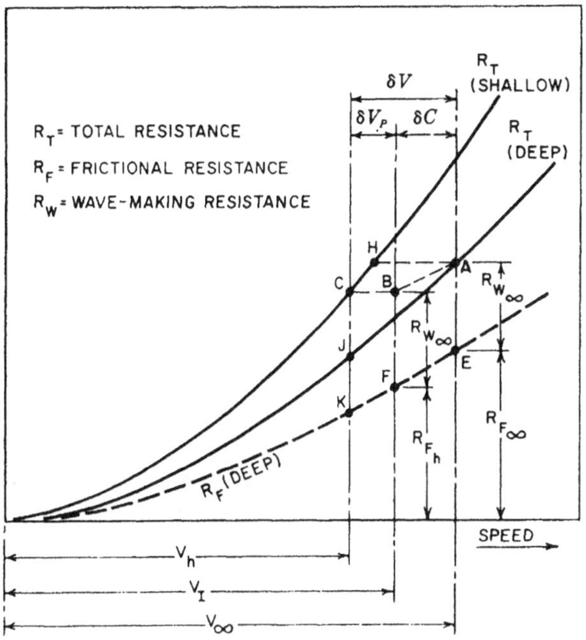

Fig. 45 Determination of shallow water resistance by Schlichting's method

These percentages are given in contour form in Fig. 47.

Schlichting's work is not theoretically rigorous, but it may be looked upon as a good engineering solution of a confused and complicated problem. In particular, the assumption of equal wave resistance in deep and

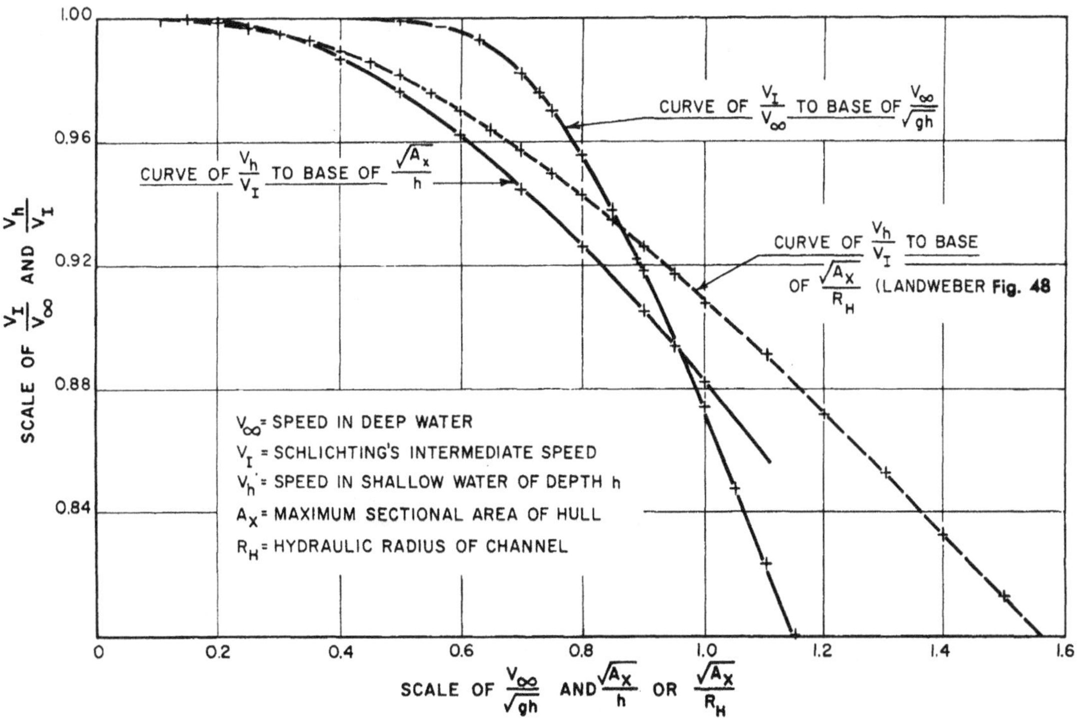

Fig. 46 Curves of velocity ratios for calculating resistance in shallow water (Schlichting)

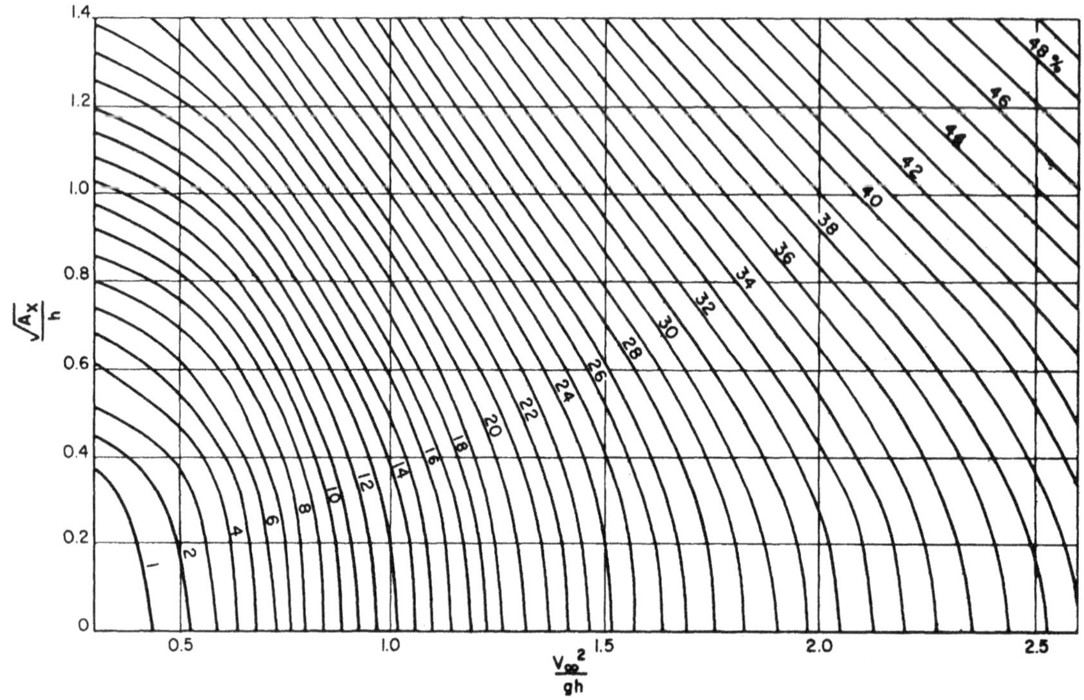

Fig. 47 Schlichting's chart for calculating reduction in speed in shallow water

Total loss of speed δV is given in contours as percentage of speed, $\frac{\delta V}{V_\infty} \times 100$

shallow water when the lengths of the ship-generated waves are the same is open to question. The waves will be steeper and the resistance therefore higher in shallow water, which means that the speed deduced for point C will be somewhat too high. This will partly offset the fact noted above that for the same total resistance the speed should be somewhat higher than that given by the point C, and with all the unknown factors in the problem, C probably gives a close estimate of the shallow-water speed.

As an example of the use of the contours, consider the ore-carrier given on the SNAME Resistance Data Sheets No. 9 (Section 61.7, p. 397, of Saunders, 1957). The ship has dimensions 112.8 m (370 ft) $\times$ 19.5 m (64 ft) $\times$ 5.3 m (17.5 ft). Assuming a deep-water speed of 13 knots, the speed in water 7.3 m (24 ft) deep, unrestricted laterally, is required

$$A_X = 103.2 \text{ m}^2; \sqrt{A_X} = 10.16 \text{ and}$$

$$\sqrt{A_X}/h = 1.392$$

$$V_\infty/\sqrt{gh} = (13 \times 0.5144)/(9.81 \times 7.3)^{1/2} = 0.790$$

and

$$V_\infty^2/gh = (0.79)^2 = 0.624$$

From the contours in Fig. 47, the speed loss $\delta V/V_\infty$ = 20.3 percent, and the ship speed = 10.35 knots.

When the shallow water is restricted laterally, as in an estuary, river or canal, the increase in resistance or the loss of speed will be enhanced. Landweber (1939) published the results of experiments on the resistance of a merchant ship model in a number of different sized rectangular channels, all at speeds below the critical speed. An analysis of the data suggested to Landweber an extension of Schlichting's method for predicting shallow-water resistance to the case of lateral restriction also, i.e., in channels.

Since the speed of translation of a wave in a channel depends only on the depth of water, h, as it does in unrestricted water, it seemed reasonable to assume that Schlichting's method of correcting for the wave-making part of the resistance would still be valid. The speed correction for the displacement flow would, however, have to be modified to take into account the additional restriction introduced by the lateral limitation of the channel walls. In shallow water of unlimited width the speed reduction is a function of $\sqrt{A_X}/h$, and Landweber sought a similar parameter which would also introduce the width of the channel, b. He found this in terms of the *hydraulic radius* of the channel, R_H.

This ratio is in common use in practice of hydraulics, and is defined as

$$R_H = \frac{\text{area of cross section of channel}}{\text{wetted perimeter}} \quad (50)$$

For a rectangular channel of width b and depth h

$$R_H = bh/(b + 2h)$$

When b becomes very large, $R_H = h$, this corresponds

to the case of shallow water of unlimited width.

When a ship or model is in a rectangular channel, the hydraulic radius is

$$R_H = (bh - A_X)/(b + 2h + p) \qquad (51)$$

where

A_X = maximum cross-sectional area of hull
p = wetted girth of hull at this section.

From the model results, Landweber was able to deduce a single curve giving the ratio V_h/V_I in terms of $\sqrt{A_X}/R_H$ for use in restricted, shallow channels. This curve is shown in Fig. 48. It is also reproduced on Fig. 46, and it will be seen that it does not quite agree with that derived by Schlichting, a fact which can be accounted for by Schlichting's neglect of the effect of the width of the tank in which his experimental data were obtained. Saunders prefers to use Landweber's curve for both unlimited shallow water or restricted water, as the case may be, entering with the value of $\sqrt{A_X}/h$ or $\sqrt{A_X}/R_H$ as necessary.

For completeness and convenience in use, therefore, the curve of V_I/V_∞ to a base of $V_\infty/\sqrt{gh}$ is repeated on Fig. 48.

To illustrate the case of resistance in restricted channels, consider the ship of the example, given previously, moving in a channel having the section shown in Fig. 49; see Saunders 1957, Vol. 2, section 61.7.

The cross-sectional area of the canal

$$= [(76.2 \times 10.67) + (10.67)^2/2 + (18.43 \times 10.67/2)]$$

$$= 968.3 \text{ m}^2$$

and the wetted perimeter is

$$= [76.2 + 10.67 \operatorname{cosec} 45° + 10.67 \operatorname{cosec} 30°]$$

$$= 112.6 \text{ m}$$

The maximum cross-sectional area of the ship is 104.0 m^2 and the girth 30.17 m. Hence

$$R_H = (968.3 - 104.0)/(112.6 + 30.17)$$

$$= 864.3/142.77 = 6.05 \text{ m}$$

was the ratio

$$\sqrt{A_X}/R_H = \sqrt{104}/6.05 = 1.69$$

The equivalent depth of the canal for calculating the critical wave speed is given by

$$\frac{\text{Cross-sectional area}}{\text{Width at water surface}} = \frac{968.3}{105.3} = 9.20 \text{ m}$$

To find the speed of the ship in the canal when the resistance is equal to that in deep water at, say, 8 knots

$$V_\infty = 8 \text{ knots} = 4.115 \text{ m/sec}$$

and

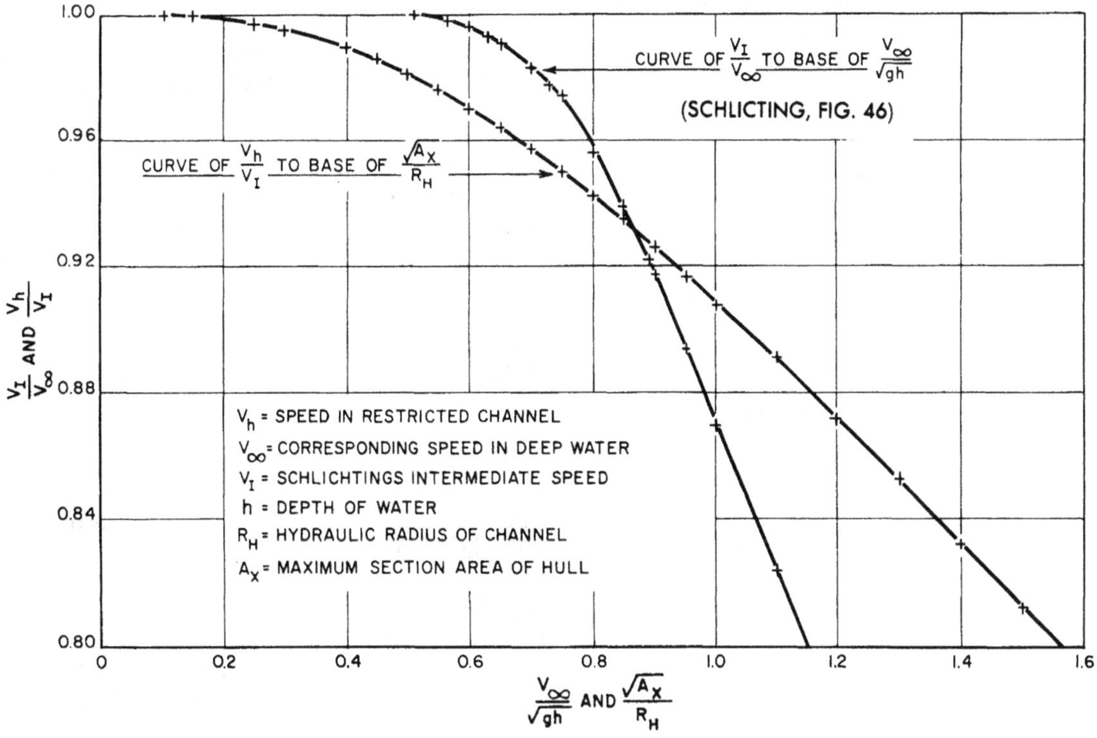

Fig. 48 Curves of velocity ratios for calculating resistance in restricted channels (Landweber)

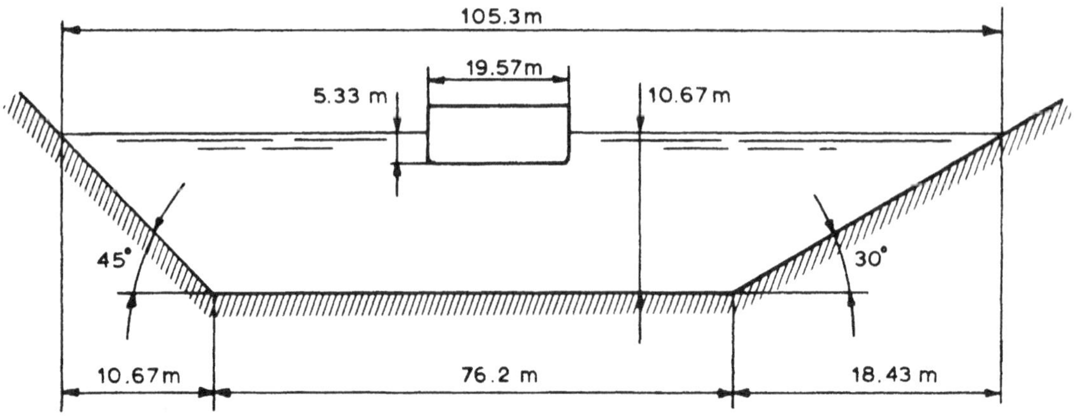

Fig. 49 Cross-section of a canal

$$V_\infty / \sqrt{gh} = 4.115/\sqrt{9.81 \times 10.67} = 0.402$$

From Fig. 46 the ratio of V_I/V_∞ is unity, and $V_I = V_\infty$, so that there is no effect of wave-making and all the speed reduction is due to potential flow changes.

From Fig. 48 for

$$\sqrt{A_X}/R_H = 1.67, \; V_h/V_I = 0.783$$

and therefore, since $V_I = V_\infty$

$$V_h = 0.783 \times V_\infty = 0.783 \times 4.115 = 3.222 \text{ m/sec}$$

$$= 6.26 \text{ knots}$$

or a reduction of 22 percent.

Incidentally, if the curve obtained by Landweber is used for the unrestricted shallow-water case, as suggested by Saunders, instead of that published by Schlichting, the speed loss for this ship in shallow water is 18.6 percent instead of the 20.3 percent found previously, the speed being 10.58 knots instead of 10.35.

When large, medium-speed ships or very high-speed ships such as destroyers have to run measured mile trials, the question often arises of finding a course over which the water is deep enough to ensure that the effects on resistance and speed will be negligible or within stated limits. Conversely, if such a course is not available in a particular case, it is desirable to be able to correct the trial results to obtain the probable performance in deep water.

Both these problems can be solved by the methods described here, and a number of such cases are given by Saunders (1957), Chapter 61, together with charts to assist in their more rapid solution.

The effect of shallow water on some typical merchant ship forms has been investigated at DTRC by running resistance and propulsion tests on models (Norley, 1948). These represented a Liberty ship, a Victory ship, a T-2 ocean-going tanker and a T-1 inland tanker.

The increase in resistance and shaft power, together with sinkage and trim, were measured over a range of speeds in a number of depths of water from 6.71 m in the case of the T-1 tanker up to deep water.

The detailed results for each model are given in the report. Figs. 50, 51 and 52 show generalized plots of the sinkage at bow and stern and the increase in P_D and revolutions per minute to a base of the ratio of ship draft to water depth. The sinkage increased with speed and with decrease in water depth, that at the bow being greater at all speeds up to the maximum reached, for which $V/\sqrt{gL} = 0.149$. There were indications that the sinkage would be greater with larger ratios of beam to draft, but further model tests would be necessary to confirm this trend. There appeared to be a real danger of the ship striking the bottom if the

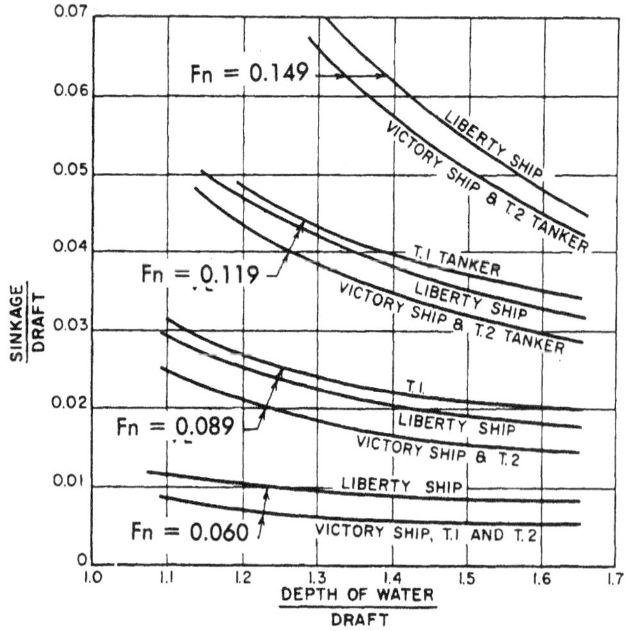

Fig. 50 Sinkage of bow at FP in shallow water

speed were too high in any given case, and the safe depths of water for values of $V/\sqrt{gL}$ of 0.149, 0.119 and 0.089 were defined by values of depth to draft ratios of 1.3, 1.2 and 1.1. In the last of these conditions the increase in P_D was about 100 percent, but at such low speeds this does not have much importance.

When the speed of a ship in shallow water approaches and passes the critical speed (the speed of translation of a wave in the depth of water in question), radical changes occur in the wave pattern around the ship, in the ship's attitude, and in its resistance. These changes in wave pattern are generally similar to those described for a pressure point under similar conditions.

The characteristic behavior of a high-speed ship in shallow water is illustrated in Fig. 53, which shows resistance and trim curves for a model of a destroyer run in different depths of water (Rota, 1900). On each curve is marked the *critical speed* where the value of $V/\sqrt{gh}$ is unity for that particular depth. Considering the curves for the shallowest depth, where the ratio (depth of water)/(draft of ship) is 3.08, it will be seen that as the speed of the model approaches the critical speed the trim by the stern and the resistance both begin to increase very rapidly. When the critical speed is passed the trim falls off very quickly and the resistance remains almost constant for a time before beginning to increase again but then at a somewhat lower rate than in deep water. Finally, at a $V/\sqrt{gL}$ of about 0.48 both the trim and resistance curves cross those for deep water, and above this speed, now well in the supercritical zone, the resistance and trim are both less than in deep water.

The typical characteristics of shallow-water behavior found in such model experiments are well borne out in full-scale trials (Watts, 1909). The speeds at which the power begins to increase, that is, where the effects of shallow water are first noticed, can be given approximately by the relation

$$V_K = 4.17\sqrt{h},\ h \text{ in meters} \qquad (52)$$

This formula may therefore be used with some confidence to calculate the speed at which the wave making resistance in shallow water begins to increase appreciably as the critical speed is approached. The critical speed is given by

$$V_K = 6.09\sqrt{h},\ h \text{ in meters} \qquad (53)$$

The ratio of the speed at which the power begins to increase (Equation (52)) to the critical speed is 4.17/6.09 or 0.68. Observations of the wave pattern on the trials of the destroyer HMS *Cossack* (Watts, 1909) showed that in deep water the stern wave increased in size continuously with increase in speed. On the shallow measured mile course, however, it increased very rapidly in size up to about 22 knots and then appeared to become unstable and fell away to about half its maximum height. At 28 knots the stern wave had practically disappeared and did not re-appear during the higher speed runs. This changing pattern is generally in accord with that described by Havelock for pressure disturbances.

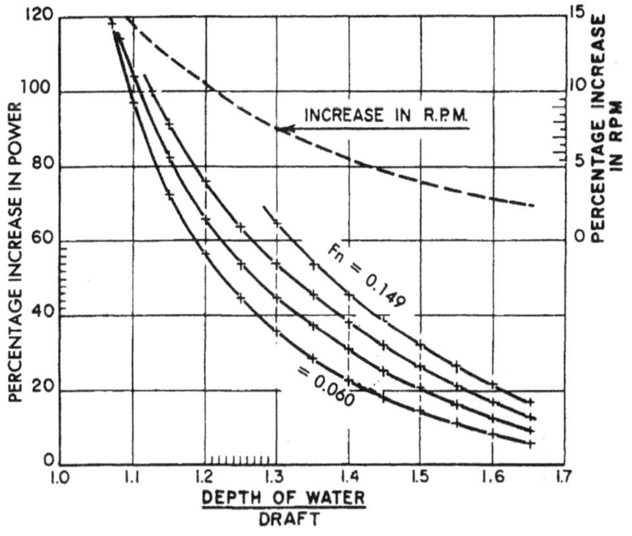

Fig. 52 Increase in P_D and RPM in shallow water

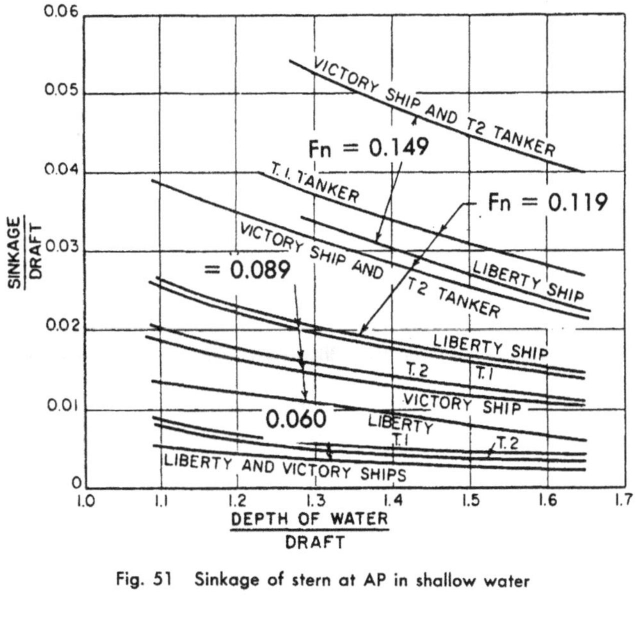

Fig. 51 Sinkage of stern at AP in shallow water

The effect of depth of water on the resistance of high-speed ships is brought out clearly when the percentage increase in resistance is plotted against the value of $V/\sqrt{gh}$. Such curves are shown in Fig. 54. The peaks occur at values of $V/\sqrt{gh}$ somewhat less than unity, that is, a little below the speed of the wave of translation. The percentage increase at the peak is greater the smaller the ratio of draft to depth of water.

There are obvious advantages in the use of models for investigating shallow-water effects, where all the variables can be carefully controlled, and most of the

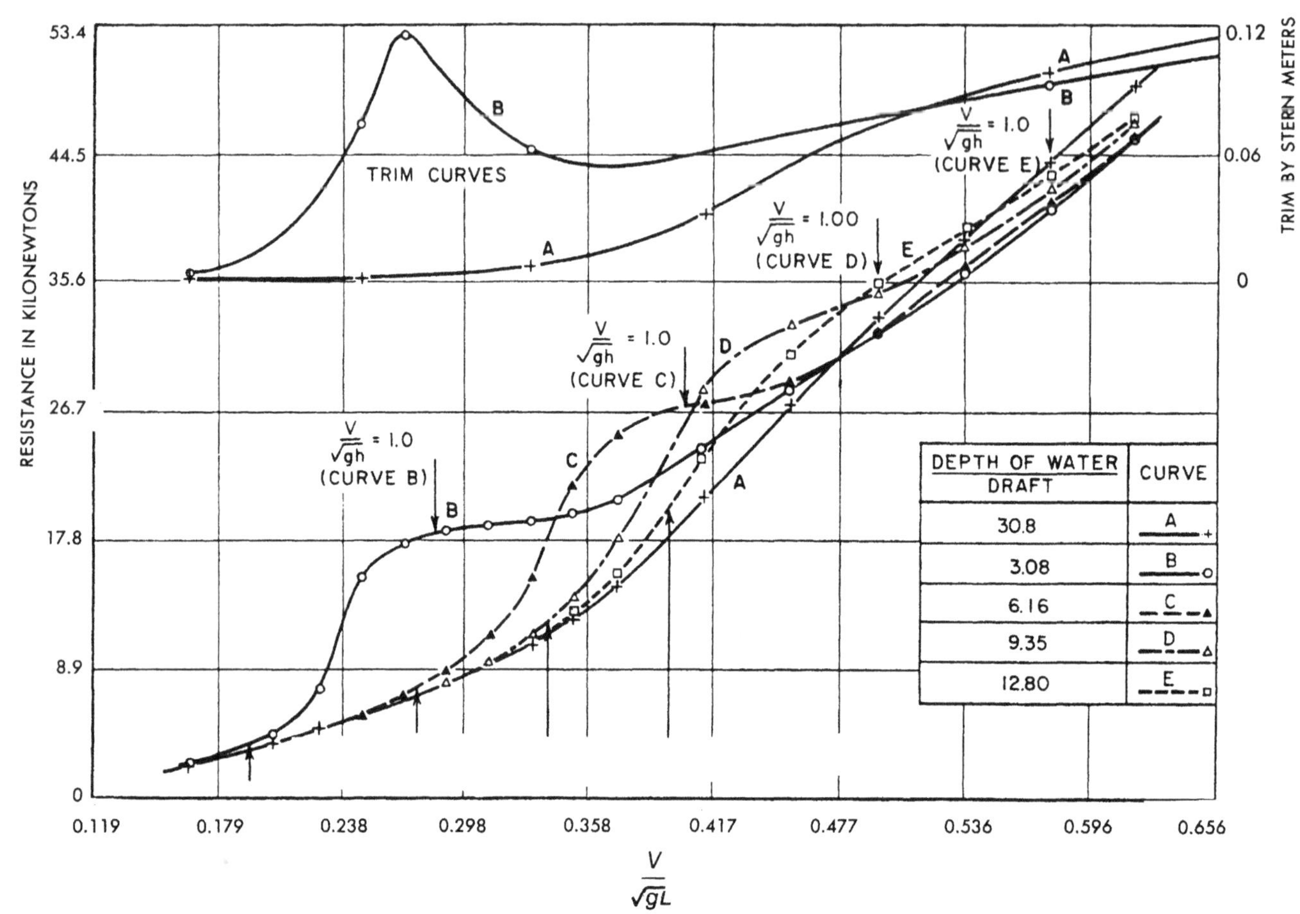

Fig. 53 Resistance of destroyer in shallow water

Model: 3.76 × 0.41 × 0.098 m, C_B = 0.435

data available have been obtained by this method. There are difficulties in applying the results to the ship, however. Owing to the accelerated flow around the model, the skin-friction correction will be different from that in deep water. The increase in resistance in the tank, where both width and depth are restricted, will be greater than that in shallow water of unlimited width, and this must be properly allowed for in any such application. If only resistance experiments are carried out, due allowance must be made for the heavier loading of the propeller when estimating the shaft power. The flow differences may also affect the elements making up the propulsive efficiency, such as wake and thrust deduction, and self-propulsion experiments are advisable. Whenever possible such tests should be carried out over a solid bottom, rather than a suspended "false" bottom which may move, either in a tank specially designed for this kind of work or by lowering the water level in a deep water tank to the required depth and making special towing rigs for the model. In general, for a variety of reasons, model tests in tanks of limited width tend to exaggerate the increase of resistance in shallow water.

5.7 Resistance increase due to leeway and heel, with special reference to sailing yachts. The increase in resistance that occurs when a ship has a fixed yaw and or a heel angle is most marked in sailing ships and yachts sailing to windward. Steady yaw angles usually occur when the vessel has to counteract some aerodynamic side force on the superstructure or sails by developing an equal and opposite hydrodynamic side force on the hull and on the keel. For a sailing vessel with completely symmetrical hull the only way hydrodynamic side forces can develop is when the hull adopts an angle of attack relative to its course through the water. In sailing yachts this angle is usually referred to as leeway. The same aerodynamic force on the superstructure or the sails will also cause a heel angle for yachts sailing to windward. Typical leeway angles are between 3 and 6 degrees, although on yachts with inefficient keels (i.e. keels which do not possess the characteristic of being able to develop the required side force at small angles of leeway) larger leeway angles may occur. Heel angles of up to 30 degrees are quite normal. For other ships yaw angles can also occur because of unsymmetrical properties of the hull, such as can occur when shaft brackets or other appendages are misaligned. In those cases a constant (usually small) rudder angle is required to compensate for the hydrodynamic side force that results. However, in that case a hydrodynamic moment is exerted on the vessel which has to be counterbalanced by the hull adopting some leeway angle. In these cases, however, angles are usually within 1 or 2 degrees.

The resistance increase due to leeway angles less than about 5 degrees is almost entirely due to the so-called induced resistance associated with the production of side force or lift on the hull. Lift is generated by deflecting a flow over an angle α_i downward (or sideways) from its undisturbed direction. The force generated by the body that induces (deflects) this flow is directed at approximately right angles to the direction of the deflected flow, as shown in Fig. 55. It follows that the component of this force, $F \sin \alpha_i$, then acts against the direction of motion. This force is called the induced drag force R_I, because it is associated with the induced flow field. From Fig. 55 it follows that $R_I = F \sin \alpha_i = L \tan \alpha_i$. Hence $C_{R_I} = C_L \tan \alpha_i$. It can be shown that the induced flow angle α_i is related to the lift coefficient and the aspect ratio according to:

$$\alpha_i = \frac{C_L}{\pi AR}, \text{ where } AR = \frac{\text{span or depth of hull (or keel)}}{\text{length of hull or keel}} \quad (54)$$

Therefore,

$$C_{R_I} = \frac{C_L^2}{\pi AR} \quad (55)$$

since for small α_i-angles, $C_{R_I} \simeq C_L \alpha_i$. Use of Equation (55), together with $R_I = \frac{1}{2} \rho V^2 C_{R_I} A$, leads to:

$$R_I = \frac{1}{2} \rho V^2 \frac{C_L^2}{\pi AR} A \quad (56)$$

where A = lateral area of hull or keel.

The induced resistance of a sailing yacht then can be approximately determined by adding the induced resistance of the hull, keel and rudder.

Equation (56) is strictly valid only for an elliptical lift distribution over the span of the lifting surface. The planforms of present-day keels and rudders rarely lead to an elliptical spanwise loading, however. Appreciable increments in induced drag are found in planforms that are either extremely tapered or close to a

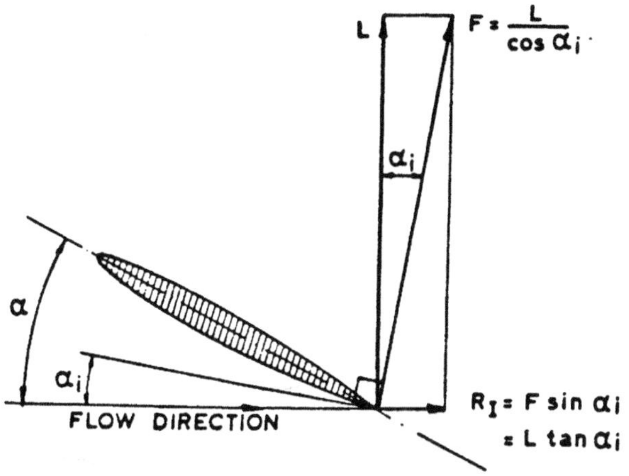

Fig. 55 Induced drag: a component of the lift force.

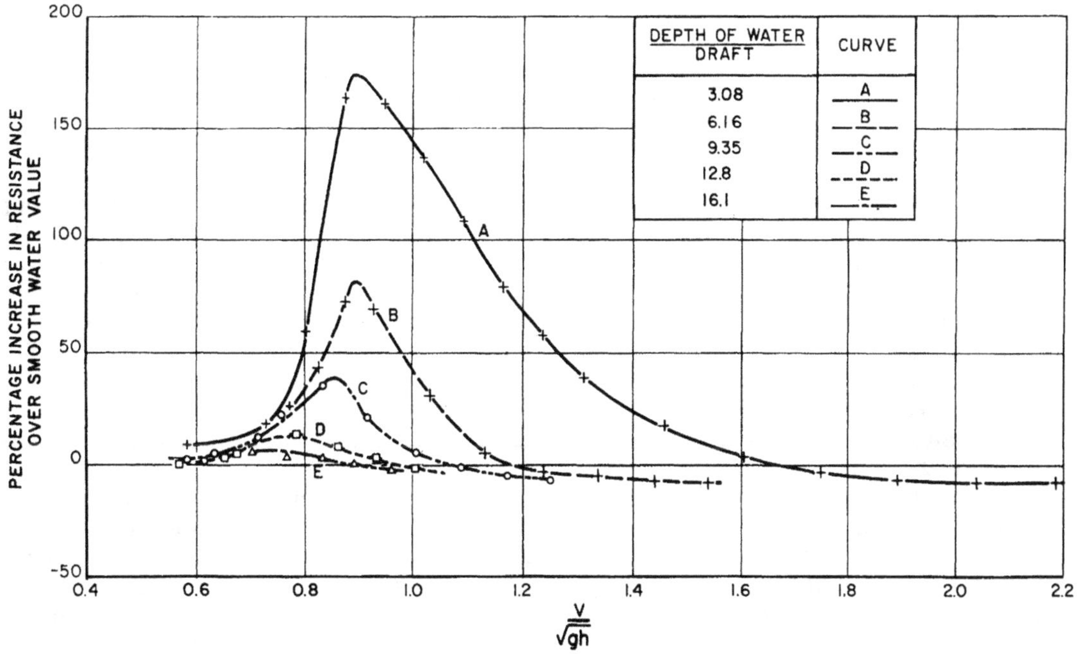

Fig. 54 Percentage increase in resistance in shallow water. Same model as Fig. 53

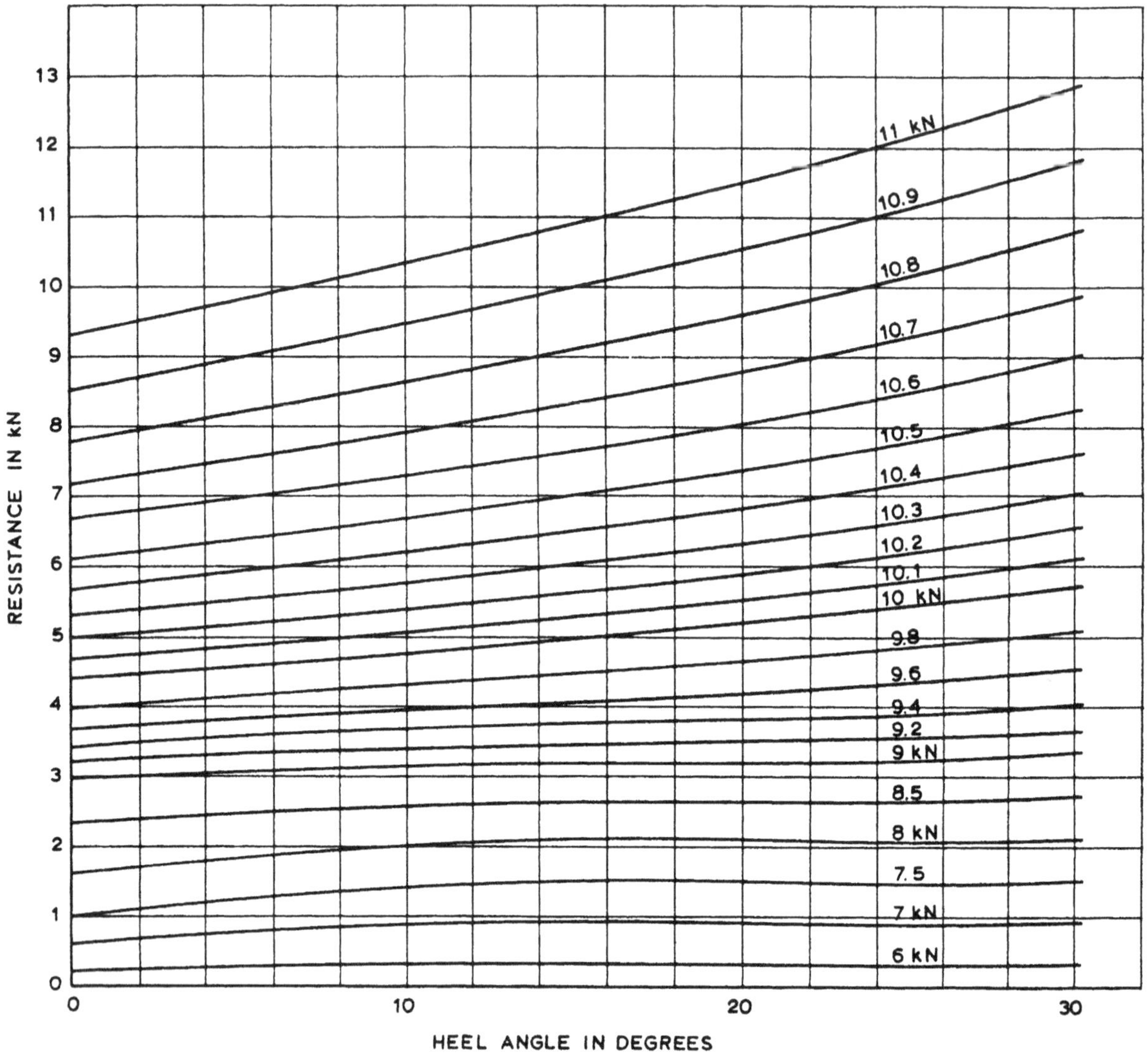

Fig. 56 Resistance of the J-class yacht *Rainbow* as a function of heel angle at zero yaw angle, as determined from model tests with a 1 to 9 scale model at MARIN (unpublished data).

rectangular shape. For taper ratios (i.e. the ratio between the lower and upper end chord lengths) between 0.3 and 0.4, an elliptical spanwise loading is nearly obtained. In that case the additional induced drag is very small (about 1 or 2 percent).

It should be noted that whereas elliptical or rounded planforms might be advantageous in minimizing induced drag, they also lead to a reduction in the total lift. The effective span of rounded planforms is less than that of rectangular planforms. A consequence of this fact is that rectangular planforms often lead to the highest lift to drag ratios.

The effect of sweep is to increase the loading near the tip of the lifting surface. According to Hoerner (1965), a sweep-back angle of 30 degrees requires a taper ratio of about 0.15 to obtain near-elliptical loading (instead of about 0.35 for zero sweep-back). Since such taper ratios are rarely practical (except in delta configurations) it follows that the spanwise loading of swept-back lifting surfaces is not often near-elliptical, leading to somewhat higher induced drag values. Also, the lift force of each chordwise segment of the lifting surface approaching the tip is tilted further "backward" because of an increasing deflection of the induced flow, leading to larger flow angles α_i (see Fig. 55). It follows that because of this the component of the lift in the direction of the undisturbed flow becomes greater with increasing sweep-back. According to Hoerner (1965) the induced drag increases proportionally with sweep angle according to $1/\cos\alpha$, where α is the sweep-back (or forward) angle of the quarter-chord line of the lifting surface. Such an increase in induced drag, however, is never found in experimental sailing yacht studies. It would appear that the increase in induced drag due to sweep is nearly completely compensated for by the favorable influence of sweep on wave resistance.

At yaw angle larger than about 5 degrees the flow along the aftbody usually separates and the resistance associated with yaw angle increases markedly. Even on sailing yachts, with block coefficients around 0.4, this occurs because the flow on the windward side of the hull forward of the rudder separates.

The immersed hull of a heeled ship will be asymmetrical, with the leeside of the vessel being considerably bluffer. This invariably leads to an increment in the wavemaking resistance. In many cases also the viscous resistance increases because of added wetted area or because of more unfavorable boundary layer development leading to flow separation, or both. In yachts with long bow and stern overhangs, this increase in resistance is compensated for to a certain extent because of the increase in effective wavemaking length of the hull as the hull heels. A typical result for such a hull is shown in Fig. 56 which gives the resistance in kN for the J-class yacht *Rainbow*, as measured at MARIN for various angles of heel. At speeds between 6 and about 9 knots it can be seen that the increase in resistance with heel angle is marginal due to the increase in wavemaking length.

Section 6
The Uses of Models for Determining Ship Resistance

6.1 Historical. Because of the complicated nature of ship resistance it was natural that early recourse should have been made to experiments, and it is recorded that Leonardo da Vinci (1452-1519) carried out tests on three models of ships having different fore-and-aft distributions of displacement (Tursini, 1953). The next known use of models to investigate ship resistance were qualitative experiments made by Samuel Fortrey (1622-1681), who used small wooden models towed in a tank by falling weights (Baker, 1937). From this time onwards there was a steady growth of interest in model experiment work (Todd, 1951). Colonel Beaufoy, under the auspices of the Society for the Improvement of Naval Architecture, founded in London in 1791, carried out between nine and ten thousand towing experiments between 1791 and 1798 in the Greenland Dock, using models of geometrical shape and flat planks (Beaufoy, 1834). Benjamin Franklin was probably the first American to make model experiments, in 1764, to verify observations he had made in Holland that resistance to motion increased in shallow water (Rumble, 1955).

Throughout this period this method of gravity towing was universally used, and William Froude made his first model experiments in 1863 in a large rainwater tank using the same type of towing mechanism. He soon became dissatisfied with the limitations of these experiments and turned his mind to the use of a larger tank, making proposals to the British Admiralty in 1868, which were accepted, and a new tank was completed near his home in Torquay in 1871 (W. Froude, 1955). This tank had a length of 84.7 m (277.8 ft), a width at the water surface of 11 m (36 ft) and a depth of water along the centerline of 3.05 m (10 ft). It was equipped with a mechanically propelled towing carriage to tow the models, in place of a gravitational device, and because of this and its size may be considered as the forerunner of the tanks so common today. At this time Froude was already 61 years of age, having put forward his law of comparison in 1868 and shown how it could be used in practice to predict ship resistance from model results.

D. W. Taylor graduated from the U.S. Naval Academy in 1885 and from there went to the Royal Naval College, Greenwich, England, where he became aware of the work done by Froude. On his appointment to the Navy Department in Washington in 1894 he advocated the building of a towing tank for the U.S. Navy. As a result, the Experimental Model Basin (EMB) was built in the Washington Navy Yard. It had a length of 143.3 m (470 ft), a breadth on the water surface of 12.8 m (42 ft) and a centerline depth of 3.06 m (10 ft), the towing carriage having a top speed of 7.7 m/sec (25.2 ft/sec) (EMB, 1925). The Basin was opened in 1900, and Taylor remained in charge of it for some 14 years, during which time much work of great value to naval architects everywhere was conducted under his inspiration and guidance.

At the end of the 19th century there were perhaps five model experiment tanks in the world. Now they number about 125 and are regarded as a necessary and important adjunct to the shipbuilding industry of every maritime nation.

6.2 Modern Facilities. There is a considerable number of small tanks generally associated with educational and research establishments, using models 1 to 2 m in length, engaged in measuring resistance in smooth water and motions, resistance, and loss of speed in waves. In some of these tanks the towing force is still provided by a falling weight, suitably geared, and which is constant during a run, the technique being to measure the speed attained for a given towing force, i.e., for a given resistance. A few facilities have monorail carriages.

The larger tanks in general employ mechanically or electrically-driven towing carriages, use models 4 to 10 or more meters in length (32.8 ft), and conduct propulsion as well as resistance tests on ship models and various other bodies. Typical dimensions of these larger tanks are 250 m (820 ft) long, 10 m (33 ft) wide, and 5 m (16 ft) deep. For investigations in shallow water, some establishments have adopted a basin in excess of 20 m (66 ft) wide and a (variable) depth of up to about 1.5 m (5 ft). For investigations involving high-speed craft, the tank needs to be extra long. In that case, often relatively narrow basins are employed, typically 4 m wide and 4 m deep, (13 ft) and the speed of the towing carriages needs to be well in excess of the maximum speed of about 10 m/sec (33 ft/sec) normally employed in other tanks. Resistance tests are carried out with the model attached to the carriage. Besides measuring the so-called towing force, the rise (or sinkage) of the model at the forward and aft perpendicular is measured and observations of the waves along the model are made.

After the first resistance tests, many models are modified and further experiments made. This is often done on the basis of observations of the flow around the hull by means of a paint-smear technique and/or wool-tufts, which set themselves in the flow lines. Wool-tufts secured to pins set normal to the model surface enable the flow to be explored at points some distance out from the hull. Photographs of the underwater hull form are then taken for analysis of the flow pattern (Fig. 57). Such observations provide useful information for judging the quality of the hull and appendages, often leading to design modifications.

Fig. 57 Model in MARIN towing tank showing stream-line flow

6.3 Model Testing Techniques. The accepted basis of predicting ship resistance from that of a model still rests on the assumption made by Froude that the total resistance can be divided into frictional and residuary components, as set out in Section 2.3. The residuary resistance coefficient C_R is assumed to be the same for model and ship at the same value of the Froude number $\mathrm{Fn} = V/\sqrt{gL}$, but the frictional resistance coefficient C_F is a function of Reynolds number and therefore bears the major responsibility for correct extrapolation. The "jump" in going from model to ship is very large; in the case of a 125-m ship and a 5-m model, used as an example in Section 2.3, the speeds of the ship and model were, respectively, 25 and 5 knots, so that the values of Reynolds number, proportional to the product VL, would be in the ratio of 1 to 125. (This is not always realized in looking at experiment plots, because it is general to use a base of log Rn, which greatly reduces the apparent degree of extrapolation.) The measurement of the model resistance must therefore be extremely accurate to minimize errors in the extension to the ship. Constancy of speed of the towing carriage is a most important basic requirement, as is the accuracy of the towing dynamometer.

The models must be made to close tolerances, the surface correctly finished and the models properly ballasted and trimmed. The choice of model length is governed by several considerations. The larger the model the more accurately it can be made and the larger are the forces to be measured, both features leading to greater accuracy in the measurement of resistance. However, the bigger the model the more expensive it is to build and handle, the larger are the facilities and instruments necessary, and some compromise in size must be reached. If the model is too large for a particular basin, interference from the walls and bottom will increase the resistance. There is still no real agreement on the proper assessment of this interference effect. Broadly speaking, the model should not have a length greater than the depth of water or than half the width of the basin in order to avoid interference with the wave resistance. The midship cross-sectional area of the model should not exceed about 1/200 of that of the basin in order to avoid setting up appreciable return flow in the water around the model, the so-called blockage effect. However, in cases where wavemaking is small, larger models can be used and corrections made for the remaining blockage effect (Comstock, et al, 1942, Telfer, 1953-4, Hughes, 1957, Emerson, 1957 Hughes, 1961, and Kim, 1962). In certain high-speed models care must also be taken to avoid the critical speed $V = \sqrt{gh}$, in the basin which would result in the formation of a wave of translation, as already described. A model run at any speed above about 0.7 of this value will give a resistance different from that appropriate to deep water.

Another precaution that must be taken in all model testing is to ensure that the flow over the model is fully turbulent, since flow around the full-scale ship is turbulent. In discussing the frictional resistance of smooth planks, it was shown that two regimes of flow are possible, laminar and turbulent, the latter giving a much higher specific resistance than the former at the same value of Reynolds number. Also, with increasing Reynolds number, the regime changes, but

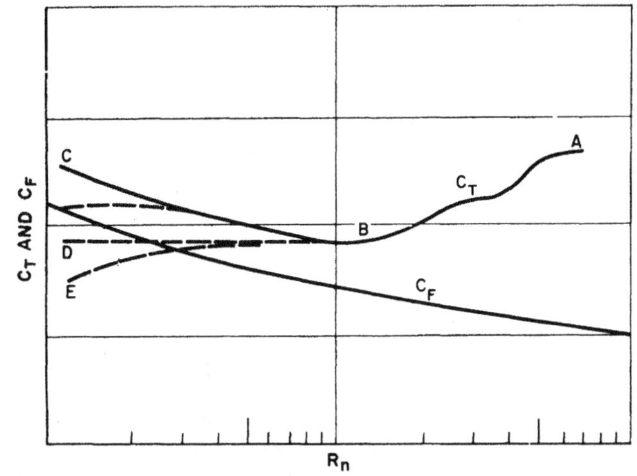

Fig. 58 Effect of laminar flow on resistance curves

the transition curve depends on the individual circumstances (Fig. 2).

The resistance curves of models show the same character of transition at low values of Reynolds number, and the resulting low resistance coefficients if scaled up to the ship on the basis of any of the turbulent friction lines would give much too low ship resistance values. It is therefore necessary to avoid this situation.

The presence of laminar flow can usually be detected from the shape of the resistance curve (Fig. 58). At low values of the Froude number, where the wave-making resistance is vanishingly small, the C_T curve should run in more or less parallel to the curve of skin friction coefficient C_F, as *ABC*. A curve which falls away in this region or even becomes horizontal, such as *ABD* or *ABE*, is at once suspect as being subject to partial laminar flow. To investigate this, Prohaska's method can be used. For low speeds the resulting value of $1 + k$ should approach a constant value.

Some typical values of Reynolds numbers for ships and models are given in Table 6 as lending a quantitative meaning to the problem. Results of plank experiments, such as the plot shown in Fig. 3, indicate that laminar-flow effects occur up to Reynolds numbers of the order of 5×10^6. On this basis the models required for the 125 m and 300 m ships considered in Table 6, would have to have minimum lengths of about 6 and 4 m, respectively, to avoid serious laminar flow at the speeds in question. Experiments with ship models have shown, however, that without special devices to stimulate turbulence, as are discussed later, even these sizes are sometimes inadequate. The persistence of laminar flow has been found to depend to a great extent on the pressure gradient along the entrance (which is absent in the plank) and on the factors which affect this, such as shape of stem profile, half-angle of entrance on the load waterline and the shape of the entrance area curve. When these features combine to give a negative pressure gradient just abaft the bow, with consequent increasing velocity, the flow is stable and laminar flow tends to persist over considerable areas, sometimes as far as to the forward shoulder, where the pressure gradient becomes constant or positive (Allan, et al, 1949).

The practical answer to the problem is to deliberately "trip" the laminar flow by some kind of roughness near the bow. Perhaps the first reference to such a practice is the use of a 0.025 mm diameter trip wire at the nose of a spheroid, 0.61 m long and 0.15 m diameter (2 × 0.492 ft) tested in air, which completely altered the character of the resistance curve (ARC, 1922).

Trip-wires some 0.9 mm diameter (0.035 in.) placed around the hull at a station 5 percent of the forward perpendicular were used in the Berlin tank as early as 1925 and came into general use there around 1933, and it is now standard practice to use such a wire or other equivalent device in most model basins. Among these other devices are struts towed ahead of the model and sandstrips, studs or pins on the hull itself. The stimulating device, if attached to the model, increases the resistance because of its own parasitic drag. If it is placed too near the stem, there is a danger of the laminar flow reestablishing itself if the pressure gradient is favorable, while when placed in the usual position, 5 percent aft of the stem, it leaves the laminar flow, if it exists, undisturbed over the first part of the length up to the stimulator. In this case the resistance of this portion of the surface will be less than the turbulent resistance desired. It is usual to assume that this defect in resistance balances the additional parasitic drag of the wire or studs. The strut has the advantage of being unattached to the model, so that its drag does not come into the measurement of model resistance, and if it successfully stimulates turbulence in the water there will be no area of laminar flow on the model. However, experiments with fine, high-speed

Table 6—Typical Reynolds Numbers

Scale		Length in m	Speed in m/sec.	Rn at 15 deg C
		125-m ship at 10 knots		
1/1	(ship)	125	5.144	540.9×10^6
1/10		12.5	1.627	17.10×10^6
1/15		8.33	1.328	9.31×10^6
1/20		6.25	1.150	6.05×10^6
1/25		5.00	1.029	4.33×10^6
1/50		2.50	0.728	1.53×10^6
1/100		1.25	0.514	0.54×10^6
		300-m ship at 30 knots		
1/1	(ship)	300	15.432	3894.3×10^6
1/30		10.0	2.817	23.70×10^6
1/50		6.0	2.182	11.01×10^6
1/75		4.0	1.782	6.00×10^6
1/100		3.0	1.543	3.89×10^6
1/125		2.4	1.380	2.79×10^6
1/250		1.2	0.976	0.99×10^6

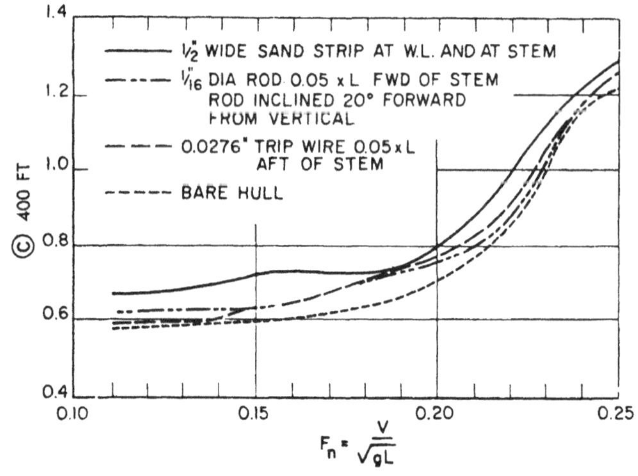

Fig. 59 Resistance curves showing effects of turbulence stimulation

models which showed no signs of laminar flow when unstimulated, have shown reduced resistance with a strut ahead, presumably due to the strut wake. The use of studs, some 3 mm in diameter and 2.5 mm high (0.12 × 0.098 in.) spaced 25 mm (1 in.) apart along a line parallel to the stem contour, was proposed by Hughes et al (1951). The distance of the line of studs from the stem was controlled by the half-angle of entrance on the waterline. The object is to ensure that the stud drag approximately balances the reduction in resistance of the laminar flow area forward. For models 5 to 6 m in length, the studs appeared to be more effective than the trip-wire, stimulating turbulent flow at an earlier speed and giving greater coverage, and they also maintained turbulent flow to lower speeds, an important factor in experiments designed to investigate methods of extrapolation.

The magnitude of the problem depends upon the size of model and type of ship. Stimulation shows little or no effect in 6 m models of high-speed warships and merchant ships of block coefficient 0.65 and below. In the fuller types of merchant ships, the effect of stimulation appears to depend to a considerable extent upon the type of bow. A heavily raked stem and pronounced V-sections seem to favor the persistence of laminar flow, while a vertical stem and U type sections seem to feel the effect much less. A typical example of the former type of hull is the *Liberty* ship, and some results of experiments on a model of this design are shown in Fig. 59. At the service speed the spread in resistance between the bare model without stimulation and the highest resistance obtained with stimulation is of the order of 20 percent.

Without turbulence stimulation, therefore, even comparative model tests may be misleading, to say nothing of the errors in ship estimates. Two models of the same design, one with V and the other with U-sections forward could give results apparently showing the former to have considerably less resistance, whereas some or all of this difference could be due simply to a greater area of laminar flow. Again, as the LCB is moved forward, the fore end becomes fuller, pressure gradients are altered, and changes in resistance attributed to the change in shape may, in truth, be purely a stimulation effect. For these reasons it is now almost universal practice to use stimulating devices on any model which is even remotely likely to suffer from laminar flow. In this connection it is well to bear in mind that many of the methodical series tests of the past were run without any stimulation devices and, indeed, before the need was even recognized. For the type of ship which is prone to such trouble, these early series results and those of individual models also should be treated with caution. Some indications of the probable magnitude of the correction necessary have been given by Dawson, et al (1949).

If self-propelled model experiments are to follow the resistance tests, as is usual, the size of the model propellers has also to be considered when choosing the scale for the hull model (Section 5. Chapter VI).

The model when completed is ballasted and trimmed to the required displacement and waterline, and attached to the resistance dynamometer of the towing carriage. A clear statement as to whether molded or total displacement is meant, should be included (see Chapter I). The model is free to take up any sinkage, rise or trim that may be dictated by the water forces, but any yawing motion is prevented by guides.

Table 7—Principal Particulars of Model and Ship

	Ship	Model
Scale, $1/\lambda$	1/1	1/21.667
Length on waterline, *m*, L_{WL}	260.0	12.000
Length between perpendiculars, *m*, L_{PP}	262.0	12.092
Beam, *m*, *B*	42.00	1.938
Draft, *m*, *T*	10.640	0.4911
Displacement volume (mld), m^3, ∇	86266	8.481
Block coefficient, C_B	0.7425	0.7425
Wetted surface, m^2, *S*	12898.9	27.476

Table 8—Model Experiment Results

Model speed, m/sec. V_M	Ship speed, knots V_k	Measured model resistance, *N* R_{TM}
0.7736	7.000	32.05
0.8842	8.000	40.89
0.9947	9.000	50.91
1.1052	10.000	62.21
1.2157	11.000	74.86
1.3262	12.000	89.17
1.4368	13.000	104.60
1.5473	14.000	121.18
1.6578	15.000	138.76

Basin water temperature, deg C = 16.2 (61 deg F)

On any given test run, the carriage is driven at the desired constant speed and records are taken of speed, resistance and trim of the model, and often the wave profile along the hull is photographed as an aid to subsequent understanding of the results. For any given displacement and trim condition, a number of test runs are made at different speeds and a curve of resistance against speed obtained.

Before proceeding to propulsion tests, a number of other experiments are often made, to determine the best line for bilge keels, the flow around the afterbody to settle the best alignment for bossings, shaft brackets and rudders and, in some cases, the flow over the whole form, as previously discussed.

6.4 Calculation of Effective Power. The estimation of ship resistance and effective power from model tests is carried out on the basis of the Froude assumption as set out in Section 2.3, with refinements based on increased understanding of resistance.

In 1978 the ITTC Performance Committee advocated

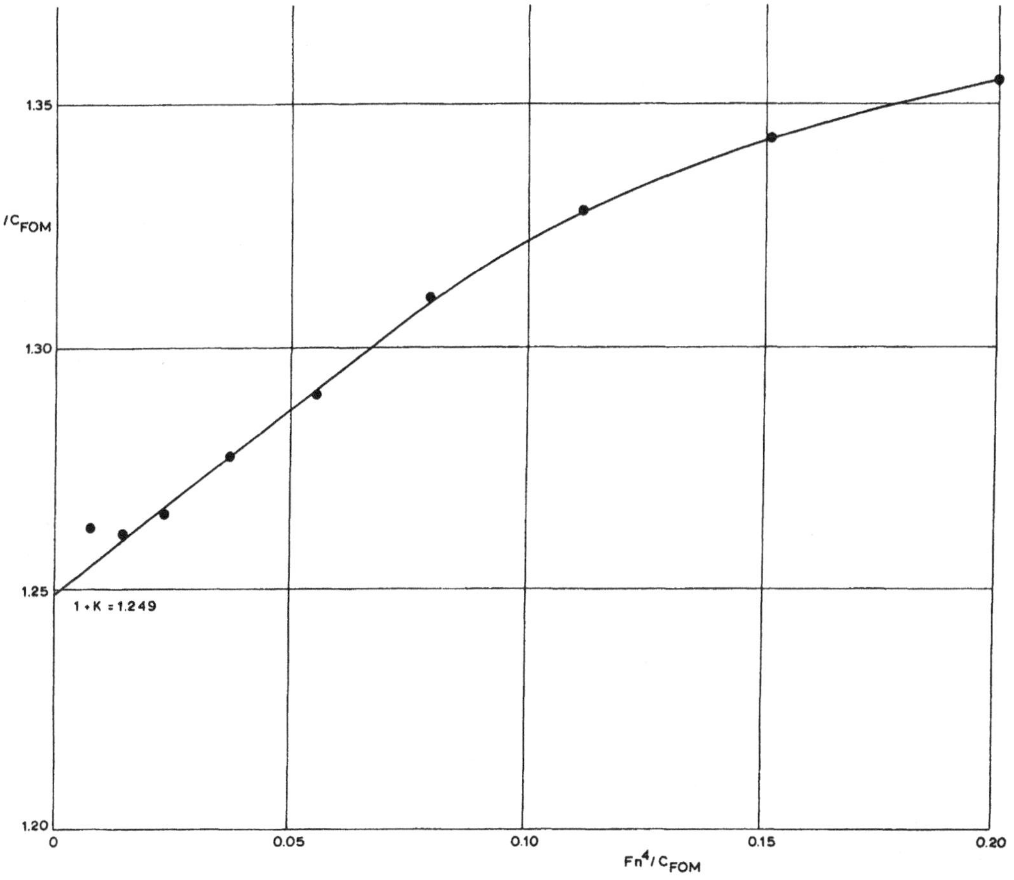

Fig. 60 Plot of Prohaska method to derive the effect of hull form on viscous resistance, based on data in Table 9.

Table 9—Values Required to Find 1 + *k* According to Prohaska

V_M	Fn	$Fn^4 \times 10^6$	$Rn \times 10^{-6}$	$C_{TM} \times 10^3$	ITTC-1957 $C_{FOM} \times 10^3$	Fn^4/C_{FOM}	C_{TM}/C_{FOM}
0.7736	0.0713	2.586	8.409	3.901	3.092	0.0084	1.262
0.8842	0.0815	4.414	9.611	3.810	3.021	0.0146	1.261
0.9947	0.0917	7.069	10.812	3.748	2.960	0.0239	1.266
1.2157	0.1221	15.773	13.215	3.690	2.860	0.0552	1.290
1.3262	0.1223	22.338	14.419	3.693	2.818	0.0793	1.310
1.4368	0.1324	30.775	15.618	3.691	2.780	0.1107	1.328
1.5473	0.1426	41.391	16.819	3.687	2.746	0.1507	1.343
1.6578	0.1528	54.543	18.020	3.678	2.715	0.2009	1.355

The values of C_{FOM} adopted in Table 9 are those according to the ITTC-1957 friction line which, together with the ATTC (Schoenherr) friction line, are the most widely used. The values of C_F according to both these friction lines are given in Table 12.

Table 10—Values of Kinematic Viscosity ν for Fresh and Salt Water Adopted by the ITTC in 1963; Salinity of Salt Water is 3.5 Percent

Temperature deg C*	Kinematic viscosity of fresh water ν, $\frac{m^2}{sec} \times 10^6$	Kinematic viscosity of salt water ν_S, $\frac{m^2}{sec} \times 10^6$
0	1.78667	1.82844
1	1.72701	1.76915
2	1.67040	1.71306
3	1.61655	1.65988
4	1.56557	1.60940
5	1.51698	1.56142
6	1.47070	1.51584
7	1.42667	1.47242
8	1.38471	1.43102
9	1.34463	1.39152
10	1.30641	1.35383
11	1.26988	1.31773
12	1.23495	1.28324
13	1.20159	1.25028
14	1.16964	1.21862
15	1.13902	1.18831
16	1.10966	1.15916
17	1.08155	1.13125
18	1.05456	1.10438
19	1.02865	1.07854
20	1.00374	1.05372
21	0.97984	1.02981
22	0.95682	1.00678
23	0.93471	0.98457
24	0.91340	0.96315
25	0.89292	0.94252
26	0.87313	0.92255
27	0.85409	0.90331
28	0.83572	0.88470
29	0.81798	0.86671
30	0.80091	0.84931

For other salinities, interpolate linearly.

the use of the form-factor approach (see Section 3.6) in determining the effective power from model tests. Calculations have been made to illustrate this modified method in detail and to show the differences in the final effective power which result from using this approach relative to the two-dimensional frictional resistance formulations (see Section 3.5), for both ATTC and ITTC (1957) friction lines. The calculation is for a 12-m (39.36 ft) model of the 120,000-m^3 methane carriers *Castor* and *Nestor* (Muntjewerf, et al, 1983). The main particulars of model and ship have been given in Table 7.

(*a*) *Calculation of resistance and effective power by three-dimension extrapolation procedure.* Following the Prohaska (1966) proposal (Section 3.6), the value of $1 + k$ can be determined from the gradient of the C_{TM}/C_{FOM} vs. $c\text{Fn}^4/C_{FOM}$ curve for $\text{Fn} \rightarrow 0$. For this particular case $c = 1$ has been taken, so as to fit the C_{TM} curve as well as possible. The values of the quantities required to find the form factor value are given in Table 9. The Froude number and Reynolds number given in Table 9 have been calculated as follows:

$$\text{Fn} = \frac{V_M}{\sqrt{gL_{WL_M}}}; \text{ for 14 knots}$$

$$\text{Fn} = \frac{1.5473}{\sqrt{9.81 \times 12}} = 0.1426$$

$$\text{Rn} = \frac{V_M L_{WL_M}}{\nu}; \text{ for 14 knots}$$

$$\text{Rn} = \frac{1.5473 \times 12}{1.1040 \times 10^{-6}} = 1.6819 \times 10^7$$

Table 11—Values of Mass Density ρ for Fresh and Salt Water Adopted by the ITTC in 1963; Salinity of Salt Water is 3.5 Percent

Temperature deg C*	Density of fresh water, ρ, kg/m^3 or 1000 kg/L	Density of salt water, ρ_S, kg/m^3 or 1000 kg/L
0	999.8	1028.0
1	999.8	1027.9
2	999.9	1027.8
3	999.9	1027.8
4	999.9	1027.7
5	999.9	1027.6
6	999.9	1027.4
7	999.8	1027.3
8	999.8	1027.1
9	999.7	1027.0
10	999.6	1026.9
11	999.5	1026.7
12	999.4	1026.6
13	999.3	1026.3
14	999.1	1026.1
15	999.0	1025.9
16	998.9	1025.7
17	998.7	1025.4
18	998.5	1025.2
19	998.3	1025.0
20	998.1	1024.7
21	997.9	1024.4
22	997.7	1024.1
23	997.4	1023.8
24	997.2	1023.5
25	996.9	1023.2
26	996.7	1022.9
27	996.4	1022.6
28	996.2	1022.3
29	995.9	1022.0
30	995.6	1021.7

* deg F = 9/5 C + 32

Table 12—Values of C_F According to the ITTC-1957 and ATTC Friction Lines

Rn	$C_F \times 10^3$ (ITTC)	$C_F \times 10^3$ (ATTC)
1×10^5	8.333	7.179
2	6.882	6.137
3	6.203	5.623
4	5.780	5.294
5	5.482	5.057
6	5.254	4.875
7	5.073	4.727
8	4.923	4.605
9	4.797	4.500
1×10^6	4.688	4.409
2	4.054	3.872
3	3.741	3.600
4	3.541	3.423
5	3.397	3.294
6	3.285	3.193
7	3.195	3.112
8	3.120	3.044
9	3.056	2.985
1×10^7	3.000	2.934
2	2.669	2.628
3	2.500	2.470
4	2.390	2.365
5	2.309	2.289
6	2.246	2.229
7	2.195	2.180
8	2.162	2.138
9	2.115	2.103
1×10^8	2.083	2.072
2	1.889	1.884
3	1.788	1.784
4	1.721	1.719
5	1.671	1.670
6	1.632	1.632
7	1.601	1.600
8	1.574	1.574
9	1.551	1.551
1×10^9	1.531	1.531
2	1.407	1.408
3	1.342	1.342
4	1.298	1.299
5	1.265	1.266
6	1.240	1.240
7	1.219	1.219
8	1.201	1.201
9	1.185	1.186
1×10^{10}	1.172	1.172

$$C_F = \frac{0.075}{(\log_{10}\mathrm{Rn} - 2)^2} \qquad \text{(ITTC-1957)}$$

$$\frac{0.242}{\sqrt{C_F}} = \log_{10}(\mathrm{Rn} \times C_F) \qquad \text{(ATTC)}$$

$$\mathrm{Rn} = \frac{VL}{\nu}$$

The value of the kinematic viscosity ν for fresh and salt water at the basin water temperature of 16.2 deg C (61 deg F) can be determined from Table 10.

C_{TM} is calculated on the basis of the measured total resistance, which for 14 knots (model speed, 1.5473 m/sec) was found to be 121.18 N:

$$C_{TM} = \frac{R_{TM}}{\frac{1}{2}\rho V_M^2 S_M} = \frac{121.18}{0.5 \times 999.25 \times (1.5473)^2 \times 27.476} = 0.003687$$

The value of the mass density, ρ (kg/m^3), for fresh and salt water at the basin water temperature of 16.2 deg C (61 F) can be determined from Table 11. From the plot in Fig. 60 the value of $1 + k$ is found to be 1.249, which is the value of C_{TM}/C_{FOM} when $\mathrm{Fn}^4/C_{FOM} = 0$.

To calculate the resistance and effective power for a smooth hull at a full-scale speed of 14 knots, for example, the following procedure is adopted. proceeding as in Section 3.6 and Fig. 5:

$$C_{TS} = C_{VS} + C_{WS} \qquad (57)$$

where $C_{VS} = (1 + k)\, C_{FOS}$

and $C_{WS} = C_{WM} = C_{TM} - C_{VM}$

in which $C_{VM} = (1 + k)\, C_{FOM}$

Following this procedure it has implicitly been assumed that $1 + k$ is independent of Rn. Following Tanaka (1979), the form factor may be defined as:

$$k = a + b\, C_F^{-0.5} + c\, C_F^{-1.0} + d\, C_F^{-1.5}$$

where a, b, c, d depend on the ship form. a is related to equivalence with a two dimensional body, b to equivalence with a body of revolution, c is related to separation and d is related to base drag. Thereby it is evident that the form factor may depend on Rn. However no quantitative information on this dependency is known. With $\mathrm{Rn}_M = 1.6819 \times 10^7$, $C_{FOM} = 2.746 \times 10^{-3}$ for the ITTC-1957 friction line and $C_{FOM} = 2.700 \times 10^{-3}$ for the ATTC friction line.

Thus $C_{VM} = 1.249 \times 0.002746 = 0.003422$ (ITTC) and
$C_{VM} = 1.249 \times 0.002700 = 0.003372$ (ATTC).

The corresponding value of $C_{TM} = 0.003687$ (see Table 9).

Thus $C_{WS} = C_{WM} = 0.003687 - 0.003422 = 0.000265$ (ITTC) and
$C_{WS} = 0.003687 - 0.003372 = 0.000315$ (ATTC).

It has been agreed by the ITTC and ATTC that for published work all ship results shall be given for a standard temperature of 15 deg C (59 deg F). The corresponding value of the kinematic viscosity for salt water is 1.1883×10^{-6} m^2/sec. (see Table 10). The full scale Reynolds number for 14 knots is then,

$$\mathrm{Rn}_S = \frac{V \times L_{WL_S}}{\nu_S} = \frac{14 \times 0.5144 \times 260}{1.1883 \times 10^{-6}} = 1.576 \times 10^9$$

At this value of the Reynolds number, the value of the friction coefficient (from the formula given in Table 12) $C_{FOS} = 0.001448$. Note that at this value of the Reynolds number the ITTC and ATTC values of C_{FO} are equal.

It follows that $C_{VS} = 1.249 \times 0.001448 = 0.001809$; thus $C_{TS} = 0.001809 + 0.000265 = 0.002074$ (ITTC) and $C_{TS} = 0.001809 + 0.000315 = 0.002124$ (ATTC).

Finally it is necessary to add a correlation allowance to the smooth-ship resistance to obtain the resistance of the actual ship, an allowance which has been given the symbol C_A, the A standing for "additional" resistance, in analogy with C_F for frictional, C_R for residuary, and so on. The significance to be attached to C_A is discussed next.

The value of C_A depends upon a number of things. From the differences which can arise in the predicted ship P_E from identical model results by using different extrapolation methods, these latter will be one of the prime factors influencing C_A. In order to build up a standard method of deriving such allowances for use in future design work, it is necessary for each establishment to use one method of extrapolation, and the resulting values of C_A will apply only to that method. C_A covers not only roughness, but also differences due to extrapolation methods, together with scale effects in such processes, as well as in all the factors making up the propulsive efficiency. The need for such a factor arises from our lack of exact knowledge in the foregoing fields.

The part of C_A due to hull roughness can be attributed to a number of causes:

1. Structural roughness, resulting from the method of construction of the shell—whether riveted, welded, or partly of each kind. Other contributors are valve openings, scoops, damage-control valves, waviness of plating between frames, and so on.

2. Paint roughness. Smooth, hard-drying paints will in general cover up some of the structural roughness, such as welding beads and rivet points, and may reduce the resistance below that of an unpainted shell. Paints of rough texture or applied badly, leaving "runs" on the surface, on the other hand, can increase resistance greatly.

3. Corrosion resistance, resulting from breakdown of the paint film in service, causing corrosion and erosion of the shell plating. This can be controlled to some extent by frequent docking, cleaning and painting and the use of cathodic protection, but there is a general long-term deterioration, as evidenced by full-scale trials carried out at intervals over a number of years. Modern methods of grit or shot-blasting may help to restore new-ship quality to the shell, but they are expensive treatments and must be weighed against possible savings in fuel and future maintenance costs.

4. Fouling resistance, caused by marine organisms depositing shell, grass, and so forth. One of the chief factors influencing this type of resistance is the percentage of time spent at sea. Atlantic liners foul very little. Cargo ships spending perhaps half their time in port foul more rapidly, and are affected also by the ports they visit and the time of year. The problem of fouling is much less important now because of the greater efficiency of modern antifouling paints.

Considerable progress in the area of hull roughness has been made by BMT (Lackenby, 1962). Measurements of shell roughness were made on a number of new ships prior to trial, records 0.762 m long (2.5 ft) being taken at some 50 points on each ship. The maximum amplitude of roughness was measured over each 0.05 m length of every record, and the mean of these taken to represent the average roughness of the hull.

The average value for 68 ships was 0.0188 cm (7 mil), 25 of them having values between 0.0165 cm and 0.0191 cm. The greatest roughness was twice the average, the least one half of the average. A statistical analysis of the BMT trial results for new ships indicated that paint roughness was the most significant factor in explaining the variation in C_A-values.

Lackenby (1962) has stated that an increase in average roughness of 0.0025 cm would increase the resistance of a large, new single-screw ship by about 2½ percent

On this basis, the roughest ship, with a value of 0.0366 cm (14 mil) or 0.0178 cm above the average, would have about 17 percent greater resistance, and the smoothest, with a value of 0.0089 cm would have some 8 percent less resistance than the average ship. The spread on new ships due only to variation in roughness could therefore be as much as 25 percent. In the ships tested by BMT, differences of as much as 20 percent have been found between sister ships, and this suggests again that paint surface is probably the major cause of roughness and increased resistance.

However, hull roughness is by no means the only factor affecting the correlation of ship resistance as predicted from the model and as deduced from full-scale trials on the ship. From what has been said in Section 3 it is clear that for a given model result the predicted ship resistance will depend on many other things; i.e., the size of the model, adequate stimulation of turbulence, corrections where necessary for tank wall and bottom effects on model resistance and, above all, upon the method of extrapolation adopted, as illustrated in the present example.

Recent analyses are aimed at determining the influence of hull roughness on the value of C_A, which has led to proposals to no longer use an overall value of C_A but instead to use separate values to account for the specific effects. Thus $C_A = \delta C_f' + \delta C_r$ where $\delta C_f'$ is the correlation allowance for roughness effects and δC_R is a correction for phenomena not accounted for elsewhere.

Bowden, et al (1974) proposed a formula for C_A:

$$10^3 \times C_A = 105 \left(\frac{k_s}{L}\right)^{0.333} - 0.64 \qquad (58)$$

where k_s is the mean apparent amplitude of the surface roughness over a 50mm wavelength (1.96 in.). In this relation L should not exceed 400m (1312 ft). The above formula has been established from an analysis of thrust measurements taken during ship trials. The equation is based on the measured roughness and should be used in conjunction with a form factor method and the 1957 ITTC line.

The above relation also includes phenomena which are not due to surface roughness. Therefore the increase due to roughness alone was calculated following Yokoo (1966) and Sasajima, et al (1955):

$$\delta C_f' = \delta C_{FM}(k_s, V)\left\{\frac{C_{FS}(V, L_S)}{C_{FM}(V, L_M)}\right\}^2 \qquad (59)$$

where δC_{FM} is the resistance increase due to roughness for a flat plate with a standard length L_M. For δC_{FM} the following relation was found (ITTC, 1981):

$$\delta C_{FM} = 1.867\ \delta C_{FM}^*(k_s)\left[1 - \left(1 - \left(\frac{V}{20}\right)^{1.9}\right)^{3.5}\right] \qquad (60)$$

where $\delta C_{FM}x^*(k_s)$ is a roughness parameter depending on the type of surface. When $\delta C_{FM}(k_s)$ was put equal to 0.3×10^{-3}, the following equation for δC_R could be established:

$$10^3 \times \delta C_R = 5.725\ L^{-0.11} - 3$$

Holtrop et al (1978) have analyzed 108 full-scale trial measurements of newly built ships without bulbs, in good condition, and compared the results with corresponding model test values. They derived the following statistical relation for the model-ship correlation factor C_A to be used in conjunction with the form factor extrapolation procedure, using the ITTC line, $C_A = 0.006\ (L_{WLS} + 100)^{-0.16} - 0.00205$ which is valid for $T_F/L_{WL} > 0.04$, where T_F is the draft at the fore perpendicular. This relation for C_A reflects MARIN's experience; model test results from other towing tanks may require different correlation allowances, for reasons which have been discussed above. In the present case $C_A = 0.006(360)^{-0.16} - 0.00205 = 0.000290$.

The final value of $C_{TS} = 0.002074 + 0.000290 = 0.002364$ (ITTC) and $C_{TS} = 0.002124 + 0.000290 = 0.002414$ (ATTC). The correlation allowance is taken the same if the ATTC line is used, in order to show the differences using the two lines. The ship resistance in salt water of 15 deg C (59 deg F) is given by:

$$\begin{aligned} R_{TS} &= \tfrac{1}{2}\rho S_S V_S^2 \times C_{TS} \\ &= 0.5 \times 1.0259 \times 12898.9 \times (14 \times 0.5144)^2 \times 0.002364 \\ &= 811.210 \text{ kN (ITTC) and} \\ R_{TS} &= 828.367 \text{ kN (ATTC)} \end{aligned}$$

The effective power P_{ES} is:

$$\begin{aligned} P_{ES} = R_{TS} \times V_S \text{ kW} &= 5842 \text{ kW (ITTC)} \qquad (61) \\ &= 5966 \text{ kW (ATTC)} \end{aligned}$$

(b) Calculation of resistance and effective power by two-dimensional extrapolation procedure using ITTC (1957) and ATTC friction coefficients. In the case of two-dimensional extrapolation (see Section 3.5 and Fig. 5), the following procedure is adopted:

$$C_{TS} = C_{FOS} + C_{RS}$$

where

$$C_{RS} = C_{RM} = C_{TM} - C_{FOM}$$

Again for the example considered before with $Rn_M = 1.6819 \times 10^7$, $C_{FOM} = 0.002746$ for the ITTC 1957 friction line and $C_{FOM} = 0.002700$ for the ATTC friction line (see Table 12). With $C_{TM} = 0.003687$, $C_{RS} = C_{RM}$ follows from

$$\begin{aligned} C_{RS} = C_{RM} &= 0.003687 - 0.002746 \\ &= 0.000941 \text{ for the ITTC friction line} \\ \text{and } C_{RS} = C_{RM} &= 0.003687 - 0.002700 \\ &= 0.000987 \text{ for the ATTC friction line.} \end{aligned}$$

The full-scale Reynolds number value is $Rn_S = 1.576 \times 10^9$ for which the ITTC and ATTC friction coefficients are equivalent, and equal to 0.001448 (see previous example). Hence

$$\begin{aligned} C_{TS} &= 0.001448 + 0.000941 \\ &= 0.002389 \quad \text{(ITTC)} \\ \text{and } C_{TS} &= 0.001448 + 0.000987 \\ &= 0.002435 \quad \text{(ATTC)} \end{aligned}$$

The value of the correlation allowance factor C_A to be used in conjunction with two-dimensional extrapolation procedures has not been analyzed as thoroughly as has been for the three-dimensional case. Keller (1973), however, has given overall values for C_A to be used in conjunction with two-dimensional extrapolation procedures adopting the ITTC friction coefficients, Table 13. These values again reflect MARIN's experience. They may also be used in conjunction with the ATTC friction coefficients provided that the Reynolds number at which the model tests are carried out, exceeds approximately 1×10^7. In that case differences between both friction lines are small, as is shown.

Table 13—Values of the Model-Ship Correlation Allowance C_A, According to Keller (1973)

Length of ship L_{WL}	Value of correlation allowance C_A
50 – 150 m	+0.0004 to +0.00035
150 – 210 m	+0.0002
210 – 260 m	+0.0001
260 – 300 m	+0
300 – 350 m	−0.0001
350 – 450 m	−0.00025

On adopting these values in the current example, for

which L_{WL} = 260 m, it follows that $C_A \approx 0$. The resistance and effective power then become:

$$R_{TS} = \tfrac{1}{2}\rho S_S V_S^2 \times C_{TS}$$
$$= 0.5 \times 1.0259 \times 12898.9 \times (14 \times 0.5144)^2 \times 0.002389$$
$$= 819.789 \text{ kN} \quad \text{(ITTC)}$$
$$\text{and } R_{TS} = 835.574 \text{ kN} \quad \text{(ATTC)}$$
$$P_{ES} = 5904 \text{ kW} \quad \text{(ITTC)}$$
$$\text{and } P_{ES} = 6017 \text{ kW} \quad \text{(ATTC)}$$

(c) Comparison between R_{TS} and P_{ES} calculated by the different methods. The values of resistance and effective power calculated by the different methods are summarized in Table 14.

In this case the three-dimensional extrapolation method with the correlation allowance value according to Holtrop corresponds satisfactorily with the values derived from the two-dimensional method with the correlation allowance of 0.

When not accounting for the correlation allowance in the three-dimensional method the respective R_{TS} and P_{ES} values are some 15 percent smaller. The differences between the ITTC-1957 and ATTC friction values only lead to about 2 percent differences in the respective resistance and effective power values for both the three dimensional and the two-dimensional procedures. These differences are small only because the adopted example concerns a model test with a 12 m (39 ft) long model. If a 2 m (6.5 ft) model had been used, these differences would have been nearly 10 percent because the large differences in the respective C_{FOM} values occur for Reynolds number values less than 1×10^7. Below this value the ITTC line has a steeper slope and hence greater values for the same Rn. On using models with a length of 7 meters (23 ft) or more, resistance predictions for the ship will not often be influenced to a great extent whether the ATTC or ITTC line is used.

The accuracy of both the two-dimensional and the three-dimensional extrapolation techniques depend to a large extent on the information available at individual model basins relative to the value of the correlation allowance to be adopted.

Table 14—Calculated Values of Resistance and Effective Power

Method	R_{TS} with allowance (in kN)	R_{TS} without allowance (in kN)	P_{ES} with allowance (in kW)	P_{ES} without allowance (in kW)
3D—ITTC-1957	811.2	711.7	5842	5125
3D—ATTC	828.4	728.9	5966	5249
2D—ITTC-1957	—	819.8	—	5904
2D—ATTC	—	835.6	—	6017

Section 7
Methods of Presenting Model Resistance Data

7.1 General. The most useful method of presenting model resistance data depends upon the particular purpose for which they are to be used. There is no unanimity of opinion in the matter, and the ITTC Committee on the presentation of data (now the Information Committee) has not recommended any generally acceptable method.

Two points may be made in this respect:

(a) It is desirable that the original model data be given, including measured speed and resistance, water temperature, method of turbulence stimulation, cross-sectional area of the tank, model dimensions and displacement and any other relevant information. The user can then convert them to any desired form. This policy has been followed by The Society of Naval Architects and Marine Engineers in its Model Resistance Data Sheets (undated).

(b) If the data are presented in coefficient form, these latter should be nondimensional, so that they will have the same numerical value in any consistent system of units. Unfortunately, this practice has not been followed in the past, with the result that the naval architect should be familiar with a number of the more commonly used presentations.

7.2 The C_T-Rn Presentation. In research problems concerned with the separation of resistance into its components, methods of extrapolation to the ship, model-ship correlation allowances and the like, the resistance coefficient (Section 2.3)

$$C_T = \frac{R_T}{\frac{1}{2}\rho S V^2}$$

is usually used, plotted to a base of the logarithm of Reynolds number Rn = VL/ν.

Curves of this kind have been used in earlier sections of this chapter. In any consistent system of units, both C_T and Rn are nondimensional.

7.3 Design Presentations. For design purposes, a method is desired which will show the relative merits of different ship forms.

Ships are usually designed to carry a given displacement at a specified speed. C_T is not suitable for such cases, since it is based on wetted surface and not on displacement, and can lead to misleading presentations. An obvious merit criterion is the resistance per unit displacement weight, R_T/W, which is nondimensional when R_T and W are expressed in the same units. This ratio is the basis of a number of presentations, which differ principally as regards the speed coefficient used as the base.

7.4 The Ⓒ-Ⓚ System. R. E. Froude (1888) adopted the R_T/W criterion in his "constant" system of notation.

In order to have a speed base which would also be nondimensional, he devised a coefficient Ⓚ which is the ratio of the ship's speed to the speed of a wave having a length equal to one half of the side of a cube of the same volume of displacement as the ship. If this volume is ∇, the wavelength is $\frac{1}{2}\nabla^{1/3}$, and the wave speed will be

$$\left(\frac{gL_w}{2\pi}\right)^{1/2} = \left(\frac{g\nabla^{1/3}}{4\pi}\right)^{1/2} = \left(\frac{g}{4\pi}\right)^{1/2} \nabla^{1/6}$$

Hence

$$Ⓚ = \frac{V}{\left(\frac{g}{4\pi}\right)^{1/2} \nabla^{1/6}} = \left(\frac{4\pi}{g}\right)^{1/2} \frac{V}{\nabla^{1/6}} \tag{62}$$

If R_T/W is plotted directly to such a base, the values increase rapidly at high speeds, and the curve becomes very steep, obscuring some of its important characteristics, such as the wave-making humps and hollows. Froude therefore divided the ordinates by Ⓚ^{2}, and introduced a factor 1000 to avoid small numerical values. The resistance "constant" is then

$$Ⓒ = \frac{R_T}{W}\frac{1000}{Ⓚ^2} = \frac{125}{\pi}\frac{R_T}{\frac{1}{2}\rho\,\nabla^{2/3}\,V^2} \tag{63}$$

Since at low speeds the resistance is mostly frictional and varies approximately as V^2, the Ⓒ-curves are nearly horizontal in this region. Any increase in the rate of variation of R_T with V is shown by a rise in the curve, and these changes are very valuable in a diagnostic sense when appraising the merits of a hull form. In the foregoing equations, both Ⓒ and Ⓚ are non-dimensional.

Since Ⓒ relates to the total resistance, its frictional component will vary with size of ship, and for presentation purposes it is usual to give the values of Ⓒ for a standard value. In the past this standard value was a length between perpendiculars of 121.92 m (400 ft). The ITTC in 1969, however, decided to adopt in addition a standard ship displacement volume of ∇ = 10,000 m^3. It was also decided at that time that for the presentation of resistance and propulsion data at least two additional curves for other ship sizes be shown.

For other ship sizes a correction must be applied. This correction depends on the ship length (for the calculation of the Reynolds number), the Froude number value and the wetted surface Ⓢ, where

$$Ⓢ = \frac{\text{wetted surface}}{(\text{volume of displacement})^{2/3}} = \frac{S}{\nabla^{2/3}}$$

7.5 The R_R/W vs. Fn or R_w/W vs. Fn System. The wave-making pattern and its associated resistance are largely dependent on the Froude number Fn = $V/\sqrt{gL}$. For many purposes, therefore, especially for ships with an important wave-making resistance component, it is useful to plot R_T/W against Fn.

Very often the results of so-called standard series of hull forms are presented in the R_R/W vs. Fn form, where R_R/W is called the specific residual resistance coefficient. Use of a form factor will allow the determination of the wave-making resistance coefficient, R_w/W, in which case a R_w/W vs. Fn representation can be given.

A comparison between the residuary or wave-making resistances of two alternative designs should be carried out with care, since it ignores differences in frictional or viscous resistance, and the total resistance has to be computed in all cases to make a proper evaluation.

7.6 The R_T/W vs. Fn System. When curves of R_T/W for a number of ships are plotted to a base of Fn for comparison, the relative merits of the designs at a given value of Fn will be shown by the order of the R_T/W curves. If we wish to introduce some function of speed into the ordinates to reduce the steepness of the curves and bring out the wave-making characteristics (which is one of the reasons for plotting on Fn), Telfer (1933) has shown that we can divide the ordinate R_T/W by $(V/\sqrt{L})^2$. To retain a non-dimensional quantity, however, it is possible to divide by $(V/\sqrt{gL})^2$ and obtain:

$$C_{TL} = \frac{gR_TL}{WV^2} \tag{64}$$

When plotted against $V/\sqrt{gL}$, this leads to what Telfer has called a "compatible" presentation, correctly preserving the relative merits of comparable hull forms.

7.7 Conversion Factors for Speed and Resistance Coefficients. In converting model resistance data from one form of presentation to another, the speed relationships given in Table 15 are useful. The factors for converting frequently used resistance coefficients are given in Table 16.

The most important coefficients are:

$C_{T\nabla}$ defined by:

$$C_{T\nabla} = \frac{R_T}{\frac{1}{2}\rho\,\nabla^{2/3}\,V^2} = \frac{\pi}{125}\,Ⓒ$$

C_{TL} defined by Equation (64),

Table 15—Relationships for Converting Frequently Used Speed Coefficients

	Fn (nondimensional)	Fn_∇ (non-dimensional)	$\textcircled{K}$ (non-dimensional)	$V/W^{1/6}$ (International units)	$V/\sqrt{L}$ (International units)
Fn (nondimensional)	Fn	$\frac{Fn_\nabla}{\sqrt{(L/\nabla^{1/3})}}$	$\frac{\textcircled{K}}{\sqrt{4\pi L/\nabla^{1/3}}}$	$\frac{\rho^{1/6}}{g^{1/3}} \frac{V/W^{1/6}}{\sqrt{L/\nabla^{1/3}}}$	$\frac{V/\sqrt{L}}{\sqrt{g}}$
Fn_∇ (nondimensional)	$\sqrt{L/\nabla^{1/3}}$ Fn	Fn_∇	$\frac{\textcircled{K}}{\sqrt{4\pi}}$	$\frac{\rho^{1/6}}{g^{1/3}} V/W^{1/6}$	$\frac{\sqrt{L/\nabla^{1/3}} \cdot V/\sqrt{L}}{\sqrt{g}}$
$\textcircled{K}$ (nondimensional)	$\sqrt{4\pi L/\nabla^{1/3}}$ Fn	$\sqrt{4\pi}\ Fn_\nabla$	$\textcircled{K}$	$\frac{\sqrt{4\pi}\ \rho^{1/6}}{g^{1/3}} V/W^{1/6}$	$\frac{\sqrt{4\pi L/\nabla^{1/3}}\ V/\sqrt{L}}{\sqrt{g}}$
$V/W^{1/6}$ (International units)	$\frac{g^{1/3}}{\rho^{1/6}} \sqrt{\frac{L}{\nabla^{1/3}}}$ Fn	$\frac{g^{1/3}}{\rho^{1/6}} Fn_\nabla$	$\frac{g^{1/3}}{\sqrt{4\pi}\ \rho^{1/6}} \textcircled{K}$	$V/W^{1/6}$	$\frac{\sqrt{\nabla^{1/3}}}{(\rho g)^{1/6}} V/\sqrt{L}$
$V/\sqrt{L}$ (International units)	$\sqrt{g}$Fn	$\frac{\sqrt{g}}{\sqrt{L/\nabla^{1/3}}} Fn_\nabla$	$\frac{\sqrt{g}}{\sqrt{4\pi L/\nabla^{1/3}}} \textcircled{K}$	$\frac{(\rho g)^{1/6}}{L/\nabla^{1/3}} \cdot V/W^{1/6}$	$V/\sqrt{L}$

$Fn = V/\sqrt{gL}$
$Fn_\nabla = V/\sqrt{g\nabla^{1/3}}$
$\textcircled{K} = \sqrt{4\pi}\ V/\sqrt{g\nabla^{1/3}}$
V = ship velocity (m/sec)
∇ = displacement volume (m^3)
W = displacement weight (kNewton); $W = \rho g \nabla$
ρ = mass density, kg/L (or t/m^3)
g = acceleration due to gravity (m/sec^2)

NOTE: ρ = 0.999 kg/L and 1.0259 kg/L for fresh and salt water respectively at 15 deg C (59 deg F) and g = 9.81 m/sec^2 (32 ft/sec^2).

NOTE: These notes also apply to Table 16.

Table 16—Relationships for Converting Frequently Used Resistance Coefficients

	C_T (non-dimensional)	$C_{T\nabla}$ (non-dimensional)	© (non-dimensional)	$\frac{R_T}{W}\left(\frac{R_R}{W}, \frac{R_W}{W}\right)$ (non-dimensional)	C_{TL} (non-dimensional)	$\frac{R_T}{W^{2/3}V^2}$ (International units)
C_T (nondimensional)	C_T	$\frac{C_{T\nabla}}{S/\nabla^{2/3}}$	$\frac{\pi}{125}\,\frac{\copyright}{S/\nabla^{2/3}}$	$\frac{2\,R_T/W}{(\mathrm{Fn}_\nabla)^2\,S/\nabla^{2/3}}$	$\frac{2\,C_{TL}}{\frac{L}{\nabla^{1/3}}\cdot\frac{S}{\nabla^{2/3}}}$	$\frac{\frac{R_T}{W^{2/3}V^2}}{\frac{1}{2}\left(\frac{\rho}{g^2}\right)^{1/3}\cdot\frac{S}{\nabla^{2/3}}}$
$C_{T\nabla}$ (nondimensional)	$\frac{S}{\nabla^{2/3}}\cdot C_T$	$C_{T\nabla}$	$\frac{\pi}{125}\,\copyright$	$\frac{2\,R_T/W}{(\mathrm{Fn}_\nabla)^2}$	$\frac{2\,C_{TL}}{L/\nabla^{1/3}}$	$\frac{\frac{R_T}{\nabla^{2/3}V^2}}{\frac{1}{2}(\rho/g^2)^{1/3}}$
© (nondimensional)	$\frac{125\,S/\nabla^{2/3}\,C_T}{\pi}$	$\frac{125}{\pi}\,C_{T\nabla}$	$\copyright$	$\frac{250}{\pi}\,\frac{R_T/W}{(\mathrm{Fn}_\nabla)^2}$	$\frac{250}{\pi}\,\frac{C_{TL}}{L/\nabla^{1/3}}$	$\frac{\frac{250}{\pi}\left(\frac{R_T}{W^{2/3}V^2}\right)}{(\rho/g^2)^{1/3}}$
$\frac{R_T}{W}\left(\frac{R_R}{W}, \frac{R_W}{W}\right)$ (nondimensional)	$\frac{(\mathrm{Fn}_\nabla)^2\,\frac{S}{\nabla^{2/3}}\,C_T}{2}$	$\frac{(\mathrm{Fn}_\nabla)^2}{2}\,C_{T\nabla}$	$\frac{\pi}{250}\,(\mathrm{Fn}_\nabla)^2\,\copyright$	$\frac{R_T}{W}\left(\frac{R_R}{W}, \frac{R_W}{W}\right)$	$\mathrm{Fn}^2\,C_{TL}$ $\left(=\frac{(\mathrm{Fn}_\nabla)^2\,C_{TL}}{\mathrm{L}/\nabla^{1/3}}\right)$	$\frac{(\mathrm{Fn}_\nabla)^2\,\frac{R_T}{\nabla^{2/3}V^2}}{(\rho/g^2)^{1/3}}$
C_{TL} (nondimensional)	$\frac{\left(\frac{L}{\nabla^{1/3}}\right)\left(\frac{S}{\nabla^{2/3}}\right)C_T}{2}$	$\frac{\left(\frac{L}{\nabla^{1/3}}\right)C_{T\nabla}}{2}$	$\frac{\left(\frac{L}{\nabla^{1/3}}\right)\copyright}{250/\pi}$	$\frac{R_T/W}{\mathrm{Fn}^2}$ $\left(=\frac{L/\nabla^{1/3}}{(\mathrm{Fn}_\nabla)^2}\,\frac{R_T}{W}\right)$	C_{TL}	$\frac{L}{\nabla^{1/3}}\,\frac{\frac{R_T}{W^{2/3}V^2}}{(\rho/g^2)^{1/3}}$
$\frac{R_T}{W^{2/3}V^2}$ (International units)	$\frac{1}{2}\left(\frac{\rho}{g^2}\right)^{1/3}\frac{S}{\nabla^{2/3}}\,C_T$	$\frac{1}{2}\left(\frac{\rho}{g^2}\right)^{1/3}C_{T\nabla}$	$\frac{\left(\frac{\rho}{g^2}\right)^{1/3}\copyright}{250/\pi}$	$\frac{(\rho/g^2)^{1/3}}{(\mathrm{Fn}_\nabla)^2}\,\frac{R_T}{W}$	$\left(\frac{\rho}{g^2}\right)^{1/3}\frac{C_{TL}}{L/\nabla^{1/3}}$	$\frac{R_T}{W^{2/3}V^2}$

$$C_T = \frac{R_T}{\frac{1}{2}\rho S V^2}$$

$$C_{T\nabla} = \frac{R_T}{\frac{1}{2}\rho\nabla^{2/3}V^2}$$

$$\copyright = \frac{125}{\pi}\,\frac{R_T}{\frac{1}{2}\rho\nabla^{2/3}V^2}$$

$$C_{TL} = \frac{gR_TL}{WV^2}$$

V = Ship velocity in m/sec.

∇ = Displacement volume in m^3

W = Displacement weight (kNewton); $= \rho g \nabla$

ρ = Mass density, kg/L (or t/m^3)

g = Acceleration due to gravity in m/sec.^{2}

$$C_{TL} = \frac{gR_TL}{WV^2}$$

The well known (but outdated) admiralty constants are defined by

$$C_a = \frac{W^{2/3}\, V^3}{P_D}$$

and

$$C_e = \frac{W^{2/3}\, V^3}{P_E}.$$

C_a is related to the power-displacement coefficient $C_{D\nabla}$

$$C_{D\nabla} = \frac{P_D}{\frac{1}{2}\rho\, V^3\, \nabla^{2/3}}$$

by

$$C_a = \frac{g^{2/3}}{\frac{1}{2}\rho^{1/3}\, C_{D\nabla}}$$

C_e is related to the resistance-displacement coefficient $C_{T\nabla}$ by

$$C_e = \frac{g^{2/3}}{\frac{1}{2}\rho^{1/3}\, C_{T\nabla}}.$$

Section 8
Relation of Hull Form to Resistance

8.1 Choice of Ship Dimensions. In merchant ships speed is seldom the dominant consideration, and the proportions and shape of the hull, as a rule, cannot be chosen solely to attain minimum resistance. Nevertheless, lower power and lower fuel costs have an important effect on the profits a ship can earn.

Some containerships are capable of speeds as high as 30 knots. Such ships have stimulated renewed interest in the design of hull forms which can achieve such speeds economically in smooth water and still have good seakeeping qualities and small loss of speed in rough weather.

At the other end of the scale are the bulk carriers, such as oil tankers and ore ships. Speed is not so important in such ships, because the minimum cost of transport per ton-mile is achieved by carrying as great a deadweight as possible in one ship at moderate speeds. Ships have been built with deadweights in excess of 500,000 t, with lengths such that even for a speed of 15 knots the Froude number is as low as 0.15. Restrictions on the drafts of such ships have increased the beam-draft ratios, and the block coefficients are in the 0.85 region. The efficient design of such ships poses many problems.

The prospective owner usually specifies that the new ship shall carry a certain deadweight at a particular speed, and the designer estimates the probable displacement and principal dimensions. The latter are usually subject to restrictions not associated with resistance and propulsion. Length is expensive in first cost, is limited by docking and navigation restrictions, while added length increases scantlings, equipment and manning scales. From a resistance point of view, greater length for a given displacement will reduce the wave-making resistance but increase the frictional resistance, so that longer lengths will be beneficial in ships running at high speeds and vice-versa. Longer lengths are also generally beneficial for behavior in rough seas (Chapter VIII, Vol. III).

An increase in draft, T, is generally beneficial for resistance, and is a cheap dimension in terms of cost. However, it may be limited by depths of harbors, canals, rivers, and dock sills.

The beam, B, is one of the governing factors in ensuring adequate stability, and a minimum value of B/T is generally necessary on this account. An increase in B will increase the resistance unless it is accompanied by a corresponding reduction in fineness coefficient. In cases of low-speed ships, however, a small reduction in length and a compensating increase in beam, because of the resulting decrease in wetted surface, may result in little or no increase in resistance. This results in a cheaper ship and also meets the need for increased stability in ships with large superstructures. This idea has been exploited in a number of large tankers.

The minimum wetted surface for a given displacement is also sensitive to the B/T ratio, the optimum value of which is about 2.25 for a block coefficient of 0.80 and about 3.0 at 0.50. However, the penalty for normal departures from these values is not very great. The effects of changes in B/T on wave-making resistance can be studied from model-experiment results. Generally, stability considerations and limiting drafts usually preclude values below 2.25 for full ships and 2.5 or even more for fine, higher speed ones.

While such considerations may be of guidance to naval architects in the choice of dimensions, they must meet many other demands, and will be influenced to a large extent by their knowledge of the particulars of existing successful ships. The process of design is essentially an iterative one, in which the various elements are changed until a proper balance is attained. In order to do this, parametric surveys have to be made on the effects of changes in dimensions, hull form, machinery types, and so on. This is an area in which the high-speed computer can play an important role, enabling the designer to consider a far greater number of possible solutions than could ever be made in the past.

8.2 Choice of Form Coefficients. The approximate relation between the block coefficient C_B and the Froude number Fn can be expressed by formulas originally given by Alexander (van Lammeren, et al, 1948),

$$\frac{V}{\sqrt{gL_{PP}}} = 0.595\ (1.08 - C_B) \text{ for trial speed}$$

$$\frac{V}{\sqrt{gL_{PP}}} = 0.595\ (1.05 - C_B) \text{ for service speed}$$

Troost (1955) has given a similar formula for *sustained sea speed* in terms of the prismatic coefficient C_P, which is more in line with design practice:

$$\frac{V_S}{\sqrt{gL_{PP}}} = 0.55 - 0.48\ C_P \quad (65)$$

where the trial speed is taken as

$$V_T = 1.06\ V_S \quad (66)$$

This *sustained sea speed*, V_S, lies very close to that at which the C_T-curve begins to rise steeply; i.e., to the speed at which the power begins to increase more rapidly than V^3. If the power over the first part of the rise is assumed to vary as V^4 then Equation (66) is equivalent to saying that the power at the trial speed is about 25 percent greater than that at the sustained sea speed under trial conditions. This is in keeping with the general design practice that the *service* speed should be attained under *trial* conditions at 80 percent of the maximum continuous power.

The above relationships are intended as rough guides to the designer and do not take the place of a careful analysis and comparison of alternative designs. For passenger liners, cross-channel ships and other craft in which high speed is important the relations in Equations (65) and (66) no longer apply. Comparative economic evaluations are essential in these cases.

Napier (1865) was one of the first who used a cost equation which he differentiated to find the optimum speed. Compared to this direct approach an iterative procedure is more versatile, requiring less simplifying assumptions and showing the penalties for departure from optimum configurations. Benford (1966, 1967) has presented an optimization method in which the costs have been split up into components making up the building and operating costs. The revenues are determined on the basis of the transport capacity with due allowances made for the bunker capacities required. Using appropriate economic criteria, the relative profitability of competing ship designs can be determined.

Fisher (1972) has presented such an optimization procedure applied to the Australian ore trade. In this paper the economic criterion used is the *Required Freight Rate.* He also investigated the impact of variations of fuel costs, interest rates, insurance costs and construction costs on the Required Freight Rate.

The above-mentioned procedures are valid and useful when costs (capital and operational) are known as general functions of the primary design parameters. These are, however, most often not known with enough accuracy. Fisher (1973) introduced a method based on the existence of a good (basic) design for which the full details are known. Optimization is carried out by varying the main parameters of this design, introducing errors of much smaller magnitude.

Economic optimization studies can yield valuable information concerning the relative merits of a design. However, as the results of these methods rely heavily on hydrodynamic knowledge, information concerning cost levels and predictions of the future economic and political situation (amount of freight, insurance rates, shipping routes and so on), care should be taken in the interpretation of the results.

The final decision on length and fullness should not be taken without considering the sea-going qualities of the ship. A short, full ship may well suffer such loss of speed in bad weather as to justify the extra cost of a longer, finer ship. The choice depends on many things, including the ocean conditions on the trade routes in question, particularly the length of the predominant waves and the frequency of their occurrence. Thus to maintain a weekly service on the North Atlantic in winter, requiring speeds of 28 or 29 knots, the length of express liners cannot well be less than 950 ft (See Chapter VIII).

Excessive fullness also promotes a tendency to bottom damage due to slamming. Flat areas on the bottom forward should be avoided. The floor lines should begin to lift immediately the parallel body ends, so as to give a V-shape which will allow the hull to enter the water smoothly when the ship is pitching (Todd, 1945). The relative qualities of U and V-sections in avoiding bottom damage have been analyzed by Townsend (1960) of the U. S. Salvage Association, who showed the dangers in vertical stems and too-pronounced U-sections forward. These questions are discussed further in Chapter VIII, but it is essential to have in mind the importance of seagoing behavior from the very inception of a new design.

Fig. 61 shows typical ©-curves for different types of ships (Todd, 1963). The wave-making resistance humps occur approximately at values of Fn equal to 0.24, 0.30 and 0.48, and their importance depends upon the speed and fullness of the ship. The coaster, with a prismatic coefficient $C_P = 0.83$ cannot be driven above Fn = 0.158 without an excessive increase in resistance, and as shown in Fig. 61 this coincides with Troost's definition of sustained sea speed. These speeds for the cargo ship and tanker also indicate the points where the resistance begins to increase rapidly.

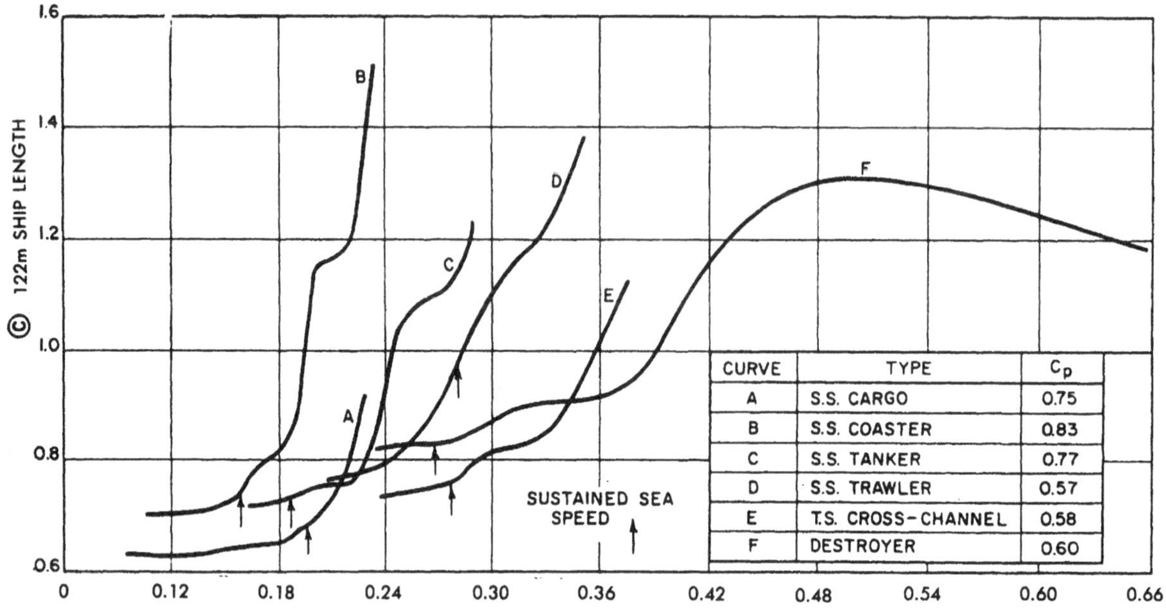

Fig. 61 Typical © curves

In the trawler, with a finer hull form of $C_P = 0.57$, the lower humps are not very marked, and a Fn value of 0.24 can be reached before the rise in the © curve begins. However, speed has great significance in these ships, to get to the fishing grounds quickly and to get home to market afterwards, and they are usually overdriven up to values of Fn = 0.30.

The cross-channel ship, of $C_P = 0.58$, can be driven to Fn = 0.33 without excessive resistance, for although the C_P is the same as in the trawler, the length is perhaps twice as great, showing the advantage of length in delaying the onset of heavy wave-making.

The destroyer, in which economy in the commercial sense is not paramount, normally has a top speed of Fn = 0.6 or more, well beyond the last hump at about Fn = 0.48.

When the principal dimensions and fullness coefficients have been chosen, the resistance then depends chiefly upon the following elements of ship form:

(a) Distribution of displacement along the length, as typified by the curve of cross-sectional areas and the LCB.

(b) Shape of the LWL, particularly in the fore body.

(c) Shape of the transverse sections, especially near the ends.

(d) Midship-section area coefficient.

(e) Type of stern; i.e., raised counter, cruiser, transom, and so on.

The midship-section coefficient C_M varies with fullness. In merchant ships with block coefficients around 0.80, it may be as high as 0.995. As the fullness decreases and the length of parallel body becomes shorter, it is necessary to ease the midship-section area somewhat to avoid too pronounced shoulders in the lower waterlines. In Series 60 the relation between C_B, C_M, and C_P is as follows:

C_B	0.800	0.750	0.700	0.650	0.600
C_M	0.994	0.990	0.986	0.982	0.978
C_P	0.805	0.758	0.710	0.661	0.614

With still finer ships, C_M is still smaller, being about 0.93 on fast passenger liners, trawlers and tugs, and 0.90 on cross-channel ships.

The choice of the shape of section area and LWL curves depends upon the values of Fn and C_P, and will also be influenced by the need to provide adequate stability. Naval architects must draw upon their own experience, with recourse to published design data, where there is much information on the best values or shapes for these elements of form for different kinds of ships. General guidance in this field has been given by Taylor, D.W. (1943), Lindblad (1961) and Todd (1945). The recommendations from the two last-named sources are summarized in Table 17. The ship types are arranged in order of decreasing block coefficients, from 0.80 for a slow-speed cargo ship to 0.52 for a cross-channel ship. As already mentioned, there is a corresponding reduction in C_M, and with the finest ships this will approximately compensate for the reduction in C_B, so that C_P tends to reach a steady value of around 0.59. Indeed, at the very highest Fn values, the C_P can be increased with advantage, as first pointed out by Taylor, and in destroyers it may be as high as 0.65. These points are illustrated in a chart given by Saunders (1957) showing the relations between speed-length ratio, prismatic coefficient, and displacement-length ratio. This is reproduced in Fig. 62. The curves were based upon data from a variety of sources, and result in two pairs of empirical curves which define two "design lanes." These apply to merchant and combatant vessels of orthodox form, and not to special types such as fishing vessels and tugs.

The load waterplane coefficient C_{WP} decreases with decreasing fullness, its value depending also to a considerable extent upon the type of transverse sections. For Series 60 it is related to the C_P by the approximate formula

$$C_{WP} = 0.18 + 0.86C_P$$

In general C_{WP} will depend also on the stability requirements and sea keeping.

In full ships considerable parallel body can be worked in with advantage, and the entrance can be short, the run being long and fine to minimize separation and form resistance. As C_P decreases, so does parallel body, and the entrance is made longer to reduce the increase in wave-making resistance, the LCB moving aft in consequence. Most of the reduction in C_P is thus accomplished by fining the entrance, the change in the coefficient of the run being much less.

The sectional area curve and load waterline follow a similar pattern. At low Fn values and high prismatic coefficients, both are slightly convex forward and aft. As Fn increases, they become straight and eventually S-shaped with a hollow near the stern. At Fn values of 0.45 and above, the hollow should disappear in the LWL, which should be straight or even slightly convex in destroyers and other high-speed types. In such ships, too, the onset of high wave-making resistance calls for as long a length as is compatible with the other design requirements.

The information given in Table 17 can only be used for general guidance in the preliminary design stage. In any particular ship design, more detailed analyses, based upon model and full-scale data for closely similar ships, must later be made to determine the most suit-

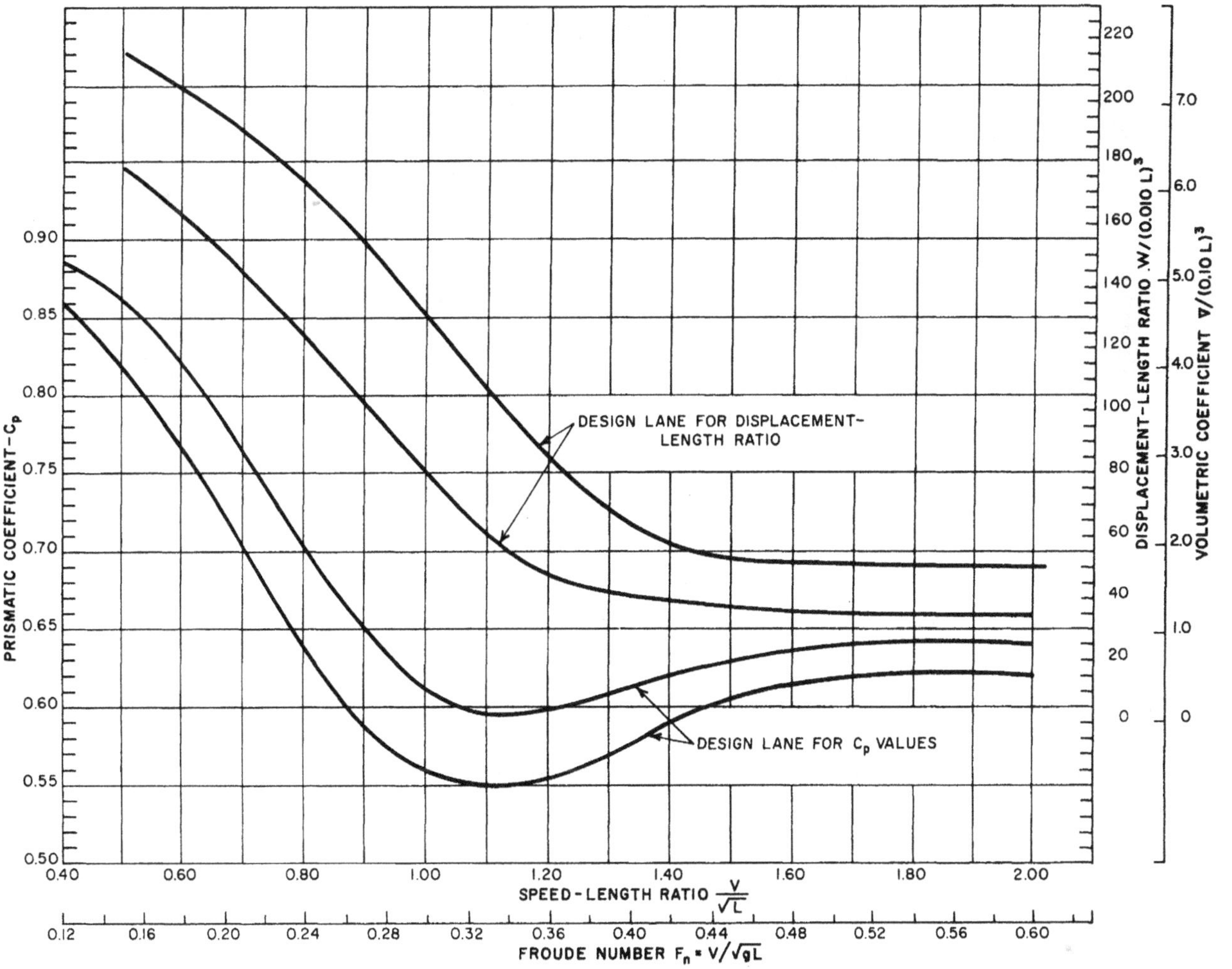

Fig. 62 Design lanes for prismatic coefficient and displacement-length ratio (Saunders, 1957) (English units).

Table 17—Variation of Form Coefficients and Elements of Hull Shape, Based on Lindblad (1961) and Todd (1945)

Type of ship		Slow speed cargo ship	Medium speed cargo ship	Cargo liners	Passenger and cargo, fruit ships	High-speed passenger liners, ferries, etc.	Fast passenger liners, trawlers, tugs	Cross-channel ships	Destroyers
C_B		0.80	0.75	0.70	0.65	0.60	0.55	0.54 –0.52	0.46–0.54
C_M		0.99–0.995	0.985–0.99	0.98	0.98	0.97	0.93	0.915–0.905	0.76–0.85
C_P		0.809–0.805	0.762–0.758	0.715	0.664	0.62	0.59	0.59 –0.575	0.56–0.64
C_{WP}		0.88	0.84	0.81	0.78	0.71	0.69	0.69 –0.675	0.68–0.76
Fn		0.15–0.18	0.18–0.19	0.21	0.24	0.24–0.30	0.24 –0.36	0.36 –0.45	0.45 and above
L_P percent		35	25	12	5	0	0	0	0
L_E/L_R		0.7	0.8	0.9	1.0	1.0–1.1	1.1–1.2	1.2	0.55
LCB percent L_{PP} from PP ⨂		1.5–2.5 fwd	1.0–2.0 fwd	0–1.0 aft	1.0–2.0 aft	1.5–2.0 aft	2.0–2.8 aft	2.0–3.0 aft	0.5–2.0 aft
Sectional area curve shape	Fwd	Straight with U sections, slightly convex with V sections and raked stem	Straight with slight hollow at extreme fore end	Straight with some hollow forward giving S-shape	S-Shape—fine entrance essential with pronounced hollow forward	S-Shape for Fn = 0.24, becoming straight above this value, with addition of bulb	S-Shape at Fn = 0.24 straight above Fn = 0.27	S-Shape with hollow forward at Fn = 0.36 becoming straighter for 0.45	Maximum area aft of midships. Straight or slightly convex area curve forward
	Aft	Straight or slightly convex with easy shoulder	Straight or slightly convex	Straight except at extreme aft end	Straight except at extreme aft end	Straight except at extreme aft end	S-Shape	Straight with hollow at extreme aft end	Good buttock lines aft and transom stern
LWL shape	Fwd	Slightly convex throughout	Slightly convex or straight	Slightly hollow forward, or straight with longer entrance	S-Shaped, hollow forward	Fine *WL*, almost straight with slight hollow	Fine *WL*, S-shaped below Fn = 0.30 straight above this value	Up to Fn = 0.30 fine *WL* with hollow, above Fn = 0.3, *WL* endings fuller and straight, or hollow with bulb	Maximum beam aft of midships. Waterline forward quite straight or even a little convex
	Aft	Slightly convex. If possible the slope should not exceed 20 deg	Slightly convex	Slightly convex	Slightly convex	Full *WL*, nearly straight	Full, straight or convex	Full *WL*, convex	*WL* aft very full to suit transom stern and cover screws
Half-angle of entrance on LWL (i_E)		35 deg	27 deg	12 deg with hollow *LWL*, 16 deg if *LWL* is straight	10 deg	6 deg	8½-10 deg	6–7 deg below Fn = 0.30. Above this speed, 9 deg with straight *WL*, 6 deg with hollow and bulb	4–11 deg

Table 17 (Continued)

Shape of sections	Fwd	Upper part U-shaped, lower V-shaped to avoid slamming damage	V-Shaped	V-Shaped, no flat area on bottom	U-Shaped, no flat area on bottom	U-Shaped, no flat area on bottom	Modified U-V sections	U sections modified near keel	U type with some bulb effect
	Aft	V-Shaped changing to U near stern	V-Shaped changing to U near stern	V-Shaped changing to U near stern	V changing to U on S.S. ships. V on T.S. ships	V changing to U on S.S. ships. V on T.S. ships	Modified U-V sections	V sections	V type becoming flat toward transom
Remarks		By eliminating deadrise on midship section bilge radius can be increased and shoulders made easier	If Fn is greater than 0.65, forward sections should be more U-type			In this range of Fn, bulbous bows are beneficial	Bulbous bows beneficial	Essential to use maximum length, compatible with other design conditions	Essential to use maximum length compatible with other design conditions and give adequate cover to screws

able form together with estimates of the probable effective and shaft powers.

8.3 Design Data. Naval architects in designing a new ship must study the effects upon power of a number of choices of hull form and proportions. Data for such comparisons are available in the publications of the technical societies and the technical press, to which they will add their own experience from past designs. Many of these data are derived from model experiments, and it is quite impossible to describe other than representative examples.

8.4 Model Resistance Data Sheets, SNAME. A valuable source of information exists in the data sheets published by The Society of Naval Architects and Marine Engineers (1953-1966). These have been compiled for some 150 ships from the results of model experiments carried out in various towing tanks. All types of ships are included, it being one of the objects of the collection to give a variety of data for the benefit of those naval architects not having access to other sources or engaging in a new basic field of design. The sheets give all the principal form coefficients, the basic model data, and predicted values of R_R/W and Ⓒ to bases of $V_K/\sqrt{L}$ and Ⓚ, respectively. Area curves and lines plans are included.

8.5 Methodical Series Experiments. The data given in the Society's sheets and in many pubished papers are valuable guides in the design of closely similar ships. On the other hand, they refer to a group of completely unrelated forms, and it is difficult to determine the trends in resistance values with changes in proportions and coefficients or, what is equally important, the penalties involved in specific changes.

Information of this kind is obtained by running a series of models in which the principal characteristics are changed in a systematic manner. The results of such methodical series can be used to plot design charts which are of inestimable value to the designer.

Such a series may be based upon a single parent form or upon a number of parents related to one another in some graphical or mathematical pattern. The prismatic coefficient can be changed by systematic variations in the curve of areas, while the proportions such as L/B and B/T can be varied by straight geometrical methods.

8.6 Taylor's Standard Series. A complete investigation of the effects of altering proportions using a single parent form was made by Admiral Taylor in the Experimental Model Basin (EMB), Washington, giving rise to the well-known Taylor's Standard Series (Taylor, D.W., 1943).

The original parent was patterned after the British cruiser *Leviathan* of 1900, which had a ram bow and twin-screw, cruiser stern. For the series parent, the ram was eliminated, the maximum section was moved to midlength, and a 3 percent bulb was adopted at the bow. The sectional-area curves and body lines for the other models were derived from the parent partly by mathematical means. The models were run at various

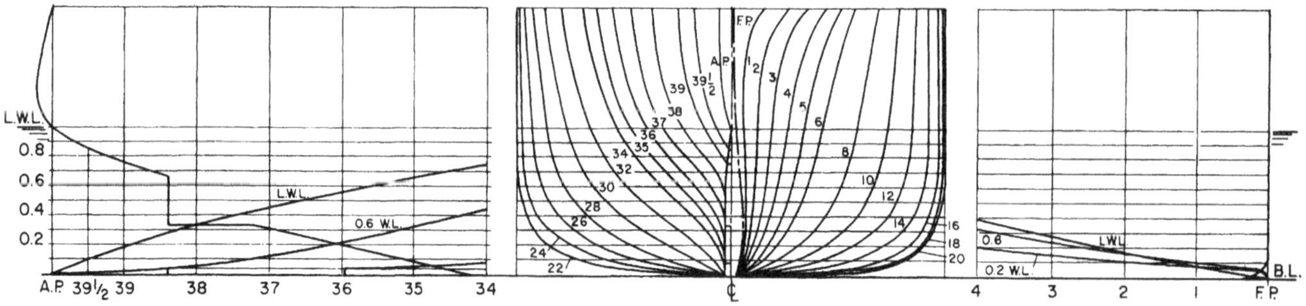

Fig. 63 Lines for the parent form of Taylor's Standard Series

periods up to 1914, and the first full presentation of results was in the 1933 edition of *Speed and Power of Ships*. The data appeared as contours of residual resistance per tonne of displacement against prismatic coefficient and displacement-length ratio, each chart being for particular values of B/T and $V_K/\sqrt{L}$.

These contours were derived by the methods in use at EMB in 1910; the model frictional resistance was determined by the use of frictional coefficients measured on 20-ft planks in the Washington tank and the full-scale frictional resistance was calculated by using Tideman's ship coefficients. After some intermediate changes, the ATTC standard method was adopted in 1947, the ship P_E being increased by a model-ship correlation allowance of +0.0004, as already described (Section 3.5). In view of this change, the Taylor data were reanalyzed, and the new contours based on the ATTC coefficients were published by Gertler (1954).

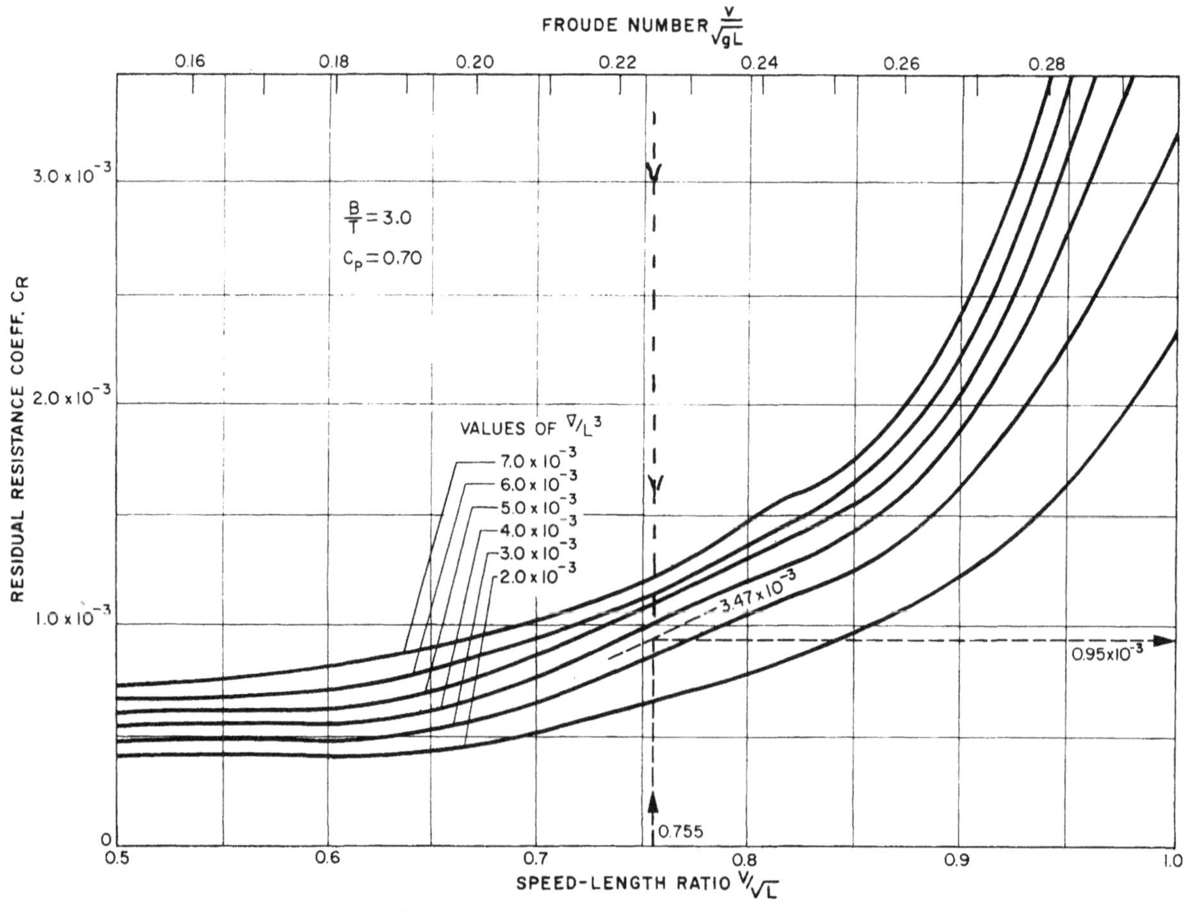

Fig. 64 Typical Taylor's Standard Series contours

The lines of the parent form are shown in Fig. 63. The midship-section area coefficient was 0.925. The prismatic coefficients of the fore-and-aft bodies were equal, and the LCB was always amidships. The quantities varied were C_P, B/T and $W/(L/100)^3$, the midship-section area coefficient C_M remaining constant.

The variations in L/B, B/T, and $W/(L/100)^3$ were obtained by selecting different ratios of B/T and L/B and varying the offsets geometrically.

The ranges of the variables covered in the Taylor's Standard Series are:

C_P	0.48 to 0.86
B/T	2.25, 3.00 and 3.75
$W/(L_{WL}/100)^3$ (English)	20 to 250
C_M	0.925
$\nabla/(L_{WL})^3$	0.70 to 8.75×10^{-3}

In presenting the reanalyzed results, Gertler used the nondimensional volumetric coefficient $\nabla/(L_{WL})^3$ in preference to $W/(L/100)^3$. Contours of the wetted surface coefficient

$$C_S = S/\sqrt{\nabla L_{WL}}$$

were derived and given for three different values of B/T.

In converting the Series results to the ATTC presentation, Gertler went back to the original model data. In doing so, he took the opportunity of making certain corrections to the data, which had been omitted in the original Standard Series presentation, including the effects of temperature, the absence of turbulence stimulation and the interference between model and tank boundaries upon the measured resistance.

To facilitate the calculation of P_E for specific ships, Gertler gave charts of C_R to a base of $V/\sqrt{gL_{WL}}$ C_R, equal to $\frac{R_R}{\frac{1}{2}\rho S V^2}$, is nondimensional in any consistent system of units, as is $V/\sqrt{gL_{WL}}$. An auxiliary scale of $V/\sqrt{L_{WL}}$ in units of knots and feet was incorporated on the charts.

The design charts give contours of C_R against $V/\sqrt{gL_{WL}}$ for various values of $\nabla/L_{WL}{}^3$, each chart being for a particular value of C_P and B/T, Fig. 64.

For merit comparisons, Gertler used the Ⓒ − Ⓚ presentation for a ship of standard length of 121.92 m (400 ft) on the L_{WL} in water of 3.5 percent salinity at 15 deg C (59 F). The values of C_F for this ship can be found from Fig. 21 of the publication by Gertler (1954), as a function of Ⓚ, $V_K/\sqrt{L_{WL}}$ or Fn. C_R is obtained from the charts, so that $C_T = C_F + C_R$, with the addition of the model-ship correlation allowance of +0.0004 if desired.

Ⓒ can then be found from the equation

$$Ⓒ = 39.78\ Ⓢ\ C_T \qquad (45)$$

Gertler gives charts for the conversion of C_T and Fn (or $V_K/\sqrt{L}$) to Ⓒ and Ⓚ. For a ship designed to carry a given displacement at a given speed, curves of Ⓒ to a base of Ⓚ will give a merit comparison between various choices of dimensions, as described in Section 7.

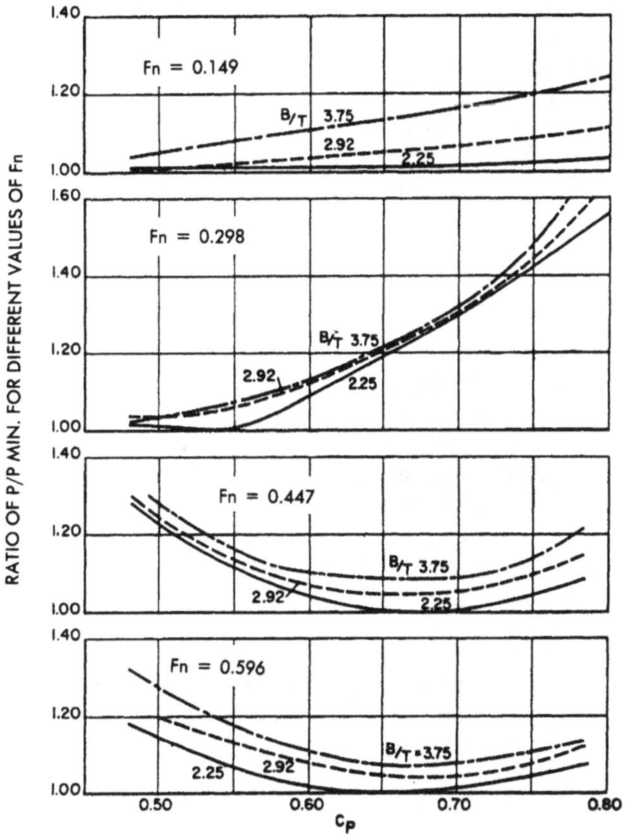

Fig. 65 Variation in Taylor's Standard Series P_E with change of C_p for a ship with $L/\nabla^{1/3} = 8.7$

Merit comparisons can also be obtained from the C_R presentation for any particular design by calculating and plotting curves of the ratio of P_E to that of the model used for reference. Such curves can also be used to find the effects of major changes in design parameters.

Such a comparison for ships of the destroyer type is shown in Fig. 65, taken from Gertler (1954). In this case, the displacement volume is 2720 m³ and the value of $L/\nabla^{1/3}$ is 8.7. For values of Fn less than 0.30, the lowest P_E is realized by using the smallest C_p value of 0.50. At high speeds, the picture is different, and at Fn = 0.60, corresponding to 40 knots, the best C_p is about 0.65 to 0.67.

Figure 65 also shows that an increase in B/T causes a moderate increase in P_E, but the effect may be larger in rough water than in smooth.

In some other experiments, Taylor, D.W. (1908) investigated the effects of shape of midship section on resistance. The models all had a $C_P = 0.56$, the same curve of areas, and the same maximum section area, the LWL curves being very nearly the same shape. The midship-area coefficient C_M varied from 0.7 to 1.1 in the five models, based upon the beam at the LWL. The fuller area coefficient had a slight advantage up to Fn well above 0.33, but the difference in R_R/W for the whole series was very small. Taylor therefore concluded that the shape of the midship section was not an important factor in determining residuary resistance.

The method of calculating the resistance and effective power from the Taylor Series charts can be illustrated as follows. Consider a vessel with the following characteristics:

$$L_{PP} = 121.92 \text{ m (400 ft)}$$
$$L_{WL} = 123.96 \text{ m (406.7 ft)}$$
$$\nabla = 6612.55 \text{ m}^3 \text{ (233,520 ft}^3\text{)}$$
$$B = 15.24 \text{ m (50 ft)}$$
$$T = 5.081 \text{ m (16.67 ft)}$$
$$B/T = 3.00$$
$$L_{WL}/\nabla^{1/3} = 6.61$$
$$\nabla/L_{WL}^3 = 3.47 \times 10^{-3}$$
$$C_B = 0.688$$
$$C_P = 0.698$$
$$S = 2363.10 \text{ m}^2 \text{ (25436 ft}^2\text{)}$$
$$V = 7.822 \text{ m/sec. (15.206 knots)}$$
$$\nabla^{1/3} = 18.770 \text{ m}$$

The full-scale Reynolds number is calculated as follows:

$\text{Rn} = VL_{WL}/\nu = 7.822 \times 123.96/1.1883 \times 10^{-6} = 8.16 \times 10^8$ for salt water at 15 deg C (59 F). and $C_{FOS} = 0.001570$ according to the ITTC-1957 friction line (see Table 12). The residual resistance coefficient C_R is determined from the appropriate Taylor's Standard Series diagram for $B/T = 3.0$ and $C_P = 0.70$ (which is close enough to the actual value of 0.698) for the appropriate Fn value of $\text{Fn} = 7.822/\sqrt{9.81 \times 123.96} = 0.224$. From Fig. 64 it follows that $C_R = 0.000950$. The correlation allowance for a ship length in between 50 m and 150 m is 0.00035 or 0.00040 (see Table 13). On adopting the value of 0.00040 the total resistance coefficient becomes $C_{TS} = 0.001570 + 0.00095 + 0.00040 = 0.00292$.

Hence
$$R_{TS} = \tfrac{1}{2}\rho SV^2 \times C_{TS}$$
$$= 0.5 \times 1.0259 \times 2363.1 \times (7{,}822)^2 \times 0.00292$$
$$= 216.56 \text{ kN}$$
and
$$P_{ES} = R_{TS} \times V = 216.56 \times 7.822$$
$$= 1693.9 \text{ kW}$$

In using the Taylor Series results it should be borne in mind that the maximum midship coefficient value of the Series is only 0.925 and that the Taylor models have a deep cruiser stern suitable for a twin screw propulsion arrangement. Also the LCB location has not been optimized but is stationed at the midships location.

Graff et al (1964) presented the results of tests with Standard Taylor Series models with higher B/T ratios and with differing LCB positions. These supplementary tests were intended for use in predicting resistance properties of slender, fast ships. Accordingly, the models were tested up to a speed corresponding to Fn = 0.90. The effects of shallow water were also studied.

8.7 Series 60. In 1948 the Society of Naval Architects, in cooperation with the ATTC, sponsored under the guidance of a technical panel the preparation of parent lines for a series of single-screw merchant ships. The parent lines were developed at the Taylor Model Basin and the model tests were carried out there under the Bureau of Ships Fundamental Hydromechanics Research Program. These models formed "Series 60" in the Model Basin sequence, and the results were published by Todd (1963).

The five parent models covered block coefficients from 0.60 to 0.80, and a range of values of L/B, B/T, displacement-length ratio and LCB position, as set out in Table 18.

The length L_{PP} is that between perpendiculars, measured from the centerline of rudder stock to the forward side of the stem on the designed load waterline. In presenting the ship results, L_{PP} was taken as 121.92 m (400 ft) and the corresponding length on the waterline, L_{WL}, was 123.96 m (406.7 ft). The models all had a vertical stem and a stern with an aperture for a single screw. There was no bulb at the bow. The models were 6.10 m (20 ft) in length and were run in the large

Table 18—Particulars of Series 60 Models

C_B	0.60	0.65	0.70	0.75	0.80
C_M	0.978	0.982	0.986	0.990	0.994
C_P	0.614	0.661	0.710	0.758	0.805
L/B	6.5–8.5	6.25–8.25	6.00–8.00	5.75–7.75	5.50–7.50
B/T	2.5–3.5	2.5–3.5	2.5–3.5	2.5–3.5	2.5–3.5
$L/\nabla^{1/3}$	5.60–7.50	5.32–7.16	5.05–6.84	4.79–6.55	4.55–6.27
Ⓢ	6.20–7.20	6.03–7.04	5.90–6.98	5.78–6.88	5.71–6.84
L_P, pct L_{PP}	0	3.5	11.9	21.0	30.0
LCB, as pct of L_{PP} from ⊗ L_{PP}	2.48A to 0.52F (optimum 1.69A)	2.46A to 1.37F (optimum 1.01A)	2.05A to 2.55F (optimum 0.25A)	0.48F to 3.46F (optimum 2.60F)	0.76F to 3.51F (optimum 2.70F)
$i_E°$	6.2–8.7	7.3–9.6	9.7–12.9	19.8–25.9	38.9–47.8

(Note: L_p is length of parallel midbody)

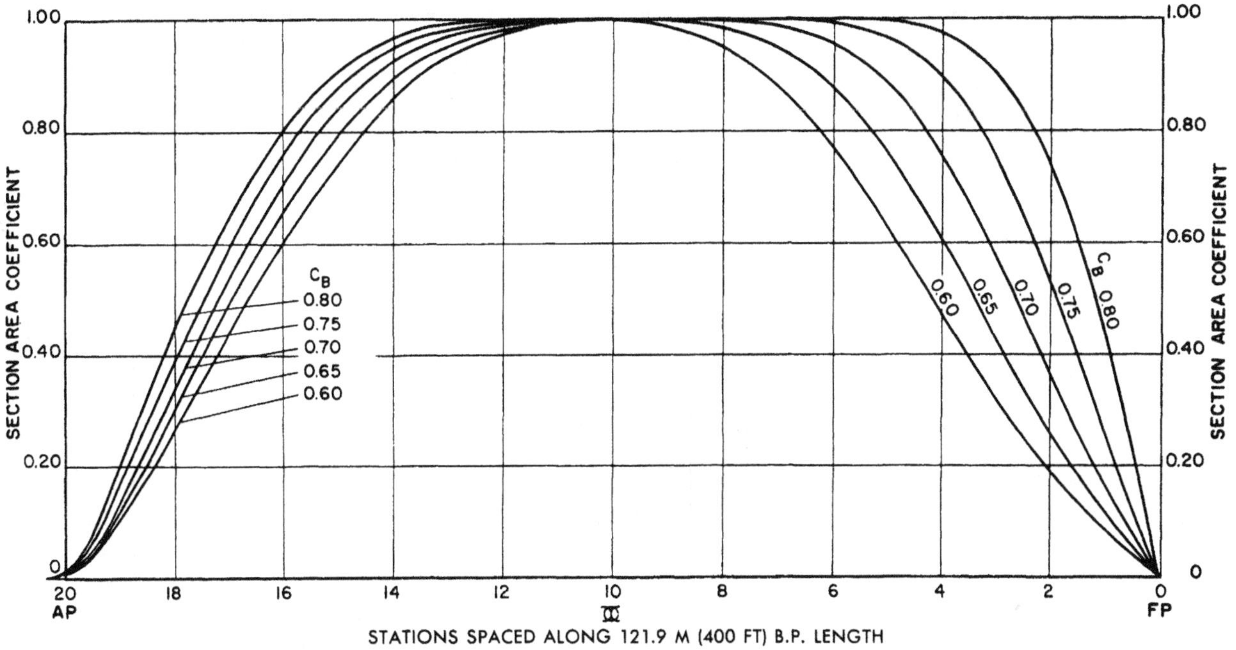

Fig. 66 Series 60—sectional area coefficient curves of parent models

TMB tank with a width of 15.5 m (50.8 ft) and a depth of 6.7 m (22 ft). Turbulence stimulators were fitted.

The models of most methodical series have been derived from a single parent form by changes in the area curve, as described for the Taylor's Series, and proportional geometrical changes. When carried to cover very different proportions and to fullness coefficients suitable to very different values of Fn, such changes led to unrealistic forms, regardless of how good the parent lines may have been for the original design conditions. In Series 60 another approach was used. Five parent forms of C_B = 0.60, 0.65, 0.70, 0.75 and 0.80 were developed, each incorporating the features considered necessary for good resistance qualities at its appropriate value as deduced from results for successful ships. The sectional area and waterline curves for these parents were plotted and faired to a base of the prismatic coefficients of entrance and run and cross-faired with the area curves and body plans. Auxiliary curves showed the lengths of parallel body for each value of C_B and the necessary lengths of entrance and run to be associated with these to give any desired position of LCB. The sectional area-coefficient curves for the five parent models are shown in Fig. 66.

A number of models were then run to determine the optimum location of LCB for each block coefficient, their lines being derived from the design charts just described. The results of these tests are summarized in Fig. 67. This shows the optimum LCB locations and the corresponding minimum Ⓒ values. For a given value of C_B, the optimum LCB location moves aft as the value of Ⓚ is increased. When C_B and Ⓚ are known, this figure will give the optimum LCB position and the corresponding minimum Ⓒ-value if the proportions and lines of the ship conform with those of the Series 60 parents. Thus, for a C_B = 0.65 and a Ⓚ = 2.1, the best position of LCB is 1.45 percent of L_{PP} aft of midships, the corresponding minimum Ⓒ value for a length of 121.92 m (400 ft) being 0.73 and Fn = 0.244 (based on L_{PP}). A point of considerable interest is the remarkable constancy of the minimum Ⓒ-value at the Troost sustained sea speeds, which varies only between 0.72 and 0.74 over the full range of block coefficient.

The optimum location of LCB was then assumed to apply to all models of a given block coefficient, regardless of L/B and B/T, and these ratios were varied on a number of models to cover the range of values shown in Table 18. The results were presented in two ways:

(a) Contours of R_R/W, lbs per long ton (2240 lb) of displacement, against C_B and L/B, each chart being for given values of B/T and $V_K/\sqrt{L_{WL}}$.

(b) Contours of Ⓒ for a length of 121.92 m (400 ft) against C_B and L/B, each chart being for given values of B/T and Ⓚ. The Ⓒ values included a model-ship correlation allowance of 0.0004, which is suitable for calculations for ship lengths up to about 150 m (492 ft) in length.

To demonstrate the use of the Series 60 results, calculations can again be carried out for the example adopted in Section 8.6 as follows:

Ⓚ $= \sqrt{4\pi}\ \mathrm{Fn}_\nabla = \sqrt{4\pi}\ V/\sqrt{g\nabla^{1/3}}$

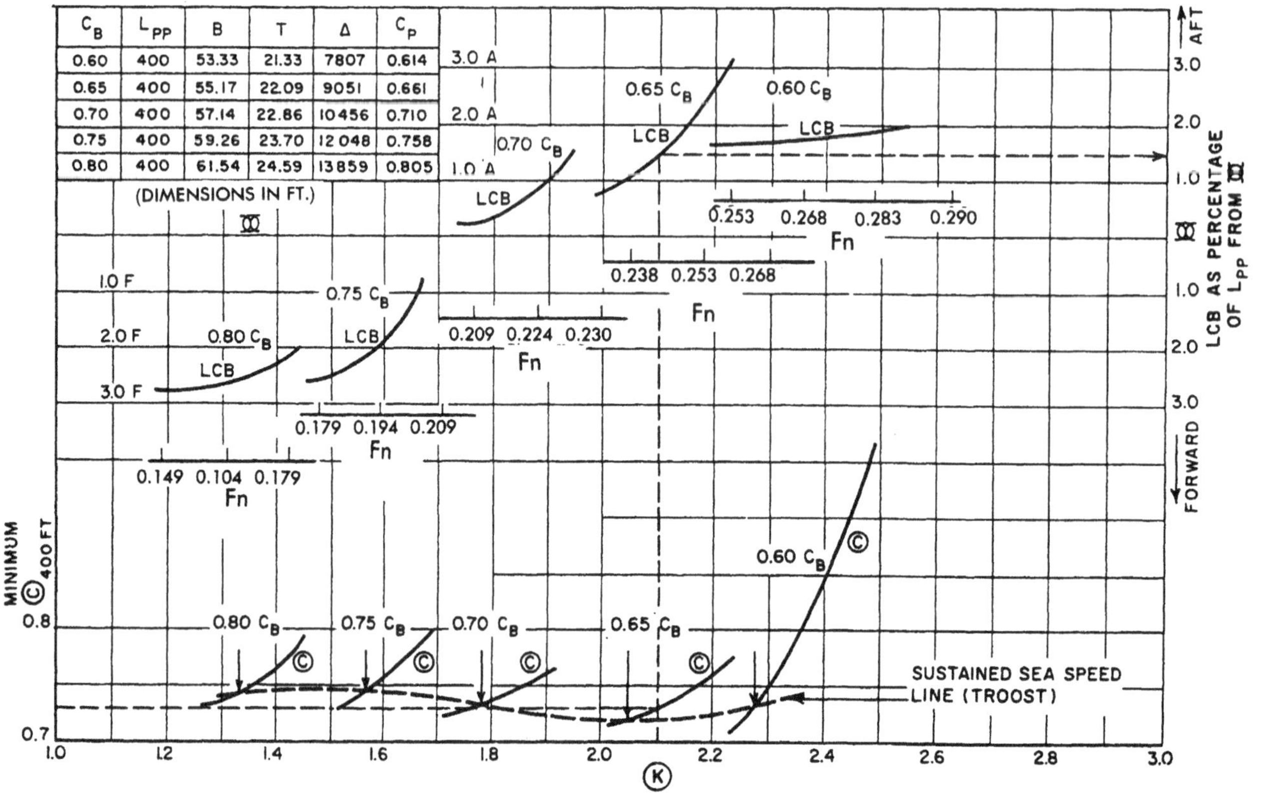

C_B	L_{PP}	B	T	Δ	C_P
0.60	400	53.33	21.33	7807	0.614
0.65	400	55.17	22.09	9051	0.661
0.70	400	57.14	22.86	10456	0.710
0.75	400	59.26	23.70	12 048	0.758
0.80	400	61.54	24.59	13859	0.805

Fig. 67 Series 60 minimum Ⓒ values and optimum LCB position

$= 3.5449 \times 7.822/\sqrt{9.81 \times 18.77}$
$= 2.043$

From charts B.61 and B.62 of Todd (1963) the following values of Ⓒ can be found (for a vessel of 121.92 m length):

Ⓚ = 2.0	Ⓚ = 2.1
Ⓒ = 0.75	Ⓒ = 0.79

By interpolation, for Ⓚ = 2.043, it follows that Ⓒ = 0.768. Accordingly,

$$R_{TS} = \frac{\pi}{125}\,\tfrac{1}{2}\rho\nabla^{2/3}\,V^2\,\text{Ⓒ}$$
$$= 0.0251 \times 0.5 \times 1.0259 \times (18.770)^2 \times (7.822)^2 \times 0.768$$
$$= 213.144 \text{ kN}$$

and $P_{ES} = R_{TS} \cdot V = 213.144 \times 7.822$
$= 1667.2$ kW

For design use, assuming that L, B, T and ∇ are known, these charts enable the resistance and effective power to be estimated over a range of speeds for a ship having Series 60 lines and an LCB in the position chosen for the parent forms. For any other position of LCB dictated by design needs, the change in R_{TS} and P_E can be estimated from the results of the auxiliary LCB variation series. The contours of sectional area and load-waterline coefficients enable a body plan to be produced quickly which will fulfill all the foregoing design conditions.

The data may also be used to explore the effects upon power of changes in principal dimensions and coefficients during feasibility studies, so as to ensure the best results within the design conditions or to determine the penalties involved in such changes.

The contours can also be used for comparative purposes. If a new design has secondary characteristics which differ from those of its Series 60 equivalent, but model results are available for some other ship which more closely resembles it in these respects, the latter may be used as a "basic ship." Calculations of P_E can be made from the contours for the Series 60 equivalents of both the new design and the basic ship. Then approximately

$$P_E \text{ for new ship} = P_E \text{ of Series 60 equivalent} \times \frac{P_E \text{ for basic ship}}{P_E \text{ for Series 60 equivalent basic ship}}$$

Fig. 68 shows the predicted P_E for a bulk carrier with $C_B = 0.78$ derived in this way, using a tanker as the basis ship, compared with the P_E derived from actual model tests.

Finally, the models of Series 60 were all run self-propelled, and Todd (1963) includes contours of wake fractions, thrust-deduction fractions and propulsive efficiencies for a number of propeller diameters and power characteristics. These data, together with the P_E values, enable close estimates to be made of the delivered horsepower, P_D, at the propeller.

Lackenby, et al (1970) presented the results of the Series 60 models in the form of © values for a 121.92 m (400 ft) ship based on both the Froude and the ITTC-1957 skin-friction lines. These © values were presented without any model-ship correlation allowance included, which will allow application to cases for which C_A is not equal to 0.0004 more directly.

8.8 Other Methodical Series of Merchant Ship Models. It is impossible to give in detail the results of the many other series of merchant ship models which have been run in different tanks. Herewith are references to some of the more important and an indication of their scope. For a full index see SNAME (1973).

(a) British Maritime Technology, (Moon, et al, 1961 and Lackenby et al, 1966). This series covers variations in block coefficient from 0.65 to 0.80, and in LCB position, for a basic, single-screw ship having the following dimensions:

$L_{PP} = 121.92$ m (400 ft)
$L_{WL} = 124.76$ m (409.32 ft)
$B = 16.76$ m (55.00 ft)
$T = 7.93$ m, 6.4 m and 4.88 m (26 ft, 21 ft and 16 ft).

Besides the influence of C_B and LCB, the influence of variations in parallel middle body and B/T were determined. The results are presented as © values to a base of speed with correction curves for movement of LCB from the standard position. The © values are based on the Froude friction coefficients, and curves are given for converting them to the basis of the ITTC-1957 friction line.

(b) Swedish State Shipbuilding Experimental Tank, Gothenburg (SSPA):

1. High-Speed, twin-screw cargo liners (Lindblad, 1951). Two series of models, Table 19, were run to investigate the effect of the position of LCB on resistance.

2. Fast, single-screw cargo ship models (Nordstrom, 1948) and (1950). Two series of models, Table 20, were run, covering the variations in length, B/T and LCB position.

3. Tanker models (Edstrand, 1953-56). The parent form represented a single-screw tanker. The experiments covered variations in forebody section shape (from extreme V to extreme U), L/B ratios between 7.2 to 8.1, B/T ratios of 2.3 and 2.5, C_B-values of 0.725 to 0.80, LCB positions from 0.3 to 3.0 percent of L_{PP} forward of amidship for the $C_B = 0.80$ model.

4. Single-screw cargo ship with $C_B = 0.525$ to 0.750 (Edstrand, et al, 1956, and Freimanis, et al, 1957-59). This work forms an extension of Nordstrom (1948) and (1950). Variations in L/B (from 6.54 to 8.14), in B/T (from 2.1 to 3.0), in LCB (from 1.0 to 4.0 percent of L_{PP} aft of amidship). Also, systematic changes in fore- and aft sections and in waterline shape were studied.

5. Coaster models (Warholm, 1953). Variations in L/B (from 4.5 to 7.5), in B/T (from 2.0 to 2.8), in C_B (from 0.6 to 0.75) and in LCB (from 1 percent of L_{PP} forward to 2 percent L_{PP} aft of amidship), were studied.

(c) NPL Coaster Models. Experiments on an extensive series of models of coasters have been carried out at the National Physical Laboratory, Teddington, England (Todd, 1931, 1934, 1938, 1940 and 1942, and Dawson 1952-60). They cover block coefficients between 0.625 and 0.81, L/B ratios from 4.44 to 8.0 and B/T ratios from 2.05 to 2.75. The results are presented as curves of © for a 60.96 m (200 ft) length to a base of Ⓚ.

(d) Trawlers. Trawlers are usually of fairly fine form and run at relatively high Froude number values of 0.3 to 0.35, while the length-displacement ratios are relatively low. This combination of high Froude numbers and low length-displacement ratios usually falls outside the range of most model series.

Maintenance of sea speed is most important, as stated previously, and it is essential to provide adequate freeboard and flare at the bow to keep the ship as dry as possible in bad weather.

Methodical series experiments on trawler forms have been made by Ridgely-Nevitt (1956) and (1963). The four parents had prismatic coefficients of 0.55, 0.60, 0.65 and 0.70 with values of the length-displacement ratio $L/\nabla^{1/3}$ ranging from 3.85 to 5.23. The lines were derived separately to suit each fullness, as in Series 60. The hulls have a cruiser stern, single-screw aperture, and no bulb. The series covered block coef-

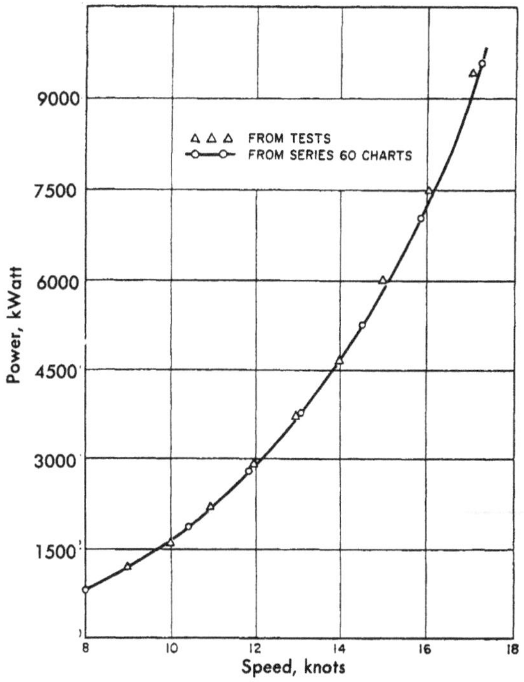

Fig. 68 Comparison of P_E from model tests of a bulk carrier with estimate from Series 60 charts.

Table 19

Series	C_B on L_{PP}	L/B	B/T	LCB as pct L_{PP} from ⊗
A	0.60	7.14	2.4	3 pct aft to 1 pct aft
B	0.65	7.14	2.4	2.7 pct aft to 1.4 pct aft

ficients from 0.40 to 0.53, L/B ratios from 4.08 to 5.55 and a B/T ratio of 2.30. The results are given as curves of Ⓒ to a base of Ⓚ.

Methodical series experiments on trawlers have also been reported on by Lackenby (1959). Variations in midship coefficient C_M (from 0.829 to 0.887), in beam-draft ratio B/T (from 2.0 to 2.75), and in length-displacement ratio $L/\nabla^{1/3}$ (from 4.35 to 5.10) were studied.

(e) Full ship forms with low L/B ratios. Systematic series experiments on full ships with low L/B ratios were reported on by Keil et al (1975). The effect of variations in L/B (from 2.73 to 4.74) and B/T (from 2.40 to 10.00) were studied. The tested hull forms are characterized by a value of the block coefficient $C_B = 0.76$ and a value of the prismatic coefficient $C_P = 0.77$ (hence the midship coefficient $C_M = C_B/C_P = 0.987$). The length-displacement ratio $L/\nabla^{1/3}$ varied from 3.73 to 5.13. The results are presented in the form $C_T = (1+k)C_{FO} + m\mathrm{Fn}^4$ as a function of Froude number Fn. The values of the form factor k and the coefficient of Fn^4 are given.

The systematic series test program and analysis sponsored by the U.S. Maritime Administration (Roseman, 1987) is also concerned with full ships. This program was carried out for a series of 16 models having the parameters defined in Figure 69. These parameters reflect the design trend toward low L/B ratios evident in the late 1960's and early 1970's for bulk-type vessels and are also applicable to restricted draft full forms with characteristically high values of breadth/draft, B/T.

The single screw hull form parent has the following general characteristics:

1. Entrance—Round waterline endings, similar to the cylindrical bow configuration of the NSMB series (Muntjewerf, 1970). The MarAd series differs from the NSMB series in that the forward stations have considerable V-form.

2. Run—The very full ends associated with high values of C_B and low values of L/B would necessarily result in blunt ends and flow separation for conventional hull forms. To minimize this effect, a buttock flow stern geometry was adopted, insuring flow along buttock planes with relatively gentle slope.

3. Longitudinal center of buoyancy—Based on the results of preliminary model tests, location of the longitudinal center of buoyancy 2.5 percent of L_{pp} forward of amidships was selected and held constant for the series.

The parent lines shown in Fig. 70 have the following principal characteristics:

C_B	0.875
L_{pp}/B	5.500
B/T	3.000
C_M	0.994
L_E/L_{pp}	0.117
L_R/L_{pp}	0.346

The test program included resistance, propulsion, and maneuvering measurements in the Tracor-Hydronautics Ship Model Basin at Laurel, MD. The publication (Roseman, 1987) contains results of the data analysis, in the form of residuary resistance coefficient, C_R, and hull efficiency factors, for systematic variations in C_B, L_{PP}/B and B/T. Maneuverability characteristics are reported in the form of non-dimensional stability and control derivatives for a similar range of parameters.

A comparison of expanded series predictions with corresponding data for conventional existing vessels of the same displacement and geometry indicates the following:

1. The expanded series model resistance tends to be lower than the resistance of equivalent conventional hull forms with the same proportions and coefficients.

2. Hull efficiency of the series models is low compared to values for equivalent prototypes of conventional form.

3. The net power required for the series models tends to be less than required power for equivalent conventional hull forms.

4. Coursekeeping characteristics of low L_{pp}/B, high B/T forms are expected to be acceptable, based on analysis of the maneuvering data, provided that appropriate auto-pilot systems are installed.

Further design data on ships of various kinds can be found in Section 8.12, Statistical Analysis of Model Data.

8.9 Bodies of Revolution, Deeply-Submerged (Submarines). The resistance tests on a methodical series of bodies of revolution have been designed and run at Taylor Model Basin by Gertler (1950). These data enable P_E to be calculated for a variety of choices of hull forms, proportions and coefficients when making feasibility studies.

The range of variables covered was as follows:

Table 20

Series	C_B on L_{PP}	$L/\nabla^{1/3}$	Variation in L	Variation in B/T	Variation in LCB position
1	0.625	5.91	−7.5 to +15%	2.2 to 2.8	3 pct aft to 1 pct fwd
2	0.575	6.07	−5 to +10%	2.1 to 3.0	3.5 pct aft to 0.5 pct aft

(a) *Without parallel body*

$$L/D = 4 \text{ to } 40$$
$$C = 0.55 \text{ to } 0.70$$
Maximum section from nose $= 0.36L$ to $0.52L$
$$L_P = 0$$
Wetted surface coefficient $C_S = S/\pi DL$
$$= 0.689 \text{ to } 0.809$$

(b) *With parallel body*

$$L/D = 7.0 \text{ to } 17.5$$
$$C_P = 0.685 \text{ to } 0.840$$

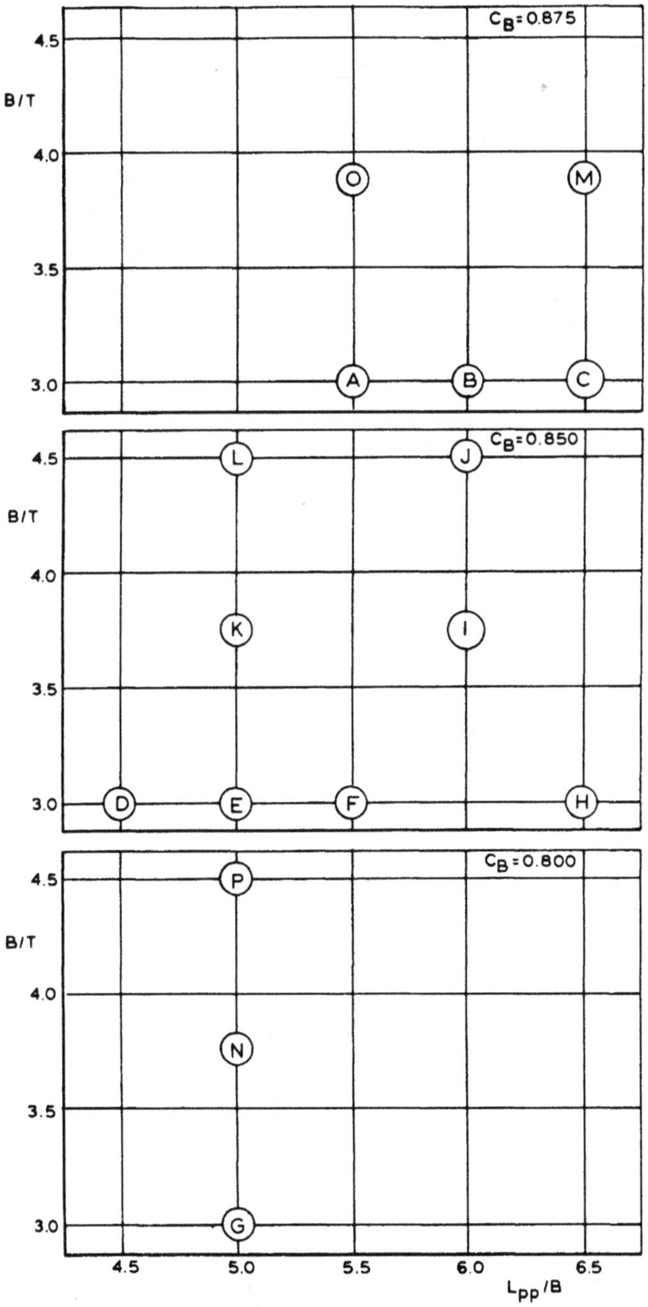

Fig. 69 Range of geometric parameters covered by MarAd Series

Maximum section from nose $= 0.28L$ to $0.16L$ (to beginning of parallel body)
$$L_P = 0.285L \text{ to } 0.60L$$
$$C_S = 0.787 \text{ to } 0.895$$
$$(L_E + L_R)/D = 5 \text{ to } 7$$

Important information is also given on the resistance of appendages and the propulsive-efficiency factors.

There have been proposals from time to time to build submarines for commercial purposes. The principal advantages claimed for such ships are the elimination of wave-making resistance and independence of weather conditions. Volume for volume, the submarine has a greater wetted surface than the ordinary ship, and so starts off with the handicap of greater frictional resistance. The absence of wave-making resistance therefore does not make itself felt until relatively high Froude numbers are reached, perhaps in excess of Fn = 0.25. Because of cargo-handling problems, most proposals have been for oil or ore-carrying submarines, and these cargoes do not call for high speeds of transport. Other suggestions have been to use such ships to open up new ocean routes under the Arctic ice cap or into ice-bound ports, and there is no doubt of their great strategic military value.

There are many difficulties in the development of such ships—the high initial cost, including that of the nuclear machinery, the need for skilled personnel and particularly the large drafts involved for building, docking, and operation. This latter feature would probably impose an elliptical or rectangular cross section rather than circular, and this would detract from the saving in resistance due to the absence of wave-making. Information on such ships and data upon their resistance and propulsion features are to be found in Todd (1960) and Russo, et al (1960).

8.10 Effect of Bulbous Bows on Resistance. D. W. Taylor, when developing the lines for his standard series from those of HMS *Leviathan*, suspected that the projecting ram bow played a definite part in its superior performance, and was also aware of the idea of a "second bow wave that would neutralize the first." He pushed the ram bow further beneath the surface, gave it a more bulbous character and thus produced the first bulb bow designed as such. A bulb of this type was incorporated in the battleship USS *Delaware* in 1907, which ship had an outstanding performance. (Saunders, 1957, Vol I, p. 368, and Saunders, 1938–41, pt. 2) and Havelock (1928) calculated the surface wave pattern around a sphere immersed in a uniform stream. The most important feature of this pattern was a wave trough just aft of the sphere, which suggested the possibility of partly cancelling the bow wave of the hull by locating a sphere below the surface in the neighborhood of the stem. Wigley (1935–36) carried out calculations of wave profiles and wave-making resistance based on Havelock's work, and published the basic theory for the bulbous bow.

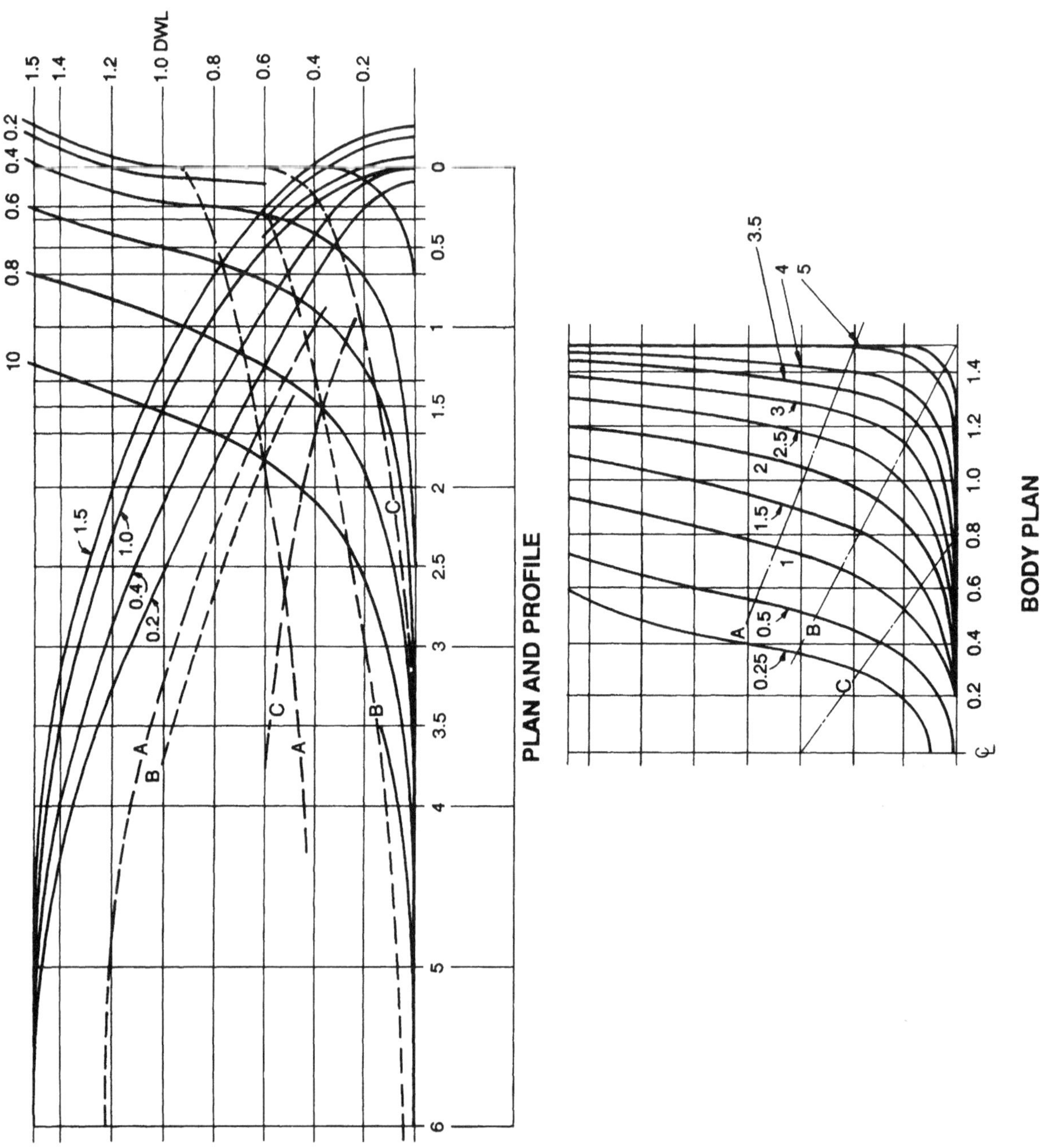

Fig. 70 Lines of basic hull form of MarAd Series

(a) Entrance

At low speeds Wigley found the total resistance to be increased owing to the additional frictional and form drag of the bulb. At high speeds, the reduction in wave resistance due to the interference between the wave systems of hull and bulb, if properly located, is more than sufficient to overcome the frictional and form drag of the bulb, and the net result is a reduction in total resistance.

Fig. 71 shows details of one bulb design in three different locations and Fig. 72 the percentage change in the calculated resistance for a ship 121.92 m (400 ft) in length (see the discussion by Todd on Inui (1962). The calculated results were confirmed by experiments on 4.88 m models, in which the sphere was faired into the hull by a cylinder. The reduction in resistance is greater the further the nose of the bulb projects forward of the stem, while in the farthest aft position there is a substantial increase in resistance.

Wigley's principal conclusions were:

(a) The useful speed range of a bulb is generally from about Fn = 0.24 to 0.57.

(b) Unless the lines are extremely hollow, the best position for the bulb is with its center at the bow; that is, with its nose projecting forward of the hull.

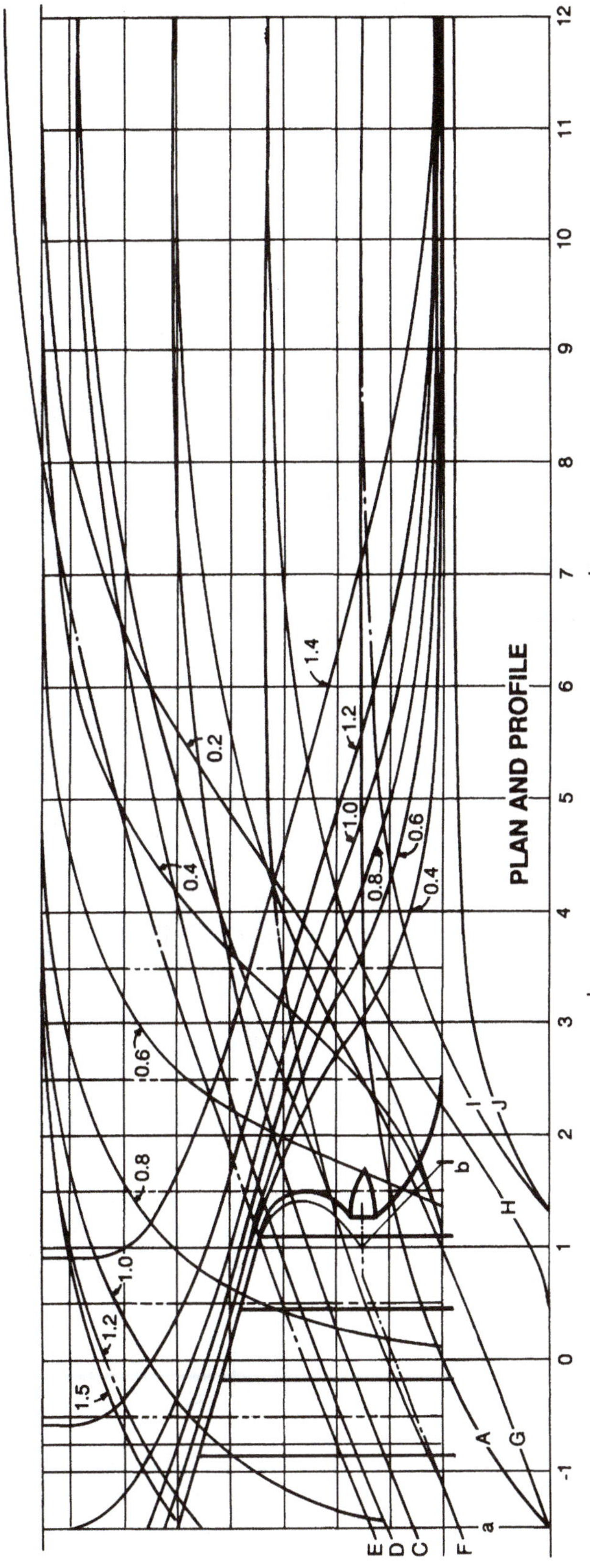

x/L_{ab}	y/t
0	1.000
0.2	0.925
0.4	0.850
0.6	0.775
0.8	0.700
1.0	0.000

$L_R/B = 1.6$

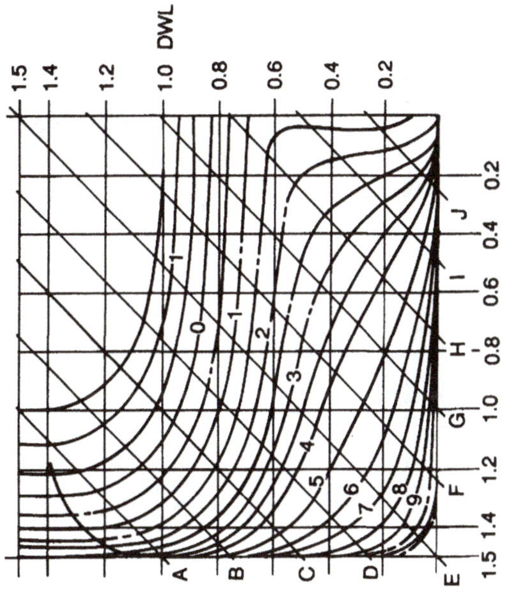

BODY PLAN

(b) Short Run

Fig. 70 (Continued)

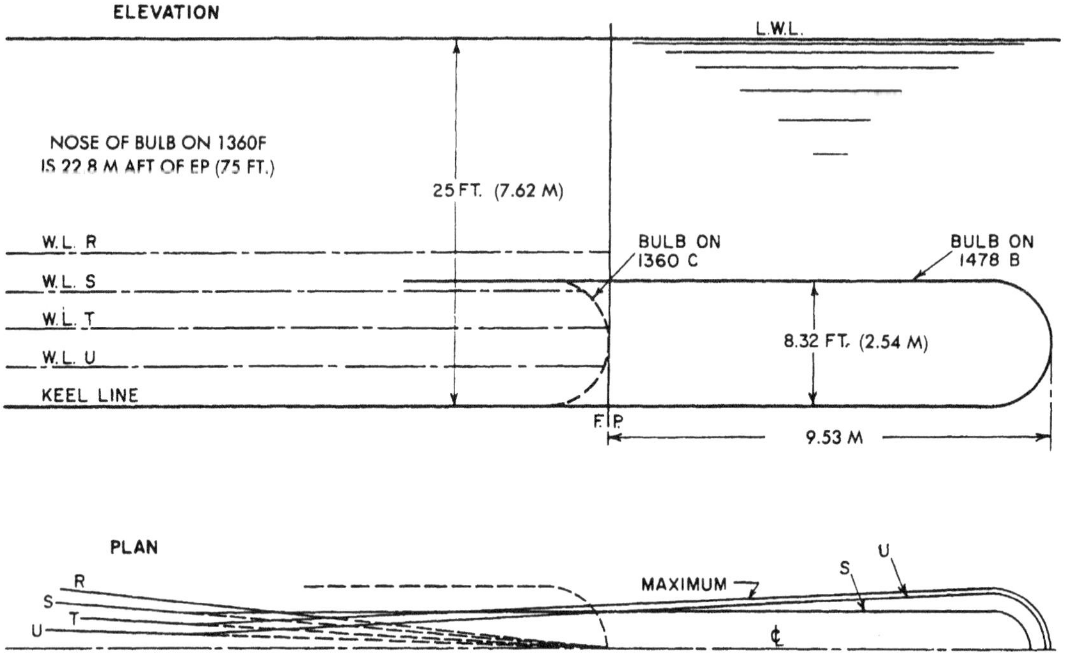

Fig. 71 Details of bulbs fitted to Models 1360 and 1478 (Wigley)
(Dimensions are for a ship 121.92 m (400 ft) in length, 11.43 m beam and 7.62 m draft)

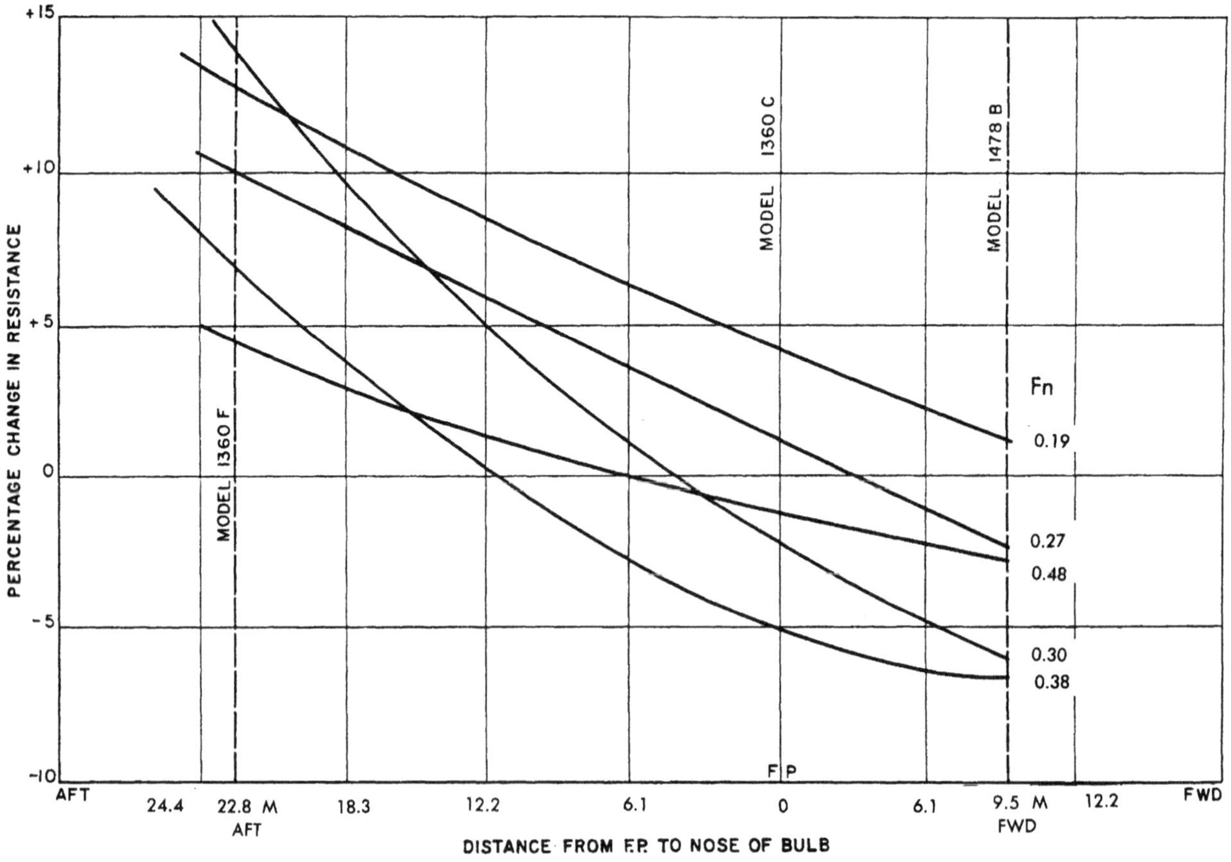

Fig. 72 Percentage change in resistance due to bulb

(c) The bulb should extend as low as possible, and should be as short longitudinally and as wide laterally as possible, consonant with fairness in the lines of the hull.

(d) The top of the bulb should not approach too near to the water surface.

As mentioned previously, two of the principal features of ship form which control the resistance are the shapes of the area and load-waterline curves, especially at the forward end. Both of these can be altered materially by the use of a bulb. The area curve can be filled out at the extreme fore end by adding displacement near the forefoot in the shape of a bulb without altering the LWL, although there will be a forward movement of the LCB. Alternatively, without changing the total displacement, the LWL can be made finer to balance the volume added by the bulb.

The forebody area curves for a ship with and without a bulb are shown in Fig. 73. The bulb is usually defined by two coefficients, f and t, introduced by Taylor, where f is a measure of the area of cross section at the forward perpendicular, FP, and t is a measure of the slope of the area curve at that point.

For a ship without a bulb, Fig. 73(a), f is zero. If a tangent is drawn to the area curve at the FP and extended to cut the midship ordinate at B, then t is defined as the ratio $AB/AC = 1.20/1.00 = 1.20$.

For a ship with a bulb, Fig. 73(b), the area curve is continued through to the FP, ignoring the final rounding of the bulb, to give a fictitious area at that point, equal to EF. The coefficient f is the ratio of this area to the midship area, and in the case shown $f = EF/AC = 0.16/1.00 = 0.16$. The ship is said to have "a 16 percent bulb." A tangent is drawn to the fictitious area curve at the FP, i.e., at the point E, and extended to cut the midship ordinate at B. Then

$$t = \frac{BD}{DC} = \frac{(AB - AD)}{(AC - AD)} = \frac{(1.36 - 0.16)}{(1.00 - 0.16)} = \frac{1.20}{0.84} = 1.43$$

The value of t is not a true tangent, but has the same value regardless of the scales on which the area curve is drawn. Saunders (1957) has called it the "terminal ratio."

Two typical bulb forms with f-values of 0.045 and 0.135 (4.5 and 13.5 percent bulbs) are shown in Fig. 74.

Many model experiments have been made to find the effects of bulbs upon resistance (Taylor, D. W., 1911; Bragg, 1930, and Lindblad, 1944 and 1948).

While the results of these experiments are not altogether consistent, there is a general confirmation of Wigleys' theoretical conclusion that the bulb is most beneficial at Fn = 0.24 to 0.57, but some tests show substantial benefits at considerably lower speeds than Fn = 0.24, due to possible effects of the bulb on the viscous resistance as well.

Trawlers run at high values of Fn (0.30 to 0.37) and have large wave-making resistance. These are conditions which should be favorable to the use of a bulbous bow, and this has been confirmed by model experiments. Doust (1961) has shown that the resistance of good conventional designs can be reduced by 10 to 15 percent through the use of a bulb, and that the propulsive efficiency is also increased by 4 to 5 percent, so that reductions in P_D of as much as 20 percent are possible in smooth water. In model tests in waves, the bulb-type trawler also experienced a smaller reduction in speed. To achieve improvements, the bulb must not be treated merely as an addition or appendage, but the whole forebody should be redesigned, a fine load waterline being used with half-angles of entrance 5 to 10 deg less than those of a normal trawler, and with the LCB as far aft as possible. Doust recommends that the bulb area should not exceed 5 percent in order to avoid risk of slamming damage.

Johnson (1958) in Sweden has tested models of trawlers with ram bulbs having values up to 8 percent. The resistance was reduced at all speeds above Fn values

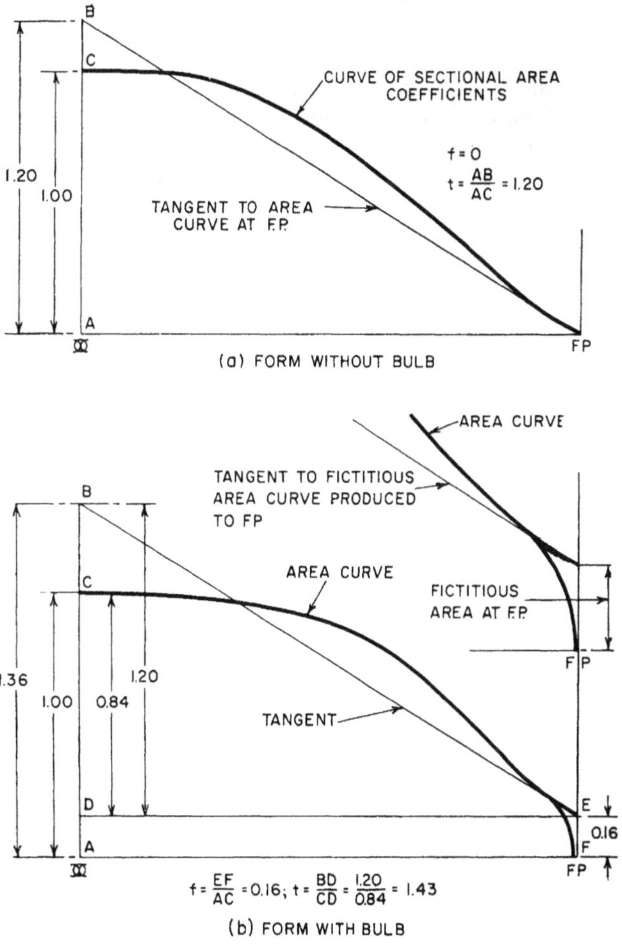

Fig. 73 Definition of bulb in terms of *f* and *t*

around 0.28 to 0.30, the decrease at Fn = 0.36 being about 12 percent at the design draft, and 20 percent when trimmed by the stern.

Professor Inui (1960, 1962) in Japan has devoted much effort to the design of bulbs, visualizing their use at both stem and stern to annul the wave systems normally arising at those points. In his analysis, he keeps the bow and stern systems separate and cancels each one by means of an additional distribution of sources and sinks or "singularities." These latter must generate a wave system such that the resulting elementary waves have approximately the same amplitude as the free waves from the bow or stern and are in inverse phase. His so-called "waveless" form is based upon this principle.

Bulbs of the Inui design have been fitted to a number of ships. Comparative trials have been carried out on two Japanese ferryboats, the *Murasaki Maru* and the *Kurenai Maru*. The second ship was fitted with a large bulb, three times the size of the ordinary one, the diameter being 3.5 m (11.48 ft) and the displacement about 40 m^3 (1412 ft^3) (JSS, 1961). The f-value for this bulb was 16.7 percent as compared with 5.5 percent for the normal size.

The two ships were run together and aerial photographs confirmed the reduction in the size of the wave pattern. At 18.5 knots the engine output with the 5.5 percent bulb was 4023 kW, whereas with the 16.7 percent bulb it was only 3494 kW, a reduction of 13.1 percent, equivalent to an increase in speed of nearly half a knot for the same power. It was estimated that the increase over a similar ship with no bulb was about 0.8 knot.

Systematic tests have been carried out for the Maritime Administration at the Davidson Laboratory to investigate the effects of different bows both in smooth water and in waves, (Dillon, et al, 1955). In their preface the authors called attention to the fact that while some ships are built with sharp stems, often heavily raked, others, designed for similar operating conditions, are built with pronounced bulbous bows and vertical stems. They concluded that the lack of agreement between naval architects is due to an absence of objective information on the subject, especially in rough-water conditions.

One of the reasons commonly given against the adoption of bulbs is the fear of rough-water effects, but the authors found little recorded complaint in this matter. However, the evidence gathered by Townsend (1960) (Section 8.2) in this context should not be overlooked.

The particulars and the results of the Maritime Ad-

Table 21—Particulars of Maritime Administration Model Tests (Bulbous bows)

L_{WL}	= 189.59 m	C_B	= 0.595
LP_P	= 183.49 m	C_M	= 0.970
B	= 26.21 m	C_P	= 0.613
T	= 8.53 m	L/B	= 7.23
∇	= 25,278 m^3	B/T	= 3.07
$L/\nabla^{1/3}$	= 6.46		

Design sea speed = 23 knots; Fn = 0.274

Bulb area, percent	0	4.5	9.0	13.5
i_E (half-angle of entrance on L_{WL}), deg	11.25	8.75	6.25	3.75
LCB, percent L_{WL} aft of amidships	0.48	0.39	0.29	0.19
P_E (kWatt)	15347	15049	14453	14043
Reduction in percent	0	2.0	5.5	8.5

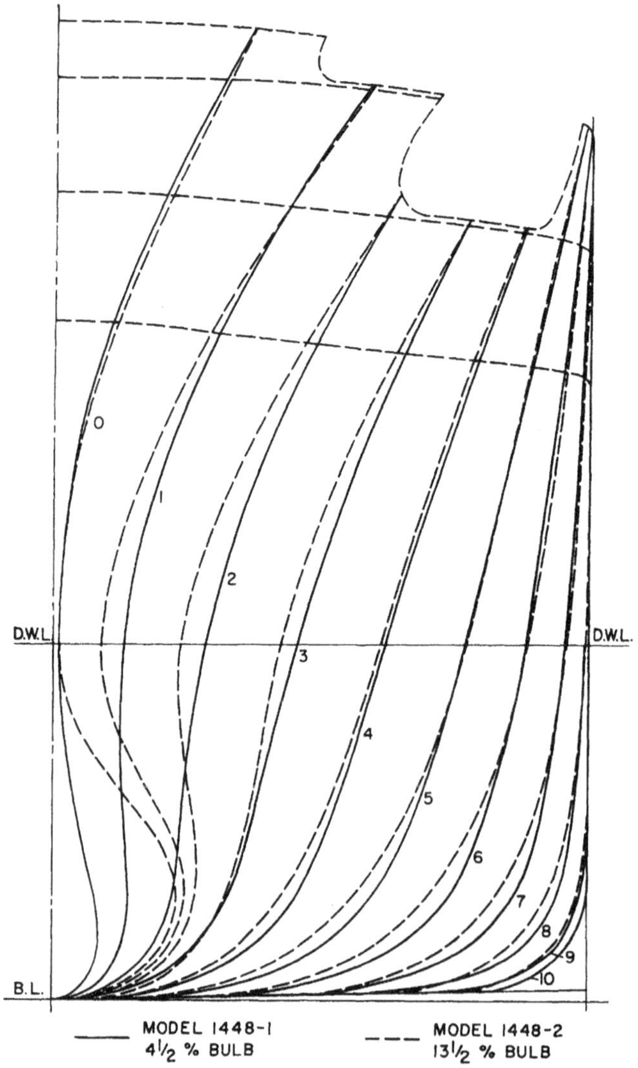

Fig. 74 Comparison of body plans. 4.5 and 13.5 percent bulbs

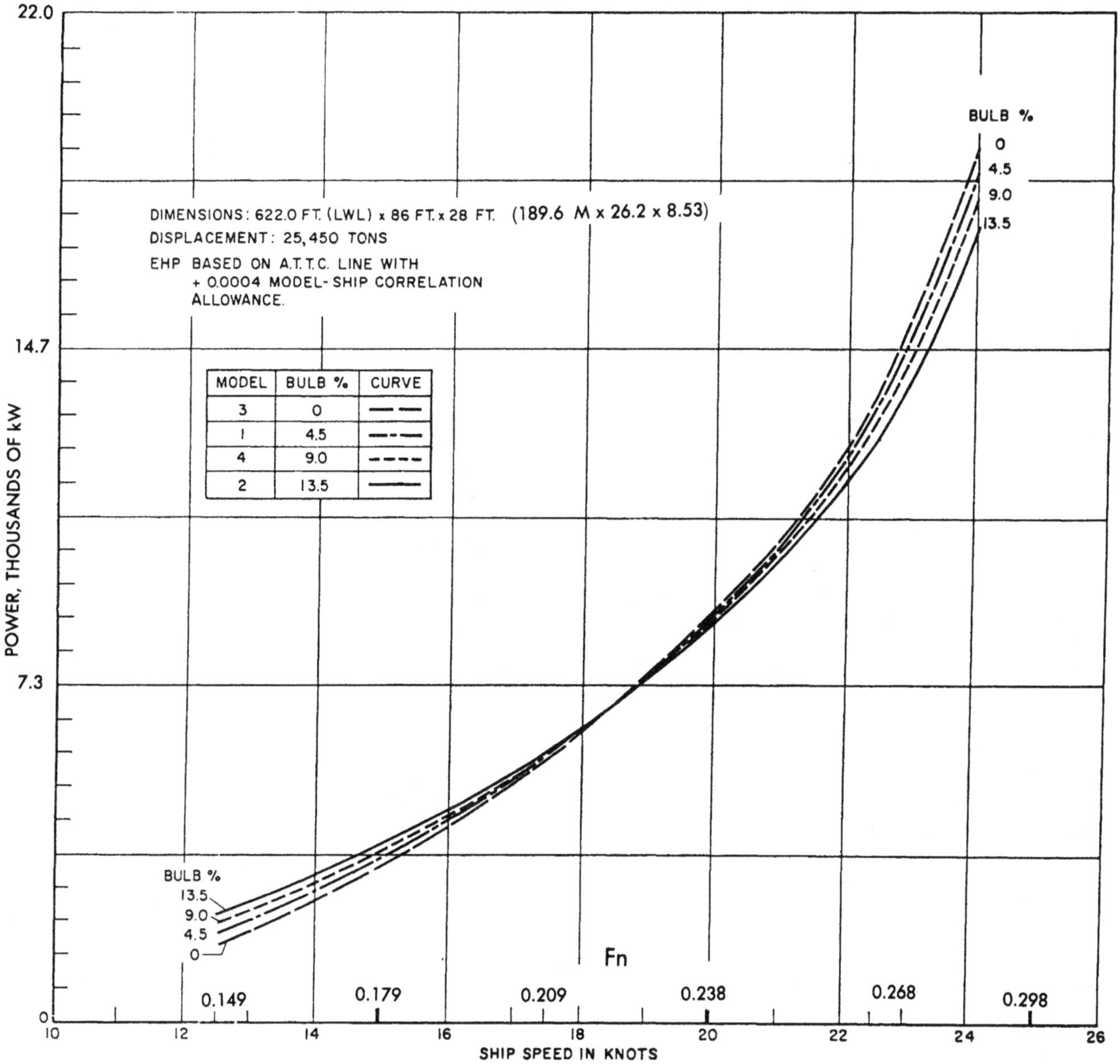

Fig. 75 Effect of bulb on P_E in smooth water

ministration models are given in Table 21, and the lines for the models with 4.5 and 13.5 percent bulb are shown in Fig. 74. The resistance results in smooth water are shown in Fig. 75. At low speeds the smaller bulb showed the better performance, at high speeds the larger, the changeover occurring between Fn = 0.21 and 0.24. The variation in bulb area was not carried sufficiently far to show a minimum resistance value.

At a constant P_E = 18,625 kW, the loss of speed in different waves with the 13.5 percent bulb is given in Table 22. The order of merit of the different bulbs in severe waves was the same as that in calm water at the same speed, i.e., around 17 knots.

The wide variation in bulb size was found to have only a small effect on power or speed and on pitching motion in head seas. The wave length L_W and the period of encounter had a much greater effect on these characteristics than did bulb size. Hence, the authors concluded that, for this design, the choice of bulb could be based on the smooth-water performance. There was

Table 22—Reduction in Speed for 13.5 Percent Bulb in Waves

Wave length-ship length ratio L_W/L_{WL}	0	0.25	0.50	0.75	0.875	1.00	1.15	1.25
V_k in knots	24.3	24.1	23.7	22.9	21.8	18.6	17.6	18.0
Loss in V_k in percent	0.0	0.8	1.7	4.6	9.2	22.1	26.3	25.0

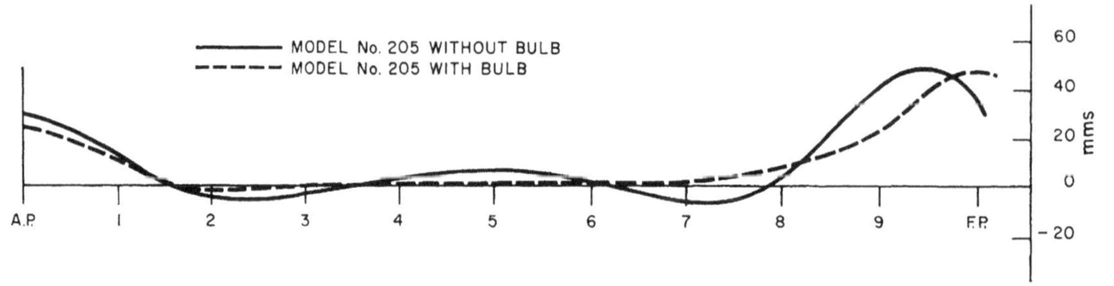

Fig. 76 Wave profiles, with and without bulb, at Fn = 0.1267 (designed speed) (Tsunodu, et al, 1963)

no slamming in regular waves, but in irregular waves some occurred with all models.

No evidence was found that large bulbs should not be used if found desirable for smooth-water performance.

Much of the research into the effects of bulbs has been devoted to resistance and powering aspects. The effects on sea-going qualities must also be investigated before the decision is taken to apply a bulb for any particular case.

The evidence seems to be that bulbs have little effect on pitching and if anything are beneficial in reducing such motions. Slamming and resultant hull damage are also feared, though in general there is little evidence that ships with bulbs have suffered any worse in this respect. Out-size bulbs do introduce problems in berthing and anchoring. Many warships today carry large sonar equipment forward, and certainly every effort should be made to house these in "bulbs" which will at least not add to the hull resistance at service speeds.

A problem that has arisen in high-speed ships with bulbs is the occurrence of cavitation on the bulb surface, resulting in erosion and noise. The nose of the bulb should be elliptical rather than circular, and calculations should be made to ensure that the curvature is nowhere sharp enough to cause cavitation. Special attention should be paid to smoothing off weld beads and other roughnesses in this area.

Large bulbs are now commonly fitted to big tankers and bulk carriers running at low Fn values, at which the wave-making resistance is relatively small. Reductions in resistance of approximately 5 percent in full load and 15 percent in the ballast condition have been obtained in model tests. These results are confirmed in full-scale trials. In general about 1 knot increase in speed in the ballast condition is realized. Such gains are apparently possible on ships with block coefficients around 0.80 and at Froude number values of about 0.18. It is significant that the most substantial improvements are found in the ballast condition when the bulb is near the surface. The draft forward appears to be critical and care should be taken in choosing the ballast operating condition.

The reasons for the large reduction in resistance are not entirely clear, but one effect of fitting such bulbs seems to be a reduction in the separation phenomena which experiments have shown to be present under the bows of some models, particularly those with strong U-shaped sections and hard bilges forward. Flow tests on certain models have shown that the presence of a large ram bulb appears to stabilize the flow in this vicinity.

It has been assumed generally in the past that the skin-friction resistance is chiefly a function of the wetted area and relatively insensitive to changes of shape of the hull. However, recent advances in the theoretical treatment of wave-making, mostly in Japan and directed towards obtaining a "waveless" form, have resulted in models which have shown quite large reductions in resistance at quite moderate speeds. The proportion of wave-making resistance at these speeds is such that these reductions in total resistance seem unlikely to be due only to wave-making. It suggests that the particular designs in question, which generally incorporate very large, protruding bulbs at the forefoot, have also affected the flow around the hull in such a way as to reduce the average velocity and so the frictional resistance. Evidence pointing in this direction also occurs in a paper giving the results of model experiments on a cargo liner (R. Tsunoda, et al. 1963). The wave profile for the normal form has a trough following the bow crest, which indicates low pressure and high velocity around the hull in this region, whereas the "waveless" bulb form, while being by no means waveless, and having a bow crest of about the same height as that of the conventional form, has no following trough, the water surface being always above the still waterline, Fig. 76. This suggests that the velocities over the hull, and so also the skin friction, are less with this type of form. This may be a further part of the explanation for the relatively large reduction in total resistance found for such designs, even in large tankers at low values of the Froude number. It points to the necessity for studying the shape of the hull in order to reduce skin friction, pressure resistance as well as wave-making resistance. Such studies are now being carried out.

Much research remains to be done to gain a full understanding of the effects of large bulbs on slow, full ships.

8.11 Cylindrical and Elliptical Bows (Spoon

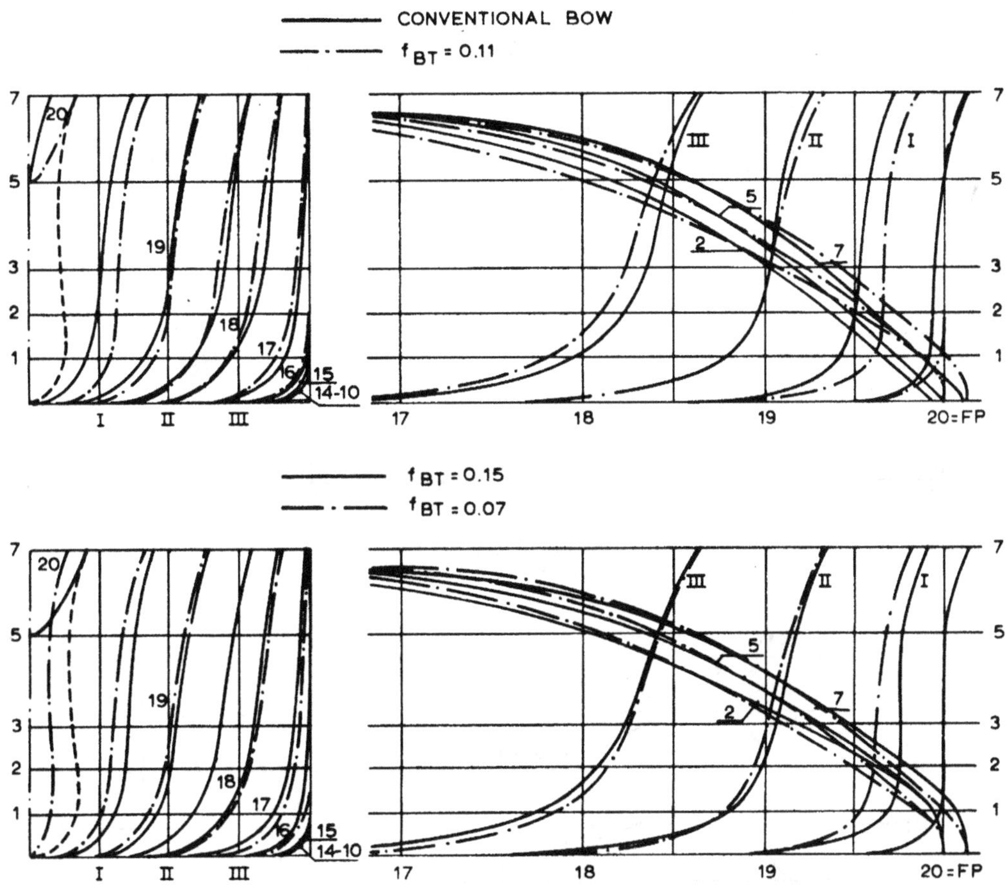

Fig. 77 Body plans for models with C_B of 0.825

Table 23—Parameter Field Chosen for the MARIN Program

L/B		6·5	
B/T		2·65	
C_B	0·800	0·825	0·850
f_{BT}	0 0·07 0·11 0·15	0 0·07 0·11 0·15	0 0·07 0·11 0·15
LCB	1·6 percent F	2·1 percent F	2·8 percent F

Bows). At MARIN, systematic experiments have been carried out on full-form ships (C_B = 0.800, 0.825 and 0.850) for which three cylindrical bows were designed (Muntjewerf, 1970). Table 23 gives the range of variables chosen for the experiments. The models represented a 183 m (600 ft) bulkcarrier and at the same time a 244 m (800 ft) tanker. The body plans for the 0.825 block models are shown in Fig. 77. The test results were extrapolated according to the ITTC-Froude method, using the ITTC (1957) line with a correlation allowance C_A = 0.00035 for the 183 m ships and C_A = 0.00020 for the 244 m ships.

Fig. 78 shows the optimum bow size as a function of Fn for different block coefficients and for two loading conditions. From this figure, it follows that for ballast as well as full load condition the optimum bow size increases with increasing Fn. Further it shows that the higher the block coefficient the larger the cylindrical bow should be. According to the results the optimum bow size for ballast condition is considerably smaller than for full load condition.

Fig. 79 shows the reduction in effective power with an optimum cylindrical bow. This figure clearly shows that the advantages are most pronounced at 100 percent displacement. The highest reductions are obtained in the range of block coefficients between 0.825 and 0.850.

Using these results one often finds that a cylindrical bow has a slight negative effect in ballast especially at lower speeds. In Muntjewerf (1970) it also turns out that the improvements in the required power at the propeller are even a little more pronounced than the improvements in bare hull resistance.

For a ship with C_B = 0.85 experiments were carried out with a number of bow forms (Luthra, 1971). Fig. 80 shows the parent form and the tested variations. The experiments were carried out at two drafts. Fig. 81 shows the resistance results for the two loading conditions. For both drafts the elliptical bow forms turned out to have remarkably better resistance char-

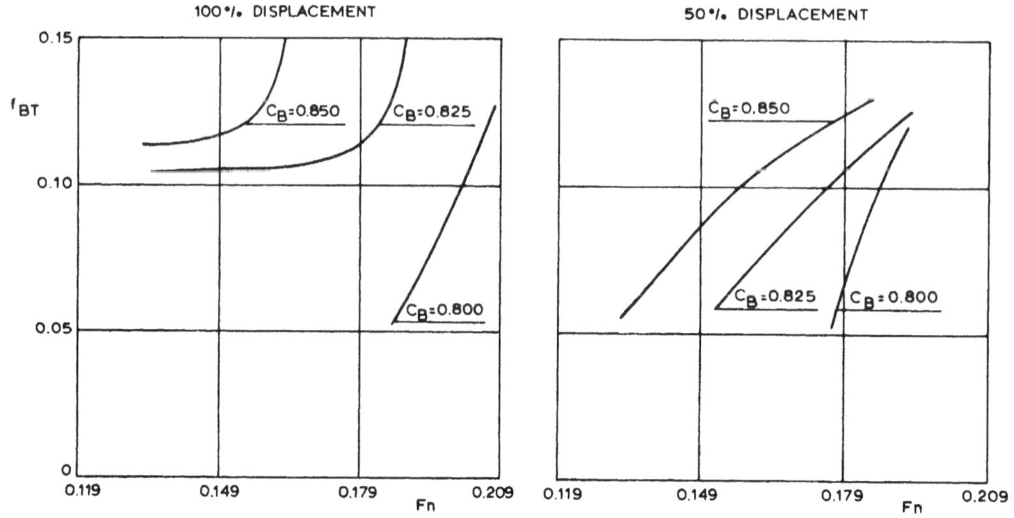

Fig. 78 Optimum bow size as a function of speed-length ratio for different block coefficients (Muntjewerf, 1970)

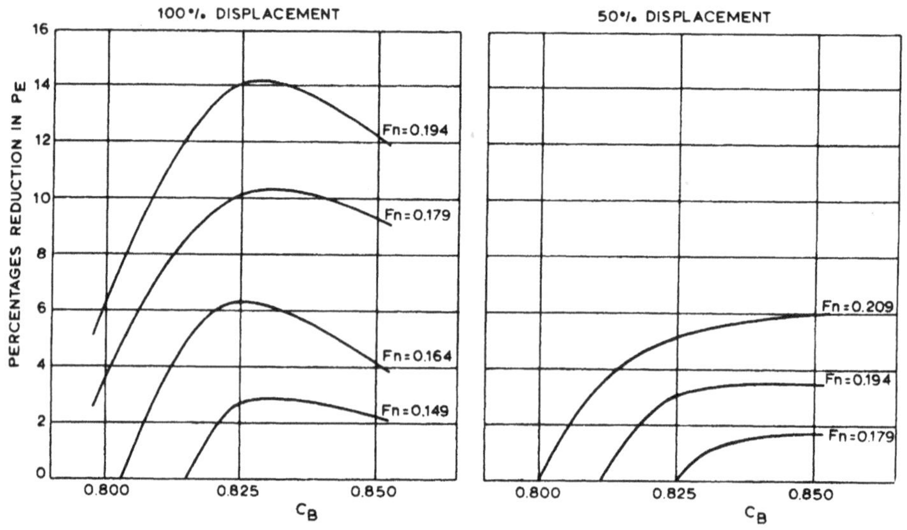

Fig. 79 Percentages reduction in effective power with optimum cylindrical bow

acteristics. The knuckled elliptical bow form has a slightly lower resistance values compared to the smooth elliptical bow, the latter having higher building costs.

In a later publication by Luthra (1975), experiments were reported in which the elliptical bow form was fitted with a bulb. This form has a lower resistance than the parent bow form without a bulb. For a ship with C_B = 0.75 experiments with a spoon bow have also been carried out but the improvements in this case turned out to be only marginal.

8.12 Statistical Analysis of Model Data. In addition to the published results for methodical model series, there exists a vast store of resistance data for the many models tested for specific designs. These are generally unrelated except in a generic way, but they contain the results of many changes made to hull forms in the effort to improve their performance. Such data might therefore be expected to yield valuable results if analyzed statistically. Attempts to do this in the past have been unsuccessful because of the arithmetical labor involved, but the advent of high-speed computers has changed the situation.

The first application to ship model data was made by Doust, et al (1958–59) at the National Physical Laboratory in England. He chose fishing trawlers for the purpose, some 150 models of which had been run in the NPL tanks, making a homogeneous set of data for a class of ship in which wave-making resistance, the part principally susceptible to hull shape, contributed

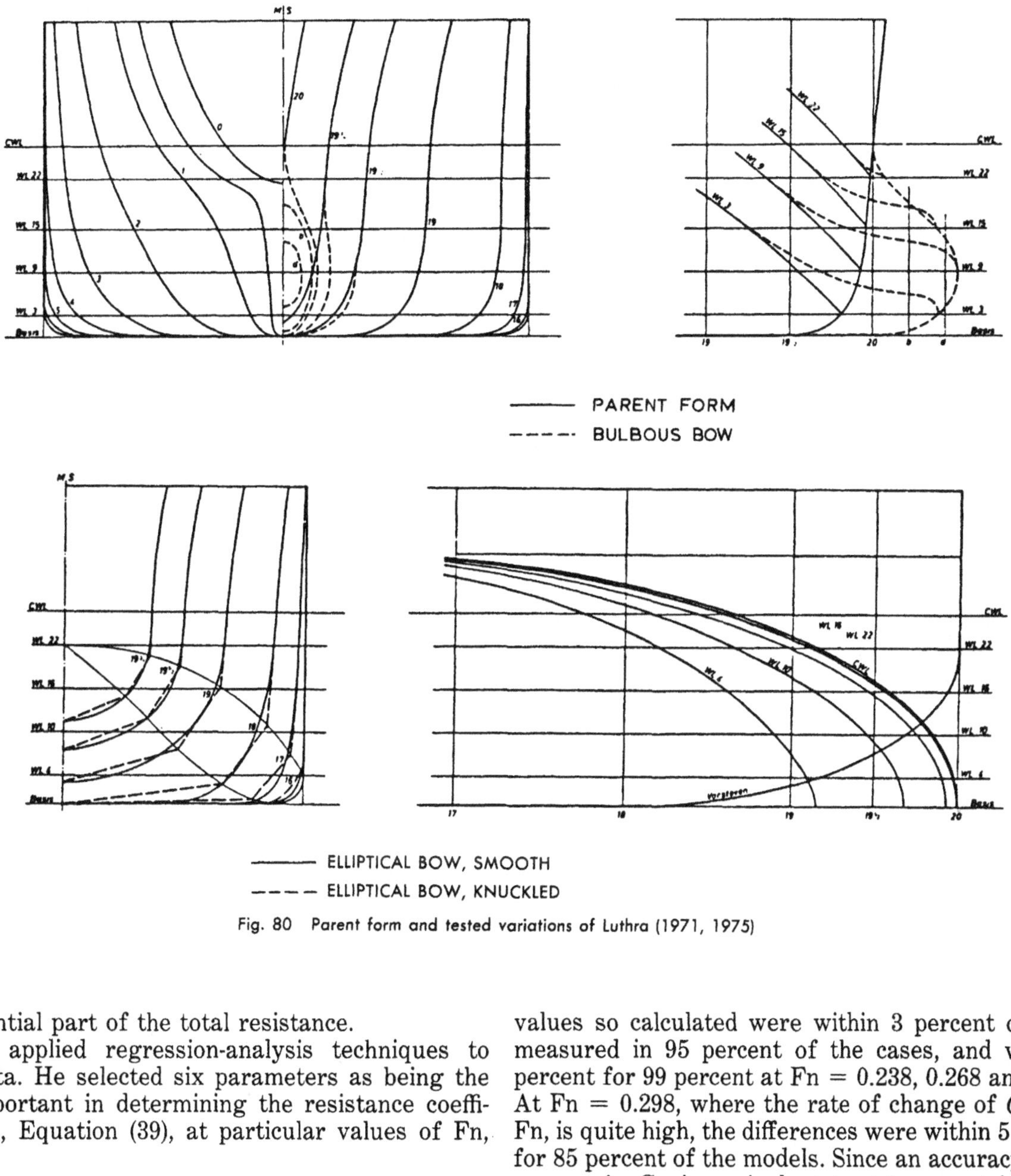

Fig. 80 Parent form and tested variations of Luthra (1971, 1975)

a substantial part of the total resistance.

Doust applied regression-analysis techniques to these data. He selected six parameters as being the most important in determining the resistance coefficient C_{TL}, Equation (39), at particular values of Fn, writing

$$
\begin{aligned}
C_{TL} &= gR_TL/WV^2 \\
&= f(L/B,\ B/T,\ C_M,\ C_P,\ \text{LCB position},\ i_E) \\
&= 0.00505\,[a_0 + a_1\,(B/T) + a_2(B/T)^2 \ldots, \\
&\quad + a_{28}\,(B/T)\,C_P + a_{29}\,(B/T)^2\,C_P]
\end{aligned}
$$

The terms contained no powers of any one parameter higher than the second.

For four values of Fn (0.238, 0.268, 0.298 and 0.328) the values of a_0—a_{29} were determined on the computer for the 150 trawler models.

The accuracy of the analysis was first tested by applying the equation to calculate the resistance of the models used to derive the coefficients—what may be called "playing back" the data. It was found that C_{TL}-values so calculated were within 3 percent of those measured in 95 percent of the cases, and within 5 percent for 99 percent at Fn = 0.238, 0.268 and 0.328. At Fn = 0.298, where the rate of change of C_{TL} with Fn, is quite high, the differences were within 5 percent for 85 percent of the models. Since an accuracy of ± 4 percent in C_{TL} is equivalent to ± 1 percent in speed, or say 1/8 of a knot, this is quite acceptable for preliminary design use.

The next step was to use the regression equation to investigate the optimum choice of parameters to suit any given design requirements. The equation was extended by two more terms, and four hulls designed to give optimum resistance performance over a range of prismatic coefficient were made and tested. The results for one of these are shown in Fig. 82, which indicates the closeness of the prediction (Doust, 1962–63).

Fig. 83 shows the results of the four new designs at a Froude number value Fn = 0.329, compared with the original data on which the analysis was based. It can be seen that the new designs, derived from the statistical analysis, all give results below the best ones

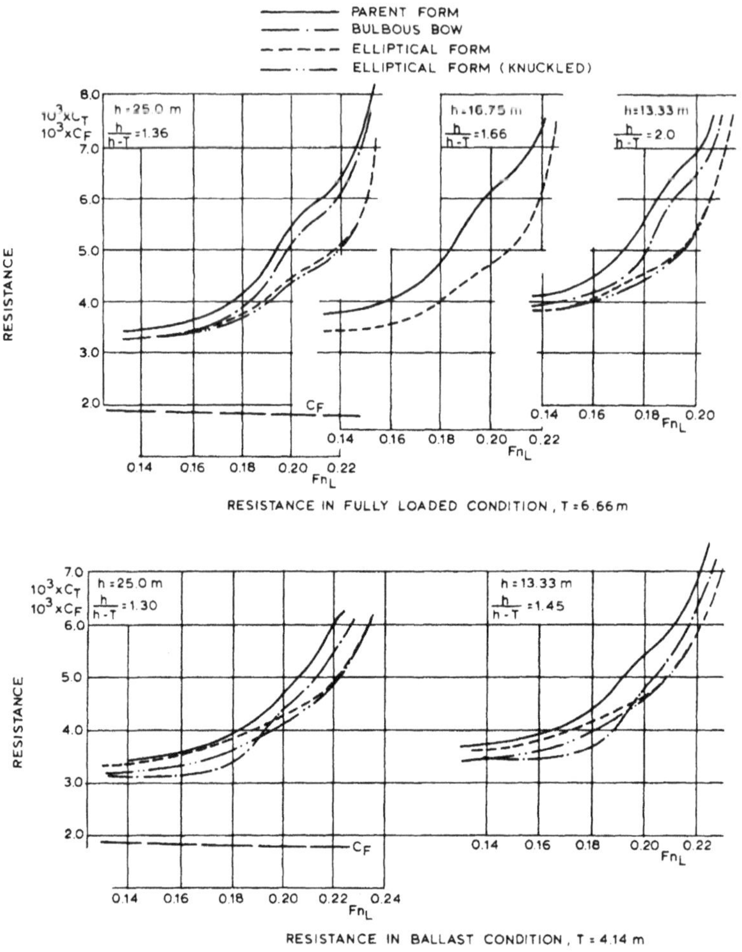

Fig. 81 Resistance coefficients for two drafts (Luthra, 1971, 1975)

previously attained, the maximum reduction for the fullest model being some 22 percent.

The same methods have been applied by Doust (1963) to fast cargoliner models and to the prediction of propulsive efficiencies for both classes of ship.

Recently, Holtrop, et al (1982) published the results of a statistical analysis of the results of resistance and propulsion tests with 191 models of various types of ship carried out at MARIN. It was found that for 95 percent of the cases the accuracy of the statistically-derived formulas is satisfactory in preliminary design work if the range of variables is within that given in Table 24.

Holtrop, et al (1984), extended their method to include the Series 64 hull forms (Yeh, 1965). Also better formulas were obtained for the higher speed ranges. The regression analysis is now based on the results for 334 models.

The viscous resistance is calculated from:

$$R_V = \tfrac{1}{2}\rho V^2 C_{FO}(1+k_1)S \qquad (67)$$

where

C_{FO} = friction coefficient according to the ITTC-1957 friction line according to the formula

$$C_{FO} = \frac{0.075}{(\log \text{Rn} - 2)^2}$$

Table 24—Range of variables for statistical power prediction method at MARIN

Ship type	Maximum Froude number	C_P min	C_P max	L/B min	L/B max	B/T min	B/T max
Tankers, bulk carriers (ocean)	0.24	0.73	0.85	5.1	7.1	2.4	3.2
Trawlers, coasters, tugs	0.38	0.55	0.65	3.9	6.3	2.1	3.0
Containerships, destroyer types	0.45	0.55	0.67	6.0	9.5	3.0	4.0
Cargoliners	0.30	0.56	0.75	5.3	8.0	2.4	4.0
Roll-on, roll-off ships; car-ferries	0.35	0.55	0.67	5.3	8.0	3.2	4.0

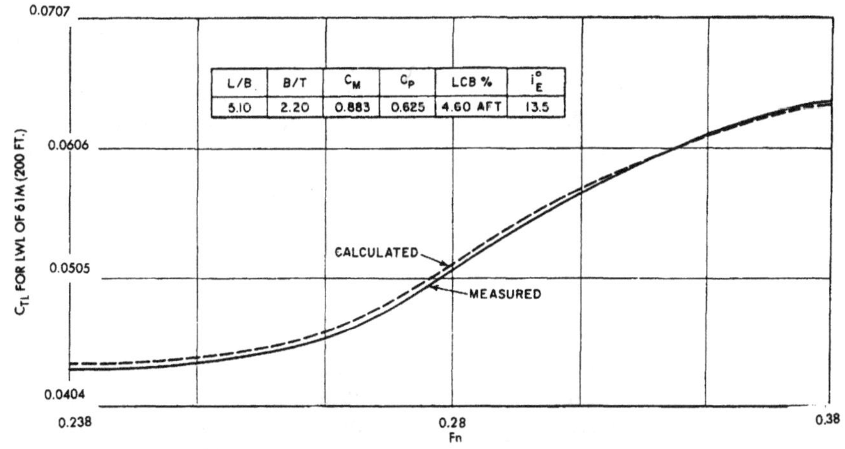

Fig. 82 Comparison of calculated and measured resistance on optimized trawler form (Doust, 1962, 1963)

k_1 = form factor of bare hull

The value of $1+k_1$ was derived statistically. The following formula was found:

$$1 + k_1 = 0.93 + 0.4871c\,(B/L)^{1.0681}\,(T/L)^{0.4611}\,(L/L_R)^{0.1216}\,(L^3/\nabla)^{0.3649}(1-C_p)^{-0.6042}$$

in which c is a coefficient accounting for the specific shape of the afterbody and is given by

$$c = 1 + 0.011\,c_{\text{stern}}$$

$c_{\text{stern}} = -25$ for pram with gondola
$c_{\text{stern}} = -10$ for V-shaped sections
$c_{\text{stern}} = 0$ for normal section shape
$c_{\text{stern}} = +10$ for U-shaped sections with Hogner stern.

L_R is the length of the run, which—if unknown—can be estimated from the following formula:

$$L_R/L = 1 - C_P + 0.06C_P\,\text{LCB}/(4C_P-1)$$

If unknown, the wetted surface of the bare hull can be estimated from the following, statistically derived, formula:

$$S = L(2T+B)C_M^{0.5}(0.4530 + 0.4425\,C_B - 0.2862\,C_M - 0.003467\,B/T + 0.3696\,C_{WP}) + 2.38\,A_{BT}/C_B$$

in which

T = average molded draft in m
L = waterline length in m
B = molded breadth in m
LCB = longitudinal center of buoyancy forward of (+), or abaft (−) midship as a percentage of L
A_{BT} = cross-sectional area of bulb in the vertical

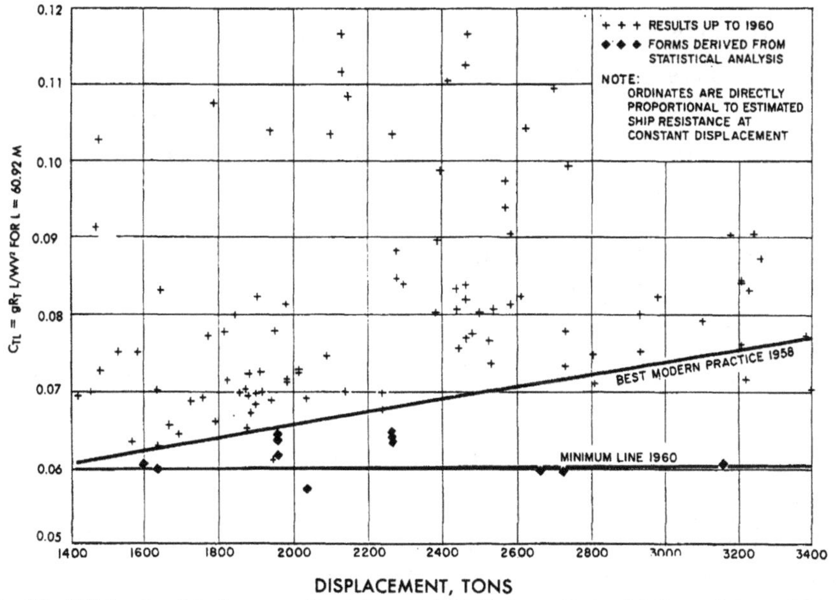

Fig. 83 NPL trawler data for conventional 61-m (200 ft) BP forms. Design (trial) speed, Fn = 0.329

Table 25—Effective Form Factor Values k_2 for Different Appendages

Type of appendage	Value of $1+k_2$
Rudder of single-screw ship	1.3 to 1.5
Spade-type rudders of twin-screw ships	2.8
Skeg-rudders of twin-screw ships	1.5 to 2.0
Shaft brackets	3.0
Bossings	2.0
Bilge keels	1.4
Stabilizer fins	2.8
Shafts	2.0
Sonar dome	2.7

plane interesecting the stem contour at the water surface.

All form coefficients are based on the length of the waterline L.

The resistance of appendages was also analyzed and the results were presented in the form of an effective form factor, including the effect of appendages, Table 25.

$$1 + k = 1 + k_1 + [1 + k_2 - (1 + k_1)] \frac{S_{app}}{S_{tot}}$$

where

k_2 = effective form factor of appendages
S_{app} = total wetted surface of appendages
S_{tot} = total wetted surface of bare hull and appendages

The effective form factor is used in conjunction with a modified form of Equation (67);

$$R_v = \tfrac{1}{2}\rho V^2 C_{FO} S_{tot} (1 + k)$$

The effective value of k_2 when more than 1 appendage is to be accounted for can be determined as follows:

$$(1 + k_2)_{\text{effective}} = \frac{\Sigma S_i (1 + k_2)_i}{\Sigma s_i}$$

in which S_i and $(1 + k_2)_i$ are the wetted area and appendage factor of the i th appendage.

For the calculation of the wave-making resistance the following equation, derived by Havelock (1913), was adopted:

$$\frac{R_W}{W} = C_1 e^{-m\mathrm{Fn}^{-2}/g} + e^{-m\mathrm{Fn}^{-2}} \{ C_2 + C_3 \cos (\lambda \mathrm{Fn}^{-2})\}$$

In this equation, C_1, C_2, C_3, λ and m are coefficients which depend on the hull form. This expression describes the wave-making resistance of two pressure disturbances of infinite width with the first term as a correction to account for the induction of the diverging waves. The distance between the centers of the disturbances λL can be regarded as the wave-making length. The interaction between the transverse waves, accounted for by the cosine term, results in the typical humps and hollows in the resistance curve.

This formula was further simplified into the following formula:

$$\frac{R_w}{W} = C_1 C_2 C_3 e^{m_1 \mathrm{Fn}^d} + m_2 \cos (\lambda \mathrm{Fn}^{-2})$$

Although the theoretical basis for this wave-resistance formula is relatively poor it has been used because the original Havelock formula is not very suitable for regression analyses. Moreover, a more elaborate theoretical expression would require more detailed information on the lines plan and would therefore be less suitable for preliminary design purposes.

It turned out that a better fit could be obtained by introducing a low-speed and a high-speed wave resistance formula.

For the low-speed range ($\mathrm{Fn} \leq 0.4$) the following coefficients were derived:

$$C_1 = 2223105\, C_4^{3.7861} (T/B)^{1.0796} (90 - i_E)^{-1.3757}$$

with:	$C_4 = 0.2296\,(B/L)^{0.3333}$	for $B/L \leq 0.11$
	$C_4 = B/L$	for $0.11 \leq B/L \leq 0.25$
	$C_4 = 0.5 - 0.0625\, L/B$	for $B/L \leq 0.25$
Further	$d = -0.9$,	
	$m_1 = 0.01404\, L/T - 1.7525\, \nabla^{1/3}/L - 4.7932\, B/L - C_5$	
with	$C_5 = 8.0798\, C_p - 13.8673\, C_p^2 + 6.9844\, C_p^3$	for $C_p \leq 0.8$
	$C_5 = 1.7301 - 0.7067\, C_p$	for $C_p \geq 0.8$
and	$m_2 = C_6\, 0.4\, e^{-0.034\, \mathrm{Fn}^{-3.29}}$	
in which	$C_6 = -1.69385$	for $L^3/\nabla \leq 512$
	$C_6 = 1.69385 + (L/\nabla^{1/3} - 8.0)/2.36$	for $512 \leq L^3/\nabla \leq 1727$
	$C_6 = 0.0$	for $L^3/\nabla \geq 1727$
Lastly	$\lambda = 1.446 C_p - 0.03\, L/B$	for $L/B \leq 12$
	$\lambda = 1.446 C_p - 0.36$	for $L/B \geq 12$

where i_E denotes the half angle of entrance of the load waterline in degrees. The value C_2 accounts for the effect of a bulb. If no bulb is fitted, $C_2 = 1$. The value of C_2 is found from:

$$C_2 = e^{-1.89} \frac{A_{BT} r_B}{BT(r_B + i)}$$

in which r_B is the effective bulb radius, equivalent to:

$$r_B = 0.56\, A_{BT}^{0.5}$$

i represents the effective submergence of the bulb as determined by:

$$i = T_f - h_B - 0.4464 r_B$$

in which T_f = molded draft at the forward perpendicular

h_B = height of the centroid of the area A_{BT} above the base line.

The final coefficient in the formula, C_3, is given by

$$C_3 = 1 - 0.8\, A_T/(BTC_M)$$

C_3 accounts for the influence of a transom stern on the wave resistance. A_T is the immersed area of the transom at zero speed.

If in the preliminary design stage the half-angle of entrance of the load waterline is unknown the following formula can be used:

$$i_E = 125.67\, B/L - 162.25\, C_p^2 + 234.32\, C_p^3 + 0.1551 \left(\text{LCB} + \frac{6.8\,(T_a - T_f)}{T}\right)^3$$

in which T_a = molded draft at the aft perpendicular.

For the high speed range (Fn $\geq$ 0.55) some of the above coefficients obey different relations:

$$C_1 = 6919.3\, C_M^{-1.3346}\, (\nabla/L^3)^{2.0098}\, (L/B-2)^{1.4069}$$
$$m_1 = -7.2035\, (B/L)^{0.3269}\, (T/B)^{0.6054}$$

For intermediate speeds (0.4 $\leq$ Fn $\leq$ 0.55) an interpolation formula is suggested:

$$\frac{R_W}{W} = \frac{1}{W}\Big[R_W(\text{Fn}{=}0.4) + (10\text{Fn} - 4)\,\{(R_W(\text{Fn}{=}0.55) - R_W(Fn{=}.4)\}/1.5\Big]$$

The formula derived for the model-ship correlation allowance C_A is:

$$C_A = 0.006\,(L_{WL} + 100)^{-0.16} - 0.00205$$

which is valid for $T_F/L_{WL} > 0.04$. If $T_F/L_{WL} < 0.04$ then the appropriate value of C_A should be determined from:

$$C_A = 0.006\,(L_{WL} + 100)^{-0.16} - 0.00205 + 0.003\,(L_{WL}/7.5)^{0.5}\, C_B^4\, C_2\,(0.04 - T_F/L_{WL})$$

where C_2 is the coefficient adopted to account for the influence of a bulb, and given above.
The total resistance is then determined from

$$R_T = \tfrac{1}{2}\rho\, V^2 S_{tot}\,[C_F(1+k) + C_A] + \frac{R_W}{W}\, W$$

Holtrop and Mennen also derived statistical results for the various propulsion coefficients. These are given in Section 5, Chapter VI.

Of course, the present regression analysis reflects the statistics on which it is based. Despite the large number of ships it will therefore contain systematic errors when applied to ships that do not reflect the characteristics of this group.

Other notable applications of regression analysis to model test results have been made by Sabit (1971), (1972), (1976) for the BMT methodical series tests, for the Series 60, and for the SSPA cargo liner series. Van Oortmerssen (1971), (1973) has derived statistical relations for small ships. A study carried out by Berger (1980) showed that no appreciable differences in accuracy occur among the results of using Gertler (1954), Guldhammer-Harvald (1974) or Holtrop-Mennen (1978).

An important recent contribution by Lin, Day and Lin (1987) describes the development of a statistical prediction technique based on results of tests on 255 models at DTRC. Although the ships in the data base are identified as naval auxiliaries, "they include designs for cargo ships, passenger liners and some amphibious types," both single and twin-screw. Statistical measures of prediction error are given for the new method in comparison with methods of Holtrop (1982, 1984) and Tagano (1974.)

Use of the statistically derived relations in a computer-aided design process forms a powerful tool in carrying out parametric studies for new designs, in order to investigate the effect of changes in size, proportions and fullness. Before such relations are put to use for such a purpose, however, one must make sure that the input data used to generate the relations are accurate and that the number of models or data points used is at least 3 times the number of terms in the resulting regression equation.

Fairlie-Clarke (1975) gives a survey of the statistical theory related to regression analysis and applies this to a few cases. He shows that using this theory, accurate equations can be obtained.

Section 9
High-Speed Craft and Advanced Marine Vehicles

Considerable research has been carried out during the last decade on the resistance of various types of high-speed craft and advanced marine vehicles. Such craft can be distinguished by the means adopted to support their weight: through buoyancy, through hydrodynamic lift, through aerostatic lift, or through combinations of these. Fig. 84 displays the sub-division that can be made in this way. The first category is composed of round-bilge mono-hulls and planing craft; the second of catamarans and Small Waterplane Area Twin Hull (SWATH) ships. The third category is composed of surface-piercing and submerged foil hydrofoil craft, while the fourth category is composed of air cushion vehicles (ACV) and surface effect ships (SES).

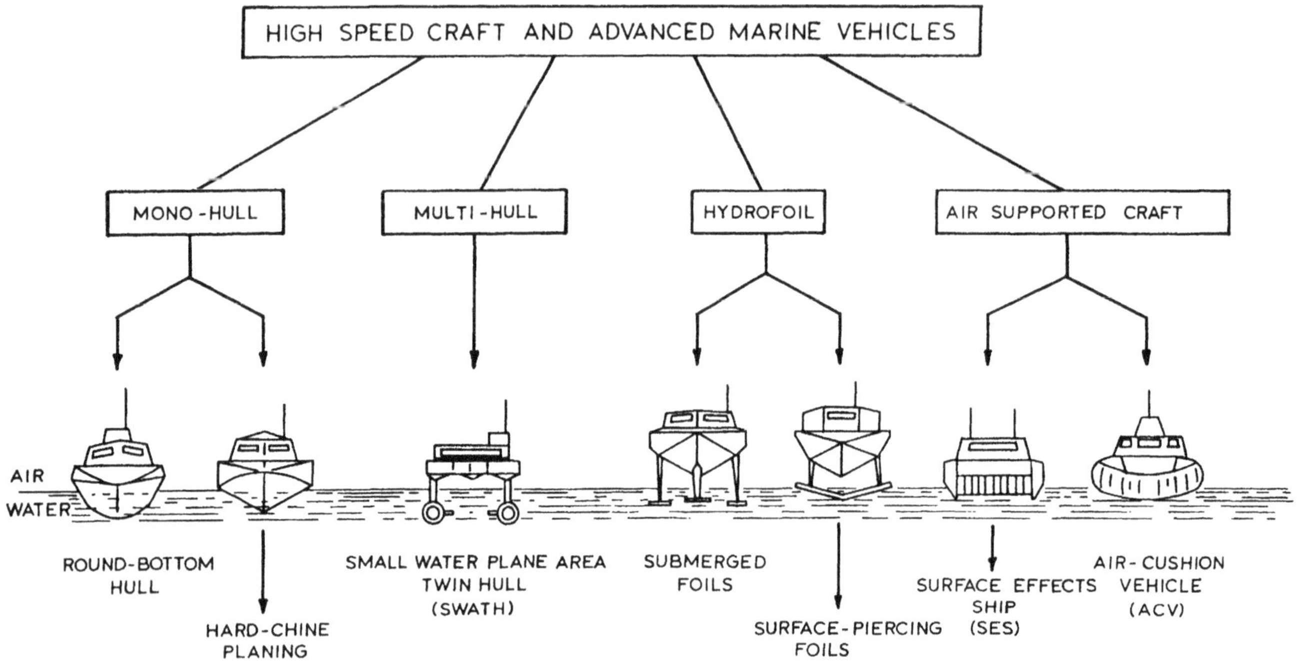

Fig. 84 Main types of high-speed craft and advanced marine vehicles

An overview of each of these craft types, including their application, technology and special attributes and limitations, is given in the Special Edition of the *Naval Engineers Journal*, "Modern Ships and Craft" (1985). A bibliography concerning model tests of high-speed marine vehicles may be found in Savitsky, et al (1984). This section will identify those published series data and analytical methods that can be used to make estimates of the resistance of high-speed craft.

9.1 Round-Bilge Semi-Displacement Craft. The geometry of these vessels is characterized by transom sterns, round bilges, a rise in after-body buttock lines and a maximum operational volume Froude number $Fn_\nabla \approx 2.5$. A significant amount of information is available on the resistance of high-speed, round-bilge craft. A summary of available data was recently presented by Van Oossanen (1982) and Muller-Graf (1980).

(*a*) *Nordstrom's Methodical Series.* Nordstrom (1951) published the results of tests carried out at the Royal Institute of Technology in Stockholm with 14 different round-bilge models, five of which were tested at more than one draft. Three of these models, each tested at three different drafts, form a small systematic series. The hull form is shown in Fig. 85, along with other series to be discussed here.

The results of resistance tests with these models, carried out in calm water, have been reanalyzed and included with the De Groot Series, following.

(*b*) *De Groot Series.* De Groot (1955) published the results of tests with 31 round-bilge, high-speed full forms, seven of which were tested at two or more drafts. These models were tested at MARIN. Four of these 31 models, each tested at four drafts, constitute a small systematic series. Tests with these four models were also carried out in the towing tank of the Delft University of Technology.

The results of resistance tests with all thirty-one models, carried out in calm water, were analyzed using the 1947 ATTC friction coefficients. De Groot derived a single graph showing the average resistance coefficients of 76 models (including the Nordstrom data) as a function of the displacement-length ratio for the speed range corresponding to $V_K/\sqrt{L} = 1.0$ to 3.5. Results are still often used as a standard to which the results of model resistance tests of fast, round-bilge hull forms are compared in order to arrive at a first approximation of the lines of a new design.

To facilitate preparing a resistance prediction using such average data, the total model resistance values have been converted at MARIN to values of R_R/W by using the average value for the wetted surface of the models given by $S = 2.75\sqrt{\nabla L}$, and the 1947 ATTC friction coefficients with $C_A = 0$. These residual resistance values given in Fig. 86 include the Nordenstrom (1951) reanalyzed data.

(*c*) *Series 63 Methodical Tests.* Results of resistance tests with models of five 15.24-m (50-ft) round-bilge utility boats were reported on by Beys (1963). These tests were carried out in the towing tank of the Davidson Laboratory of the Stevens Institute of Technology. These models form a methodical series in that the body plans of all five models are geometrically similar. The parent model has a nominal length-beam ratio of four. The hull form is shown in Fig. 85. All

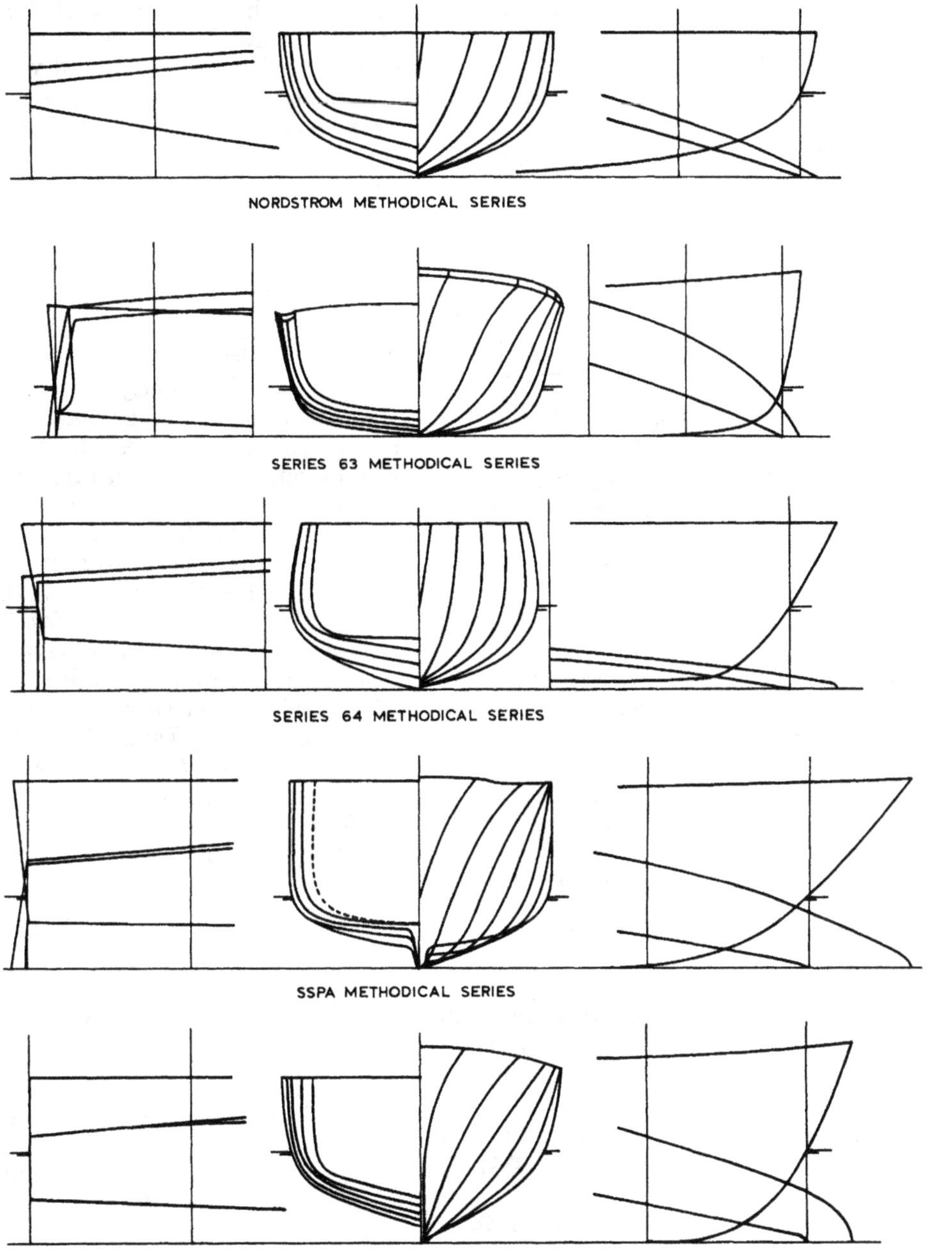

Fig. 85 Body plans, stem and stern outlines and design waterline of high-speed round-bilge methodical series

other models were derived from this parent model by multiplying the waterline and buttock spacings of the parent model by a constant so as to obtain nominal L/B values of 2.5, 3.0, 4.0, 5.0 and 6.0.

Results were analysed and plotted by van Oossanen (1982) in the same format as Fig. 86. The residual resistance values presented there were derived from the measured total resistance by means of the 1947 ATTC friction line with $C_\Delta = 0$.

(*d*) *Series 64 Methodical Tests.* The Taylor Standard Series models were run only up to a Froude number Fn = 0.60. The continually increasing speeds demanded of naval ships eventually made it desirable to explore the resistance to higher values of Fn, and in 1959 DTRC began a new series of methodical model experiments.

Series 64 (Yeh, 1965) consisted of low-wave-drag, displacement-type hulls, tested up to a speed corresponding to Fn = 1.50. Three parameters were chosen as the primary variables: block coefficient, C_B, length-displacement ratio, $L/\nabla^{1/3}$, and beam-to-draft ratio, B/T. The prismatic coefficient C_P was kept constant at

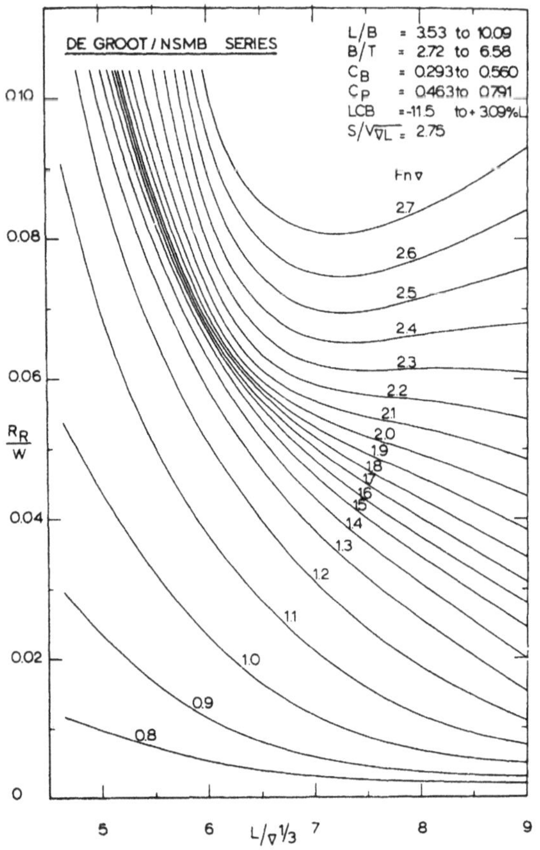

Fig. 86 Residual resistance-displacement weight ratio R_R/W derived from the averaged, total resistance of 76 models compiled by De Groot (1955)

0.63. The range of variables and other data are given in Tables 26 and 27. The models had a heavily raked stem, no bulb, fine entrance angles, and a transom stern with a round knuckle; see Fig. 85. The maximum area and maximum beam were at 60 and 70 percent of the length from the forward perpendicular, respectively, and the LCB was at 56.6 percent of the length from forward. Twenty-seven models were made, all 3.048 m in length, and towed without appendages or turbulence stimulation in the large tank at DTRC, which has a cross section 15.5 m × 6.7 m.

Table 26—Dimensions and Coefficients for Series 64 Models of $B/T = 3.0$

For all models:

$B/T = 3.0$ $\quad C_P = 0.63$

$LCB/L_{WL} = 0.566$ $\quad L_E/L = 0.60$ $\quad C_{WP} = 0.761$

C_B	C_M	$L/\nabla^{1/3}$	L/B	$\frac{S}{\sqrt{\nabla L}}$	i_E
0.55	0.873	8.04	9.762	2.622	6.7
		8.94	11.447		5.8
		10.45	14.479		4.5
0.45	0.714	8.59	9.762	2.675	6.7
		9.58	11.487		5.8
		11.26	14.643		4.5
0.35	0.556	9.35	9.762	2.882	6.7
		10.45	11.551		5.8
		12.40	14.913		4.5

Above a value of Fn = 0.90, the wave resistance is no longer an important factor, frictional resistance being dominant. At such high values of Fn it is therefore necessary to keep the wetted surface to a minimum. Due to the rather extreme type of hull forms in this series (see Fig. 85), the resistance results for the individual models are not often used or referred to. Average resistance values for the complete series, however, are frequently adopted for use in parametric studies for slender ships and other purposes. For this reason the average residual resistance-displacement weight ratio R_R/W, as a function of $L/\nabla^{1/3}$ and Fn_∇, is shown in Fig. 87. In preparing this figure the data for $L/\nabla^{1/3} = 11.26$ and $C_B = 0.45$ were not considered because of the "inconsistency" of these data with the results for the other models of the series.

(*e*) *SSPA Methodical Series.* Lindgren, et al (1968) presented the results of resistance tests with a methodical series of nine models of high-speed, round-bilge displacement vessels carried out at the Swedish State Shipbuilding Tank (SSPA). The hull form parameters varied were $L/\nabla^{1/3}$ (values of 6, 7 and 8) and B/T (values of 3.0, 3.5 and 4.0). The value of the block coefficient C_B was kept equal to 0.40 for all models, resulting in L/B values ranging from 4.62 to 8.20, since:

$$\frac{L}{B} = \sqrt{\frac{C_B\,(L/\nabla^{1/3})^3}{B/T}}$$

The speed range covered corresponds to a range in the Froude number Fn from 0.4 to 1.2 (equivalent to a range in $V_K/\sqrt{L}$ from 1.34 to 4.0).

The residuary resistance values of the models were obtained by using the 1957 ITTC frictional resistance coefficients. Up to an Fn-value equal to about 0.90, the results for the three B/T values are almost identical, leading to the observation that in the speed range between Fn = 0.4 and about 0.9 the length-displacement ratio is the only significant parameter. For this speed range the residuary resistance-displacement weight ratio R_R/W of this series is shown in Fig. 88, as a function of $L/\nabla^{1/3}$ and Fn_∇. The hull form of this series is also shown in Fig. 85.

(*f*) *NPL Methodical Series.* Very useful resistance data on high-speed, round-bilge displacement forms have been published by Marwood, et al (1969) and by Bailey (1976), based on the tests carried out at the Ship Division of the British National Physical Laboratory (now the National Maritime Institute) with a systematic series of 22 models in which L/B and B/T were varied. Five models were tested having a L/B value of 3.33 with B/T values ranging from 3.19 to 10.21,

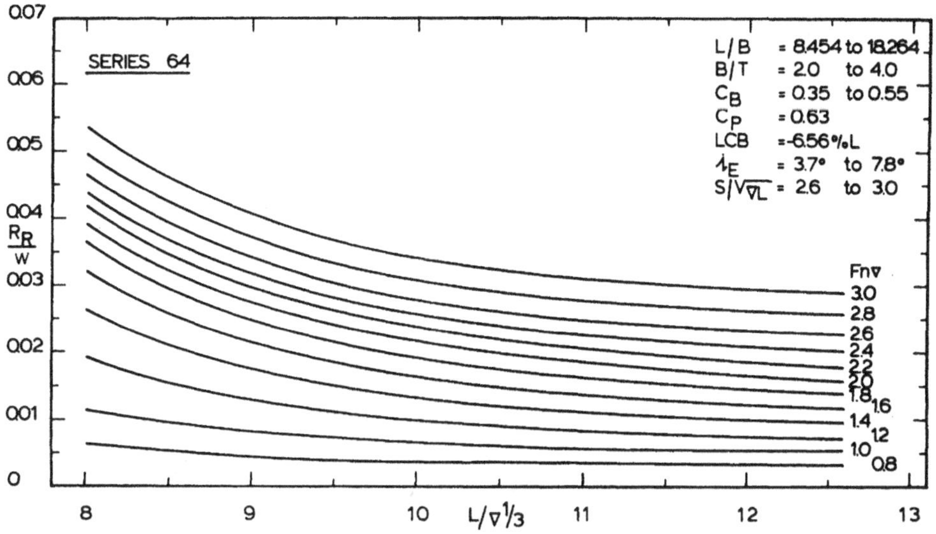

Fig. 87 Averaged value of the residual resistance-displacement weight ratio R_R/W of the Series 64 methodical models.

six with L/B equal to 4.54 with B/T values ranging from 1.72 to 6.87, four with L/B equal to 5.41 with B/T values ranging from 1.94 to 4.86, four with L/B equal to 6.25 with B/T values ranging from 1.93 to 5.80 and three models with $L/B = 7.50$ with B/T values ranging from 2.01 to 4.02. Other main hull form parameters were kept constant ($C_B = 0.397$, $C_P = 0.693$ and the longitudinal center of buoyancy LCB was positioned 6.4 percent L aft of the midship section). The speed range covered corresponds to values of the Froude number ranging from 0.3 to 1.20 (equivalent to a range in $V_K/\sqrt{L}$ from 1.0 to 4.0).

The residuary resistance values were calculated from the measured model resistance by subtracting the frictional resistance as determined by means of the 1957-ITTC skin friction formulation. The residuary resistance-displacement weight ratio was then plotted against $L/\nabla^{1/3}$ for various Fn_∇-values for each L/B value. These graphs are reproduced here as Fig. 89. The hull form is shown in Fig. 85.

(*g*) *Statistical Resistance Prediction Method Derived by Mercier and Savitsky.* Mercier, et al (1973) carried out a regression analysis of the resistance results obtained by Nordstrom (1951) for a small systematic series (nine models), by De Groot (1955) for a small systematic series (12 models), by Beys (1963) for the series 63 (21 models), by Yeh (1965) for the series 64 (27 models), by Lindgren, et al (1968) for the SSPA series (nine models), and by Marwood, et al (1969) for the NPL series (23 models). The results obtained by Clement, et al (1963) for the Series 62 hard-chine hull forms (17 models) were also incorporated in the data base.

Formulas were derived for the total resistance-displacement weight ratio R_T/W for eleven values of the volumetric Froude number, 1.0, 1.1–2.0, for a displacement weight of 100,000 lb (445.0 kN). Four parameters were selected for inclusion in the resistance equations as independent variables. These are the displacement-length ratio, $\nabla^{1/3}/L$, the beam-loading coefficient, $C_\nabla = \nabla/B^3$, the square root of the angle of entrance of the load water line in degrees $\sqrt{2i_E}$, and the ratio of transom area to maximum section area A_T/A_X.

Although the formulas derived by Mercier and Savitsky were originally intended for predicting the resistance of planing craft in the pre-planing, displacement mode, they can also be successfully used for predicting the resistance of displacement hulls,

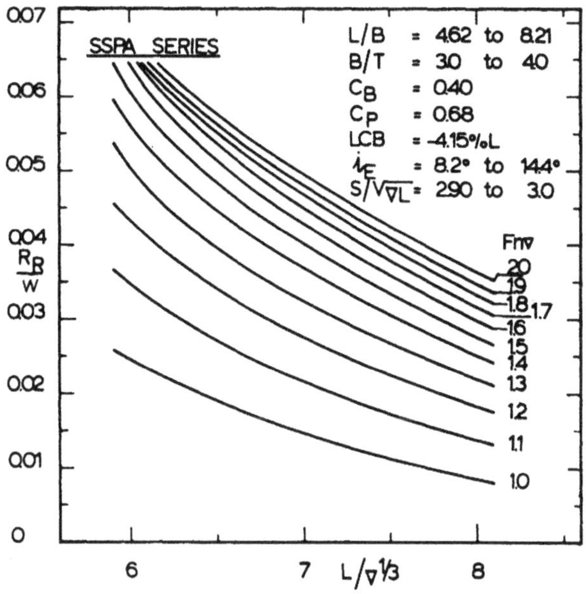

Fig. 88 Averaged value of the residual resistance-displacement weight ratio R_R/W of the SSPA methodical series.

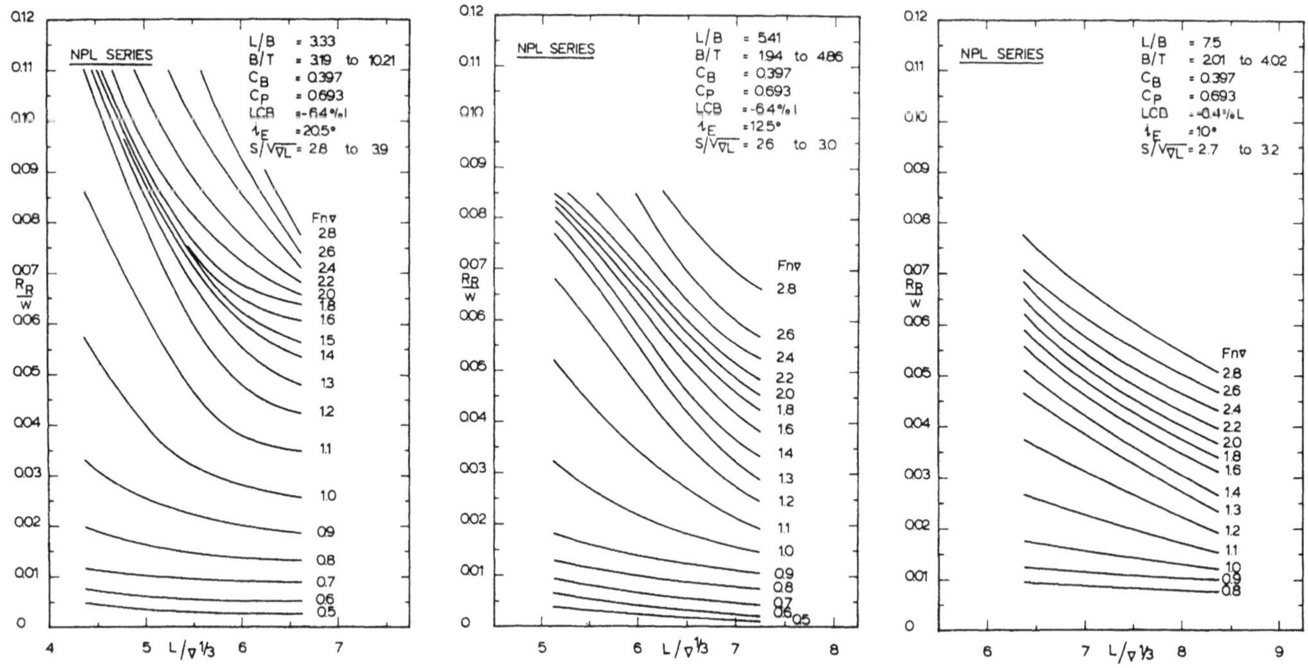

(a) $L/B = 3.33$ (b) $L/B = 5.41$ (c) $L/B = 7.5$

Fig. 89 Averaged value of the residual resistance-displacement weight ratio R_R/W of the NPL methodical series with L/3 equal to 7.5.

Table 27—Series 64 Residuary Resistance Coefficients C_R for $B/T = 3.0$

$$C_R = \frac{R_R}{\frac{1}{2}\rho s V^2}$$

C_B	0.55			0.45			0.35		
	$L/\nabla^{1/3}$			$L/\nabla^{1/3}$			$L/\nabla^{1/3}$		
Fn	8.04	8.94	10.45	8.59	9.58	11.26	9.35	10.45	12.40
0.060	3.155	2.775	2.340	3.419	3.017	3.847	1.799	2.129	1.374
0.119	2.760	1.387	1.462	2.350	2.011	1.602	1.124	1.064	1.030
0.179	2.629	1.233	1.300	2.278	1.788	1.424	1.499	1.300	1.221
0.238	2.711	1.561	1.316	2.457	1.571	1.122	1.574	1.131	1.202
0.297	2.524	2.072	1.497	2.564	1.368	.974	1.726	1.192	1.099
0.357	2.914	2.543	1.657	2.872	1.648	.926	1.798	1.212	1.030
0.416	3.219	2.624	1.600	3.069	1.785	.889	1.835	1.194	.953
0.476	3.228	2.601	1.535	3.004	1.728	.841	1.812	1.147	.858
0.535	3.067	2.409	1.430	2.774	1.614	.760	1.710	1.077	.780
0.595	2.823	2.127	1.298	2.487	1.508	.718	1.546	1.000	.714
0.654	2.542	1.880	1.179	2.210	1.379	.678	1.397	.915	.704
0.714	2.267	1.670	1.080	1.970	1.250	.623	1.261	.850	.687
0.773	2.025	1.510	.983	1.765	1.130	.561	1.149	.800	.683
0.833	1.819	1.387	.907	1.596	1.026	.510	1.069	.765	.687
0.892	1.647	1.299	.853	1.466	.943	.467	1.011	.742	.696
0.952	1.528	1.236	.827	1.368	.888	.461	.969	.740	.692
1.011	1.441	1.197	.814	1.313	.863	.456	.949	.748	.694
1.071	1.375	1.179	.820	1.274	.850	.459	.938	.755	.704
1.130	1.333	1.168	.830	1.250	.850	.476	.932	.766	.715
1.190	1.301	1.158	.842	1.231	.857	.497	.928	.779	.721
1.249	1.298	1.162	.849	1.221	.869	.520	.930	.788	.729
1.309	1.343	1.160	.853	1.216	.881	.546	.934	.796	.738
1.368	1.421	1.156	.858	1.212	.891	.572	.937	.805	.743
1.428	1.524	1.206	.859	1.209	.899	.601	.935	.809	.746
1.487	1.628	1.258	.861	1.209	.907	.628	.932	.812	.749

Table 28—Values of the Coefficients in Equation (88) as a Function of Fn_∇ for W = 100,000 lb in Sea Water at 15 °C using the 1947 ATTC Friction Coefficients with C_A = 0.

	$Fn_\nabla = 1.0$	$Fn_\nabla = 1.1$	$Fn_\nabla = 1.2$	$Fn_\nabla = 1.3$	$Fn_\nabla = 1.4$	$Fn_\nabla = 1.5$
A_1	0.06473	0.10776	0.09483	0.03475	0.03013	0.03163
A_2	−0.48680	−0.88787	−0.63720	0.0	0.0	0.0
A_4	−0.01030	−0.01634	−0.01540	−0.00978	−0.00664	0.0
A_5	−0.06490	−0.13444	−0.13580	−0.05097	−0.05540	−0.10543
A_6	0.0	0.0	−0.16046	−0.21880	−0.19359	−0.20540
A_7	0.10628	0.18186	0.16803	0.10434	0.09612	0.06007
A_8	0.97310	1.83080	1.55972	0.43510	0.51820	0.58230
A_9	−0.00272	−0.00389	−0.00309	−0.00198	−0.00215	−0.00372
A_{10}	0.01089	0.01467	0.03481	0.04113	0.03901	0.04794
A_{15}	0.0	0.0	0.0	0.0	0.0	0.08317
A_{18}	−1.40962	−2.46696	−2.15556	−0.92663	−0.95276	−0.70895
A_{19}	0.29136	0.47305	1.02992	1.06392	0.97757	1.19737
A_{24}	0.02971	0.05877	0.05198	0.02209	0.02413	0.0
A_{27}	−0.00150	−0.00356	−0.00303	−0.00105	−0.00140	0.0

	$Fn_\nabla = 1.6$	$Fn_\nabla = 1.7$	$Fn_\nabla = 1.8$	$Fn_\nabla = 1.9$	$Fn_\nabla = 2.0$
A_1	0.03194	0.04343	0.05036	0.05612	0.05967
A_2	0.0	0.0	0.0	0.0	0.0
A_4	0.0	0.0	0.0	0.0	0.0
A_5	−0.08599	−0.13289	−0.15597	−0.18661	−0.19758
A_6	−0.19442	−0.18062	−0.17813	−0.18288	0.20152
A_7	0.06191	0.05487	0.05099	0.04744	0.04645
A_8	0.52049	0.78195	0.92859	1.18569	1.30026
A_9	−0.00360	−0.00332	−0.00308	−0.00244	−0.00212
A_{10}	0.04436	0.04187	0.04111	0.04124	0.04343
A_{15}	0.07366	0.12147	0.14928	0.18090	0.19769
A_{18}	−0.72057	−0.95929	−1.12178	−1.38644	−1.55127
A_{19}	1.18119	−1.01562	0.93144	0.78414	0.78282
A_{24}	0.0	0.0	0.0	0.0	0.0
A_{27}	0.0	0.0	0.0	0.0	0.0

since 101 of the 118 sets of model data used are those of round-bilge, displacement hulls.

The general form of the resistance equation adopted by Mercier and Savitsky (1973) is as follows:

$$R_T/W = A_1 + A_2X + A_4U + A_5W + A_6XZ + A_7XU + A_8XW + A_9ZU + A_{10}ZW + A_{15}W^2 + A_{18}XW^2 + A_{19}ZX^2 + A_{24}UW^2 + A_{27}WU^2 \quad (68)$$

where $X = \nabla^{1/3}/L$, $Z = \nabla/B^3$, $U = \sqrt{2i_E}$ and $W = A_T/A_X$.

The values of the coefficients A_1 to A_{27} are given in Table 28 for the eleven values of the Froude number and a displacement weight of 100,000 lb (445.0 kN).

In using Equation (68) it is essential to remain within the range of values of the independent variables used in the data base. Gross errors can occur otherwise. For other displacement values, water temperatures, friction coefficients or C_A-values, Equation (68) can be corrected according to the following expression:

$$(R_T/W)_{\text{corr}} = (R_T/W)_{\text{Eq.}} + \left(C'_f - C_{F_{\text{Eq.}}} + C_A\right)\frac{1}{2}\frac{S}{\nabla^{2/3}}\text{Fn}_\nabla^2 \quad (69)$$

where $(R_T/W)_{\text{corr}}$ is the corrected value of R_T/W, $(R_T/W)_{\text{Eq.}}$ is the value of R_T/W according to Equation (88), C_F' is friction coefficient for alternative displacement, water temperature or friction formulation, $C_{F_{\text{Eq.}}}$ is friction coefficient according to the 1947 ATTC friction formulation, C_A is appropriate value of the model-ship correlation factor, S is wetted surface.

An analysis of the still water values of the wetted surface of the models comprising the data base resulted in the following formula, with an accuracy of ±9 percent for 95 percent of the cases comprising the data base.

$$S/\nabla^{2/3} = 2.262\sqrt{L/\nabla^{1/3}}\left[1 + 0.046\,B/T + 0.00287\,(B/T)^2\right] \quad (70)$$

9.2 Planing Craft. The planing hull evolved to overcome the inherent hydrodynamic limitations associated with high-speed operation of the traditional displacement hull. The interest in planing craft was given great impetus by the development of high-power light-weight engines which are an essential requirement for this hull form.

Briefly, the lines of traditional displacement ships have longitudinal and transverse curvature in order to minimize flow separation at the stern and bilges and

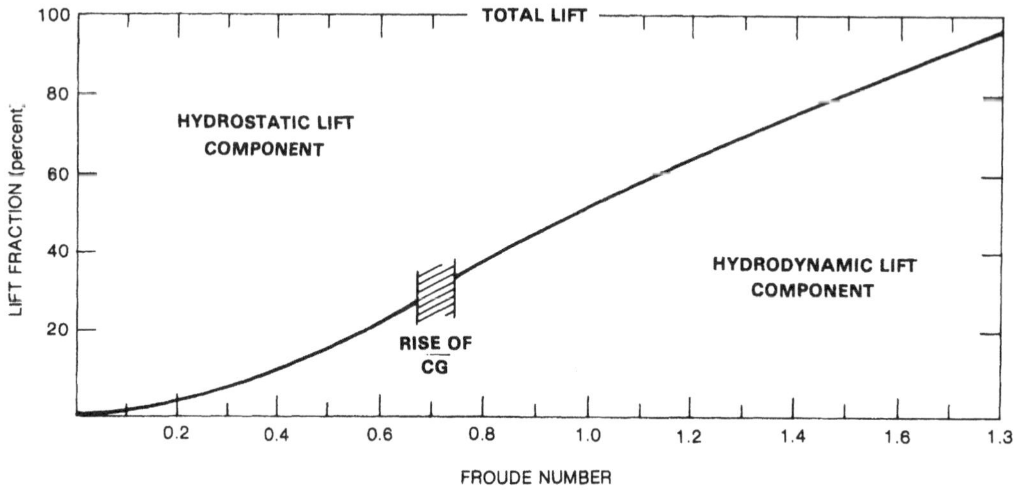

Fig. 90 Distribution of hydrostatic and hydrodynamic lift

thus reduce drag. When run at high speeds, the dynamic pressure on these convex surfaces is negative, resulting in large trim by the stern, and substantial increases in resistance.

In contrast, the planing hull form is configured to develop positive dynamic pressures so that its draft decreases with increasing speed, enabling it to run higher and higher on the wave it is generating, thus avoiding the large drag increases experienced by displacement hulls when run at high speed. To attain positive dynamic pressures the planing hull avoids convex curvature of both the buttocks and transverse sections. Whereas, in the displacement hull, all means are taken to reduce flow separation, in a planing boat the straight buttock lines are cut off cleanly by the transom stern so as to induce early flow separation. The transverse section is typically a deadrise section with sharp intersection of the bottom and sides to form a hard chine from which the flow will also separate. Fig. 90 shows the relation between hydrostatic and hydrodynamic lift versus Fn for a typical planing hull.

The *step,* sometimes adopted in planing hulls, is a sharp discontinuity across the bottom of the boat, either straight or V-shaped in plan form. The idea is that the boat will plane on two small areas, one just ahead of the step, one just ahead of the transom. By this means the total wetted surface may be reduced by 60 percent or more, but owing to the greater complexities of such designs practically all boats, other than racing craft, now have stepless hulls.

Small, high-speed planing craft became possible in the 1930's as a result of the development of power plants with a high power-weight ratio. The size of planing craft has steadily increased from the 17 m (55 ft) long Coastal Motor Boat in England with a displacement volume of some 14 m^3 and a speed of 46 knots to the Russian *Nanuchka* class missile corvette displacing nearly 1000 m^3 (35,280 ft^3). An excellent summary of the development of the planing hull form has been given by Savitsky and Gore (1979) and Savitsky (1964).

Information on resistance of planing hulls can be found in SNAME Bulletin 1-23, providing the data included in the "Small Craft Data Sheets" issued by the Society and explaining how they can be used in design (Clement, 1963). Whereas the resistance of round-bilge craft depends mainly on one parameter, $L/\nabla^{1/3}$, that of high-speed planing boats depends on several, including LCG location and the relation between projected planing bottom area and volume of displacement at rest, $A_P/\nabla^{2/3}$. Consequently, the practice of testing at standard values of these parameters is usually adhered to, and these conditions have also been used, where possible, in the SNAME sheets. They are: $A_P/\nabla^{2/3} = 7.0$ and LCG 6 percent L_P abaft the centroid of A_P. Here L_P is defined to be the projected chine length.

The principal parameters affecting the performance of planing hulls, including the above, are:

(*a*) Length-beam ratio, L_P/B_{PA}, where the mean beam over chines $B_{PA} = A_P/L_P$.

(*b*) Size-weight ratio, defined by the coefficient $A_P/\nabla^{2/3}$.

(*c*) Longitudinal position of CG from center of area of A_P.

(*d*) Deadrise and its variation along the length.

(*e*) Longitudinal curvature of buttock line $B_{PA}/4$ from CL.

(*f*) Shape of chine in plan.

(*g*) Shape of sections.

The SNAME data sheets list the foregoing particulars, together with hull lines, coefficients of form, loading, measured model results, dimensionless performance characteristics and P_E and V_K in knots for a boat of gross weight $W = 44.5$ kN (10,000 lb).

Since displacement is the most important factor in the design of a boat, it is desirable to compare hull

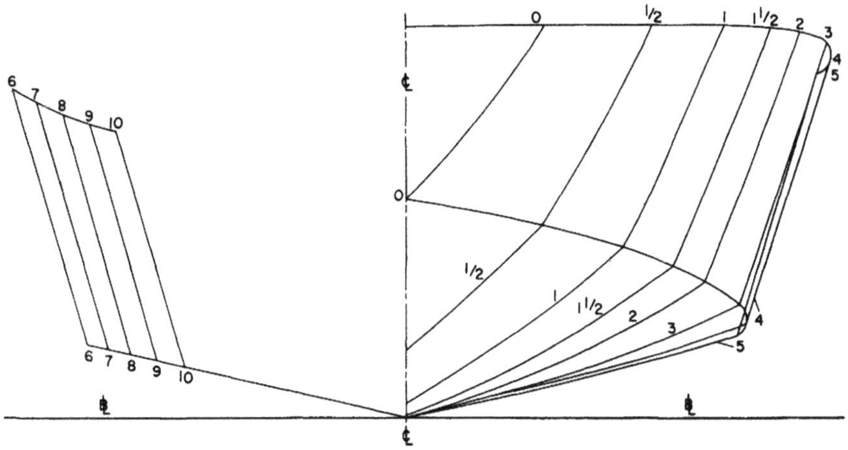

Fig. 91 Lines of stepless planing hull, Series 62

forms on a basis of equal W or ∇ by using the dimensionless ratios

$$V/\sqrt{g\nabla^{1/3}},\ R/W,\ S/\nabla^{2/3},\ L/\nabla^{1/3} \text{ and } CG_{rise}/\nabla^{1/3}$$

Wetted surface and trim angle are included in the data, because frictional resistance varies with wetted surface and wave-making resistance with trim angle.

In cases where rough-water performance must be considered in the design of small high-speed craft, this consideration may require modification of hull dimensions and characteristics selected on the basis of minimum resistance in smooth water (McGown, 1961).

(*a*) *Series 62 Methodical Tests.* Results of tests on a systematic series of models of planing hulls made at the DTRC have been given by Clement, et al (1963). Typical lines of these Series 62 models are shown in Fig. 91. The parent form was based upon an analysis of previous designs, including the features desirable for good steering qualities and good rough-water performance. Tests showed that the design had less resistance than any of the conventional stepless designs previously tested there.

The range of parameters covered was:

$A_P/\nabla^{2/3} = 4.0$ to 8.5
$L_P/B_{PX} = 2.0$ to 7.0
$L_P/B_{PA} = 2.36$ to 8.56
Centroid of A_P, as percent
L_P forward of transom = 47.5 to 48.8

The results for five models of Series 62 are shown in Fig. 92, where R/W and trim angle α are plotted against Fn_∇. These curves indicate the important effects of L_P/B_{PX} ratio on resistance—the form with the highest value of this ratio had the least drag both at low and very high speeds.

The high-speed data from the tests of Series 62 were found to collapse into a single graph when plotted in the form of contours C_{Lb} against R/W and l_{cp}/b, where

R is total resistance
l_{cp} is distance of center of gravity (or pressure) forward of transom
b is breadth over spray strips at longitudinal position of CG
C_{Lb} is lift coefficient
$= W/\tfrac{1}{2}\rho V^2 b^2$ in consistent units

This graph is shown in Fig. 93 for a displacement

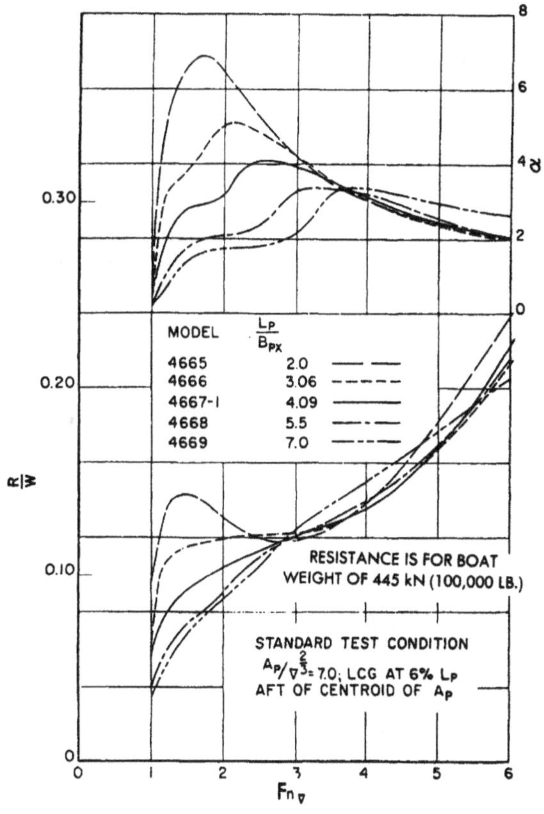

Fig. 92 Resistance/weight ratio and trim angle versus speed coefficient for five models of Series 62

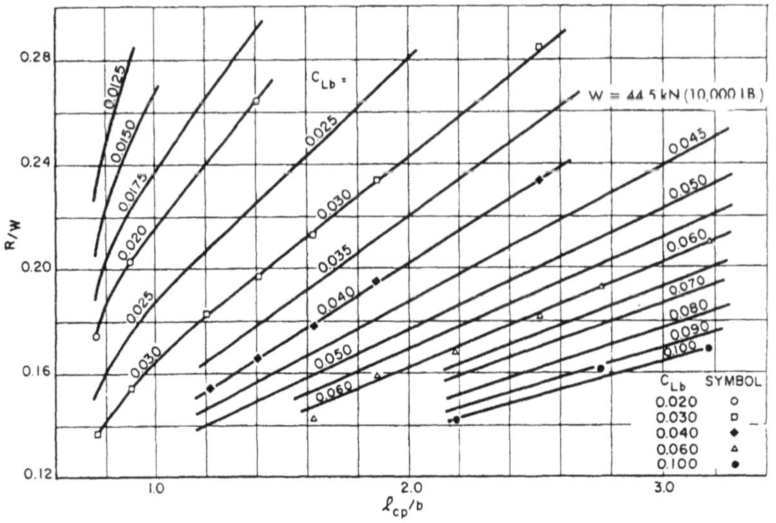

Fig. 93 *R/W* versus ℓ_{cp}/b with C_{Lb} as parameter, Series 62

weight of 44.5 kN (10,000 lb).

(*b*) *Series 65 methodical tests.* To study the resistance characteristics of planing hulls particularly suitable for hydrofoil craft, the Series 65 planing hulls were designed and tested on a model scale at DTRC. The results of these tests were presented by Holling, et al (1974) and by Hadler, et al (1974). The Series 65 planing hulls are based on the AG(EH) hydrofoil hull. This hull has moderate deadrise, fine bow entrance for easy wave penetration, hard chines for effective flow separation and minimal convex curvature to avoid the development of negative bottom pressures. The geometric characteristics of the resulting hulls are shown in Fig. 94 and compared for $W = 10{,}000$ lb with those of Series 62 in Table 29 (for other displacements see Hubble, 1974). The Series 65 hull form lends itself easily to tandem and canard foil configurations (Section 9.6) by terminating the lines at 70 and 50 percent of the length, respectively, and increasing the station spacing. As a consequence of the need to test both types of hulls for these hydrofoil applications, seven models were developed around the airplane (full length) type configuration and designated Series 65-A and nine models around the canard configuration which are designated Series 65-B. Systematic variations in length-beam ratio, beam-draft ratio, and deadrise were developed for each configuration by multiplying the parent length, beam, and/or draft by factors of 1, $1/\sqrt{2}$, $\sqrt{2}$, or 2.

The results of the Series 62 and Series 65 model tests show that the LCG location has a marked effect on the resistance-displacement weight ratio R/W. The minimum R/W for values of $Fn_\nabla < 2.5$ usually occurs when the LCG location results in an initial trim value (for zero speed) close to zero. For values of $Fn_\nabla > 3.0$ the minimum drag occurs as the LCG is shifted aft. Results also show that at the lower speeds, $Fn_\nabla \leq 2.0$ there is a tendency for the data to collapse into a very narrow band showing that the slenderness ratio, $L/\nabla^{1/3}$ is the dominant characteristic in this speed range overshadowing other parameters.

(*c*) *Theoretical approach to planing craft design.* Savitsky (1964) has given formulas for the lift and drag forces on planing hulls. These formulas are based on a large number of resistance tests with prismatic, or wedge-type surfaces, in which the trim angle, deadrise angle, wetted length and length-beam ratio were varied systematically.

Hadler (1966) presented a method to predict the performance of a planing hull. He used Savitsky's formulas to calculate the hydrodynamic forces on the hull and an open water diagram to evaluate the propeller forces. The solution of the three equations of equilibrium (sum of forces in X- and Z-direction and the sum of moments must be equal to zero) yields the three unknowns: trim angle, wetted length and rate of revolutions. The following approach is followed.

Given the speed V, the maximum beam over the chines or spray strips b, the displacement volume at rest ∇, the following values are calculated:

- the displacement Froude number, $Fn_\nabla = V/\sqrt{g\nabla^{1/3}}$
- the Froude number based on b: $C_v = V/\sqrt{gb}$
- the equivalent flat plate lift coefficient, $C_{Lb} = \dfrac{\rho g \nabla}{\frac{1}{2}\rho V^2 b^2}$

The following Savitsky formula can then be applied to determine the trim angle for equilibrium:

$$C_{Lb} = \tau^{1.1}\,(0.0120\sqrt{\lambda} + .0055\lambda^{5/2}/C_V^2) \qquad (71)$$

where τ is trim angle, deg

λ is mean wetted length-beam ratio, L_m/b

V is speed, m/sec

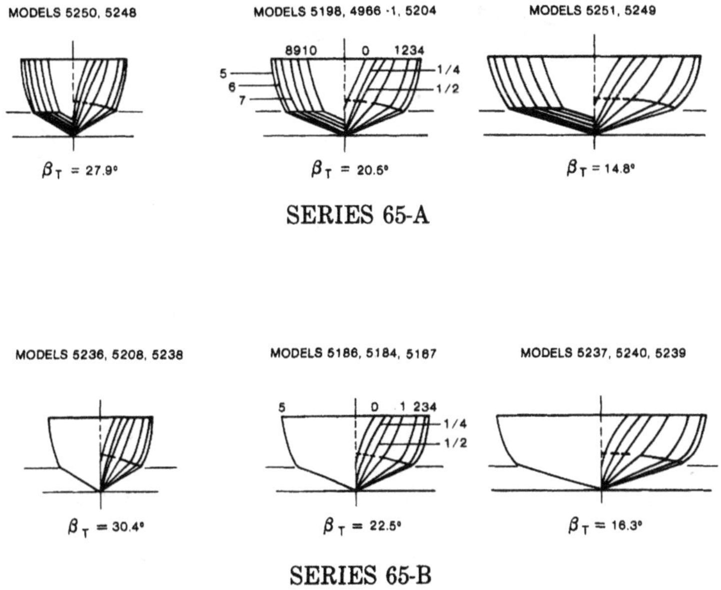

Fig. 94. Body plans of series 65 hulls

Table 29—Hull Characteristics of Series 62, Series 65 A and Series 65 B. (W = 10,000 lb = 44.5 kN)

MODEL	L_P	A_P	B_{PA}	B_{PT}	B_{PX}	$\frac{L_P}{B_{PA}}$	$\frac{L_P}{B_{PX}}$	$\frac{B_{PX}}{B_{PA}}$	$\frac{B_{PT}}{B_{PX}}$	$\bar{x}_p$	β (deg)	β_T (deg)	LCG/L_P for $\tau_o \simeq 0°$
SERIES 65-A													
5251	1.861	0.761	0.409	0.198	0.564	4.55	3.30	1.38	0.35	0.151	16.0	14.8	0.52
5249	2.632	1.076	0.409	0.198	0.564	6.44	4.66	1.38	0.35	0.151	16.0	14.8	0.52
5198	2.226	0.769	0.346	0.167	0.477	6.44	4.66	1.38	0.35	0.151	22.1	20.5	0.52
4966-1	2.632	0.761	0.284	0.140	0.399	9.10	6.60	1.38	0.35	0.151	22.1	20.5	0.52
5204	2.632	0.538	0.205	0.099	0.282	12.87	9.34	1.38	0.35	0.151	22.1	20.5	0.52
5250	2.632	0.761	0.289	0.140	0.399	9.10	6.60	1.38	0.35	0.151	29.9	27.9	0.52
5248	2.632	0.538	0.205	0.099	0.282	12.87	9.34	1.38	0.35	0.151	29.9	27.9	0.52
SERIES 65-B													SERIES 65-B
5237	1.872	1.101	0.588	0.792	0.798	3.18	2.35	1.36	0.99	0.121	21.2	16.3	0.38
5240	1.872	0.779	0.416	0.560	0.564	4.50	3.32	1.36	0.99	0.121	21.2	16.3	0.38
5239	2.648	1.101	0.416	0.560	0.564	6.36	4.69	1.36	0.99	0.121	21.2	16.3	0.38
5186	1.872	0.779	0.416	0.560	0.564	4.50	3.32	1.36	0.99	0.121	28.7	22.5	0.38
5184	1.872	0.550	0.294	0.396	0.399	6.36	4.69	1.36	0.99	0.121	28.7	22.5	0.38
5167	2.648	0.779	0.294	0.396	0.399	9.00	6.64	1.36	0.99	0.121	28.7	22.5	0.38
5236	1.872	0.550	0.294	0.396	0.399	6.36	4.69	1.36	0.99	0.121	37.4	30.4	0.38
5208	2.648	0.779	0.294	0.396	0.399	9.00	6.64	1.36	0.99	0.121	37.4	30.4	0.38
4238	2.648	0.550	0.208	0.280	0.282	12.73	9.38	1.36	0.99	0.121	37.4	30.4	0.38
SERIES 62													SERIES 62
4665	1.192	0.601	0.504	0.477	0.596	2.36	2.00	1.18	0.80	0.145	13.0	12.5	0.41
4666	1.825	0.903	0.495	0.422	0.596	3.69	3.06	1.21	0.71	0.147	13.0	12.5	0.42
4667-1	2.438	1.189	0.488	0.381	0.596	5.00	4.09	1.22	0.64	0.149	13.0	12.5	0.43
4668	2.438	0.884	0.363	0.285	0.443	6.72	5.50	1.22	0.64	0.149	13.0	12.5	0.43
4669	2.438	0.695	0.285	0.224	0.348	8.56	7.00	1.22	0.64	0.149	13.0	12.5	0.43

b is beam of planing area, m
g is acceleration of gravity, m/sec^2

The first term in this equation represents the dynamic component of lift while the second term is the buoyant component of lift. At speed coefficients $C_V >$ 10, there is little buoyant lift, so that, all other conditions being equal, lift varies as the speed squared.

The lift coefficient for a finite deadrise $C_{L\beta}$ can then be calculated from $C_{L\beta} = C_{Lb} - 0.0065\beta C_{Lb}^{0.6}$ where β is the deadrise angle at the mid-chine position (deg.)

Savitsky (1964) gives a formula for locating the distance, p, of the center of pressure forward of the transom. However, in many cases it may be assumed that the resultant normal force on the planing bottom, N, passes through the CG, i.e., p = LCG (Fig. 95). With appropriate values of C_V and p/b = LCG/b, the corresponding λ and $C_{Lb}/\tau^{1.1}$ are then read off the

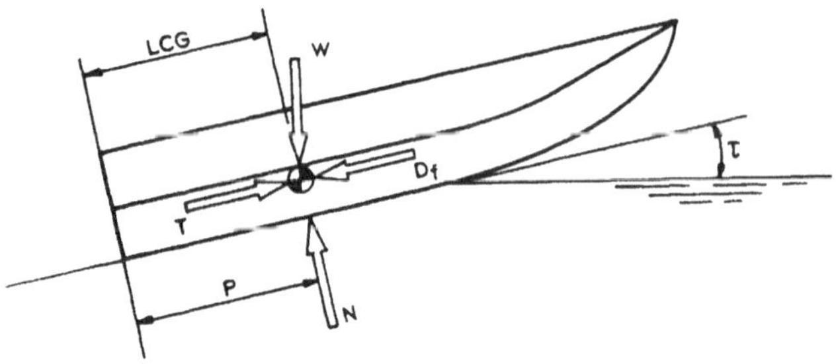

Fig. 95 Resultant normal force on planing bottom usually passes through CG

nomograph, Fig. 96. (This graph by Koelbel is valid when the propeller thrust, the resistance force and the resultant of the planing force all act through the CG) Hence, the mean wetted length, L_m, and trim angle, τ, can be determined.

Savitsky also gives a formula to correct the mean wetted length ratio, λ, to the keel wetted length ratio, λ_k if desired,

$$\lambda_k = \lambda - 0.03 + \tfrac{1}{2}\left(0.57 + \frac{\beta}{1000}\right)\left(\frac{\tan\beta}{2\tan\tau} - \frac{\beta}{167}\right) \tag{72}$$

where the value of β should be taken at the mid-chine-length position.

The value of λ_k should now be compared with the value of L_{WL}/b. If $\lambda_k \geq$, L_{WL}/b then the bow is not clear of the water and the craft is not fully planing. In that case the resistance may be calculated with the help of the formulas of Section 8.12. When $\lambda_k \leq L_{WL}/b$, the bow is essentially clear of the water and the resistance can be predicted from the following equation:

$$R_T = W\tan\tau + \tfrac{1}{2}\rho V^2\lambda b^2 C_{FO}/(\cos\tau\cos\beta) \tag{73}$$

where C_{FO} is the friction coefficient according to the ITTC 1957 or the ATTC friction lines as a function of the Reynolds number, $\text{Rn}_b = \dfrac{V_1\lambda b}{\nu}$

Here V_1 is the average bottom velocity, which is less than the forward planing velocity V owing to the fact that the planing bottom pressure is larger than the free-stream pressure. The following equation applies for zero deadrise:

$$V_1 = V\left(1 - \frac{0.0120\tau^{1.1}}{\sqrt{\lambda}\cos\tau}\right)^{1/2} \tag{74}$$

For other deadrise angles Savitsky (1964) gives con-

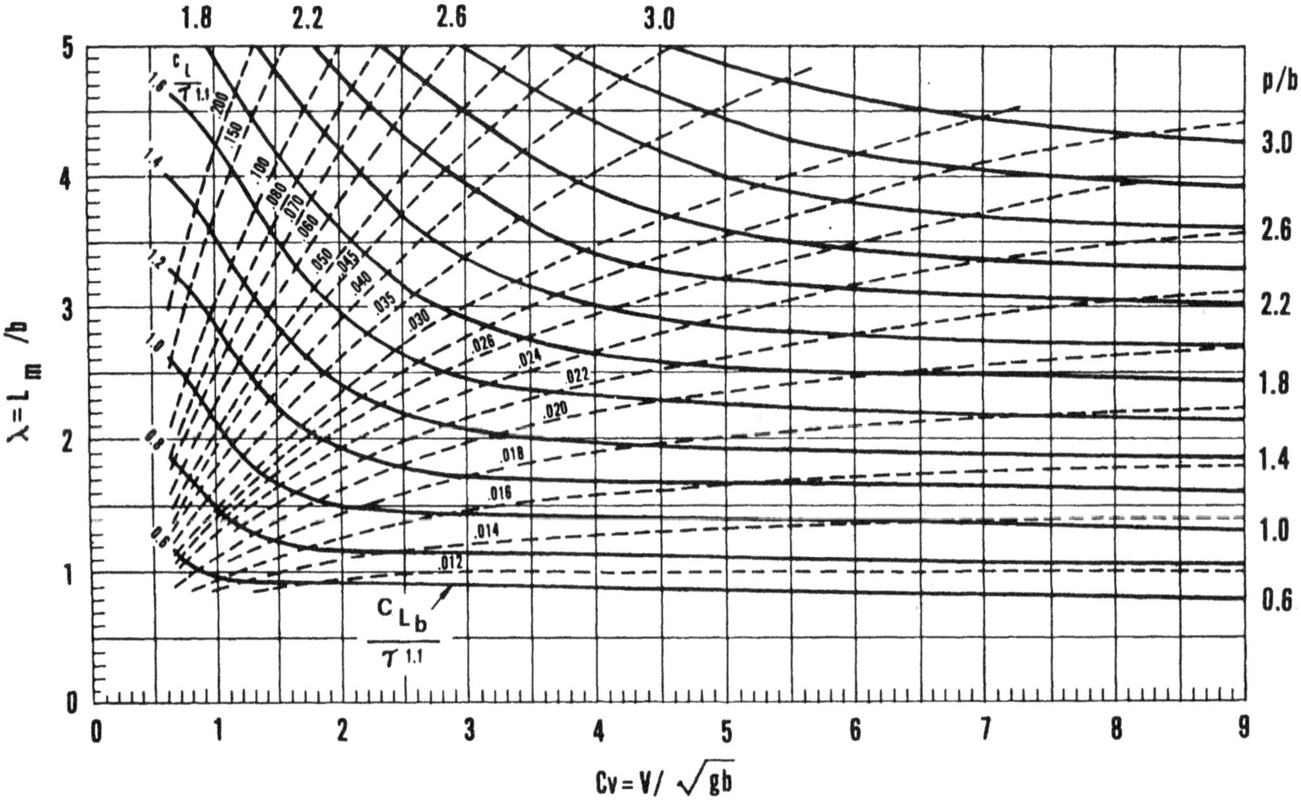

Fig. 96 Nomogram for equilibrium conditions when all forces act through CG: λ and $C_{L_b}/\tau^{1.1}$ for given values of C_v and p/b (Koelbel)

venient graphs for the average bottom velocity.

The first term of Equation (73) is the total pressure drag, while the second term is the frictional drag in the pressure area. There is additional frictional drag in the spray area.

When the propeller thrust is not parallel to the viscous force vector and the lift force does not pass through the center of gravity (Fig. 95) the detailed procedure given by Savitsky (1964) should be used.

(d) Example: consider a hard-chine hull with the following particulars:

L_{WL} = 25 m (82 ft)
b = 7.5 m (24.6 ft) (maximum chine beam)
β = 15 deg (deadrise at mid-chine length)
∇ = 90 m^3 (3,175 cu ft) (displacement volume)
LCG = 10 m (32.8 ft) from the transom.

For volumetric Froude number values of 1.5, 2.0, 3.0 and 4.0, the following values are calculated for C_V, C_{Lb} and $C_{L\beta}$.

Fn_∇	$V (m/s)$	V_k (knots)	C_v	C_{Lb}	$C_{L\beta}$
1.5	9.946	19.335	1.160	0.317	0.268
2.0	13.261	25.780	1.546	0.179	0.144
3.0	19.891	38.668	2.319	0.079	0.058
4.0	26.522	51.559	3.092	0.045	0.030

If all the forces on the hull pass through CG then the graph of Fig. 97 can be used. The value of p/b is

$$p/b = \text{LCG}/b = 10/7.5 = 1.333$$

From Fig. 97 the following values for λ and $C_{Lb}/\tau^{1.1}$ are found from which τ can be calculated. Subsequently, R_T may be computed ($W = \rho g \nabla = 905$ kN):

Fn_∇	C_V	p/b	λ	$C_L/\tau^{1.1}$	τ (deg)	V_1	Rn (*10^{-8})	C_{FO}	R_T (kN)
1.5	1.160	1.333	3.25	0.100	2.85	9.8	2.02	0.001886	63.2
2.0	1.546	1.333	2.60	0.044	2.93	13.1	2.15	0.001870	71.9
3.0	2.319	1.333	2.15	0.024	2.23	19.7	2.67	0.001816	81.4
4.0	3.092	1.333	1.85	0.019	1.51	26.3	3.08	0.001781	93.1

In the calculations the kinematic viscosity was taken as 1.188×10^{-6} m^2/s (Table 10, Section 6.4).

(e) Controllable transom flaps. Transom flaps have become accepted as a means of controlling the trim of power boats so as to optimize their performance. A study of flap effectiveness by Savitsky and Brown (1976) resulted in simple expressions for the increase in lift, drag and moment due to flaps and for the flap hinge moment. The following equations provide the flap-induced forces and moments. The lift increment δF (in N) due to the flap deflection is given by:

$$\delta F = 0.205\, L_F\, d\, \sigma\, b \frac{\rho}{2} V^2 \quad (75)$$

Here the symbols denote:

L_F is flap chord, m
σ is flap span-beam ratio
d is flap deflection, deg.

The drag increment due to the flap is:

$$D_F = 0.0052\, \delta F\, (\tau + d) \quad (76)$$

The flap moment about the trailing edge of the flap may be represented by:

$$M_F = 0.305\, \delta F\, [0.6\, b + L_F\, (1 - \sigma)]$$

whereas the flap hinge moment is:

$$H_F = 0.0424\, \delta F L_F$$

The flap effects are combined with the hull hydrodynamic equations to obtain the equilibrium conditions for a planing hull with transom flaps. A detailed procedure for this calculation is presented by Savitsky and Brown (1976).

9.3 Catamarans The catamaran or twin-hull concept has been employed in high-speed craft design for several decades, and both sailing as well as powered catamarans are in use. For commercial purposes semi-planing type catamarans are predominant. The component hulls (demihulls) are of the planing type, featuring V-type sections and a cut-off transom stern.

The division of displacement and waterplane area between two relatively slender hulls results in a large deck area, good stability qualities and consequently a small rate and angle of roll. Yermotayev, et al (1977) found that seakeeping qualities in terms of angle and rate of pitch are poor compared to a conventional hull. Active control of pitching motions by means of fins may eliminate this problem.

The resistance of a catamaran is mainly affected by the wetted surface ratio ($S_w/\nabla^{2/3}$), the slenderness ratio ($L/\nabla^{1/3}$), and the hull spacing (s/L). The wetted surface ratio is relatively high compared with planing monohulls of the same displacement. Consequently, catamarans show poor performance at low speeds (Fn $\leq$ 0.35) where skin friction is predominant. At higher

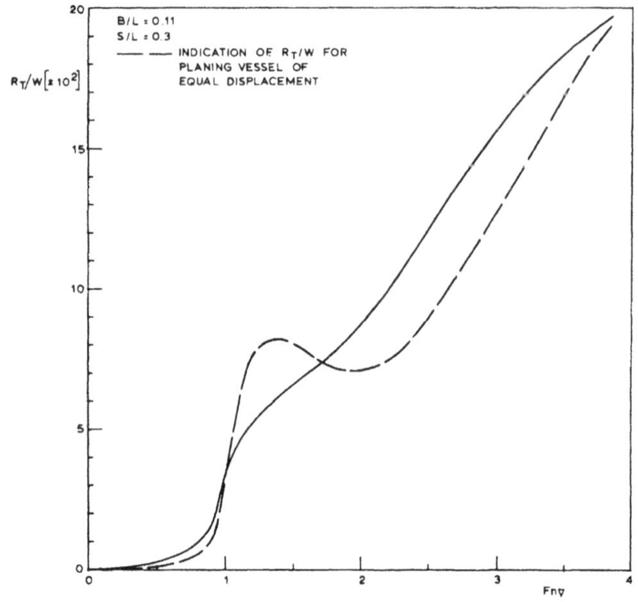

Fig. 97 Typical resistance curve for catamaran and planing craft of equal displacement

speeds (in the hump region, Fn $\approx$ 0.5) the low trim angles associated with the slender demihulls of the catamaran lead to a favorable performance (Michell, 1961, and Fry, et al 1972). This effect is clearly visible in Fig. 97, no pronounced hump being present in the resistance curve. At planing speeds (Froude numbers around 1.0) the equivalent monohull (of equal displacement) will show an advantage, as the hydrodynamic performance decreases with decreasing aspect ratio (the ratio of the wetted breadth of the demihull to its length) (Michell, 1961). The case is similar to that of a pair of biplane wings, and a single wing with the same total area and chord length but with twice as high an aspect ratio.

Concerning the performance at high planing speeds, results by Clement (1961) of a comparison between a specific catamaran and a monohull may be mentioned. He found that the catamaran had less resistance at speeds in excess of Fn_∇ = 4.6. At Fn_∇ = 6.0 the catamaran had some 30 percent less resistance, this reduction increasing to about 45 percent at Fn_∇ = 7.0. This advantage is due to the fact that at such high speeds the conventional boat is operating at a very small trim angle and high resistance, while the catamaran operates at a higher trim angle nearer to that for minimum resistance. An indication of the relative performance of catamarans and planing vessels is given in Fig. 97.

The hull spacing ratio is associated with interference effects between the component hulls. These effects consist of wave interference effects and body interference effects. Wave interference effects are due to the superposition of the two wave systems, each associated with a component hull in isolation. The body interference effects are caused by the change of flow around one demihull due to the presence of the other demihull. Several studies on interference effects on resistance have been undertaken, e.g. Fry et al (1972), Sherman, et al (1975), Yermotayev, et al (1977) and Ozawa, et al (1977).

The main component of the changed velocity field associated with body interference effects results from the induced flow of one demihull at the location of the other one. This induced flow is due partly to thickness effects and partly to lift effects. Consequently, the resulting flow around a symmetrical demihull will be composed of a symmetrical and an asymmetrical part. Vollheim (1968) and Myazawa (1979) have carried out velocity studies by means of pressure measurements. These results referred to a displacement type of catamaran with symmetrical demihulls. Myazawa found an increase of the mean velocity both between the demihulls and on the outer sides. He also concluded that the asymmetrical contribution to the local velocity field was small. His results apparently do not agree with those of Vollheim, however.

The asymmetrical onflow of one demihull and the possibly asymmetrical shape of that demihull will lead to hydrodynamic lift forces. On account of the finite aspect ratio, trailing vortices are shed leading to induced velocities around the other hull. This effect is believed to be of less importance.

The wave interference may influence the resistance to a large extent. Everest (1968) showed from a wave pattern analysis that beneficial wave interference is achieved by the cancellation of part of the divergent wave systems of each demihull, whereas adverse wave interference arises on interaction of the transverse wave system.

Fig. 98 shows the influence of the wave interference effects on the resistance obtained by Tasaki (1962) for a mathematical hull form. Here the wave interference factor is expressed by

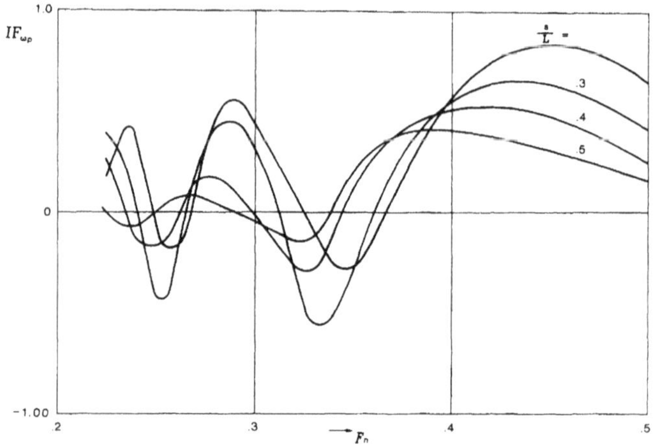

Fig. 98 Interference-factor as a function of Froude number

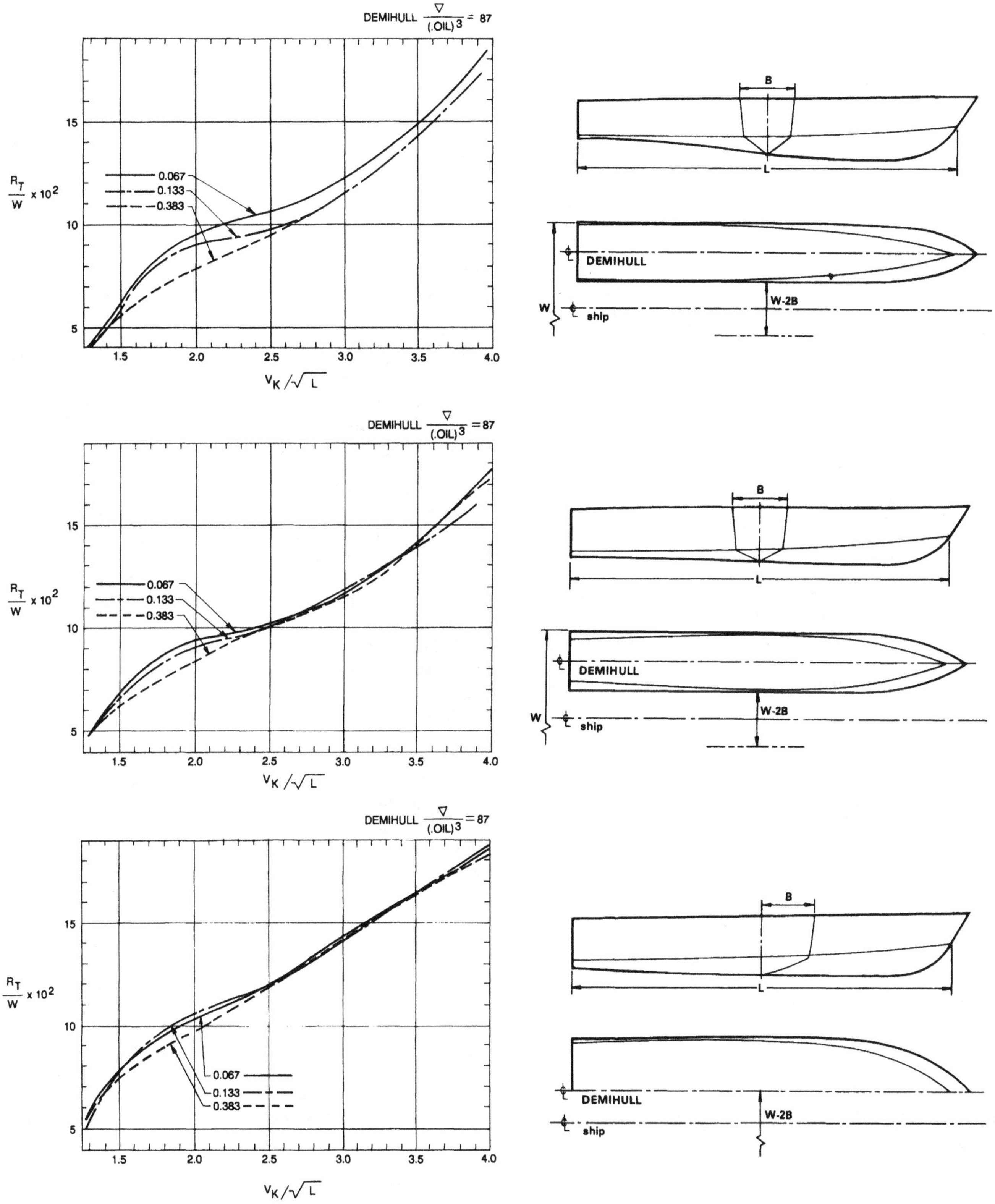

Fig. 99 Resistance of semi-planing catamarans as a function of speed-coefficient and hull spacing (s/L)

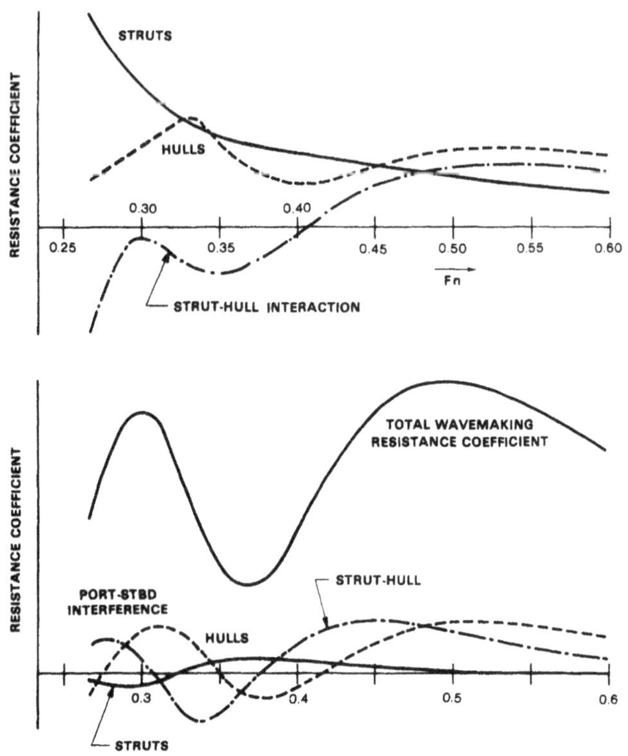

Fig. 100 Typical wavemaking resistance coefficients versus Froude number single-strut-per-hull SWATH

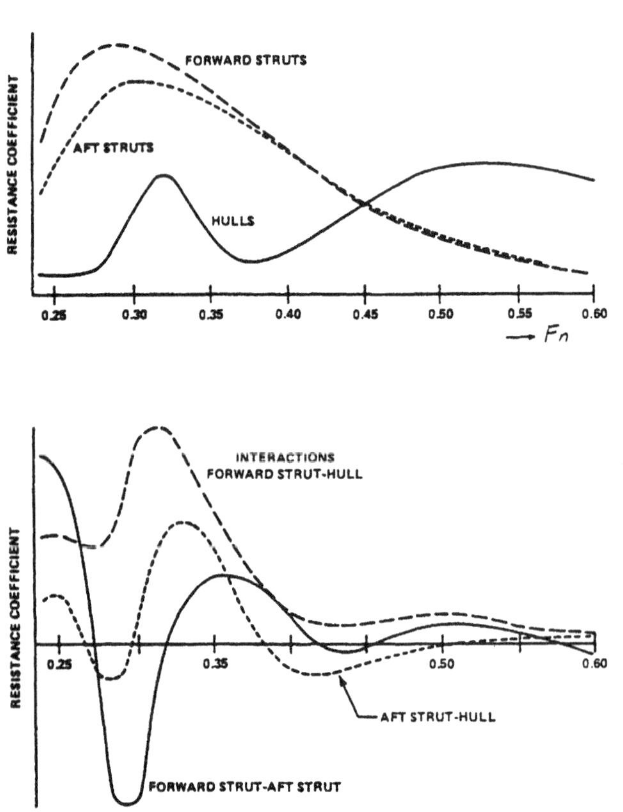

Fig. 101 Wavemaking resistance coefficients versus Froude number tandem strut SWATH

$$IF_{wp} = \frac{R_{wp}{}^{-2} R_{wpd}}{2R_{wpd}} \tag{77}$$

where R_{wp} is the wave pattern resistance of the catamaran and R_{wpd} is the wave pattern resistance of one demihull.

In general, experiments confirm this behavior but smaller beneficial and adverse effects occur (Everest, 1968). Theoretical and experimental evidence for symmetrical demihulls indicates that wave interference becomes significant at Fn-values of 0.2. Maximum beneficial effects occur at Fn $\approx$ 0.32 and $s/L \approx 0.3$ whereas adverse effects are most pronounced around Fn = 0.4. For asymmetrical demihulls, Everest (1969) and Turner, et al (1968) have made measurements.

The generation of vertical hydrodynamic lift and the associated change of hull form because of trim and rise of the center of gravity, may have a significant effect on the interference effects. Therefore, for the semi-planing speed range other tendencies may be expected, see Fig. 99. Fry, et al (1972) show model test results from which it may be concluded that the interference effects are small at speeds exceeding Fn = 0.8. Only for small hull spacings do the effects still seem to be significant. These conclusions are in contradiction with those of Sherman, et al (1975) for which there is no satisfactory explanation presently. For measurements of planing catamarans with asymmetrical demihulls the work of Ozawa, et al (1977) and Sherman, et al (1975) may be mentioned.

Design charts for planing catamarans have been published by Clement (1961). These are based on model tests. Application of these design charts is restricted to:

- low-aspect-ratio hulls, i.e., $0.1 \leq AR \leq 0.3$,
- small deadrise angles, i.e., 0 deg $\leq \beta \leq$ 10 deg.,
- high planing speeds, where buoyant forces are small.

Furthermore, the effects of interference between the hulls and of spray on the tunnel roof were not included.

Sherman, et al (1975) modified Savitsky's (1964) planing performance prediction method for catamarans. The program does not include interference effects on drag and trim. Resistance due to spray interfering with the tunnel roof is again not included.

9.4 Small Waterplane Area Twin Hull (SWATH) Ships. The SWATH concept is a type of catamaran that features two fully submerged hulls, each connected to an above-water, box-like deck structure by one or more relatively thin struts. See Fig. 84. Usually control surfaces are employed to reduce pitching motions of the vessel.

In the late sixties, the 190-ton 27-m (88 ft) SWATH SSP *Kaimalino* was designed as a workboat for the US Naval Ocean Systems Center to have a speed of 25 knots. This ship has been used to demonstrate the SWATH concept and to validate some of the technology. Results of theoretical and experimental studies

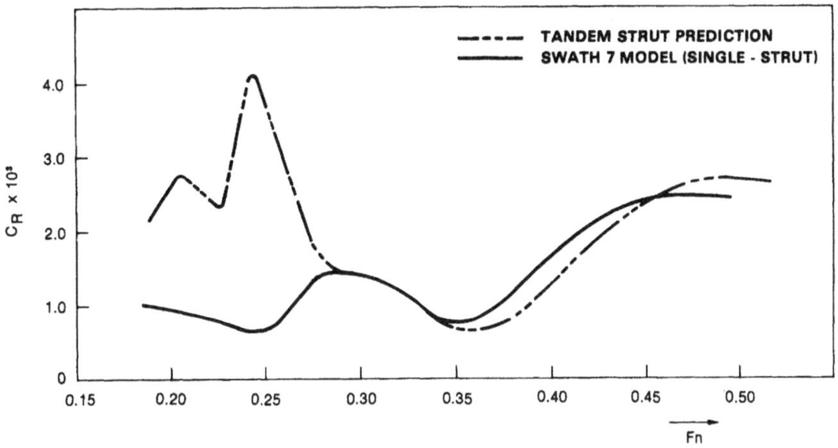

Fig. 102 Representative residuary drag coefficients for high speed single and tandem-strut SWATH designs

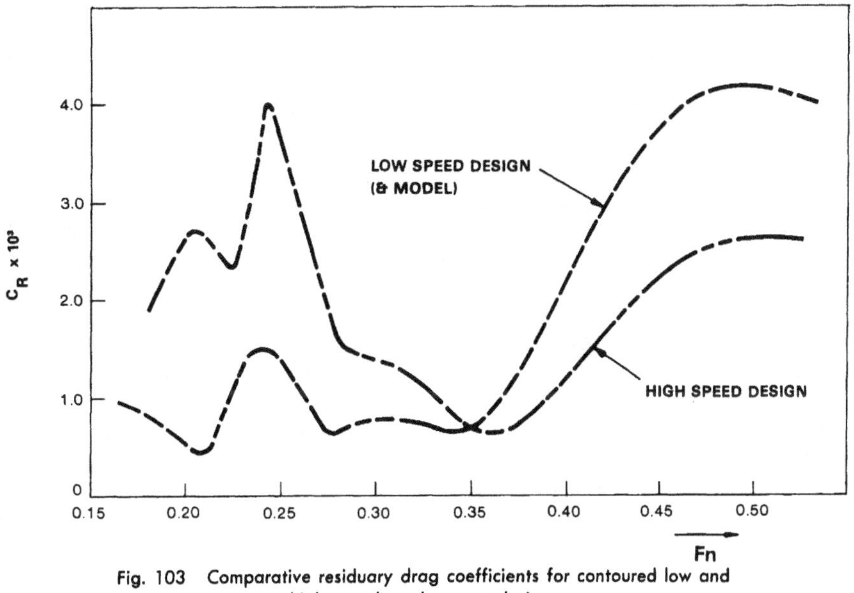

Fig. 103 Comparative residuary drag coefficients for contoured low and high speed tandem-strut designs

have revealed interest in displacements of up to 30,000 tons and in speeds of up to 40 knots (Savitsky, et al, 1981).

The benefits of the SWATH concept are derived from their low motion responses and sustained speed capability in waves. This behavior is similar to a conventional surface ship three times larger, due to the significantly reduced waterplane area, long natural pitching period and the submergence of the main bulk of the displacement volume (Savitsky, et al, 1981).

A SWATH ship has about a 75 percent larger wetted surface area compared to a single screw monohull frigate of equal displacement. Consequently SWATH ships incur a substantial frictional resistance penalty relative to conventional surface ships (Lamb, et al, 1979). This has to be compensated for (at least partly) by a reduced wave resistance, which can be obtained on account of the small waterplane area and the slenderness of the struts. However, careful attention has to be directed to the required depth of submergence of the hulls and to possible unfavorable interactions between the wave systems produced by the struts.

The induced drag of control surfaces becomes significant as soon as appreciable forces are developed to counteract any pitch instabilities of the vessel. At speeds above the primary resistance hump, the spray resistance of struts becomes significant.

Figs. 100 and 101 show typical curves taken from Numata (1981) for the components that make up the total wavemaking resistance. Fig. 100 holds for a single-strut per hull configuration, Fig. 101 for a tandem-strut configuration. The upper part of Fig. 100 shows the resistance curves disregarding interference between starboard and port sides. It is clear from this figure that interactions between hulls and struts and between port and starboard can make up a large part

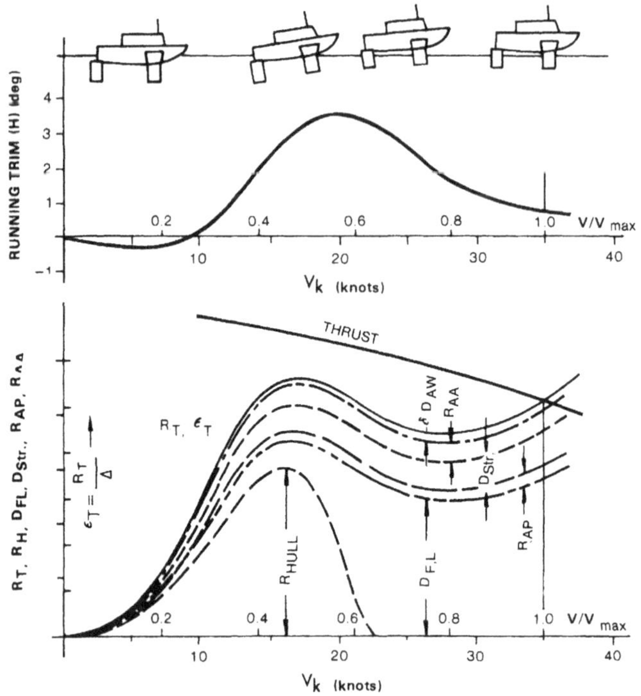

Fig. 104 Distribution of Resistance Components, Running Trim, Attitude of Hydrofoil Craft

of the total resistance. Fig. 101 presents the wavemaking resistance for a tandem-strut configuration. The interference between forward and aft struts is seen to be considerable. Port-starboard interference is not taken into account.

Fig. 102 gives representative residuary drag coefficients for high speed single- and tandem-strut SWATH designs. The existence of two peaks in the curve for the tandem-strut configuration has to be attributed to interference between the forward and aft struts. For higher Froude numbers the differences between the two configurations turn out to be small.

It is possible to reduce the peaks at low Froude numbers by appropriate contouring of the lower hulls and by adjusting the dimensions and separations of the struts. However, the design will show an increased wavemaking resistance for higher Fn in that case (Lamb, et al, 1979). Fig. 103 shows the improvements for lower speeds thus obtained and the large penalties that result for higher speeds.

Because of the unavoidable increased frictional resistance, large efforts have been directed towards the reduction of wavemaking resistance. Therefore several analytic wavemaking resistance prediction programs for SWATH forms have been developed. These programs employ linearized potential theory with thin ship approximations. Lin, et al (1974) have developed a program that requires the full input of ship geometry, a level of detail usually not available in early design stages. Chapman (1972) describes the theory for a similar program that admits a selection of an arbitrary number of hull sections of different diameters and lengths. In addition the program can select either an elliptical or a parabolic nose for the struts and the hulls, especially important for avoiding cavitation of Froude numbers below 0.35.

When making analytic predictions of full-scale resistance, it is common practice to add a correlation allowance with a constant value of 0.0005 to the predicted wavemaking drag coefficient to account for form drag and eddy-making. There is, however, insufficient experimental material to correlate the calculated wavemaking resistance and to arrive at better values for the correlation allowances for SWATHS. Some limited results of a series of systematic model tests carried out at DTRC have been published by Lamb, et al (1979). Otherwise very little information or systematic series is yet available in open literature.

A new computational method for design of small SWATHs was presented by Salvesen, et al (1985). The computer programs generate mathematically faired hull forms, compute the calm water resistance, minimize the wave resistance by contouring the lower hulls and predict the nominal wake.

9.5 Hydrofoil Craft *(a) General Considerations.* A hydrofoil craft has two modes of operation: the normal slow-speed hullborne mode and, with increasing speed through the take-off, the flying or foilborne mode. At hullborne mode the craft thus behaves more or less like a planing vessel with its characteristic hump resistance. Hulls for hydrofoil craft should be optimized for the relative amount of time spent in the cruising and the foilborne modes. For small craft planing hulls are generally used, but as size or cruising time increases, semi-planing or even displacement hulls may be more beneficial.

Take-off may occur at speeds near the maximum hullborne drag (hump) with subsequent acceleration to cruise speed, which typically may be twice the take-off speed. The drag-speed characteristic of hydrofoil craft has a broad minimum between hump speed and cruise speed beyond which resistance begins to climb rapidly, as shown in Fig. 104. The various resistance components in this figure are denoted by:

R_T is total resistance
R_H is hull resistance
D_{FL} is drag of foils
D_{St} is drag of struts
R_{AP} is drag of appendages
R_{AA} is wind resistance
δD_{AW} is added resistance in waves

In the foilborne mode the effective lift-drag ratio must be adequate for the intended operation, and at the same time the craft must be able to fly in a stable, controllable manner. These requirements are somewhat conflicting and the contributing factors are very speed dependent. As a result, a number of designs have emerged which provide different solutions to these general requirements.

Many different configurations are possible, although there always must be lifting surfaces forward and aft for longitudinal stability. The load may be divided more or less evenly between these foils in a *tandem* arrangement, or most of the load may be carried on the bow foil in what is termed an airplane or *conventional* configuration. Alternatively, the aft foil may support most of the weight in the *canard* configuration; see Fig. 105. Each of these fore or aft lifting surfaces may be split into two independent foils for purposes of design or retraction in shallow water; see Fig. 106.

There is a basic distinction between different types of craft in the method used for stabilization and control once foilborne operation is achieved, and in obtaining the modulation of dynamic lift needed between take-off and cruise. An early and still successfully used principle is that of variable area stabilization with surface piercing hydrofoils. These hydrofoils may consist of a member in the shape of a "V" or "W" or several members arranged in a ladder, see Fig. 106. Requirements of take-off and stability can be satisfied with either arrangement, although the resulting system is sensitive to waves, and therefore additional lift control devices are often needed.

An alternative scheme is to submerge both lifting surfaces, thereby avoiding area changes with changes of relative wave height. Submerged-foil, as well as surface-piercing systems experience orbital wave motions which may be appreciable, and because the lifting surface area is fixed the submerged foil system is inherently less stable. Lift modulation can be obtained by such means as flaps, variable submersion and air injection. In the following paragraphs, these principal systems will be described in more detail. The reader is also referred to Mantle (1976) in this respect.

(b) Surface-piercing (SP) hydrofoil craft. Hydrofoil craft with SP foils are designed to operate in a speed range of approximately 25 to 45 knots and in exceptional cases up to 60 knots. The take-off speed varies between 0.4 and 0.5 V_{max}. The resistance of the vessel in foil-borne mode, comprising foil, appendage and aerodynamic drag, is less than 50 percent of an ordinary ship with comparable weight.

SP-foil systems are characterized by their inherent static and dynamic stability in pitch, roll, yaw and heave. The foilborne craft is automatically stabilized by area stabilization. A deviation from the equilibrium position causes a change of lift-producing wetted area which creates restoring forces and moments. However, when running with waves, a foil is acted upon by downward orbital velocities when climbing up in a wave, and in the crest the orbital velocities tend to reduce the velocity of flow over the foil. These effects tend to cause the craft to "sit down" just when a rise is needed. This is opposite to the situation when heading into the sea, where the orbital velocities tend to cause the craft to rise in a wave. The SP hydrofoil craft may suffer from heaving instability, as the surface penetration of

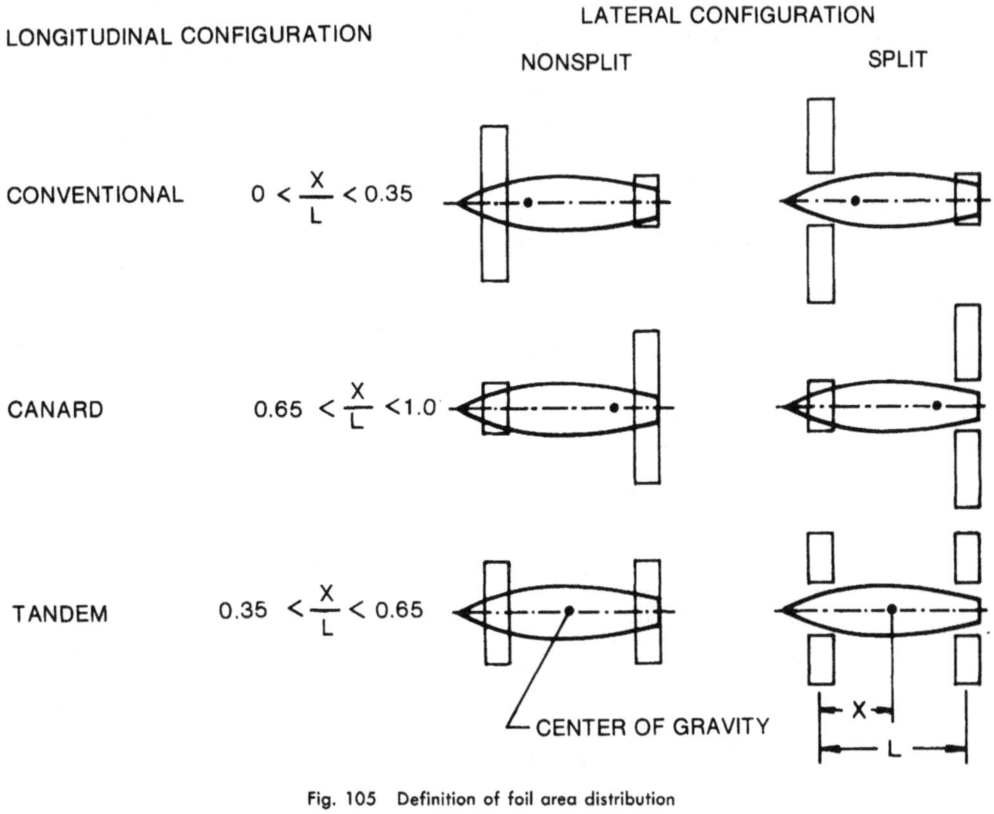

Fig. 105 Definition of foil area distribution

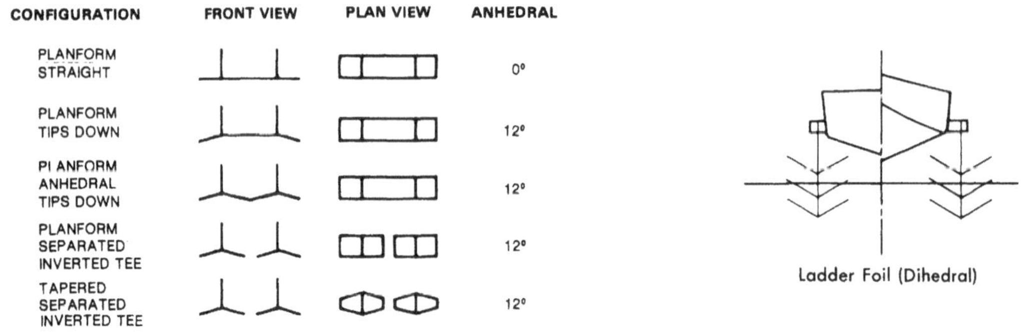

Fig. 106 Foil systems

the foils makes them prone to ventilation. This leads to a lift loss and a subsequent sinkage of the vessel, which causes it to rise again because of the increased foil area and to repeat the cycle. A detailed account of subjects related to surface-piercing hydrofoils may be found in Savitsky (1984).

(c) Fully-submerged hydrofoil craft. Hydrofoil craft with fully submerged foil systems are designed to operate at speeds above approximately 30 knots. The foils can be either deeply (DS) or shallowly (SS) submerged. The former run deeper than one chord length and the latter about one-half chord length below the water surface. Hull design, foil configuration, stability characteristics, take-off speed and seakeeping qualities of the craft depend mainly upon this submergence.

The main difference between DS and SS types of foil systems is their stability characteristics. DS foil systems do not have natural or inherent static and dynamic stability. The required stability in pitch, roll, yaw and heave of the craft is obtained by a continuous modification of the lift forces in relation to the momentary position of the vessel. Stability, maneuverability and seakeeping qualities can be guaranteed only by means of an automatic control system, which controls the lift forces either by incidence control, by actuating trailing edge flaps or by air feed into the suction side of the foil, thus reducing lift.

Hydrofoil craft with DS foils are designed to operate at speeds of 40 to 50 knots; experimental craft have attained 80 knots. Typical for these craft is the relatively high take-off speed, $\geq 0.5\ V_{max}$. These craft have excellent seakeeping qualities. Thus, they are used for high-speed passenger transport in coastal waters, as well as for patroling and military purposes in offshore areas.

SS foil systems are characterized by their inherent stability, which is based on the effect of the free surface on the lift curve slope. At foil submergences smaller than one chord length, the lift decreases with decreasing submergence, which provides an altitude control in calm water.

Hydrofoil craft with SS foils are designed to operate foilborne at speeds of 30 to 80 knots, predominantly in shallow, calm inland waterways. Take-off speed is generally smaller than $0.5\ V_{max}$. On some special designs, the rear foil is replaced by a planing surface. This type of craft is used exclusively for high-speed passenger transport on rivers and lakes, mostly in the USSR. These foils operate with low lift coefficients, which precludes generation of the necessary lift for take-off and thus an auxiliary set of subfoils in the form of planing ski-surfaces are added near the bow foil to increase the trim and consequently the angle of attack of the foils. In this way sufficient lift is generated for take-off.

(d) Hydrofoil Section Types. A third major distinction between hydrofoil configurations relates to the maximum speed. The major obstacle to achieving high sustained speeds in water is the occurrence of cavitation with its detrimental effects. Above a critical speed of about 50 knots, a radically different approach must be taken in designing the foil system. A distinction must be made, therefore, between subcavitating and supercavitating foil sections.

1. Subcavitating foil sections. The hydrodynamic characteristics of subcavitating hydrofoils are very similar to the subsonic aerodynamic characteristics of aircraft wings. Thus it has been possible to adopt much of the airfoil theory and techniques in their design. For hydrofoils that are to operate in the subcavitating regime the problem is to develop foil configurations having sufficient strength, minimum weight and maximum lift-to-drag ratio, while at the same time raising the critical cavitation speed as high as practical. The achievement of high cavitation-inception speeds is made more difficult by flow interactions at foil, strut and pod intersections, surface roughness and discontinuities, and the craft motions in a seaway combined with the orbital wave velocities.

In addition, there are severe problems due to effects of air venting from the free surface. Ventilation of submerged foils can occur through surface piercing struts and through tip vortex tubes that occasionally vent to the free surface. Ventilation problems have been dealt with primarily by the use of *fences* to block the passage of air from the free surface to regions of low pressure in the fluid. Forward sweep and dihedral of surface piercing struts are also used to reduce the tendency to ventilate.

Thus far, most foil and strut section shapes have been those selected from the NASA design literature, such as the 16 or 63 series. These sections have characteristically flat pressure distributions and provide maximum lift within limits of cavitation inception. From this point of view, bearing in mind the 800 to 1 ratio of water and air densities, it appears that under ideal conditions a foil loading of about 10^5 N/m^2 (2090 lb/ft^2) is the maximum attainable. In practice, considerably lower loadings in the order of 6.10^4 N/m^2 (1280 lb/ft^2) must be employed if cavitation is, in fact, to be avoided. Furthermore, it appears that speeds above 50 knots will always show some cavitation unless extreme care is taken in the design and fabrication of the foil system.

2. Base-vented and supercavitating foil sections. At speeds above 60 knots a new design area is entered. At present, there are two basic approaches to high-speed foil design. One involves the use of so-called supercavitating sections and the other the use of fully wetted base-vented sections. See Fig. 107.

In the supercavitating foil design the sharp leading edge causes the formation of a fully developed cavity over the entire upper surface of the foil. Cavity collapse occurs well abaft the trailing edge, and problems of buffeting and erosion are thus avoided. There are a number of difficulties with supercavitating foil design yet to be resolved:

- A high angle of attack is needed to generate a reliable, steady cavity compared with subcavitating hydrofoils, leading to relatively large drag values;
- at lower speeds, the lift-to-drag ratio is relatively small;
- the effects of proximity to the free surface are not clear;
- obtaining structural strength of the thin leading edge is difficult; and
- there is a problem of generating high lift at slow speed for take-off.

In general, it appears that true, supercavitating foils offer the greatest promise at speeds above 80 knots. At speeds below 80 knots it is felt that fully wetted base-vented designs may be more practical. The success of base-vented foils depends to a large degree on

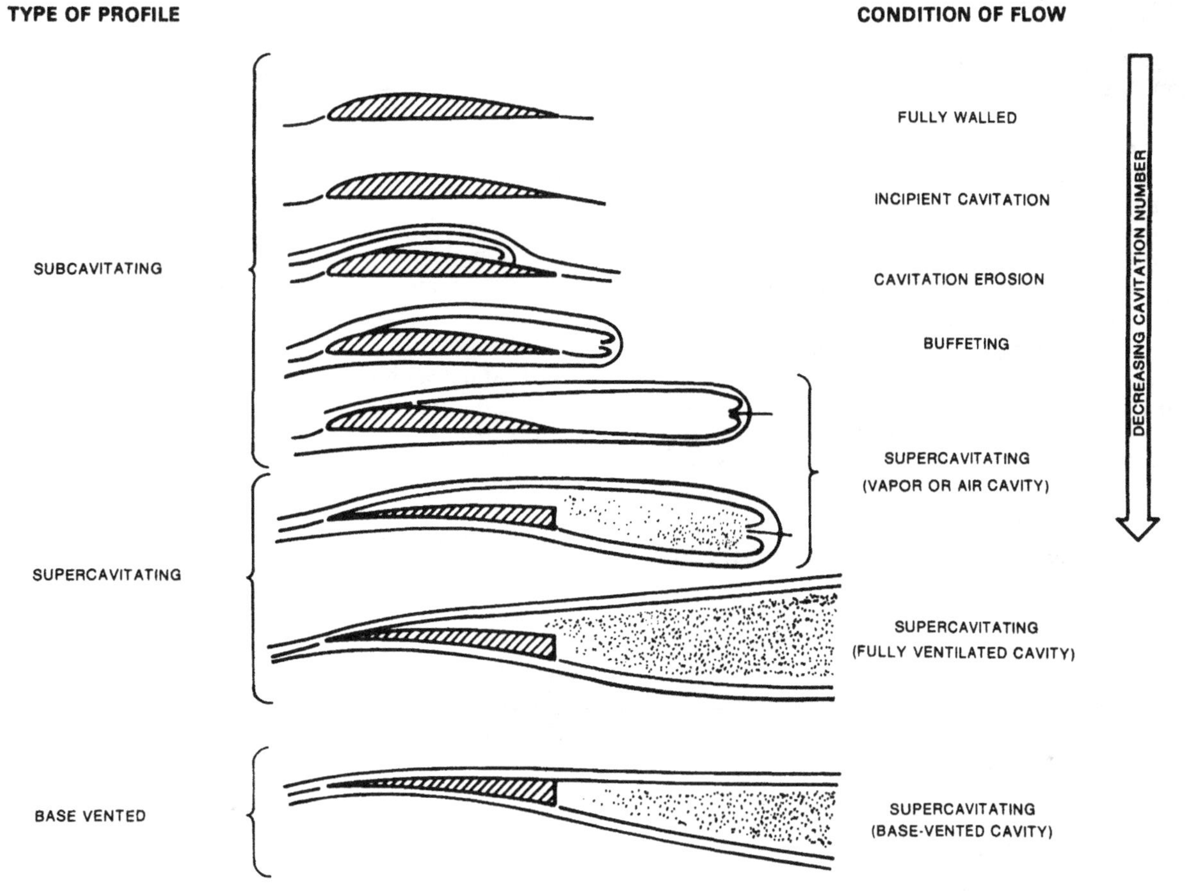

Fig. 107 Hydrofoil sections and flow patterns

the development of sections with decreased cavitation and ventilation-sensitivity to angle of attack. This is achieved by natural ventilation along the cavitating blunt trailing edge of surface-piercing struts. Fig. 107 gives a survey of the several types of sections.

(e) Hydrodynamic Characteristics. The basis of the performance prediction of hydrofoil craft is the solution of a set of equations, comprising horizontal, vertical and moment equilibrium of forces. The forces involve different contributions that will be discussed in the following:

- Hull characteristics may be obtained from regression formulas or systematic series such as Series 65 planing hulls (see Section 9.2). This series provides data on resistance, hull moment, wetted surface and dynamic draft as a function of speed, trim, loading and hull form parameters. Therefore, these results are suitable for predicting the hull characteristics for hullborne and take-off conditions.
- Drag forces on the appendages of hydrofoil craft can be as much as 20 percent of the total craft resistance. Traditionally, these forces have been predicted by using empirical equations or by subscale model data. The appendage drag forces are then added to the bare hull resistance, together with some interaction effects.

A survey of empirical formulations is given by ITTC (1984). They are believed to give reasonable results for speeds up to $Fn_\nabla = 1.5$. For higher ship speeds the application of such formulations normally leads to serious underprediction of the appendage drag. Reasons for discrepancies of this nature can be sought in the adopted value of the inflow velocity, in lack of alignment of shaft struts, rudders etc. to the flow, the occurrence of cavitation and ventilation, and interaction effects between appendages, between appendages and the hull and between appendages and the propulsor. Full and subscale experimental results for high-speed operation are given by Gregory (1973) and Gregory, et al (1979).

Hydrofoil lift and drag forces of subcavitating foil sections are determined according to aeronautical practice, including corrections for the presence of the free surface. The following formula for the 3-dimensional lift-curve slope, $C_{L\alpha}$, is a combination of formulas found in Johnson (1965), Kaplan, et al (1955) and Korvin-Kroukovsky, et al (1951):

$$C_{L\alpha} = \frac{2\pi P\, AR \cos\Lambda}{AR + 2P(1+\sigma)(1+\zeta)\cos\Lambda\left[1+\left\{1+\left(\frac{AR}{2P\cos\Lambda}\right)^2\right\}^{0.5}\right] - (1+\sigma)(1+\zeta)\,AR} \quad (78)$$

where

α is foil angle of attack
α_o is zero-lift angle of attack
AR is aspect ratio
Λ is sweep angle

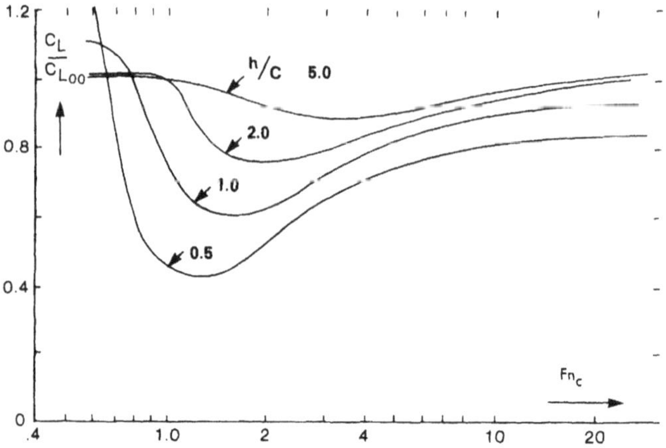

Fig. 108 Chord Froude number and depth of chord ratio effect on the lift ratio of a submerged flat plate

σ is Munk's interference factor
ζ is foil planform correction factor
P is a depth correction, estimated for high speeds by

$$P = \left(16\left(\frac{i}{c}\right)^2 + 1\right)\Big/\left(16\left(\frac{i}{c}\right)^2 + 2\right) \quad (79)$$

where

i is submergence
c is chord.

But for the lower chord Froude numbers another correction is needed. See Fig. 108. The chord Froude number is defined as $Fn_c = \frac{V}{\sqrt{gc}}$

The 3-dimensional lift coefficient is obtained from

$$C_L = C_{L\alpha}(\alpha - \alpha_o)$$

Lift and drag characteristics of surface piercing foils may be determined by a stripwise integration of depth effects or by a fully-submerged replacement foil at one half of the apex submergence. The foil drag is given by

$$C_D = C_{DP} + \delta C_{DP} + C_{Di} + C_{Dw} + C_{Ds} \quad (80)$$

where

C_{DP} is skin friction and profile pressure drag at zero lift. In Section 5.4 the following formula was given (36)

$$C_{DP} = C_F[1 + 2(t/c) + (60t/c)^4]$$

where (t/c) is foil (maximum) thickness-to-chord ratio.

C_F is skin friction coefficient according to the ITTC-friction line.

δC_{DP} is profile-drag increment due to angle of attack, which may be taken to be approximately $\delta C_{DP} = 0.005\ C_L^2$.

C_{Di} is induced-drag coefficient

$$C_{Di} = \frac{2P(1+\sigma)(1+\zeta)\cos\Lambda\left[1+\left(1+\left(\frac{AR}{2P\cos\Lambda}\right)^2\right)^{0.5}\right] - AR(1+\sigma)(1+\zeta)}{2\pi P\, AR \cos\Lambda\ C_L^{-2}} \tag{81}$$

which formula is again a combination of formulas found in Kaplan, et al (1955), Korvin-Kroukovsky, et al (1951) and Johnson (1965), where

C_{Dw} is free-surface wave-drag coefficient.

For this drag Kaplan, et al (1955) have derived theoretical expressions. For higher Froude numbers the following expression based on the results of Kaplan and Breslin may be used:

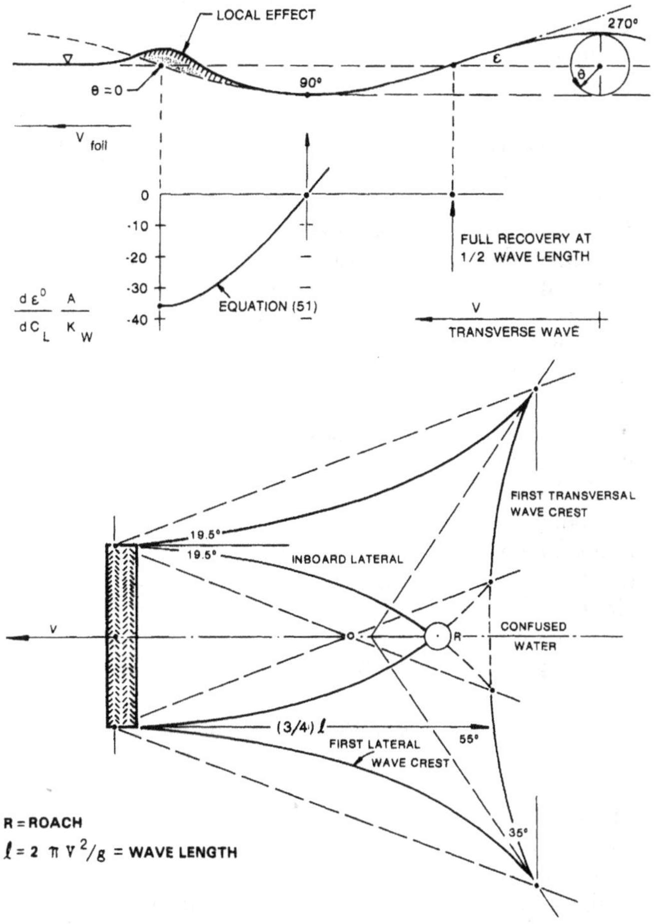

Fig. 109 Wave pattern and downwash angle behind submerged lifting hydrofoil

$$C_{Dw} = \frac{0.5 C_L^2}{F_h^2 e^{2/F_i^2}} \tag{82}$$

where $F_h = \frac{V}{\sqrt{gi}}$, a modified depth Froude number.

C_{Ds} is the spray-drag coefficient of surface piercing struts and foils. The following formula is based on model test results of Chapman (1972):

$$C_{Ds} = [7.68 - 6.40\,(t/c)]\, C_F \tag{83}$$

Lift and drag are also influenced by interference effects. These effects occur where different parts of the foil system join each other, for instance at strut-hull bottom junctions and foil-strut junctions. Hoerner (1965) gives empirical data for assessing these effects. They have also been discussed in Section 5.4. Finally, front foil-aft foil interaction may significantly influence the craft's performance.

The trailing vortices and wave system associated with the forward foil are the sources of a free surface elevation and an angle of attack variation at the aft foil. Kaplan, et al (1955) describe an approximate theoretical model for the determination of these effects. Some less accurate, but much easier to apply, empirical formulas are given below (Li Bai-Qi, 1981):

$$\delta h = -C_L \cdot c\left\{\exp\left[-\,i/(cF_c^{2\nu})\right] + \frac{1}{2\lambda}\left(F_c^{2\nu} - 2\right)\right\}\sin\left[k/(cF_c^{2\nu})\right] \tag{84}$$

$$\delta\alpha = \frac{3C_L}{4F_c^{2\nu}}\left\{\exp\left[-\,i/(cF_c^{2\nu})\right] + \frac{1}{2\lambda}\left(F_c^{2\nu} - 2\right)\right\}\cos\left[x/(cF_c^{2\nu})\right] \tag{85}$$

where

δh is free surface elevation (positive upward)

$\delta\alpha$ is downwash angle (positive if reckoning the angle of attack)

ν is $\exp(-0.73/\lambda^{0.2})$

λ is s/c

s, c, i are front-foil span, chord and submergence, resp.

C_L is front-foil lift coefficient

x is front-aft foil distance

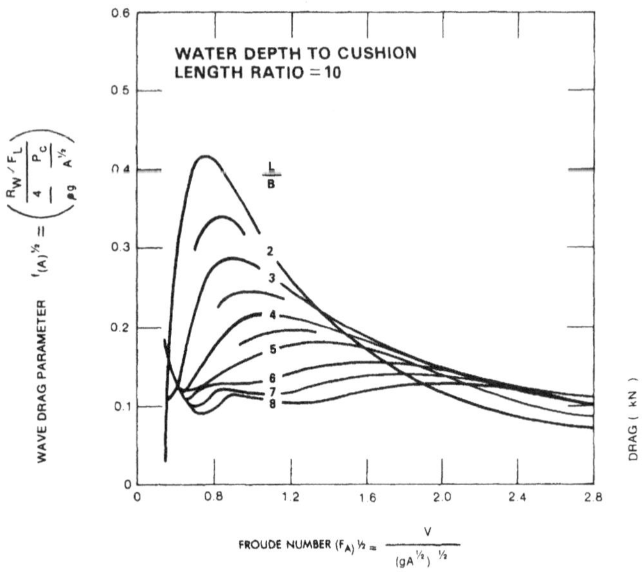

Fig. 110 Wave drag

Fig. 109 shows a hydrofoil developed wave pattern.

9.6 Air-Supported Craft. The first man-carrying air-supported craft was launched in 1960. The problem

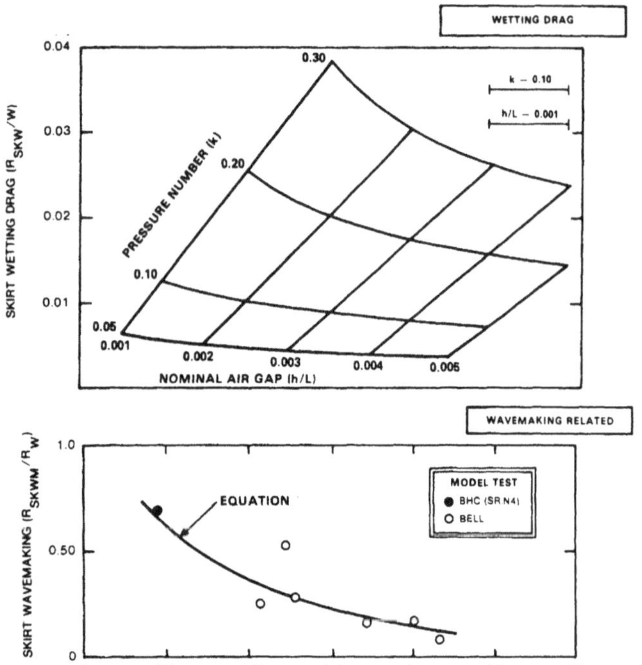

Fig. 111 Calm-water skirt drag components

of arriving at acceptable power levels to obtain a specified lift has been solved in two ways. The first was the introduction of flexible skirts by means of which a fully amphibious craft was obtained: the Air Cushion Vehicle (ACV). The other solution of the lift power problem employed rigid sidehulls and flexible seals at the bow and stern of the vessel: the Surface Effect Ship (SES).

Applications of air-supported craft are both commercial and military. Commercial applications concentrate mainly on fast passenger ferries or low speed hoverbarges (ACV's) for support of civil engineering projects. Military versions are used as patrol craft or mine countermeasure craft. The amphibious ACV is also suitable as a landing craft. Because of the aerostatic lift, these craft have a very small wetted surface, which has a significant positive influence on the resistance at higher speeds where the wavemaking resistance of the cushion is small.

The amphibious ACV has an air cushion contained by a flexible skirt around the whole perimeter of the craft. The air for the cushion is supplied by fans. ACV's usually employ air propulsion, normally either ducted propulsors or airscrews, making them amphibious. Typical service speeds are in the range of Fn (based on cushion length) between 0.8 and 1.5. Cushion length-to-beam ratios are generally in the order of 1.5 to 2.7, and the cushion pressures on modern marine vehicles tend to vary up to 5000 N/m^2 (104 lb/ft^2).

For the Surface Effect Ship thrust is provided by propellers or waterjets located on each sidehull at the transom. Cushion length-to-beam ratios may vary from 2 to 8. The speeds of interest for an SES with an L/B of 2 to 3.5 are normally at Froude numbers up to 2.0. Usually, SES's of L/B of 5 and greater have design Froude numbers of 0.6 to 0.8 (Savitsky, et al. 1981).

As compared to an ACV, the SES hull, which pierces the water surface, has less air leakage, better longitudinal and transverse stability and an acceptable form for utilizing water propulsion systems, which, at speeds up to about 120 knots, are more efficient than air propulsion (Butler, 1985). The ACV on the other hand has the advantage of being amphibious.

For both types basically two sorts of power can be discerned: a propulsion power to overcome the total drag and a lift power to raise the craft. The total drag of ACV- and SES-craft can be divided into two main components: aerodynamic drag (profile drag and momentum drag) on the one hand and hydrodynamic drag (cushion wavemaking drag, skirt or seal drag and sidehull drag for SES) on the other hand. The aerodynamic momentum drag is the drag due to the rate of change of momentum required for accelerating the cushion air to craft velocity.

The cushion wavemaking drag relates to the interaction between the cushion and the calm water surface over which it is traveling. Theoretical treatments of this drag component have been given among others by Newman and Poole (1962), Barratt (1965) and Doctors (1970). Fig. 110 gives the wavemaking drag-to-cushion-lift ratio R_W/F_L as a function of the Froude number (based on the square root of the cushion area

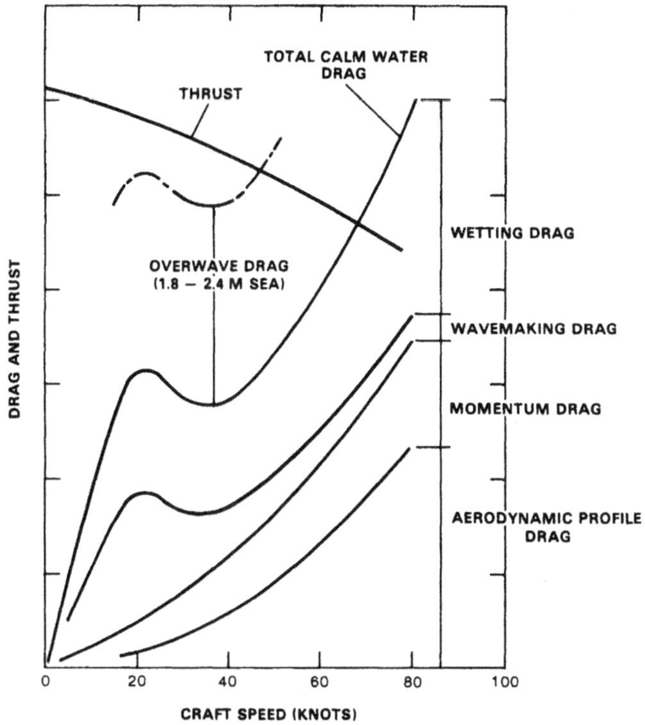

Fig. 112 Drag breakdown of typical large air cushion craft

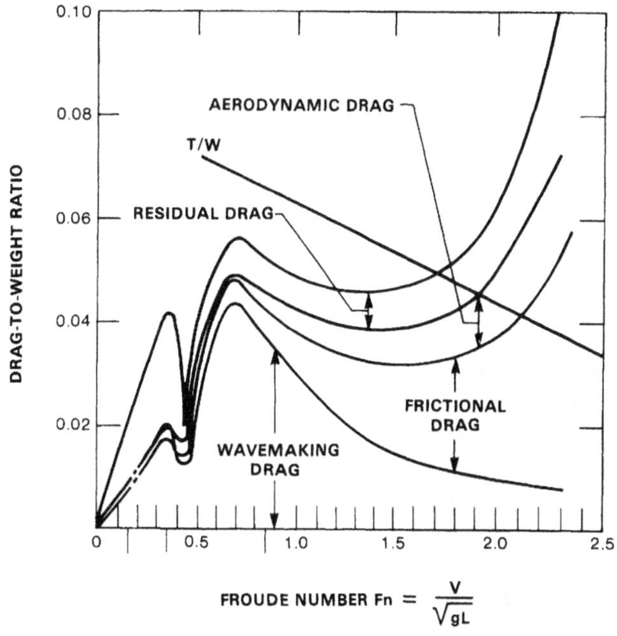

Fig. 113 Drag components of an SES of length-to-beam LIB = 2.0

A). This figure is useful in illustrating the wavemaking drag for different length-to-beam ratios. It shows that for craft of higher *L/B* (but the same displacement, cushion area and cushion pressure), the wavemaking drag is substantially lower than that for low *L/B*, except at very high velocities.

Little information about the skirt or seal drag is available from open literature. Bell and British hovercraft researchers showed that good agreement with model tests could be achieved if the calm water skirt drag of an ACV is considered to consist of a wetting drag R_{SKW} and of a cushion wavemaking related drag R_{SKWM} (Mantle, 1980). The wetting drag is related to craft speed, peripheral hemline length and the volume flow through the air gap. Fig. 111 shows the dependence of both drag contributions on the mentioned parameters. In this figure h denotes the height of the air gap, k represents the ratio of the dynamic forces generated by the oncoming air stream to the static forces generated by the air cushion pressure p_c:

$$k = \frac{\rho V^2}{2p_c} \qquad (86)$$

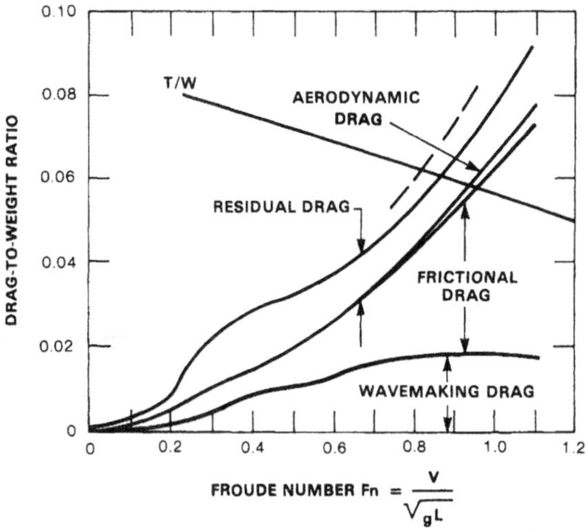

Fig. 114 Drag components of an SES of length-to-beam LIB = 6.0

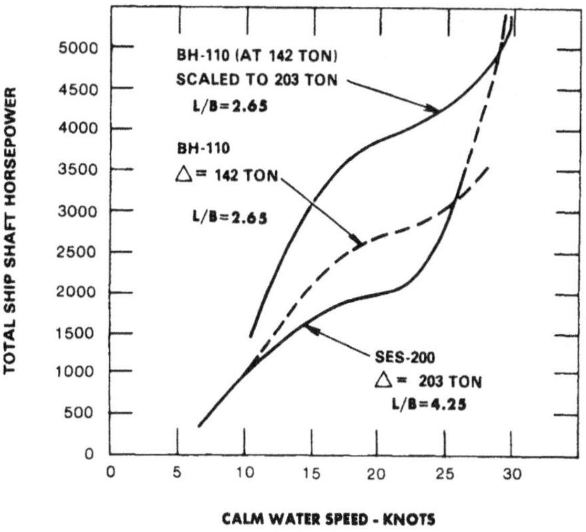

Fig. 115 Comparison of BH-110 and SES-200 performance

The sidehull drag of an SES mainly consists of frictional resistance in the high-speed region due to the high L/B ratio of the sidehull and its small displacement.

Fig. 112 illustrates the contribution of the different drag components in both calm and rough water for a typical large ACV. Figs. 113 and 114 show typical drag components for a SES with an L/B of 2 and a SES with an L/B of 6 respectively. These two sets of drag components are significantly different. In the L/B = 2 case there is a pronounced drag hump at a Froude number of about 0.7, which is primarily due to the wave-making drag. In the L/B = 6 case, the frictional drag (of the sidehulls primarily) plays a more dominant role. From these figures it is clear that the low L/B SES has a much smaller hump thrust margin (being the difference between thrust and drag) and therefore is more critical to hump drag increases due to rough water or weight increase for example. Adams, et al (1984) give a performance comparison of an L/B = 2.65 and an L/B = 4.25 craft (BH110 and SES200), based on full-scale performance data, Fig. 115. The SES200 is a lengthened version of the BH110 with all other systems retained. As can be seen in Fig. 115 the longer version requires substantially less power for speeds up to 27 knots. Apparently, the reduction in wavemaking drag achieved by increasing the L/B ratio and the reduction of the cushion loading more than balances the increase in hull drag from the 15-m extension and the weight increase.

Few publications give methods to estimate the total drag on a theoretical-empirical basis, e.g. Mantle (1980). Another method of determining the resistance of an ACV-craft is to perform model tests and reduce the data to individual components. These components are then approximately Froude-scaled or Reynolds-scaled to obtain full scale values (Barratt, et al, 1969), (Wilson, et al, 1978).

Apart from the propulsion system, the lift system is another large power consumer. On cushion, the cushion supports 80–95 percent of a SES weight. Because of air leakage underneath the skirts or seals, the cushion needs to be supplied with air to maintain full cushion pressure. Additional lift will still lower drag, at first significantly and then more gradually. The optimum airflow is defined as the flow at which the power required to increase the flow is greater than the reduction in propulsion power needed to maintain a constant speed. In general the optimum flow increases modestly with speed and sea state and requires 15–30 percent of the total power.

References

Adams, J.D. and Beverly, W.F. (1984), "Technical Evaluation of the SES-200 High Length to Beam Surface Effect Ship," *Naval Engineers Journal,* Vol. 96.

Allan, J.F. and Conn, J.F.C., (1950), "Effect of Laminar Flow on Ship Models," *Trans.* INA, Vol. 92.

Allan, J.F. (1950), "Some Results of Scale Effect Experiments on a Twin-Screw Hull Series," *Trans.* IESS.

ATTC (1953), Minutes of the 10th ATTC, MIT, Appendix XX, p.3, item 4(b).

ATTC (1956), *Trans.,* 11th General Meeting of the ATTC, Washington, D.C., p.28, TMB Report 1099.

Baba, E. (1969), "Study on Separation of Ship Resistance Components," Mitsubishi Technical Report No. 59.

Baba, E. (1972), "An Application of Wave Pattern Analysis to Ship Form Improvement," *Journal* of Society of Naval Architects of Japan, Vol. 132.

Bai, K.J. and McCarthy, J.H. (1979), "Proceedings of the Workshop on Ship Wave-Resistance Computations," DTNSRDC Report.

Bailey, D. (1976), "The NPL High-Speed Round-Bilge Displacement Hull Series. Resistance, Propulsion, Manoeuvring and Seakeeping Data," RINA, Maritime Technology Monograph No. 4.

Baker, G.S. (1915), "Notes on Model Experiments," NECI Report.

Baker, G.S., (1937), "Development of Hull Form of Merchant Vessels," *Trans.* NECI, Vol. 54.

Barratt, M.J. (1965), "The Wave Drag of a Hovercraft," *Journal of Fluid Mechanics,* Vol. 22, part 1.

Barratt, M.J., Everest, J.T., Hogben, N., Shipway, J.C. and Wheatley, J.H.W. (1969), "Estimation of Power and Drag for Marine Hovercraft," Hovercraft Unit Report 10, National Physical Laboratory.

Beaufoy, M.G. (1834), *Nautical and Hydraulic Experiments,* edited and published by H. Beaufoy, London.

Benford, H. (1966), "On the Rational Selection of Ship Size," Pan American Congress of Naval Architecture and Maritime Transportation, Rio de Janeiro.

Benford, H. (1967), "The Practical Application of Economics to Merchant Ship Design," *Marine Technology,* Vol. 4.

Berger, G. (1980), "Korrelation des Schiffswiderstandes nach Taylor-Gertler, Guldhammer-Harvald und Holtrop mit Modellversuchsergebnissen der Hamburgischen Schiffbau-Versuchsanstalt," *Schiff und Hafen,* Vol. 32.

Berkelom, W.B. van (1981), "Wind Forces on Modern Ship Forms—Effects on Performance," *Trans.* NECI, Vol. 97.

Beys, P.M. (1963), "Series 63—Round Bottom Boats," Stevens Institute of Technology, Davidson Laboratory, Report 949.

Blasius, H. (1908), "Grenzschichten in Flüssigkeiten

mit kleiner Reibung," *Zeitschrift für Mathematik und Physik,* band 56.

Bowden, B. and Davison, N. (1974), "Resistance Increments Due to Hull Roughness Associated with Form Factor Extrapolation Methods," NPL Ship Division Report TM 380.

Bragg, E.M. (1930), "Results of Experiments Upon Bulbous Bows," SNAME *Transactions,* Vol. 38.

Butler, E.A. (1985), "The Surface Effect Ship," *Naval Engineers Journal,* Vol. 97.

Chapman, R.B. (1972), "Spray Drag of Surface Piercing Struts," AIAA/SNAME/USN Advance Marine Vehicles Meeting, Paper No. 72-605.

Chapman, R.B. (1972), "Hydrodynamic Drag of Semisubmerged Ships," *Journal of Basic Engineering,* ASME, Vol. 94.

Clement, E.P. (1957), "Scale Effect on the Drag of a Typical Set of Planing Boat Appendages," DTMB Report 1165.

Clement, E.P. (1961), "Graphs for Predicting the Ideal High Speed Resistance of Planing Catamarans," DTNSRDC Report No. 1573.

Clement, E.P. and Blount, D.L. (1963), "Resistance Tests of a Systematic Series of Planing Hull Forms," SNAME *Transactions,* Vol. 71.

Clement, E.P. (1963), "How to Use the SNAME Small Craft Data Sheets for Design and for Resistance Prediction," SNAME Technical and Research Bulletin No. 1-23.

Comstock, J.P. and Hancock, C.H. (1942), "The Effect of Size of Towing Tank on Model Resistance," SNAME *Transactions,* Vol. 50.

Dagan, G. and Tulin, M.P. (1969), "Bow Waves Before Blunt Ships," Hydronautics Inc., Technical Report 117-14.

Daube, O. (1980), "Contribution au Calcul Non Linéaire de la Résistance de Vagues d'un Navire," Doctoral Thesis, Université de Paris VI.

Daube, O. and Dulieu, A. (1981), "A Numerical Approach of the Non-linear Wave Resistance Problem," 3rd International Conference on Numerical Ship Hydrodynamics, Paris.

Dawson, J. and Lindblad, A. (1949), Discussion on Allan et al (1950).

Dawson, J. (1952–60), "Resistance of Single-Screw Coasters," Parts 1 to 4, *Trans.* IESS, Mar. 1953, Nov. 1954, Feb. 1956, Mar. 1959.

DeGroot, see Groot, de.

Dillon, E.S. and Lewis, E.V. (1955), "Ships With Bulbous Bows in Smooth Water and in Waves," SNAME *Transactions,* Vol. 63.

Doctors, L.J. (1970), "The Wave Resistance of an Air Cushion Vehicle," University of Michigan, Ann Arbor.

Doust, D.J. and O'Brien, T.P. (1958–59), "Resistance and Propulsion of Trawlers," *Trans.* NECI, Vol. 75.

Doust, D.J. (1961), "Bulb Trawlers," *Ship and Boat Builder,* Vol. 14. Also "Fishing Boats of the World-2," published by Fishing News (Books), Ltd., London, p. 445.

Doust, D.J. (1962–63), "Optimised Trawler Forms," *Trans.* NECI, Vol. 79.

Doust, D.J. (1963), "Statistical Analysis of Resistance and Propulsion Data," National Physical Laboratory, Ship Report 42, Feltham, England.

Dyne, G. (1977), "A Theoretical Scale Effect Study on the Propulsion Coefficient of a Body of Revolution," Det norske Veritas Symposium on Hydrodynamics, Oslo.

Edstrand, H., Freemanis, E. and Lindgren, H. (1953–56), "Experiments with Tanker Models," Publication Nos. 23, 26, 29, 36 and 37, SSPA; also NECI, Vol. 72, 1955–56 for Summary.

Edstrand, H. and Lindgren, H. (1956), "Systematic Tests with Models of Ships with C_B = 0.525," Publication No. 38, SSPA.

Eggers, K. (1962), "Uber die Ermittlung des Wellenwiderstandes eines Schiffsmodells durch Analyse seines Wellensystems," Symposium on Theoretical Naval Architecture, University of Hamburg. Also *Schiffstechnik,* band 9, heft 46.

Eggert, E.F. (1939), "Further Form Resistance Experiments," SNAME *Transactions,* Vol. 47.

EMB (1925), "A Description of the US Experimental Model Basin," EMB Report No. 118, DTRC.

Emerson, A. (1959), "Ship Model Size and Tank Boundary Correction," *Trans.* NECI, Vol. 76.

Everest, J.T. (1968), "Some Research on the Hydrodynamics of Catamarans and Multi-Hulled Vessels in Calm Water," *Trans.* NECI, Vol. 84.

Everest, J.T. (1969), "Some Comments on the Performance of a Single-Hull Trawler Form and Corresponding Catamaran Ships made up from Symmetrical and Asymmetrical Hulls," National Physical Laboratory, Report No. 129.

Fairlie-Clarke, A.C. (1975), "Regression Analysis of Ship Data," ISP, Vol. 22.

Fisher, K.W. (1972), "Economic Optimisation Procedures in Preliminary Ship Design (Applied to the Australian Ore Trade)," *Trans.* RINA, Vol. 114.

Fisher, K.W. (1973), "The Inclusion of IMCO Tanker Design Constraints in General Optimization Procedures," SNAME *Transactions,* Vol. 81.

Freimanis, E. and Lindgren, H. (1957–59), "Systematic Tests with Ship Models," Publication Nos. 39, 41, 42 and 44, SSPA.

Froude, R.E. (1881), "On the Leading Phenomena of the Wave-Making Resistance of Ships," *Trans.* INA, Vol. 22.

Froude, R.E. (1888), "On the 'Constant' System of Notation of Results of Experiments on Models Used at the Admiralty Experiments Works," *Trans.* INA, Vol. 29.

Froude, R.E. (1889), "On the Part Played in the Operation of Propulsion by Differences in Fluid Pressure," *Trans.* INA, Vol. 30. See also Vol. 33 (1892) and Vol. 53 (1911).

Froude, W. (1872, 1874), "Experiments on Surface Friction," British Association Reports.

Froude, W. (1874), "On Experiments with HMS *Greyhound*," *Trans.* INA, Vol. 15.

Froude, W. (1877), "On Experiments Upon the Effect Produced on the Wave-Making Resistance of Ships by Length of Parallel Middle Body," *Trans.* INA, Vol. 18.

Froude, W. (1955), "Observations and Suggestions on the Subject of Determining by Experiment the Resistance of Ships," The Papers of William Froude, 1810–1879, *Trans.* INA, 1955.

Froude, W. (1955), "Memorandum to Mr. E.J. Reed, Chief Constructor of the Navy," The Papers of William Froude, 1810–1879, *Trans.* INA, 1955.

Fry, E.D. and Graul, T. (1972), "Design and Application of Modern High-Speed Catamarans," *Marine Technology,* Vol. 9.

Gadd, G.E. and Hogben, N. (1962), "An Appraisal of the Ship Resistance Problem in the Light of Measurements of the Wave Pattern," NPL Ship Division Report No. 36.

Gebers, F. (1919), "Das Aehnlichkeitsgesetz bei im Wasser geradlinig fortbewegter Platten," *Schiffbau,* Vol. 22.

Gertler, M. (1947), "The Prediction of the Effective Horsepower of Ships by Methods in Use at the David Taylor Model Basin," TMB Report 576, DTRC.

Gertler, M. (1950), "Resistance Experiments on a Systematic Series of Stream-lined Bodies of Revolution, for Application to the Design of High-Speed Submarines," TMB Report C-297 (declassified) DTRC.

Gertler, M. (1954), "A Re-Analysis of the Original Test Data for the Taylor Standard Series," TMB Report 806 DTRC.

Graff, W., Kracht, A. and Weinblum, G.P. (1964), "Some Extensions of D.W. Taylor Standard Series," SNAME *Transactions,* Vol. 72.

Granville, P.S. (1977), "Drag and Turbulent Boundary Layer of Flat Plates at Low Reynolds Numbers," *Journal of Ship Research,* Vol. 21, No. 1.

Gregory, D. (1973), "The Performance of High Speed Rudders in a Cavitating Environment," SNAME, April 2–4.

Gregory, D. and Beach, T. (1979), "Resistance Measurements of Typical Planing Boat Appendages," DTNSRDC Report No. SPD-0911-01.

Groot, D. de (1955), "Resistance and Propulsion of Motor-Boats," *International Shipbuilding Progress,* Vol. 2.

Guilloton, R. (1951), "Potential Theory of Wave Resistance of Ships with Tables for its Calculation," SNAME *Transactions,* Vol. 59.

Guldhammer, H.E. and Harvald, S.A. (1974), "Ship Resistance. Effect of Form and Principal Dimensions (Revised)," *Akademisk Forlag,* Copenhagen.

Hadler, J.B. (1966), "The Prediction of Power Performance on Planing Craft," SNAME *Transactions,* Vol. 74.

Hadler, J.B., Hubble, E.N. and Holling, H.D. (1974), "Resistance Characteristics of a Systematic Series of Planing Hull Forms—Series 65," SNAME, Chesapeake Section.

Harrington, R.L. (1971), *Marine Engineering,* SNAME.

Havelock, T.H. (1908), "The Propagation of Groups of Waves in Dispersive Media, With Application to Waves on Water Produced by a Travelling Disturbance," *Proceedings* of the Royal Society, London, Vol. 81.

Havelock, T.H. (1913), "Ship Resistance, the Wave-making Properties of Certain Travelling Pressure Disturbances," *Proceedings* of the Royal Society, London, Vol. 89.

Havelock, T.H. (1923), "Studies in Wave Resistance—Influence of the Form of the Water-Plane Section of the Ship," *Proceedings* of the Royal Society, London, Vol. 138.

Havelock, T.H. (1926), "Some Aspects of the Theory of Ship Waves and Wave Resistance," *Trans,* NECI. Vol. 42.

Havelock, T.H. (1928), "The Wave Pattern of a Doublet in a Stream," *Proceedings* of the Royal Society, London, Vol. 141.

Havelock, T.H. (1935), "Ship Waves—The Relative Efficiency of Bow and Stern," *Proceedings* of the Royal Society, London, Vol. 149.

Havelock, T.H. (1951), "Wave Resistance Theory and its Application to Ship Problems," SNAME *Transactions,* Vol. 59.

Hoerner, S.F. (1965), "Fluid Dynamic Drag," Hoerner Fluid Dynamics, Bricktown, New Jersey.

Holling, H.D. and Hubble, E.N. (1974), "Model Resistance Data of Series 65 Hull Forms Applicable to Hydrofoils and Planing Craft," DTNSRDC Report 4121, Annapolis, MD.

Holtrop, J. and Mennen, G.G.J. (1978), "A Statistical Power Prediction Method," ISP, Vol. 25.

Hughes, G. (1930), "Model Experiments on the Wind Resistance of Ships," *Trans.* INA, Vol. 72.

Hughes, G. (1932), "The Air Resistance of Ships' Hulls with Various Types and Distributions of Superstructures," *Trans.* IESS.

Hughes, G. and Allan, J.F. (1951), "Turbulence Stimulation on Ship Models," SNAME *Transactions,* Vol. 59.

Hughes, G. (1952), "Frictional Resistance of Smooth Plane Surfaces in Turbulent Flow," *Trans.* INA, Vol. 94.

Hughes, G. (1954), "Friction and Form Resistance in Turbulent Flow and a Proposed Formulation for Use in Model and Ship Correlation," *Trans.* INA, Vol. 96.

Hughes, G. (1957), "The Effect of Model and Tank

Size in Two Series of Resistance Tests," *Trans.* INA, Vol. 99.

Hughes, G. (1961), "Tank Boundary Effects on Model Resistance," *Trans.* RINA, Vol. 103.

Inui, T., Takahei, T. and Kumano, M. (1960), "Tank Experiments on the Wave-Making Characteristics of the Bulbous Bow, Part 1," *Trans.* JSNA, Vol. 108.

Inui, T. (1962), "Wave-making Resistance of Ships," SNAME *Transactions*, Vol. 70.

Inui, T., Kajitani, H. and Miyata, H. (1979), "Experimental Investigations on the Wave-Making in the Near-Field of Ships," *Journal* of the Kansai Society of Naval Architects, Vol. 173.

Inui, T. (1980), "From Bulbous Bow to Free Surface Shock Wave—Trend of Twenty Years Research on Ship Waves at the Tokyo University Tank," 3rd George Weinblum Memorial Lecture.

ITTC (1957), Proceedings of the 8th ITTC, Madrid, Spain, published by Canal de Experiencias Hidrodinamicas, El Pardo, Madrid.

ITTC (1963), Proceedings of the 10th ITTC, London, England, published by the National Physical Laboratory, England.

ITTC (1978), Proceedings of the 15th ITTC, The Hague, The Netherlands, published by the Netherlands Ship Model Basin, Wageningen, Netherlands.

ITTC (1981), Proceedings of the 16th ITTC, Leningrad, USSR. Published by the Krylov Shipbuilding Research Institute, USSR.

ITTC (1984), "Report of the High Speed Marine Vehicle Committee," Proceedings 17th ITTC, Göteborg, Sweden.

Johnson, N.V. (1958), "Test With Bulbous Bow on Trawlers," Chalmers Univ. Gothenburg, Report No. 1.

Johnson, R.S. (1965), "Prediction of Lift and Cavitation Characteristics of Hydrofoil-Strut Arrays," *Marine Technology,* Vol. 2.

Jφrgensen, H.D. and Prohaska, C.W. (1966), "Wind Resistance," *Proceedings* 11th ITTC, Tokyo.

JSS (1961), *Japan Shipping and Shipbuilding,* p.15.

Kaplan, P. and Breslin, J.P. (1955), "Evaluation of the Theory for the Flow Pattern of a Hydrofoil of Finite Span," Stevens Institute of Technology, Report No. 561.

Kayo, Y. and Takekuma, K. (1981), "On the Free Surface Shear Flow Related to Bow Wave Breaking of Full Ship Models," *Journal* of the Society of Naval Architects of Japan, Vol. 149.

Keil, U. and Schenzle, P. (1975), "Modellversuche mit Extrem Breiten Schiffsformen—Widerstandsversuche mit Extrem Breiten Schiffsformen (C_p = 0.77)," Institut für Schiffbau der Universität Hamburg, Bericht Nr. 333.

Keller, J. Auf'm (1973), "Extended Diagrams for Determining the Resistance and Required Power for Single Screw Ships," ISP, Vol. 20.

Kelvin, Lord (1887, 1904), "Ship Waves," *Trans.* IME, London. "On Deep Water 2-Dimensional Waves Produced by Any Given Initiating Disturbance, On the Front and Rear of a Free Procession of Waves in Deep Water, Deep Water Ship Waves," *Proceedings* of The Royal Society of Edinburgh, Vol. 25.

Kempf, G. and Sottorf, W. (1928), "Probefahrtsmessungen," *Werft, Rederei und Hafen.*

Kempf, G. (1929), "Results Obtained in Measuring Frictional Resistance," *Trans.* INA, Vol. 71.

Kerczek, C.H. von, Stern, F., Scragg, C.A. and Sandberg, W. (1983), "Total Resistance Calculations of Appended Destroyer Forms," *Proceedings,* 20th ATTC, Stevens Institute of Technology.

Kim, H.C. (1962), "Blockage Correction in Ship Model Testing," ATTC, University of Michigan, Ann Arbor.

Kirkman, K.L. and Kloetzli, J.W. (1980), "Scaling Problem of Model Appendages," ATTC, University of Michigan, Ann Arbor.

Korvin-Kroukovsky, B.V. and Wenninck, R.J. (1951), "A Note on the Lift and Drag of Hydrofoils; Application of Theory to Experimental Results," Stevens Institute of Technology Report No. 438.

Lackenby, H. (1952), "On the Acceleration of Ships," IESS.

Lackenby, H. (1955), "BSRA Resistance Experiments on the *Lucy Ashton,* Part III—The Ship-Model Correlation for the Shaft Appendage Conditions," *Trans.* INA, Vol. 97.

Lackenby, H. (1959), "Review of BSRA Trawler Research—Resistance of Trawlers," FAO 2nd World Fishing Boat Congress, Rome.

Lackenby, H. (1962), "The Resistance of Ships With Special Reference to Skin Friction and Hull Surface Condition," 34th Thomas Lowe Gray Lecture, IME, London.

Lackenby, H. (1963), "The Effect of Shallow Water on Ship Speed," *Shipbuilder and Marine Enginebuilder,* Vol. 70.

Lackenby, H. and Parker, M.N. (1966), "The BSRA Methodical Series—An Overall Presentation. Variation of Resistance with Breadth-Draught Ratio and Length-Displacement Ratio," *Transactions* INA, Vol. 108.

Lackenby, H. and Milton, D. (1970), "DTMB Standard Series 60—A New Presentation of the Resistance Data for Block Coefficients, LCB, Breadth-Draught Ratio and Length-Breadth Ratio Variations," RINA Paper W8.

Lamb, G.R. and Fein, J.A. (1979), "The Developing Technology for SWATH Ships," AIAA/SNAME 5th Conference on Advanced Marine Vehicles, Paper 79-2003.

Lammeren, W.P.A., Troost, L. and Koning, J.G. (1948), "Resistance, Propulsion and Steering of Ships," The Technical Publishing Company H. Stam, Haarlem, Holland. Also NSMB Publication No. 75, MARIN.

Landweber, L. (1939), "Tests of a Model in Restricted Channels," EMB Report 460, DTRC.

Lap, A.J.W. (1956), "Scale Effects in the Resistance of Bossings and Shaft Brackets," NSMB Publication 134a. Also Association Technique Maritime et Aeronautique, Paris.

Li Bai-Qi (1981), "A Prediction Method of Foilborne Performance Characteristics of Hydrofoil Craft in Calm Sea," Conference on High Speed Surface Craft, London.

Lin, C-W, Day, W. G. and Lin, W-C (1987), "Statistical Prediction of Ship's Effective Power Using Theoretical Formulation and Historic Data, *Marine Technology*, vol. 24, no. 3, July.

Lin, W.C. and Day, W. (1974), "The Still Water Resistance and Propulsion Characteristics of SWATH Ships," AIAA/SNAME Advanced Marine Vehicles Conference, Paper 74-32.

Lindblad, A. (1944), "Experiments with Bulbous Bows," SSPA Publication No. 3.

Lindblad, A. (1948), "Further Experiments with Bulbous Bows," SSPA Publication No. 8.

Lindblad, A. (1951), "Some Experiments With Self-Propelled Models of Twin Screw Ships," *Trans.* Chalmers Univ. No. 110, Gothenburg. Also INA (Vol. 88, 1946) (Vol. 91, 1949) and (Vol. 92, 1950).

Lindblad, A. (1961), "On the Design of Lines for Merchant Ships," *Trans.* Chalmers Univ., No. 240, Gothenburg.

Lindgren, H. and Williams, A. (1968), "Systematic Tests with Small, Fast Displacement Vessels, Including a Study of the Influence of Spray Strips," SNAME.

Lunde, J.K. (1957), "The Linearized Theory of Wave Resistance, etc.," SNAME Bulletin No. 1-18.

Luthra, G. (1971), "Untersuchung Widerstandsgünstiger Bugformen für Seeschiffe sehr Grösser Völligkeit," *Schiff und Hafen,* Vol. 23.

Luthra, G. (1975), "Untersuchung Widerstandsgünstiger Bugformen für Seeschiffe Sehr Grosser Völligkeit," *Schiff und Hafen,* Vol. 27.

Mandel, P. (1953), "Some Hydrodynamic Aspects of Appendage Design," SNAME *Transactions,* Vol. 61.

Mantle, P.J. (1976), "Cushions and Foils," SNAME Spring Meeting, Philadelphia.

Mantle, P.J. (1980), "Air Cushion Craft Development (first revision)," DTNSRDC Report No. 80/012 (4727 revised).

Maruo, H., Kasahara, K., Takusagawa, Z. and Okada, H. (1977), "A Trial of Determining High Speed Practical Hull Forms by the Aid of the Minimum Wave Resistance Theory," *Journal* of the Society of Naval Architects of Japan, Vol. 141.

Marwood, W.J. and Bailey, D. (1969), "Design Data for High-Speed Displacement Hulls of Round-Bilge Form," National Physical Laboratory, Ship Division, Report 99.

McGown, S. (1961), "Seaworthiness Problem in Small High-Speed Craft," SNAME Section paper, NY.

Mercier, J.A. and Savitsky, D. (1973), "Resistance of Transom-stern Craft in the Pre-planing Regime," Stevens Institute of Technology, Report 73-1667.

Michell, J.H. (1898), "The Wave Resistance of a Ship," *Philosophical Magazine,* London.

Michell, W.H. (1961), "The Seagoing Catamaran—Its Features and its Feasibility," *International Shipbuilding Progress,* Vol. 8.

Miyata, H., Inui, T. and Kajitani, H. (1980), "Free Surface Shock Waves Around Ships and Their Effects on Ship Resistance," *Journal,* JSNA, Vol. 147.

Müller-Graf, B. (1980), "Leistungsvorhersage schnellen Rund- und Knickspantboote," *Jahrbuch* der Schiffbautechnischen Gesellschaft, Vol. 74.

Muntjewerf, J.J. (1970), "Methodical Series Experiments on Cylindrical Bows," *Trans.* RINA, Vol. 112.

Muntjewerf, J.J. (1983), "Propulsion Research by Model Testing," *Jahrbuch* der Schiffbautechnischen Gesellschaft, Vol. 77. Also NSMB Publication No. 752.

Muntjewerf, J.J. and Jourdain, M. (1983), "Correlation Mer-Bassin; Application de la Méthode ITTC 1978 a Deux Méthaniers de 122000 m^3," AdMAR.

Myazawa, M. (1979), "A Study on the Flow Around a Catamaran," *Journal* of the Society of Naval Architects of Japan, Vol. 145.

Napier, J.R. (1865), "On the Most Profitable Speed for a Fully Laden Cargo Steamer for a Given Voyage," *Proceedings* of the Royal Philosophical Society of Glasgow.

Newman, J.N. and Poole, F.A.P. (1962), "Wave Resistance of a Moving Pressure Distribution in a Canal," DTNSRDC Report 1619, DTRC.

Nordstrom, H.F. (1948), "Some Systematic Tests With Models of Fast Cargo Vessels," Publication No. 10, SSPA.

Nordstrom, H.F. (1950), "Systematic Tests With Models of Cargo Vessels with Block Coefficients 0.575," Publication No. 16, SSPA.

Nordstrom, H.F. (1951), "Some Tests with Models of Small Vessels," Publication No. 19, SSPA.

Norley, W.H. (1948), "Shallow Water Effects on the Performance of SS Vessels of the US Maritime Commission, etc.," TMB Report 640, DTRC.

Numata, E. (1981), "Predicting Hydrodynamic Behavior of Small-Waterplane-Area Twin-Hull Ships," *Marine Technology,* Vol. 18.

Ogilvie, T.F. (1973), "The Wave Generated by a Fine Ship Bow," University of Michigan, Report 175.

Oortmerssen, G. van (1971), "A Power Prediction Method and its Application to Small Ships," ISP, Vol. 18.

Oortmerssen, G. van (1973), "A Power Prediction Method for Motor Boats," 3rd Symposium Yacht Architecture, The Netherlands, also NSMB publication 429.

Oossanen, P. van (1982), "Geavanceerde Scheepstypen," *Tweede Tideman Herdenking,* The Netherlands.

Ozawa, H., Shimada, K., Saito, T., Kobayashi, M., Nojiri, T., Yamashita, S. and Daimon, Y. (1977), "Experimental Study on the Resistance of Twin Hulled Air Cushion Vehicle," *Japan Shipbuilding and Marine Engineering,* Vol. 11.

Payne, M.P. (1936), "Historical Note on the Derivation of Froude's Skin Friction Constants," *Trans.* INA, Vol. 78.

Peck, R.W. (1976), "The Determination of Appendage Resistance of Surface Ships," AEW Technical Memorandum 76020.

Pien, P.C. and Lee, C.M. (1972), "Motion and Resistance of a Low-Waterplane Area Catamaran," 9th ONR Symposium on Naval Hydrodynamics, Paris.

Prandtl, L. (1904) "Ueber Fluessigkeitsbewegung bei Kleiner Reibung," International Mathematical Congress, Heidelberg.

Prandtl, L., (1921), "Ergebnisse der Aerodynamischen Versuchsanstalt zu Goettingen," Abhandlungen aus dem Aerodynamischen Institut, Aachen.

Prohaska, C.W. (1966), "A Simple Method for the Evaluation of the Form Factor and Low Speed Wave Resistance," *Proceedings,* 11th ITTC.

Reynolds, O. (1883), *Philosophical Transactions* of the Royal Society, London.

Ridgely-Nevitt, C. (1956), "The Resistance of Trawler Hull Forms of 0.65 Prismatic Coefficient," SNAME *Transactions,* Vol. 64.

Ridgely-Nevitt, C. (1963), "The Development of Parent Hulls for a High Displacement-Length Series of Trawler Forms," SNAME *Transactions,* Vol. 71.

Roseman, D.P. (1987) *The MARAD Systematic Series of Full-form Ships,* SNAME.

Rota, G. (1900), "On the Influence of Depth of Water on the Resistance of Ships," *Trans.* INA, Vol. 42.

Rumble, H.P. (1955), Discussion on paper "The Admiralty Experiment Works, Haslar," by R.W.L. Gawn, *Trans.* INA, Vol. 97.

Russo, V.L., Turner, H. and Wood, F.W. (1960), "Submarine Tankers," SNAME *Transactions,* Vol. 68.

Sabit, A.S. (1971), "Regression Analysis of the Resistance Results of the BSRA Series," ISP, Vol. 18.

Sabit, A.S. (1972), "An Analysis of the Series 60 Results, Pt 1. Analysis of Forms and Resistance Results," ISP, Vol. 19.

Sabit, A.S. (1972), "An Analysis of the Series 60 Results, Pt 2. Regression Analysis of the Propulsion Factors," ISP, Vol. 19.

Sabit, A.S. (1976), "The SSPA Cargo Liner Series. Regression Analysis of the Resistance and Propulsion Coefficients," ISP, Vol. 23.

Salvesen, N., Kerczek, C.H. von, Scragg, C.A., Cressy, C.P. and Meinhold, M.J. (1985) "Hydro-Numeric Design of SWATH Ships," SNAME *Transactions,* Vol. 93.

Sasajima, H. and Yoshida, E. (1955), "Frictional Resistance of Wavy Roughened Surfaces," ISP, Vol. 2.

Saunders, H.E. (1938–41) "The David W. Taylor Model Basin," Part 1, Part 2, and Part 3, SNAME *Transactions,* Vol. 46, 48, 49.

Saunders, H.E. (1957), *Hydrodynamics in Ship Design,* Vol. I, Vol. II, SNAME.

Savitsky, D. and Dingee, D.A. (1984), "Some Interference Effects Between Two Flat Surfaces Planing Parallel to Each Other at High Speed," *Journal of the Aeronautical Sciences,* Vol. 21.

Savitsky, D. (1964), "Hydrodynamic Design of Planing Hulls," *Marine Technology,* Vol. 1.

Savitsky, D. and Brown, P.W. (1976), "Procedures for Hydrodynamic Evaluation of Planing Hulls in Smooth and Rough Water," *Marine Technology,* Vol. 13. Also Stevens Institute of Technology, Davidson Laboratory Report No. SIT-DL-75-1859, 1975.

Savitsky, D. and Gore, J.L. (1979), "A Re-evaluation of Planing Hull Forms," AIAA/SNAME Advanced Marine Vehicles Conference.

Savitsky, D., Stevens, M.J., Balquet, R.J., Müller-Graf, B., Murakami, T., Prokhorov, S.D., Oossanen, P. van and Wilson, R.A. (1981), "Status of Hydrodynamic Technology as Related to Model Tests of High-Speed Marine Vehicles," DTNSRDC Report 81/026, Bethesda, MD.

Savitsky, D., Knowles, P.C., Aláez, J.A., Müller-Graf, B., Murakami, T., Oossanen, P. van, Prokhorov, S.D., Rutgersson, O. and Wilson, R.A. (1984), "Bibliography and Proposed Symbols on Hydrodynamic Technology as Related to Model Tests on High-Speed Marine Vehicles," SSPA Public Research Report No. 101.

Schlichting, O. (1934), "Ship Resistance in Water of Limited Depth—Resistance of Sea-Going Vessels in Shallow Water," *Jahrbuch der STG,* Vol. 35 (see also EMB Translation 56, 1940).

Schoenherr, K.E. (1932), "Resistance of Flat Surfaces Moving Through a Fluid," SNAME *Transactions,* Vol. 40.

Shearer, K.D.A. and Lynn, W.M. (1960), "Wind Tunnel Tests on Models of Merchant Ships," *Trans.* NECI, Vol. 76.

Sherman, J. and Fisher, P. (1975), "A Study of Planing Catamaran Hull and Tunnel Interactions," University of Michigan, Ann Arbor, MI.

SNAME (1948), "Uniform Procedure for the Calculation of Frictional Resistance, etc., "SNAME Bulletin 1-2; replaced by Bulletin 1-25, (SNAME (1964).

SNAME (1962), "Standardization Trials Code," SNAME, 1949 reprinted 1962.

SNAME (1976), "Tables of Coefficients for ATTC Model-Ship Correlation and of Kinematic Viscosity and Density of Fresh and Salt Water," SNAME Bulletin 1-25.

SNAME (1973), "Index of Methodical Series Ship Model Resistance Tests," T & R Bulletin 1-31.

SNAME (undated), Project 2 of the Hydromechanics Sub-Committee of SNAME: "Model and Expanded Resistance Data Sheets," available from the Society.

Stanton, T.E. and Pannell, J.R. (1952), "Similarity of Motion in Relation to the Surface Friction of Fluids," *Trans.* of the Royal Society, London, Vol. 214.

Tagano, H. (1973), "Form Effects on Viscous Resistance of Full Ships," Kansai Society of Naval Architects of Japan.

Tagano, H. (1974), "Prediction of the Wave Resistance of Ships by Statistical Analysis," Mitsubishi *Technical Bulletin No. 90*, Tokyo.

Tanaka, I. (1979), "Scale Effects on Wake Distribution and Viscous Pressure Resistance for Ships," *Journal* of the Society of Naval Architects of Japan, Vol. 146.

Taneda, S. and Amamoto, A. (1969), "On the Necklace Vortex," Bulletin of the Research Institute Applied Mechanics, Kyushu University, No. 31.

Taniguchi, K., Watanabe, K. and Tamura, K. (1966), "On a New Method of Designing the Hull Form of a Large Full Ship, Based on the Separability Principle of Ship Form," BSRA Translation No. 2661.

Tasaki, R. (1962), "A Note on Wavemaking Resistance of Catamarans," Technical Report of the University of Michigan, Ann Arbor, MI.

Taylor, D.W. (1908), "The Influence of Midship Section Shape Upon the Resistance of Ships," SNAME *Transactions*, Vol. 16.

Taylor, D.W. (1911), "Some Model Basin Investigations of the Influence of Form of Ships Upon Their Resistance," SNAME *Transactions*, Vol. 19.

Taylor, D.W. (1943), "*The Speed and Power of Ships*," Second Revision, U.S. Maritime Commission.

Taylor, E.S. (1974), "Dimensional Analysis for Engineers," Clarendon Press, Oxford, England.

Taylor, G.I. (1950), "The Instability of Liquid Surfaces when Accelerated in a Direction Perpendicular to Their Plane," *Proceedings* Royal Society, Vol. 201.

Telfer, E.V. (1927, 1950, 1951, 1952), "Ship Resistance Similarity," *Trans.* INA, Vol. 69, Vol. 92, Vol. 93, Vol. 94.

Telfer, E.V. (1933), "Miscellaneous Notes," ICSTS, The Hague, published by NSMB, Wageningen.

Telfer, E.V. (1953), "Ship Model Correlation and Tank Wall Effect," *Trans.* NECI, Vol. 70.

Telfer, E.V. (1963), The Design Presentation of Ship Model Resistance Data," NECI, Vol. 79, MARIN.

Tideman, B.J. (1876), "Results of Resistance Tests with Ship Models," *Memoriael van de Marine II.* (A list of the frictional coefficients is given in Taylor, D.W. (1943), Table V, p. 34).

Timman, R. and Vossers, G. (1955), "On Michell's Expression for the Velocity Potential of the Flow Around a Ship," *International Shipbuilding Progress*, Vol. 2.

Todd, F.H. (1931), (1934), (1938), (1940), (1942), "Further Model Experiments on the Resistance of Mercantile Ship Forms-Coaster Vessels," *Trans.* INA 1931 and 1938; NECI 1934 and 1942 and IME, 1940.

Todd, F.H. (1934), "The Effect of a Fin upon the Efficiency of Ship Propulsion," Liverpool Engineering Society.

Todd, F.H. (1945), "The Fundamentals of Ship Form," *Trans.* IME, London, Vol. 57.

Todd, F.H. (1951), "The Fundamentals of Ship Model Testing," SNAME *Transactions*.

Todd, F.H. (1960), "Submarine Cargo Ships and Tankers," 3rd Symposium on Naval Hydrodynamics, Office of Naval Research.

Todd, F.H. (1963), "Series 60-Methodical Experiments with Models of Single-Screw Merchant Ships," TMB Report No. 1712, DTRC.

Townsend, H.S. (1960), "Some Observations on the Shape of Ship Forebodies With Relation to Heavy Weather," SNAME Section paper.

Troost, L. (1955), "A Simplified Method for Preliminary Powering of Single-Screw Merchant Ships," SNAME Section paper.

Tsunoda, R., Kasahara, K. and Takekuma, K. (1963), "A High-Speed Cargo-Liner Design Based Upon the 'With-Bulb' Waveless Concept," International Seminar, University of Michigan, Ann Arbor, MI.

Turner, H. and Taplin, A. (1968), "The Resistance of Large Powered Catamarans," SNAME *Transactions*, Vol. 76.

Tursini, L. (1953), "Leonardo da Vinci and the Problems of Navigation and Naval Design," *Trans.* INA, Vol. 95.

Van Berkelom, see Berkelom, van.

Van Oortmerssen, see Oortmerssen, van.

Van Oossanen, see Oossanen, van.

Vollheim, R. (1968), "Über Formgebung und Widerstand von Katamaranen," *Schiffbauforschung*, Vol. 7.

Von Kerezek, see Kerezek, von

Wagner, B. (1967), "Windkrafte an Ueberwasserschiffen," *Jahrbuch* Schiffbautechnische Gesellschaft, Band 61.

Ward, L.W. (1962), "A Method for the Direct Experimental Determination of Ship Wave Resistance," thesis, Stevens Institute of Technology.

Warholm, A.O. (1953), "Systematic Tests With Models of Coasters," SSPA Publication No. 24.

Wehausen, J.V. (1973), "The Wave Resistance of Ships," *Advances in Applied Mechanics*, Vol. 13.

Weinblum, G.P. (1950), "Analysis of Wave Resistance," TMB Report 710, DTRC.

Wieghardt, K. (1973), "On Wind Resistance," International Jubilee Meeting on the Occasion of the 40th Anniversary of NSMB.

Wieghardt, K. (1976), "Remarks to the Viscous Ship Resistance," International Seminar on Ship Technology, Hydrodynamics Session, Seoul.

Wigley, C. (1931), "Ship Wave Resistance," *Trans.* NECI, Vol. 47.

Wigley, C. (1934), "A Comparison of Experiment and Calculated Wave Profiles and Wave Resistance for a Form Having Parabolic Waterlines," *Proceedings* of the Royal Society, London.

Wigley, C. (1935–36), "The Theory of the Bulbous Bow and Its Practical Application," *Trans.* NECI, Vol. 52.

Wigley, C. (1938), "Effect of Viscosity on Wave-Making of Ships," *Trans.* IESS.

Wigley, C. (1942), "Calculated and Measured Wave-Making Resistance for a Series of Forms Defined Algebraically, the Prismatic Coefficient and Angle of Entrance Being Varied Independently," *Trans.* INA, Vol. 84.

Wigley, C. and Lunde, J.K. (1948), "Calculated and Observed Wave Resistance for a Series of Forms of Fuller Midship Sections," *Trans.* INA, Vol. 90.

Wigley, C. (1962), "The Effect of Viscosity on Wave Resistance," *Schiffstechnik,* band 9.

Wilson, C.J. and Roddy, R.F. (1970), "Estimating the Wind Resistance of Cargo Ships and Tankers," NSRDC Report 3355.

Wilson, R.A., Wells, S.M. and Heber, C.E. (1978), "Powering Prediction for Surface Effect Ships Based on Model Results," AIAA / SNAME Advanced Marine Vehicles Conference, Paper No. 78-744 (Also *Journal of Hydronautics* 1979).

Yeh, H.Y.H. (1965), "Series 64 Resistance Experiments on High-Speed Displacement Forms," *Marine Technology,* Vol. 2.

Yokoo, K. (1966), "Roughness of Hull Surface and its Effect on Skin Friction," Appendix 2, Performance Committee Report, 11th ITTC.

Yermotayev, S.G., Aframeyev, E.A., Teder, L.A. and Rabinowich, Ya.S., (1977), "Hydrodynamic Features of High Speed Catamarans," *Hovercraft and Hydrofoil,* Vol. 16.

Young, A.D. (1939), "The Calculation of the Total and Skin Friction Drag of Bodies of Revolution at Zero Incidence," Aeronautical Research Committee, R & M 1874, London.

CHAPTER VI

J. D. van Manen
P. van Oossanen

Propulsion

Section 1
Powering of Ships

1.1 Historical. A moving ship experiences resisting forces from the water and air which must be overcome by a thrust supplied by some thrust-producing mechanism. In the earliest days this consisted of manually-operated oars, which gave place in turn to sails and then mechanical devices such as jets, paddle-wheels and propellers of many different forms (Todd, 1946).[1]

The earliest propulsive device to use mechanical power seems to have been of the jet type, using a prime-mover and a pump, patents for which were granted to Toogood and Hayes in Great Britain in 1661. Water is drawn in by the pump and delivered sternwards as a jet at a higher velocity, the reaction providing the thrust. At the speeds so far attained by ships, the jet is materially less efficient than other forms of propellers, and its use has been restricted to special types of craft.

In 1801 there appeared the first steam-driven side-paddle ship, the *Charlotte Dundas;* built by Symington for service on the Forth-Clyde Canal in Scotland. Six years later came the famous *Clermont,* constructed by Robert Fulton for passenger service on the Hudson River in New York.

The period from this time until about 1850 was the heyday of the paddle steamers. The first of them to cross the Atlantic was the American *Savannah* in 1819—a full-rigged ship with auxiliary steam power—and then followed a line of familiar names, including the Canadian *Royal William,* the famous first Cunarder *Britannia* in 1840, culminating in the last Cunard liner to be driven by paddles, the *Scotia,* in 1861.

These side paddle-wheels were far from ideal for sea-going ships. The immersion varied with ship displacement, the wheels came out of the water when the ship rolled, causing erratic course-keeping, and they rine engineer's point of view, they were too slow-running, involving the use of large, heavy engines. Because of the slow rate of turning they were reasonably efficient as a propulsive device, but their other operational weaknesses ensured their rapid decline from popularity once the screw propeller was proved to be an acceptable alternative. They have had a useful field among pleasure steamers and tugs plying in rivers and other protected waters. In such craft the draft does not change much and restrictions of draft due to shallow water prohibit the use of large screw propellers. Side paddles also give good maneuvering characteristics, but these latter can now be obtained by other means of propulsion which do not suffer from the drawbacks of paddle-wheels.

Paddles have also been fitted at the sterns of many ships, as in the well-known river boats on the Mississippi and other American rivers. Such "stern-wheelers" are still in use, mainly as passenger carriers.

The first proposal to use a screw propeller appears to have been made in England by Hooke in 1680, and its first actual use is generally attributed to Colonel Stevens in a steam-driven boat at New York in 1804. In 1828 a vessel 18 m (60 ft) long was successfully propelled by a screw propeller designed by Ressel, of Trieste, obtaining a speed of 6 knots, but this success was not followed by the Trieste engineers or shipowners (Baker, 1944). The first practical applications came in 1836 by Ericsson in the U.S. and Pettit Smith in England.

The screw propeller has many advantages over the paddle-wheel. It is not materially affected by normal changes in service draft, it is well protected from damage either by seas or collision, it does not increase the overall width of the ship, and it can be made to run much faster than paddles and still retain as good or better efficiency so that smaller, lighter, faster-running engines can be used. It rapidly superseded the paddle-wheel for all ocean-going ships, the first screw-pro-

[1] Complete references are listed at end of chapter.

pelled steamer to make the Atlantic crossing being the *Great Britain* in 1845.

From that time the screw propeller has reigned supreme in the realm of marine propulsion. It has proved extraordinarily adaptable in meeting the incessant quest for propellers to deliver more and more thrust under increasingly arduous conditions. While other devices have been adopted for certain particular types of ships and kinds of service, the screw propeller still has no real rival in the field of ship propulsion.

Among the more common variants of the propeller, the use of a shroud ring or nozzle has been shown to have considerable advantages in heavily loaded propellers, the ring or nozzle being shaped so as to deliver a forward thrust to the hull. The principal advantage is found in tugs, where the pull at the bollard for a given horsepower may be increased by as much as 40 percent or more as compared with that given by an open propeller. At low towing speeds, a considerable advantage is still found, but this disappears with increasing speed, and when free-running the drag of the nozzle results in a loss of speed. In ships other than tugs, the advantage can be extended to higher speeds by using thinner nozzles, with some loss of thrust at the low speeds, and such arrangements in association with special forms of stern lines have been claimed to give good propulsive efficiencies. Good maneuverability can be obtained in such craft by arranging for the nozzle to swivel, and so act as a very efficient rudder by controlling the direction of the propeller race.

Another type of propeller was used in the USS *Alarm* as long ago as 1874 (Goldsworthy, 1939). This ship carried a fixed bow gun and had to be turned to aim the gun. To keep the ship steady in a tideway, where a rudder would be useless, a feathering paddle-wheel rotating about a vertical axis, invented by Fowler in Great Britain in 1870, was fitted at the stern, completely submerged (White, 1882). It was quite successful as a means of maneuvering the ship, but its propulsive efficiency was low. The modern version of this propeller consists of a large disk set flush with the lower surface of a flat counter and carrying a number of projecting vertical blades rather resembling spade rudders. As the disk revolves about a vertical axis, each of these blades turns about its own vertical axis, being so adjusted to the flow that the total thrust from all the blades is concentrated in one direction. This resultant "thrust-direction" can be controlled by varying the blade motions, so as to drive the ship ahead, astern or sideways. The device therefore lends itself essentially to craft which need to have great ability to maneuver. It also enables the equivalent of a large diameter, orthodox propeller to be fitted to ships which have to operate in shallow water, and the propeller can be driven through gearing by relatively light, high-speed diesel engines. Although its efficiency is not as high as that of the orthodox propeller, and its maintenance is probably more costly, the foregoing advantages have resulted in many applications to river steamers, tugs, and ferries. The vertical axis propeller is discussed further in Section 10.5.

1.2 Types of Ship Machinery. In selecting the propelling machinery for a given vessel, many factors must be taken into consideration, such as the weight, the space occupied, its first cost, reliability, length of life, flexibility and quietness of operation, cost of upkeep, cost of fuel consumed and last, but not least, its suitability for the type of propeller to be used. It is beyond the scope of this text to consider all the various drives which have been developed to meet these factors, but a brief review of their advantages and disadvantages will not be out of place.

The reciprocating steam engine with two, three, or four cylinders dominated the field of ship propulsion until about 1910. Since then it has been almost entirely superseded by the steam turbine in the very high and intermediate-power ranges, and by the diesel engine in intermediate and low ranges.

The steam reciprocating engine has exceptional controllability at all loads, is easily reversed (an important consideration in ships) and its most efficient range of revolutions per minute (RPM) matches that of the screw propeller. On the other hand, the complete plant is relatively heavy, occupies much space, and the output of power per cylinder is limited. Also, the steam cannot be expanded effectively to the low pressures obtainable in modern condensing apparatus, so that the fuel consumption is rather high, an average figure for a triple-expansion engine utilizing superheated steam being about 0.70 kg of oil per kWhr (1.15 lb per hphr).

The first marine turbine was installed by Sir Charles Parsons in the *Turbinia* in 1894, a torpedo boat which attained a speed of 34 knots. Thereafter turbines made rapid progress and by 1906 were used to power the epoch-making battleship HMS *Dreadnought* and the famous Atlantic liner *Mauretania*.

The turbine delivers a uniform turning effort, is eminently suitable for large-unit power output, and can utilize very high-pressure inlet steam over a wide range of power to exhaust at very low pressures. The thermal efficiency is consequently reasonably high and the fuel consumption of large turbines is as low as 0.30 kg of oil per kWhr (0.49 lb per hphr). Under overload conditions a turbine delivers approximately constant power for a given throttle setting.

On the other hand, the turbine is nonreversible and its rotational speed for best economy is far in excess of the most efficient rpm of usual propeller types. These drawbacks make it necessary to install separate reversing turbines and to insert gears between the turbines and the propeller shaft to reduce the speed of the latter to values more suitable to the propeller.

The mechanical geared drive has been used most widely up to the present. It permits the operation of engine and propeller at their most economical speeds with a power loss in the gears of only 2 to 4 percent. A separate astern turbine is still required, which adds to cost, complexity, and power loss.

The reduction in RPM between turbine and propeller shaft can also be attained by electrical means. The turbine in such an installation is directly coupled to a generator, both running at the same high speed for efficient operation. The generator supplies a motor directly mounted on the propeller shaft, driving the latter at the RPM most desirable for high propeller efficiency. This system eliminates any direct shafting between turbines and propeller, and so gives the naval architect greater freedom in laying out the general arrangement of the ship to best advantage. In twin-screw ships fitted with two sets of turboalternators, considerable economy can be achieved when using half power, such as when a passenger ship is cruising, by supplying both propulsion motors from one turbine. The turboelectric drive also eliminates the reversing turbine, gives great flexibility and rapidity of maneuvering, and prevents racing of the propeller.

These advantages are gained, however, at the expense of rather high first cost and somewhat greater transmission losses.

Internal-combustion engines used for ship propulsion are generally reciprocating engines operating on the diesel[2] principle (compression ignition) which have taken their name from the man who first developed them for practical use. They are built in all sizes, from those fitted in small pleasure boats to the very large types fitted in modern supertankers and passenger liners. The engines in the latter ships develop over 2500 kW per cylinder, giving output as high as 30,000 kW in 12 cylinders (40,200 hp). They are directly reversible, occupy relatively little space, and have a very low fuel consumption, an average figure being around 0.20 kg of oil per kWhr (0.328 lb per hphr). They are used in large single units directly coupled to the propeller or in sets of small units driving the propeller through electric or gear transmissions. Opposed to these advantages are the facts that diesel engines are usually heavier and more expensive, both in first cost and in upkeep than steam plants of corresponding size.

The torque produced by a diesel engine is limited by the maximum pressure that may be developed in each cylinder. Therefore, when the engine is producing maximum torque, it produces maximum power only at maximum rpm. Consequently a diesel may produce a power directly proportional to the RPM for any throttle setting.

This limitation leads to the problem of matching a diesel engine and a propeller. The resistance will increase with time because of fouling and the propeller thrust decreases for the same reason. Therefore the load on the prime mover will increase to maintain the same speed. This requires the designer to select the adequate propeller particulars (such as pitch) so that later, in the life of the vessel, the engine does not become overloaded or that it never produces its full capabilities, see Kresic et al (1983).

More recently, gas turbines have been developed in which the fuel is burned in compressed air and the resulting hot gases passed through the turbine. The gas turbine originated in aeronautical applications, and its progress has depended mostly upon the development of metals which could withstand the high pressures and temperatures. It has the advantages of dispensing with boilers, being light in weight and giving a smooth, continuous drive. It is expensive in the quantity of fuel burned. One good operational characteristic is that it can quickly be brought on to full load without a long, warming-up period, some 15 min usually being sufficient after the warning to "raise steam" from cold. Marine gas turbines were fitted to a small number of merchant ships. But they are now frequently used in naval ships, sometimes associated with a diesel, steam turbine or smaller gas turbine. The latter are used for general cruising purposes, and the gas turbine is available at little or no notice when there is a demand for full power, both plants being connected to a common propeller shaft by clutches and gearing. The principal marine application so far has been to small and large destroyers and frigates and to smaller, high-speed craft, such as patrol craft and hydrofoils.

Nuclear reactors have been installed on many naval ships and in a few merchant ships and ice breakers. They replace the boilers being used, through a heat exchanger, to raise steam which is then passed to a turbine in the normal way. They also eliminate most of the weight and volume of fuel oil. The reactor can operate at full load indefinitely during the life of the charge of nuclear fuel, which enables the ship to maintain high speed at sea without carrying a large quantity of consumable fuel. The weight saved, however, cannot as a rule be devoted to increase dead-weight earning capacity, for the weight of reactor and shielding will equal or exceed that of the boilers and fuel for the normal ship.

1.3 Definition of Power. The various types of marine engines are not all rated on the same basis, inasmuch as it is inconvenient or impossible to measure their power output in exactly the same manner. Steam reciprocating engines are usually rated in terms of indicated power (P_I), internal-combustion engines in terms of indicated power or brake power (P_B), and turbine in shaft power (P_S). The term *horsepower* is still sometimes used, where 1 hp = 0.7457 kW. In English units 1 hp = 550 ft-lb per sec.

Indicated power is measured in the cylinders by means of an instrument (an *indicator*) which records continuously the steam or gas pressure throughout

[2] After Rudolf Diesel, a German engineer (1858–1913).

the length of the piston travel. From the resultant indicator card the mean effective pressure is measured and P_I is calculated for top end and bottom end separately:

$$P_I - p_M L A n \text{ in kW}$$

where

p_M is mean effective pressure, kN/m²
L is length of piston stroke, m
A is effective piston area, sq m
n is number of working strokes per sec

The total P_I of the engine is found by adding those calculated separately for all the cylinders.

Brake power is the power measured at the crankshaft coupling by means of a mechanical, hydraulic or electrical brake. It is determined by a shop test and is calculated by the formula

$$P_B = 2\pi Q n \text{ in kW}$$

where

Q is brake torque, kN-m
n is revolutions per sec

Shaft power is the power transmitted through the shaft to the propeller. It is usually measured aboard ship as close to the propeller as possible by means of a torsionmeter. This instrument measures the angle of twist between two sections of the shaft, which angle is directly proportional to the torque transmitted. For a solid, circular shaft the shaft power is given by

$$P_S = \frac{\pi (d_s)^4\, G\, \theta\, n}{32\, L_S} \text{ in kW}$$

where

d_s = shaft diameter, m
G = shear modulus of elasticity of shaft material. kN/m²
θ = measured angle of twist, deg
L_S = length of shaft over which θ is measured, m
n = revolutions per sec

The shear modulus G for steel shafts is usually taken as 8.35×10^7 kN/m².

For exact results, particularly with bored shafting, it is customary to calibrate the shaft by setting up the length of shafting on which the torsionmeter is to be used, subjecting it to known torques and measuring the angles of twist, and determining the calibration constant $K = Q L_S / \theta P_S$ can then be calculated directly from any observed angle of twist and revolutions per second as

$$P_S = K \times \frac{\theta}{L_S} \times 2\pi n$$

There is some power lost in the stern tube bearing and in any shaft tunnel bearings between the stern tube and the site of the torsionmeter. The power actually delivered to the propeller is therefore somewhat less than that measured by the torsionmeter. This *delivered power* is given the symbol P_D.

As the propeller advances through the water at a speed of advance V_A, it delivers a thrust T, and the thrust power is

$$P_T = TV_A$$

Finally, the *effective power* is

$$P_E = RV$$

1.4 Propulsive Efficiency. The efficiency of an engineering operation is generally defined as the ratio of the useful work or power obtained to that expended in carrying out the operation.

In the case of a ship the useful power obtained is that used in overcoming the resistance to motion at a certain speed, which is represented by the effective power P_E.

The power put in to achieve this result is not so easily defined. In a ship with reciprocating engines, it can be measured by the power developed in the cylinders themselves as given by the indicated power, P_I. The overall propulsive efficiency in this case would be expressed by the ratio P_E/P_I.

In the case of turbines it is usual to measure the power in terms of the shaft power delivered to the shafting abaft the gearing, and the overall propulsive efficiency is P_E/P_S.

Since mechanical efficiencies, gear losses and shaft-transmission losses all vary from ship to ship, according to the type of machinery and general layout, and even in a given ship with the load at which the machinery is operating at a particular time, it is difficult to define the hydrodynamic efficiency of a hull-propeller combination in terms of such an overall propulsive efficiency.

A much more meaningful measure of efficiency of propulsion is the ratio of the useful power obtained, P_E, to the power actually delivered to the propeller, P_D. This ratio has been given the name *quasi-propulsive* coefficient, and is defined as

$$\eta_D = \frac{P_E}{P_D} \tag{1}$$

The shaft power is taken as the power delivered to the shaft by the main engines aft of the gearing and thrust block, so that the difference between P_S and P_D represents the power lost in friction in the shaft bearings and stern tube. The ratio P_D/P_S is called the *shaft transmission* efficiency.

In this text, the propulsive efficiency is defined as follows:

Propulsive efficiency = quasi-propulsive coefficient times shaft transmission efficiency

or

$$P_E/P_S = P_E/P_D \times P_D/P_S$$

or

$$\eta_P = \eta_D \times \eta_S \tag{2}$$

The shaft transmission loss is usually taken as about 2 percent for ships with machinery aft and 3 percent for those with machinery amidships, but no very exact information exists on this point. It must be remembered also that when using the power measured by torsionmeter, the answer will depend on the position of the meter along the shaft. To approach as closely as possible to the power delivered to the propeller, it should be as near to the stern tube as circumstances permit. It is often assumed that $\eta_S = 1.0$.

The necessary brake power or indicated power in the turbines, diesel or steam-reciprocating engine, as the case may be, can be estimated in a particular design from the proper values of gear efficiency, mechanical efficiency and load factors. Values of these will be found in textbooks, handbooks and papers on marine engineering.

Section 2
Theory of Propeller Action

2.1 Momentum Principle. Propellers derive their propulsive thrust by accelerating the fluid in which they work. This action is in accordance with Newton's laws of motion, which state that force is required to alter the existing state of motion of any material body in magnitude or direction, and that the action of any two bodies upon one another is equal and opposite.

Newton's first law is expressed by the equation

$$F = m\frac{dv}{dt} \tag{3}$$

where

F = force exerted on body
m = mass of body
$\frac{dv}{dt}$ = resulting acceleration of body

Integrating between 0 and t seconds, we get

$$\int_0^t F dt = mv_2 - mv_1 \tag{4}$$

where v_1 and v_2 are the velocities at the beginning and end of the time interval.

The expression

$$\int_0^t F dt$$

is called the impulse of the force in the time interval zero to t, and the product of mass and velocity is called the momentum. The equation states that the impulse of the force in a given time interval is equal to the whole change in momentum produced by the force during this interval. In the special case when F is constant during the time interval, Equation (4) reduces to

$$Ft = mv_2 - mv_1$$

Furthermore, when the time interval is 1 sec,

$$F = mv_2 - mv_1$$

Hence in the case of a constant force the change in momentum in unit time is equal to the force which produced it.

Momentum and impulse are vector quantities, and to determine the direction and magnitude of the final velocity when the direction and magnitude to the force and of the initial velocity are given, the rules of vector composition must be applied.

2.2 General Discussion of Propeller Theories. The physical explanation of propeller action can be stated in rather simple terms, as shown in the preceding section, but the precise mathematical analysis presents considerable difficulties. As a result a satisfactory propeller theory which could explain all the observed facts and be useful for practical calculations was not developed until comparatively recent times.

The early propeller theories followed two independent lines of thought. In the first of these, the momentum theories, the production of thrust was explained entirely by momentum changes taking place in the fluid. In the second, the blade-element theories, the propeller thrust was obtained by analyzing the forces acting on the various sections of the blades and then integrating these over the propeller radius.

The momentum theories were based on correct fundamental principles, but gave no indication of the propeller form which would produce the calculated thrust. The propeller was idealized as an "actuator disk" or some similar conception, which could cause an instantaneous increase in pressure in the fluid passing through it. They led, however, to the important conclusion that the efficiency of an ideal propeller has an upper limit which varies with the loading. The blade element theories, on the other hand, were capable of predicting the effects of various changes in propeller form, but led to the incorrect result that the efficiency of an ideal propeller was unity.

The difference between the two groups of theories was not dispelled until the circulation theory developed by Lanchester in aerodynamic research was applied to the propeller problem by Betz and Prandtl. This theory showed the relation between the momentum changes in the medium and the forces acting on the blade elements, and its subsequent development finally led to the point where it is not only in good agreement with experimental results but also is flexible enough for practical design work.

2.3 The Momentum Theory of Propeller Action. The momentum theories were originally due to Rankine (1865), Greenhill (1888) and Froude, R. E. (1889). In the ideal conception of the propeller, it is regarded as a "disk" or mechanism capable of imparting a sudden increase of pressure to the fluid passing through it, the method by which it does so being ignored.

It is assumed that:

(*a*) The propeller imparts a uniform acceleration to all the fluid passing through it, so that the thrust thereby generated is uniformly distributed over the disk.

(*b*) The flow is frictionless.

(*c*) There is an unlimited inflow of water to the propeller.

The first assumption involves a contraction of the race column passing through the disk, and since this contraction cannot take place suddenly at the disk, the actual acceleration must occur outside the disk and be spread over a finite distance fore and aft.

Consider a propeller disk of area A_0 advancing with uniform velocity V_A into undisturbed fluid. The hydrodynamic forces will be unchanged if we replace this system by a stationary disk in a uniform flow of the same velocity V_A, as shown in Fig. 1.

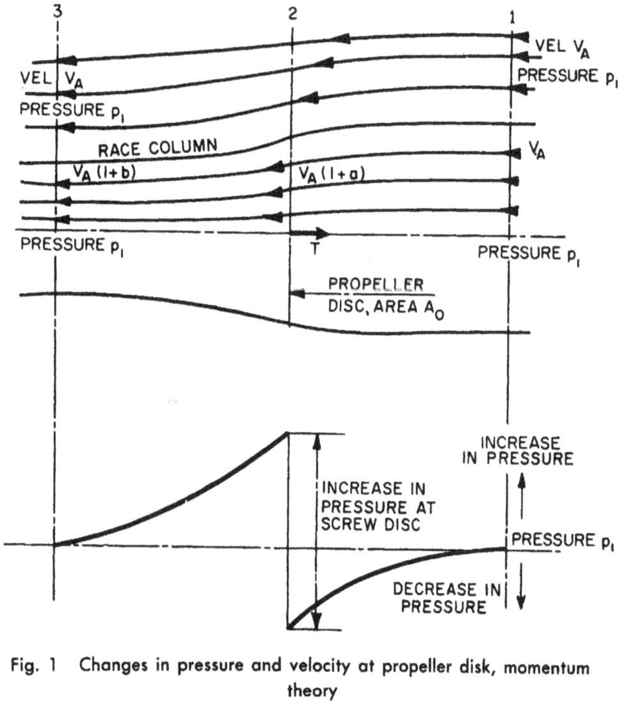

Fig. 1 Changes in pressure and velocity at propeller disk, momentum theory

At the cross section 1, some distance well ahead of the disk, the velocity of the flow is V_A and the pressure in the fluid is p_1. Well behind the screw, at section 3, the race column, i.e., the fluid which has passed through the screw disk and been acted upon by the pressure or thrust-producing mechanism there, will have some greater sternward velocity, which we may write as $V_A\,(1 + b)$. The fluid must acquire some of this increased velocity before it reaches the disk, and the velocity through it, at section 2, will be greater than V_A, and we may write it as $V_A\,(1 + a)$, where a is an axial-inflow factor.

The pressure in the race column, which is p_1 well ahead of the disk, will be reduced as the fluid approaches the disk, since by Bernoulli's law an increase in velocity is accompanied by a decrease in pressure. At the disk, the pressure is suddenly increased by some unspecified mechanism to some value greater than p_1, and then decreases again with the further acceleration in the race. If section 3 is so far aft of the disk that the contraction of the race may be assumed to have ceased, and if there is no rotation in the race, the pressure in the race at section 3 will be p_1, equal to that in the fluid outside the race.

The quantity of water passing through the disk in unit time will be

$$Q = V_A(1 + a)A_0$$

Neglecting any effect of rotation which may be imparted to the fluid, the change of momentum in unit time is

$$\rho Q[V_A(1 + b) - V_A]$$

and this must be equal to the thrust T on the disk. Hence

$$\begin{aligned} T &= \rho Q V_A b \\ &= \rho A_0 (V_A)^2 (1 + a) b \end{aligned} \tag{5}$$

The total work done per unit time (or the power expended) is equal to the increase in kinetic energy of the fluid, since we are neglecting friction, and if there is no rotation of the race the increase in kinetic energy in unit time is given by

$$\begin{aligned} &\tfrac{1}{2}\rho Q[(V_A)^2(1 + b)^2 - (V_A)^2] \\ &= \tfrac{1}{2}\rho Q((V_A)^2 b^2 + 2bV_A{}^2) \\ &= \rho Q(V_A)^2 b\,(1 + b/2) \\ &= TV_A\,(1 + b/2) \end{aligned}$$

This increase in kinetic energy has been provided by the work done on the water by the thrust, which is $TV_A\,(1 + a)$ in unit time.

Hence we have

$$TV_A\,(1 + a) = TV_A\,(1 + b/2)$$

or

$$a = b/2$$

That is, one half of the sternward increase in velocity is acquired by the fluid before it reaches the disk.

The useful work obtained from the screw, i.e., the work done upon the disk, is TV_A, and so the power lost in the screw is

$$TV_A\,(1 + a) - TV_A = TV_A \cdot a$$
$$= TV_A \cdot b/2$$

The ideal efficiency η_I will be

$$\eta_I = \frac{\text{useful work obtained}}{\text{work expended}}$$
$$= TV_A/TV_A\,(1 + a)$$
$$= 1/(1 + a) \qquad (6)$$

For many purposes, it is convenient to express the velocity increase bV_A or, as we may also call it, the slip velocity, as a fraction of the speed through the fluid, V_A. Denoting this slip ratio by s, we have

$$s = bV_A/V_A = b = 2a$$

Hence[3]

$$\eta_I = \frac{1}{1 + s/2} = \frac{2}{2 + s} \qquad (7)$$

Also, from Equation (5) and putting $a = b/2$, we find

$$T = \rho A_0 (V_A)^2\,(1 + b/2)b$$
$$= \rho A_0 (V_A)^2\,(1 + s/2)s$$

If the thrust loading coefficient is defined as

$$C_T = \frac{T}{\frac{1}{2}\rho A_0 (V_A)^2} \qquad (8)$$

then

$$C_T/2 = (1 + s/2)s \qquad (9)$$

[3] In some texts, slip ratio s is defined as the ratio of the increase of the velocity in the race, bV_A, to the final velocity in the race, $(V_A + bV_A)$. In this case,

$$s = \frac{bV_A}{V_A + bV_A} = \frac{b}{1 + b}$$

whence

$$a = \frac{b}{2} = \frac{s}{2(1 - s)}$$

The ideal efficiency is then

$$\eta_I = \frac{1}{1 + a} = \frac{1 - s}{1 - s/2}$$

Eliminating s from (7) and (9), we find

$$\eta_I = \frac{2}{1 + (C_T + 1)^{1/2}} \qquad (10)$$

This equation is of great practical importance, since it furnishes a simple criterion for the comparative efficiencies of different propellers. It shows that a propeller working at a high load coefficient C_T is less efficient than one working at a low coefficient:

C_T	0	1	2	3	4
η_I	1.00	0.827	0.732	0.667	0.618

It follows that the propeller with the largest disk area is in general the most efficient, other things being equal.

When the speed of advance is zero, the efficiency is also zero, but the propeller still delivers thrust and absorbs power. The relation between thrust and power at zero speed of advance can be derived for an ideal propeller.

The power P will be given by

$$P = \frac{\text{useful work obtained}}{\text{ideal efficiency}}$$
$$= \frac{TV_A}{\eta_I}$$
$$= TV_A \times \frac{1 + (C_T + 1)^{1/2}}{2} \qquad \text{[from Equation (10)]}$$

When V_A is very small C_T will be very large in comparison with unity, and we can write approximately

$$P = TV_A \times \frac{C_T^{1/2}}{2}$$

Putting

$$C_T = \frac{T}{\frac{1}{2}\rho A_0 (V_A)^2}$$

this reduces to

$$\frac{T}{P} \cdot \left(\frac{T}{A_0 \rho}\right)^{1/2} = \sqrt{2}$$

The value $\sqrt{2}$ applies to an ideal propeller, but for an actual propeller it is much smaller. The value can be easily determined by a dock trial and serves as a convenient measure of the relative thrusting ability of various propellers at zero speed.

2.4 The Momentum Theory, Including Angular Motion. In the simple momentum theory developed in the preceding section, the actuator disk was assumed to be capable of accelerating the fluid only in an axial direction. If we now assume a disk propeller which is capable of accelerating the fluid both axially and rotationally, we have the idealized form of the screw propeller.

For angular motion there exists a momentum theorem similar to that for linear motion. If Q is the torque or moment of a force acting on a body about an axis of rotation 0, I_P the mass polar moment of inertia of the body with respect to 0, and $d\omega/dt$ the resulting angular acceleration of the body, then the equation equivalent to (3) is

$$Q = I_P d\omega/dt$$

or

$$Qdt = I_P d\omega$$

Qdt is the angular impulse and $I_P dw$ the change in angular momentum, and the equation states that the angular impulse is equal to the change in angular momentum.

If we consider now a time interval of 1 sec during which the torque remains constant, the equation can be integrated to give

$$Q = I_P(\omega_2 - \omega_1) \tag{11}$$

where ω_1 and ω_2 are the initial and final angular velocities.

To develop an expression for the efficiency of an ideal screw propeller with rotation of the race, we assume that the fluid has a translational velocity V_A far ahead of the propeller and no rotational velocity, i.e., $\omega_1 = 0$. The disk has a rotational velocity of ω, and in passing through it the fluid will acquire some angular velocity in the same direction as the disk. Well behind the screw, the race will have a translational velocity $V_A(1 + b)$, as before, and a rotational velocity ω_2, which we may write, by analogy, in the form

$$\omega_2 = \omega(1 - b')$$

Some of this rotational velocity will be acquired by the fluid before it enters the screw disk, just as in the case of the sternward acceleration, and we can define a rotational inflow factor a' similar to the axial inflow factor a. The angular velocity of the disk relative to the water will be reduced in consequence from ω to $\omega(1 - a')$.

The total kinetic energy in the race will be increased by the energy of rotation, so that the effect will be to reduce the ideal efficiency.

Both velocity components impressed on the fluid are assumed to be uniformly distributed over the disk. Dividing the latter into concentric annular elements of width dr and area dA_0, and assuming each element works independently of all the others, the thrust dT developed by any element is given, by analogy with Equation (5) as

$$dT = \rho dA_0(V_A)^2(1 + a)b$$
$$= \rho dA_0(V_A)^2(1 + b/2)b$$

The torque dQ absorbed by the element is, by Equation (11),

$$dQ = dI_P(\omega_2 - 0)$$
$$= dMr^2\omega_2$$

where

dM = mass of fluid passing through area dA_0 in unit time = $\rho dA_0 V_A(1 + a)$

dI_P = moment of inertia of dM

and

r = radius of annular element

Thus

$$dQ = \rho dA_0 V_A(1 + a)r^2\omega_2$$

The useful work performed by the element is dTV_A. The power absorbed by the element is $dQ\omega$, which must be equal to the sum of the useful work and the energy losses. The kinetic-energy loss in translation

$$= \tfrac{1}{2}dM(bV_A)^2$$
$$= \tfrac{1}{2}dTbV_A$$

since

$$dM \times bV_A = \text{change of momentum in fluid}$$
$$= dT$$

The kinetic-energy loss in rotation

$$= \tfrac{1}{2}dI_P(\omega_2)^2 = \tfrac{1}{2}dQ\omega_2$$

The energy-balance equation then gives

$$dQ\omega = dTV_A + \tfrac{1}{2}dTbV_A + \tfrac{1}{2}dQ\omega_2$$

or

$$dTV_A\,(1 + b/2) = dQ\,(\omega - \omega_2/2)$$

This shows that one half of the angular velocity is acquired by the fluid before it enters the disk, and by definition $\omega_2/2 = a'\omega$, so that

$$dTV_A(1 + a) = dQ\omega(1 - a')$$

remembering that $a = b/2$.

The efficiency of the element is

$$\eta_I = \frac{\text{useful work performed}}{\text{power absorbed}}$$
$$= dTV_A/dQ\omega$$
$$= (1 - a')/(1 + a) \tag{12}$$

The ideal efficiency for the simple actuator disk is Equation (6),

$$\eta_I = 1/(1 + a)$$

The factor $(1 - a')$ is always less than unity. It can be shown that Equation (12) is not only the ideal ef-

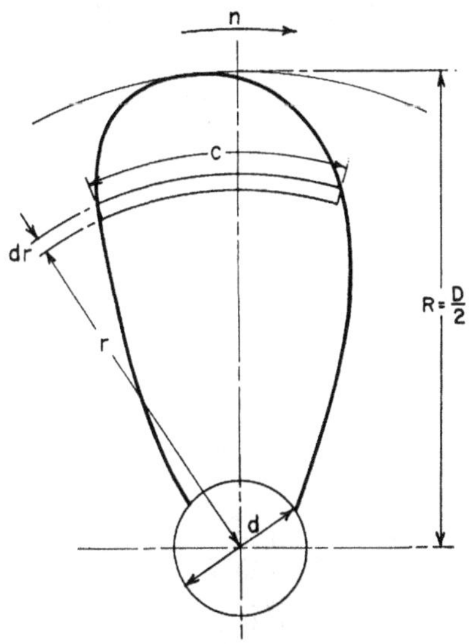

Fig. 2 Propeller-blade definitions

ficiency of the blade element, but also the expression for the ideal efficiency of a screw propeller having the minimum kinetic-energy losses (Bienen, et al, 1924). It follows that the efficiency of an ideal screw propeller is less than that of an actuator disk propeller by the fraction $(1 - a')$ or

$$(\omega - \omega_2/2)/\omega$$

2.5 Blade Element Theory of Screw Propeller. In the momentum theories of previous sections, the propeller was considered as a mechanism for increasing the momentum of the race, but no attempt was made to explain how this was done.

In the blade-element theory, the propeller is considered to be made up of a number of separate blades, which in turn can be divided into successive strips across the blades from leading to trailing edge, Fig. 2. The forces acting on each strip are evaluated from a knowledge of the relative velocity of the strip to the water and the characteristics of the section shape. The elementary forces are then resolved into the elements of thrust dT in the forward direction and of torque dQ in the plane of propeller rotation. By plotting curves of dT and dQ along the blade from boss to tip, Fig. 3, curves of thrust and torque loading are obtained which on integration will give the total thrust T and torque Q on the whole propeller. The efficiency is then

$$\eta_0 = \frac{TV_A}{2\pi nQ} \tag{13}$$

The force on a blade section set at an angle of incidence to the flow can be resolved into two components, the lift L and drag D, respectively, normal to and along the line of incident flow, Fig. 4.

The angle between the face of the section and the incident flow is the *angle of incidence* α.

The forces are usually expressed in the form of nondimensional coefficients:

$$\text{Lift coefficient, } C_L = \frac{L}{\frac{1}{2}\rho A V^2}$$

$$\text{Drag coefficient, } C_D = \frac{D}{\frac{1}{2}\rho A V^2}$$

where

ρ is mass density of fluid
A is area of plan form of section
= (chord × span) for rectangular shapes
V is velocity of incident flow

The efficiency of the section as a lifting device is measured by the ratio

$$\frac{\text{Lift}}{\text{Drag}} = \frac{L}{D} = \frac{C_L}{C_D} = 1/\tan\gamma \quad \text{(see Fig. 4)}$$

The basic data on lift and drag are generally derived from tests with airfoils of constant cross-section in wind tunnels. These foils are arranged in the test section of the wind tunnel so that they span the section. In this way the measured lift and drag forces at various angles of attack are representative of the so-called two-dimensional case corresponding to a foil with infinite span, for which the distribution of the lift and drag force along the span is uniform.

A common airfoil shape used today is the NACA 66 (modified) thickness distribution, superimposed on the NACA, $a = 0.8$, mean line. Coordinates of these thickness and mean line distributions are given in Table 30. This modification of the NACA 66 section has been

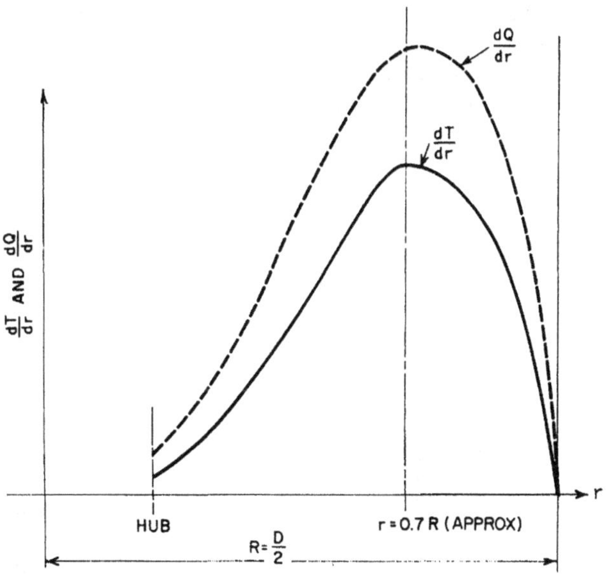

Fig. 3 Blade-loading curves

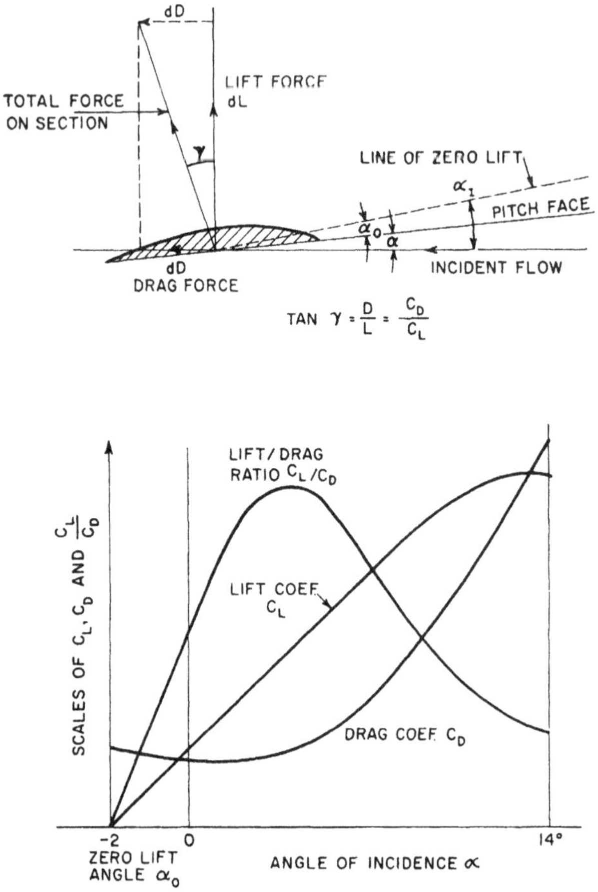

Fig. 4 Forces on a blade section

Table 1—Ordinates for NACA 66 (Mod) Thickness Distribution and NACA α = 0.8 Camber Distribution

Station, x/c percent	Thickness Ordinate, t/t_{max}	Camber Ordinate, f/f_{max}
0	0	0
0.5	0.0665	0.0423
0.75	0.0812	0.0595
1.25	0.1044	0.0907
2.5	0.1466	0.1586
5.0	0.2066	0.2712
7.5	0.2525	0.3657
10.0	0.2907	0.4482
15.0	0.3521	0.5869
20.0	0.4000	0.6993
25.0	0.4363	0.7905
30.0	0.4637	0.8635
35.0	0.4832	0.9202
40.0	0.4952	0.9615
45.0	0.5	0.9881
50.0	0.4962	1.0
55.0	0.4846	0.9971
60.0	0.4653	0.9786
65.0	0.4383	0.9434
70.0	0.4035	0.8892
75.0	0.3612	0.8121
80.0	0.3110	0.7027
85.0	0.2532	0.5425
90.0	0.1877	0.3588
95.0	0.1143	0.1713
97.5	0.748	0.0823
100.0	0.0333	0

published by Brockett (1966). Fig. 5 shows the meaning of the symbols used in Table 1.

When the results of tests on such sections are plotted, a number of interesting points emerge, Fig. 4:

(*a*) The lift coefficient C_L for small angles of incidence is a linear function of the angle of incidence α.

(*b*) At some larger value of α the lift coefficient ceases to increase linearly with α.

(*c*) Zero lift does not occur at zero incidence, but at a small negative angle, called the angle of zero lift, α_0. This is equal to -2 deg in Fig. 4. We can thus draw a *zero lift line* from the tail passing above the pitch face at an angle α_0 such that when the incident flow is along this line there will be no lifting force exerted on the section normal to the flow. When the angle of incidence to the pitch face is α, the hydrodynamic angle of incidence α_I is given by

$$\alpha_I = \alpha_0 + \alpha$$

(*d*) The drag coefficient remains small and more or less constant for small angles of incidence, but when the lift coefficient begins to fall off, the drag coefficient increases rapidly.

(*e*) The lift/drag ratio is a maximum at a small angle of incidence, and for such sections to work efficiently the angle of incidence should be small.

The ratio of span to chord is called the *aspect ratio* (AR). If this ratio were infinite, the flow past a section would be two-dimensional, and the lift distribution along the span would be uniform. With a finite span, a certain amount of "spilling" takes place at the ends, and the lift falls off to zero at those points. The results can be corrected from one aspect ratio to another, and are usually given for an AR of either 6 or infinity.

One other feature of section behavior is of importance in propeller work—the distribution of pressure around a section. An example for an airfoil shape is shown in Fig. 6. On the face of the section the pressure is increased above that in the free stream, being greatest quite close to the nose. On the back the pressure is decreased and has a marked peak some little distance from the nose. The lift force generated is the result of the differences in pressure on the two faces, and

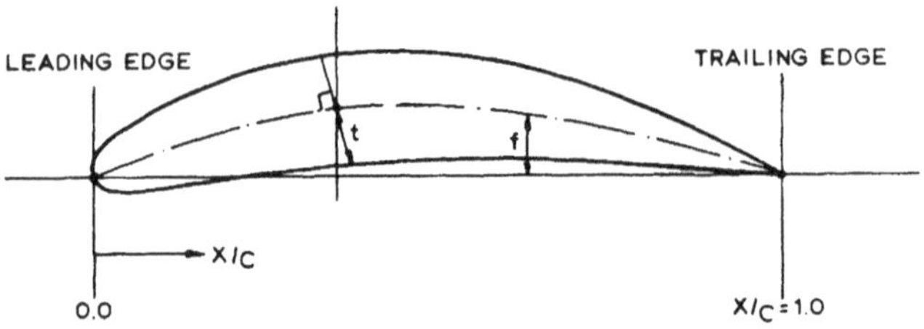

Fig. 5 Symbols defining shape of airfoil

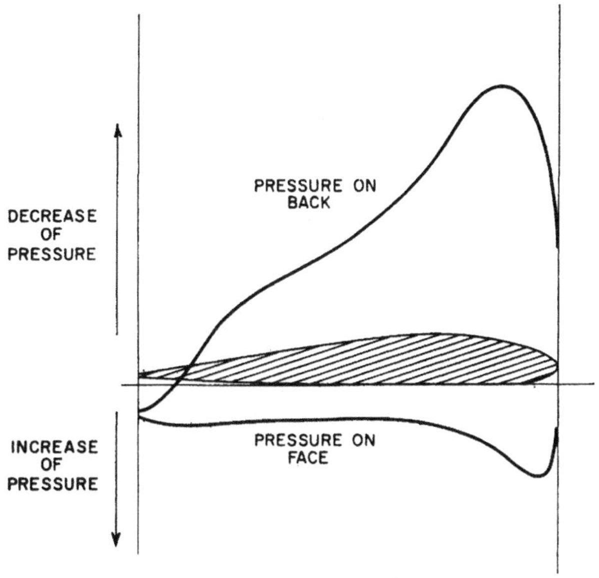

Fig. 6 Pressure distribution on blade section

for the type of pressure distribution shown in Fig. 6 it is clear that they reinforce one another and that the reduction on the back contributes more to the lift than does the increase on the face.

In a marine propeller, the surface of the blade facing aft, which experiences the increase in pressure when propelling the ship ahead, is called the *face* of the blade, the forward side being the *back*. In the simplest case, the face of a propeller blade is a portion of a true helical surface, i.e., a surface swept out by a straight line AB, Fig. 7, one end of which, A, advances at uniform speed along an axis OO', while the line itself rotates about the point A with uniform angular speed ω. When the generating line has made a complete revolution and is in the position $A'B'$, the distance it has advanced, AA', is called the face or geometrical pitch, P.

Any cylinder coaxial with OO' will cut the helical surface in a helix, and the angle between any such helix and a surface normal to the axis, such as SS, is called the pitch angle ϕ. The angle ϕ will be constant for a given helix, i.e., at a given radius, but will increase in value from the tip of the blade inwards to the hub. In practice the pitch is not always the same at all radii, it being fairly common to have a reduced pitch towards the hub and, less usually, towards the tip. In such cases the pitch at $0.7R$ is often taken as a representative mean pitch, as this is approximately the point where the maximum lift is generated, Fig. 3.

The shapes of blade outlines and sections vary greatly according to the type of ship for which the propeller is intended and to the individual designer's ideas. Fig. 8 shows a typical design and defines many of the terms in common use.

Here *skew* is defined as the angular measure from the center of the chord of each section to the reference line. This line extends from the center of the hub through the center of the chord of the section at $r = 1/2\ d$, the hub radius.

If we consider a section of the propeller blade at a radius τ with a pitch angle ϕ and pitch P, Fig. 9, and imagine the blade to be working in an unyielding medium, then in one revolution of the propeller it will advance from A to A', a distance P. If we unroll the cylinder of radius r into a flat surface, the helix traced out by A will develop into the straight line AM, and

$$\tan \phi = \frac{P}{2\pi r}$$

If the screw is turning at n-revolutions in unit time, then in that time it will advance a distance Pn and we can obtain a velocity diagram for the section, Fig. 10.

In a real fluid, there will be a certain amount of yielding when the propeller is developing thrust and the screw will not advance a distance LM, equal to Pn, in unit time, but some smaller distance LS, the distance MS being called the slip, and the ratio $MS/ML = s_R$ is called the *real slip ratio* and MAS the *slip angle* or geometrical slip angle.

From Fig. 10, it is seen that

$$s_R = \frac{Pn - V_A}{Pn} = 1 - \frac{V_A}{Pn}$$

As in the case of the actuator disk, the working of the propeller blades and the development of thrust result in an acceleration of the water ahead of the propeller, so that the total axial inflow velocity at a particular blade section is increased from V_A to $V_A(1 + a)$, while the total rotational inflow velocity is decreased from $2\pi nr$ to $2\pi nr\,(1 - a')$, Fig. 11.

It will be seen from the velocity diagram that both of the inflow factors a and a' result in a decrease in the angle of incidence at which the water meets the section to a value considerably below that which would obtain if they were neglected (from $<AOC$ to $<BOC$). The angle $<BOC$ is always small in an effi-

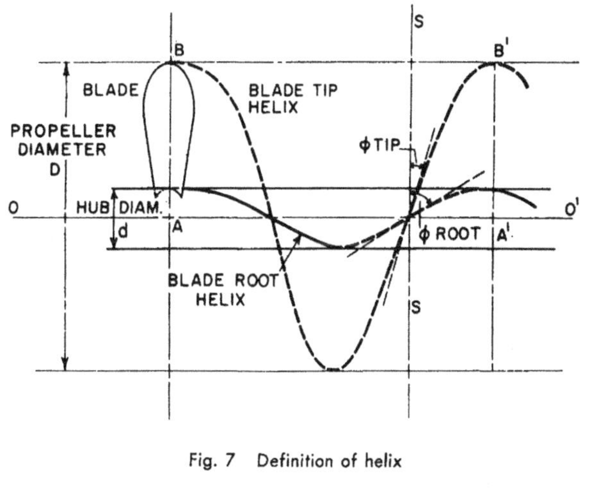

Fig. 7 Definition of helix

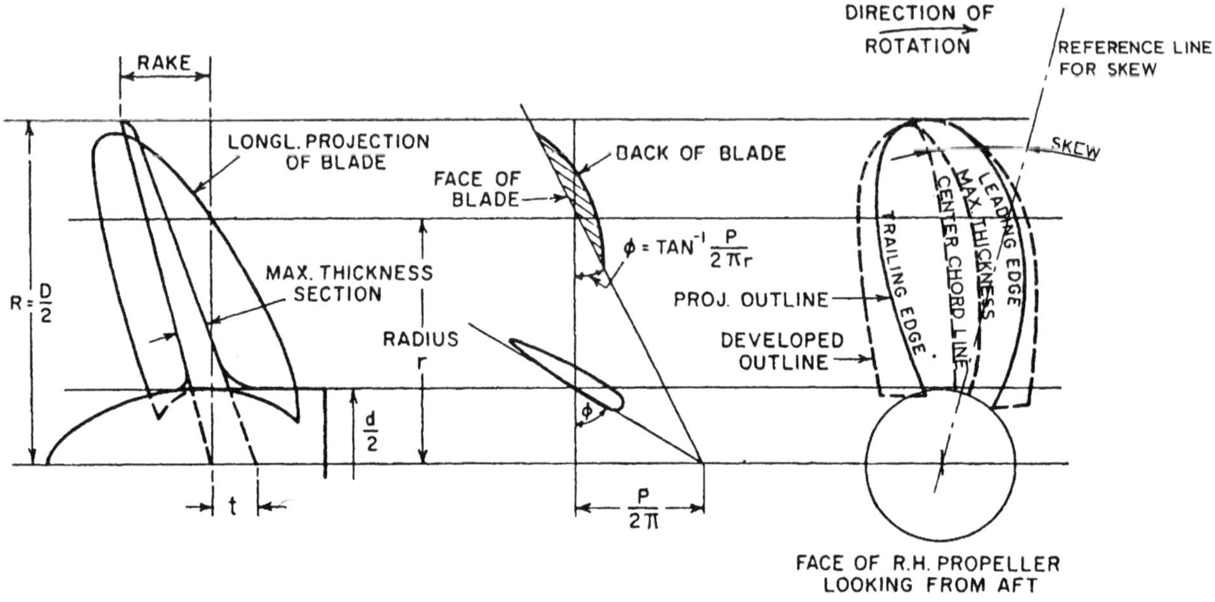

Fig. 8 Typical propeller drawing

Diameter D Pitch ratio $= \frac{P}{D}$

Pitch P Blade thickness ratio $= \frac{t}{D}$

No. of blades 4 Pitch angle $= \phi$

Disk area = area of tip circle $= \frac{\pi}{4} D^2 = A_O$

Developed area of blades, outside hub $= A_D$

Developed area ratio $= DAR = \frac{A_D}{A_O}$

Projected area of blades (on transverse plane) outside hub $= A_P$

Projected area ratio $= PAR = \frac{A_P}{A_O}$

Blade width ratio $= BWR = \frac{\text{Max. blade width}}{D}$

Mean width ratio $= MWR = \frac{A_D/\text{length of blades (outside hub)}}{D}$

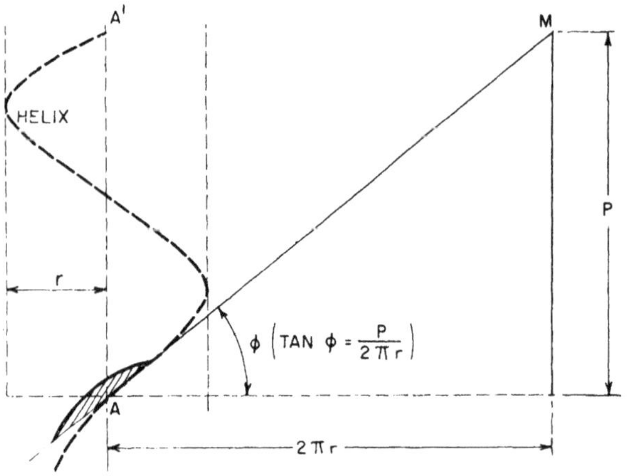

Fig. 9 Definition of pitch angle

ϕ = Pitch angle of screw propeller

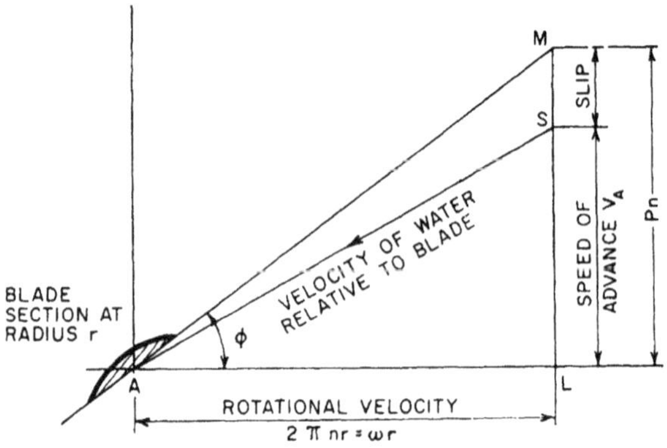

Fig. 10 Definition of slip

$$\tan \phi = \frac{Pn}{2\pi nr} = \frac{P}{2\pi r}$$

$$\text{Real slip ratio } s_R = \frac{\text{MS}}{ML} = \frac{Pn - V_A}{Pn} = 1 - \frac{V_A}{Pn}$$

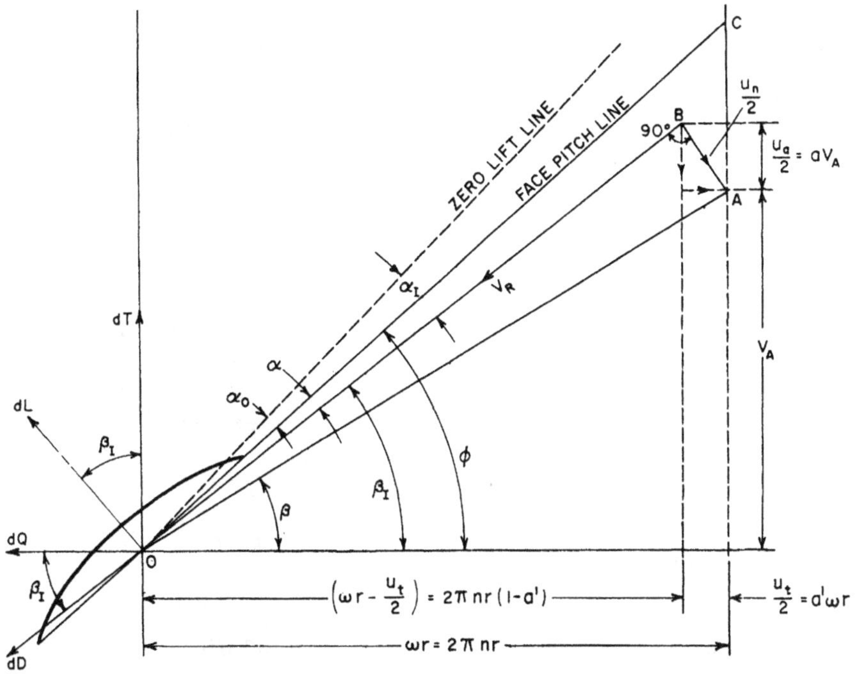

Fig. 11 Blade velocity diagram

cient propeller, usually in the region of 3 to 6 deg, at which angles the L/D ratio is approximately at its maximum value. Thus although the induced velocities defined by a and a' may be small in absolute terms as compared with the speeds of advance and rotation of the section, yet they have a major effect upon the angle of incidence and therefore upon the conditions under which the blade section works. From Fig. 11 we see that

$$\tan \beta_I = \frac{V_A(1 + a)}{2\pi nr(1 - a')} = \frac{1 + a}{1 - a'} \tan \beta \quad (14)$$

and the velocity of the water relative to the blade section is given by the vector V_R, where

$$V_R = \frac{V_A(1 + a)}{\sin \beta_I} \quad (15)$$

The angle of incidence α is given by $(\phi - \beta_I)$, where ϕ is the geometrical pitch angle.

Suppose that the propeller has Z blades, that the chord of the blades has a value c at radius n, and that the section of the blade at this point has lift and drag coefficients C_L and C_D at the angle of incidence α. Let the resultant lift and drag of an element of the propeller blade of length dr along the blade be dL and dD, respectively. Then

$$dL = \tfrac{1}{2}\rho \times \text{area} \times (\text{velocity})^2 \times C_L$$

$$= \tfrac{1}{2}\rho c dr Z \frac{(V_A)^2(1 + a)^2}{\sin^2 \beta_I} C_L$$

and

$$dD = \tfrac{1}{2}\rho c dr Z \frac{(V_A)^2(1 + a)^2}{\sin^2 \beta_I} C_D$$

Since dL and dD are, respectively, normal to and along the direction of the relative velocity V_R, the thrust and torque contributed by these elements will be

$$dT = dL \cos \beta_I - dD \sin \beta_I$$

$$dQ = (dL \sin \beta_I + dD \cos \beta_I)r$$

The first expression can be written in the form

$$dT = dL\left(\cos \beta_I - \frac{dD}{dL}\sin \beta_I\right)$$

$$= dL\left(\cos \beta_I - \frac{C_D}{C_L}\sin \beta_I\right)$$

$$= dL\,(\cos \beta_I - \tan \gamma \sin \beta_I)$$

where

$$\tan \gamma = \frac{C_D}{C_L} \text{ in Fig. 4}$$

Hence

$$dT = dL\left(\frac{\cos \beta_I \cos \gamma - \sin \beta_I \sin \gamma}{\cos \gamma}\right)$$

$$= dL \frac{\cos(\beta_I + \gamma)}{\cos \gamma}$$

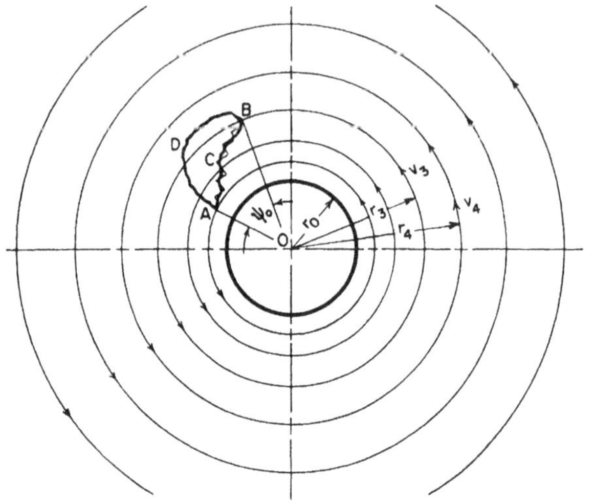

Fig. 12(a) Circulation flow

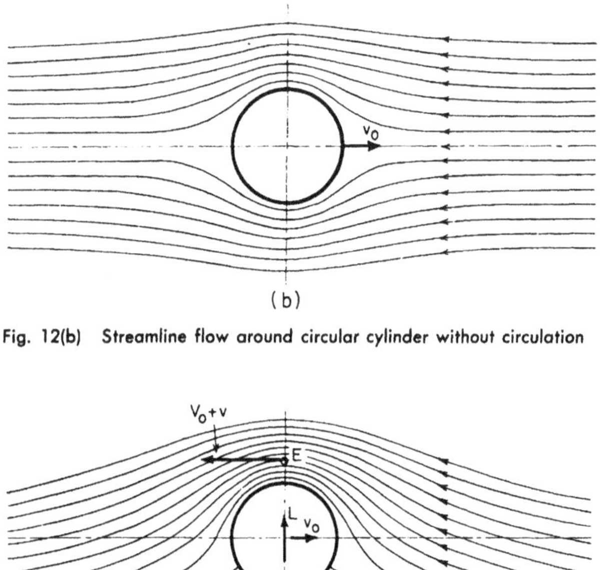

Fig. 12(b) Streamline flow around circular cylinder without circulation

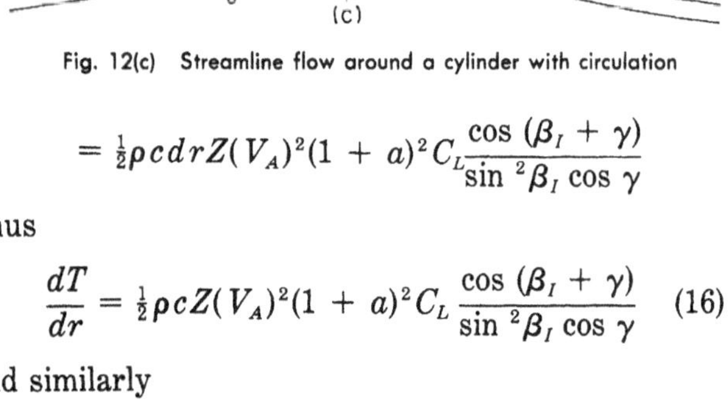

Fig. 12(c) Streamline flow around a cylinder with circulation

$$= \tfrac{1}{2}\rho c\,dr\,Z(V_A)^2(1+a)^2 C_L \frac{\cos(\beta_I+\gamma)}{\sin^2\beta_I\cos\gamma}$$

Thus

$$\frac{dT}{dr} = \tfrac{1}{2}\rho cZ(V_A)^2(1+a)^2 C_L \frac{\cos(\beta_I+\gamma)}{\sin^2\beta_I\cos\gamma} \quad (16)$$

and similarly

$$\frac{dQ}{dr} = \tfrac{1}{2}\rho cZr(V_A)^2(1+a)^2 C_L \frac{\sin(\beta_I+\gamma)}{\sin^2\beta_I\cos\gamma} \quad (17)$$

dT/dr and dQ/dr may now be plotted on a base of radius r and the total thrust T and torque Q obtained by integration, Fig. 3. Such curves show that most of the thrust and torque are developed over the outer part of the blade, the maxima occurring at about $r = 0.7R$.

The efficiency of the blade element is given by

$$\eta = \frac{dT\,V_A}{2\pi n\,dQ}$$

$$= \frac{V_A dL \dfrac{\cos(\beta_I+\gamma)}{\cos\gamma}}{2\pi n r dL \dfrac{\sin(\beta_I+\gamma)}{\cos\gamma}}$$

$$= \frac{V_A}{2\pi n r}\frac{1}{\tan(\beta_I+\gamma)}$$

$$= \frac{\tan\beta}{\tan(\beta_I+\gamma)}$$

$$= \frac{1-a'}{1+a}\frac{\tan\beta_I}{\tan(\beta_I+\gamma)} \quad \text{[by Equation (14)].}$$

The efficiency of the whole propeller will be

$$\eta_0 = \frac{TV_A}{2\pi n Q}$$

The performance of each blade element can only be determined when values of a, a', C_L, and γ are known. C_L and γ can be found from test data on the particular blade sections chosen. To find a and a', it is necessary to equate the thrust to the fore-and-aft momentum put into the race and the torque to the change in rotational momentum, as in the momentum theory.

Writing

$$F = \frac{cZC_L\cos(\beta_I+\gamma)}{8\pi r\sin^2\beta_I\cos\gamma}$$

Equation (16) becomes

$$\frac{dT}{dr} = F\rho(V_A)^2(1+a)^2 \times 4\pi r \quad (18)$$

From momentum considerations, the thrust developed by the blade element is given by

$$\left.\begin{array}{l}\text{Mass of fluid passing}\\ \text{through annular element}\\ \text{of disk per unit time} \times\\ \text{change in velocity}\end{array}\right\} = \rho 2\pi r\,dr\,V_A(1+a)\,bV_A$$

or

$$\frac{dT}{dr} = 2\pi r\rho(V_A)^2(1+a)b \quad (19)$$

This neglects any rotational momentum imparted to the race.

Equating (18) and (19), we have

$$2F(1 + a) = b$$

A similar expression can be derived for the rotational inflow factor a'.

If we knew the ratio between a and b, i.e., what proportion of the ultimate race velocity is acquired at the position of the blade section, a could be determined. For the particular value of $a = b/2$ derived from momentum considerations, we find

$$a = \frac{F}{1 - F}$$

In the early days a simplified blade-element theory was used in which the induced velocities were ignored, and the calculated thrusts, torques, and efficiencies differed considerably from those found in actual propeller performance. The comparison was improved when the induced velocity effects were included, but discrepancies still remained, owing principally to the neglect of the mutual interference between the propeller blades and the failure to allow for the falling off of the lift towards the blade tips. Later developments in propeller theory have enabled these factors to be largely accounted for in modern design methods; see Section 8.4.

2.6 Circulation Theory of Screw Propeller. The modern theoretical methods of propeller design are based upon the vortex theory first enunciated by F. W. Lanchester in his treatise *Aerial Flight* published in 1907.

Consider the type of streamline flow shown in Fig. 12(a), which is defined by the equation

$$rv = c = \text{constant} \tag{20}$$

where

r = radius vector drawn from O to any point in the field
v = velocity at any point, which is everywhere normal to radius vector

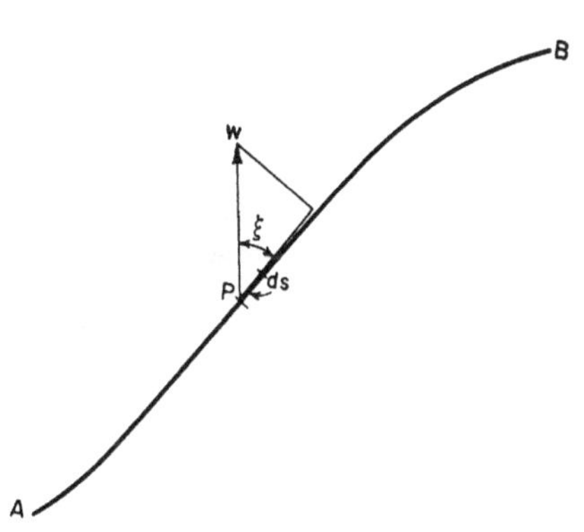

Fig. 13 Line integral

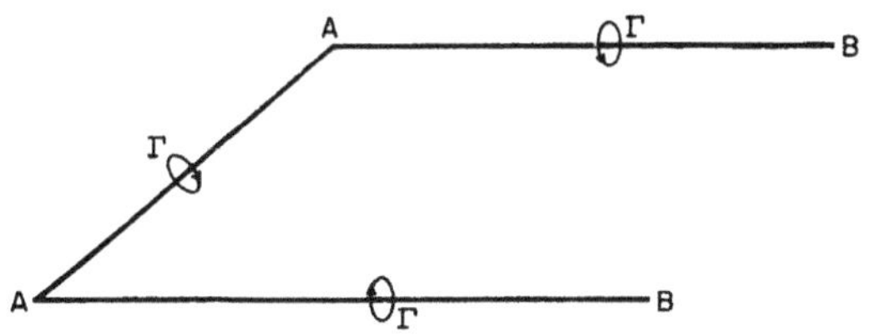

Fig. 14 Vortex of airplane wing with constant circulation

An inner streamline of radius r_0 can be considered as representing the wall of a cylinder whose axis is normal to the plane of the flow and around which the fluid circulates. When the radius r_0 is very small, we have what is known as a vortex tube or filament, because the law of velocity distribution expressed by Equation (20) applies with good approximation to the exterior field of the familiar vortex motions found in nature. Vortex filaments in ideal fluids have interesting properties, among which may be mentioned that any given vortex filament is permanently composed of the same fluid particles and that it cannot terminate abruptly in the interior of the fluid but must either return on itself or terminate on the boundary of the fluid region.

If the cylinder is placed in a uniform stream in such an ideal fluid, but without any such circulation flow, the streamlines will be symmetrical about the flow axis, and no force will be exerted upon the cylinder, Fig. 12(b).

If now a circulation flow is imposed around the cylinder, the flow pattern is greatly changed, becoming asymmetrical as shown in Fig. 12(c). At the point E the velocity parallel to the flow axis is $(V_0 + v)$ while at F it is $(V_0 - v)$. This asymmetry of velocity distribution gives rise to a similar asymmetry in pressure distribution, the pressure at F being greater than that at E. As a result, a force is exerted on the cylinder at right angles to the direction of the uniform stream flow. The production of such a force on a rotating cylinder in a stream is called the Magnus effect after its discoverer, Magnus (1853). It has been used to propel ships in place of sails in the Flettner rotor ship.

To define the mathematical concept of circulation more clearly, let A and B in Fig. 13 be two points connected by any plane curve, and let ω be a vector at the point P on the curve which makes the angle ξ with the direction of the line element ds. Then the line

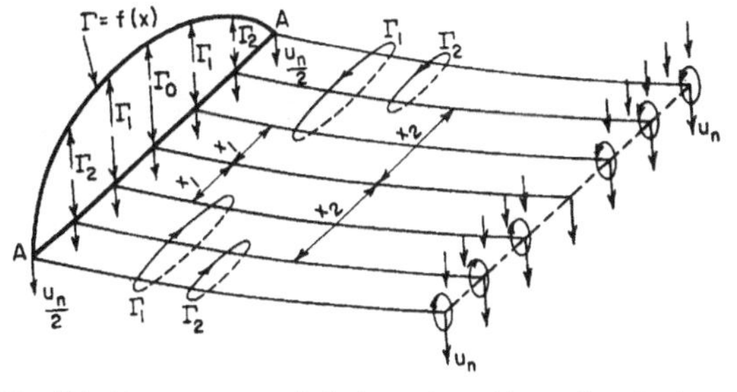

Fig. 15 Vortex system of airplane wing with varying circulation

integral between A and B is defined by the expression

$$\int_A^B \omega \cos \xi \, ds$$

In the special case when the vector ω denotes a velocity and the integration is performed around a closed curve, the line integral is called the circulation, Γ, and

$$\Gamma = \oint \omega \cos \xi \, ds$$

where the symbol $\oint$ indicates integration around a closed curve.

This type of flow has the peculiarity that when a closed curve is drawn in the field and the line integral along this closed curve is evaluated, the circulation is zero when the curve does not surround the origin O, but has the constant value $2\pi c$ when the curve surrounds the origin.

Consider the two points A and B in Fig. 12(a), which are connected by any curve whatsoever. By definition, the line integral along the curve is

$$\Gamma_{AB} = \int_A^B \omega \cos \xi \, ds$$

In order to evaluate this integral, we replace the chosen curve by a stepped line consisting of short radial lines and circular arcs. The integration along the radial lines contributes nothing to the final value, since the line element and the velocity vector are normal to each other, and consequently cos ξ is zero. The integration along the circular arcs, however, yields a definite value, since in this case cos ξ is unity.

Reasoning in this manner, we find for the value of the integral along the arbitrarily chosen curve

$$\Gamma_{AB} = \int_A^B \omega \, ds = \int_A^B \omega r \, d\psi = \pm c\psi_0 \qquad (21)$$

where ψ_0 is the angle included between the radii through A and B. The + sign applies when the integration is made in one direction, the − sign when made in the reverse direction. It appears from this equation that the value of the line integral is independent of the path and depends only on the end points. It follows therefore that if we integrate from A to B along an arbitrary path such as ACB in Fig. 12(a), and then integrate in the reverse direction along any other path not surrounding the origin, such as BDA, the value of the integral around the closed loop $ACBDA$ will be zero. On the other hand, if we integrate around a closed curve surrounding the origin O, so that the angle ψ_0 in Equation (21) has the value 2π, the line integral around the loop, or the circulation, will have the value $2\pi c$.

The transverse force L acting on the cylinder with circulation in a uniform flow, as shown in Fig. 12(c), may be shown to be given by the equation

$$L = \rho \Gamma V_0 \qquad (22)$$

where

ρ is mass density of fluid
$\Gamma = 2\pi c$ is strength of circulation flow
V_0 is velocity of uniform stream

Equation (22) is known as the Kutta-Joukowski Equation, and is one of the great generalizations of mechanics, since it applies to all bodies regardless of their shape, the shape factor being contained in the circulation factor Γ. By the aid of this equation the mathematical discussion of propeller action is greatly simplified, because we do not have to consider the shape of the propeller blades until the very end, in the meantime regarding them merely as vortex filaments or *lifting lines* endowed with circulation. These lifting lines are regarded as having finite lengths, corresponding to the lengths of the blades, not terminating abruptly at the tips, however, but having continuations, so-called tip vortices, at the free ends. Such continuations do exist at the tips of airplane wings and at the tips and roots of propeller blades, as is readily shown by wind-tunnel or water-tunnel experiments. The lift produced by an aircraft wing or a propeller blade is the result of an increased pressure on the face and a decreased pressure on the back. Since the fluid follows the pressure gradient, it tends to spill over the free ends from the face to the back, creating powerful vortices downstream, the axes of which are practically at right angles to the axis of the wing or blade, and which form the boundaries of the fluid layer which has been in contact with the blade.

The simpler case of an aircraft wing in flight through still air is illustrated in Fig. 14. The equivalent bound vortex in this case travels in a straight line at right angles to its axis. If the circulation of this bound vortex is assumed to be constant along its length, we have the simple system in which AA is the bound vortex and AB the free tip vortices already mentioned. This simple system is a useful concept and helps us to visualize the phenomenon, but does not express adequately the actual flow conditions around the wing. In reality the lift of the wing decreases from a maximum value at midspan to zero at the ends, and so the circulation around the wing must vary likewise. Assuming that the circulation around the bound vortex AA varies continuously as shown by the curve $\Gamma = f(x)$ in Fig. 15, then it can be shown by interpreting circulation in terms of its original definition as a line integral that free vortices flow not only off the free ends but also all along the trailing edge of AA, forming together a vortex sheet. The strength of any individual vortex in the sheet is equal to the change in circulation at that point on AA. Thus, if at distances x_1 and x_2 from midspan the circulation strengths are Γ_1 and Γ_2, respectively, the free vortex formed between x_1 and x_2 will have a strength equal to $(\Gamma_1 - \Gamma_2)$.

Considering now two individual vortices in the sheet, located at opposite sides of the midspan, each lies in the velocity field of the other and thus must assume the downward velocity existing at that point of the field. Inasmuch as this is true for all the vortices in the sheet, it follows that the sheet as a whole assumes a downward velocity. It was shown by Prandtl (1979) that this downward velocity is constant across the sheet when the distribution function $f(x)$ represents an ellipse. Along the vortex sheet, however, from AA to infinity at the right, the induced downward velocity is not constant but varies from the value u_n at a very large distance from AA to the value $u_n/2$ at AA. This can be proved rigorously by a theorem on vortex motion which has an equivalent in electrodynamics and is known as the Biot-Savart law. In a general way its truth can be perceived by the following reasoning. Let the vortex system in Fig. 15 be supplemented by a like system extending from AA to infinity at the left. The whole infinitely long vortex system would then move downward with the velocity u_n in accordance with the foregoing discussion. At AA this velocity would be composed in equal measure of that induced by vortices belonging to the supplementary system and by vortices belonging to the original system. Removing the supplementary system, and so reverting to the system shown in the figure, leaves only the value $u_n/2$ at the location of the bound vortex.

Similar conclusions were reached in the case of an advancing propeller blade (Prandtl, et al, 1927). The vortex sheet in this case is the helicoidal layer of fluid trailing behind the blade, and the induced velocity, which is normal to the helicoidal layer and so tends to push the sheet astern along the propeller axis and to rotate it about this axis, is identical with the previously defined slip velocity. A theorem analogous to that just mentioned holds in this case also—that the induced velocity at the position of the bound vortex, i.e., at the propeller disk, is $u_n/2$, or one half that at a great distance behind the propeller.

Betz further developed the important theorem that a propeller blade will have the smallest energy losses resulting from the induced velocities when the helicoidal vortex sheet is pushed astern along the shaft axis and rotated about this axis as though it were a rigid sheet. This theorem furnishes a simple and definite rule for the design of the propeller blade in practice; in order to obtain the maximum propeller efficiency, which is usually the aim of the designer, the blades must be so designed that the inflow velocity is the same for every blade element.

The application of the circulation theory to propeller design enables various refinements to be made to the simple blade-element theory already described. In particular, it enables the induced velocity $u_n/2$ to be calculated, an so the axial and radial inflow factors a and a'. These questions, and other developments, are discussed in detail in Section 8.4.

Section 3
Law of Similitude for Propellers

Much of the knowledge about the performance of propellers has been gained from experiments with models. To study the laws governing their behavior, the model propeller is run without any hull ahead of it. These are referred to as *open-water* conditions. In the towing tank this is done by running the propeller on a long shaft projecting well ahead of a narrow propeller "boat," containing the driving apparatus and attached to the towing carriage. The propeller advances into undisturbed water, so that the speed of advance V_A is known and the inflow is uniform over the disk. Records of thrust, torque, revolutions and speed are taken automatically over a range of values of the last two quantities.

3.1 Dimensional analysis As in the case of resistance, we can obtain guidance on the laws governing model and ship similitude by applying dimensional analysis.

The thrust of the propeller T could depend upon:

(*a*) Mass density of water, ρ.
(*b*) Size of propeller, represented by diameter D.
(*c*) Speed of advance, V_A.
(*d*) Acceleration due to gravity, g.
(*e*) Speed of rotation, n.
(*f*) Pressure in the fluid, p.
(*g*) Viscosity of the water, μ.

Writing

$$T = f(\rho^a D^b V_A{}^c g^d n^e p^f \mu^g)$$

and introducing the proper dimensions, we have

$$\frac{ML}{T^2} = \left(\frac{M}{L^3}\right)^a (L)^b \left(\frac{L}{T}\right)^c \left(\frac{L}{T^2}\right)^d \left(\frac{1}{T}\right)^e \left(\frac{M}{LT^2}\right)^f \left(\frac{M}{LT}\right)^g \qquad (23)$$

whence

$$\begin{aligned} a &= 1 - f - g \\ b &= 1 + 3a - c - d + f + g \\ c &= 2 - 2d - e - 2f - g \end{aligned}$$

and substituting a and c in the expression for b:

$$b = 2 + d + e - g$$

Then from Equation (23):

$$T = \rho D^2 (V_A)^2 f\left[\left(\frac{gD}{(V_A)^2}\right)^d \left(\frac{nD}{V_A}\right)^e \left(\frac{p}{\rho (V_A)^2}\right)^f \left(\frac{\nu}{V_A D}\right)^g\right]$$

where $\nu = \mu/\rho$.

The expressions in the square brackets are all nondimensional, and there is therefore no restriction dimensionally on the exponents d, e, f, and g. The form of the function f must be found by experiment, and may be different for each of the four terms.

The equation may be written in the form

$$\frac{T}{\frac{1}{2}\rho D^2 (V_A)^2} = f\left[\frac{gD}{(V_a)^2}, \frac{nD}{V_A}, \frac{p}{\rho (V_A)^2}, \frac{\nu}{V_A D}\right] \quad (24)$$

Note that since the disk area of the propeller, $A_0 = (\pi/4)D^2$, is proportional to D^2, the thrust coefficient can also be written in the form

$$\frac{T}{\frac{1}{2}\rho A_0 (V_A)^2}$$

Equation (24) states in effect that if all the parameters on the right-hand side have the same values for two geometrically similar but different sized propellers, the flow patterns will be similar and the value of $T/\frac{1}{2}\rho D^2 (V_A)^2$ will be the same for each.

If the model and ship quantities are distinguished by the suffixes M and S, respectively, and if λ is the linear scale ratio, then

$$D_S/D_M = \lambda$$

If the model propeller is run at the correct Froude speed of advance, then also

$$\frac{V_{AS}}{V_{AM}} = \lambda^{1/2}$$

Under these circumstances, the first term in Equation (24) will be the same for model and ship, so that the first condition for similarity of flow is that the speed of advance of the model and ship propellers should be in accordance with Froude's law of comparison.

The slip ratio has been defined as $(1 - V_A/Pn)$. For geometrically similar propellers, therefore, the second condition of Equation (24) that nD/V_A must be the same for model and ship, means that the slip ratio must be the same for each. Just as in the case of resistance, the third quantity in Equation (24) is not the same for model and ship propellers when the former is run in a towing tank, because the atmospheric pressure is not scaled down in the latter case. However, since the forces on the propeller blades are caused by differences in pressure, they will not be affected by this fact unless cavitation occurs, in which case other kinds of tests must be made (see Section 7.3). The last term, $\nu/V_A D$, is a Reynolds number, and it cannot be made the same if the model and ship speeds of advance follow Froude's law. It is concerned with the frictional resistance on the propeller blades, but as this is only a very small part of the total force on the blade, we can neglect the effect of viscosity in the first instance. However, it is necessary to make the model propeller as large as feasible within the other limitations of the hull model scale, measuring apparatus, and so on, in order to avoid as far as possible any laminar flow over the blades so as to reduce such Reynolds-number effect on the blade section drag to a minimum.

With these reservations in mind, we can say that as long as $gD/(V_A)^2$ and nD/V_A are the same in ship and model

$$T \propto D^2 (V_A)^2$$

The following relationships then hold:

$$\frac{D_S}{D_M} = \lambda; \quad \frac{V_{AS}}{V_{AM}} = \lambda^{1/2}$$

$$\frac{T_S}{T_M} = \frac{(D_S)^2}{(D_M)^2} \cdot \frac{(V_{AS})^2}{(V_{AM})^2} = \lambda^3$$

$$\frac{n_S D_S}{V_{AS}} = \frac{n_M D_M}{V_{AM}}$$

or

$$\frac{n_S}{n_M} = \frac{D_M}{D_S} \times \frac{V_{AS}}{V_{AM}} = \frac{1}{\lambda} \times \lambda^{1/2} = \frac{1}{\lambda^{1/2}}$$

or

$$n_M = n_S \times \lambda^{1/2}$$

i.e., the model revolutions are higher than those of the full-scale ship propeller.

The thrust power is given by $P_T = TV_A$, so that

$$\frac{P_{TS}}{P_{TM}} = \frac{T_S}{T_M} \times \frac{V_{AS}}{V_{AM}} = \lambda^{3.5}$$

and

$$\frac{Q_S}{Q_M} = \frac{P_{TS}}{n_S} \times \frac{n_M}{P_{TM}} = \lambda^{3.5} \times \lambda^{0.5} = \lambda^4$$

If the model results were plotted as values of

$$C_T = \frac{T}{\frac{1}{2}\rho D^2 (V_A)^2}$$

and

$$C_Q = \frac{Q}{\frac{1}{2}\rho D^3 (V_A)^2}$$

to a base of V_A/nD or J, therefore, the values would be directly applicable to the ship, apart from any scale effects such as mentioned. This method is often used, but the coefficients have the disadvantage that they become infinite for zero speed of advance, a condition

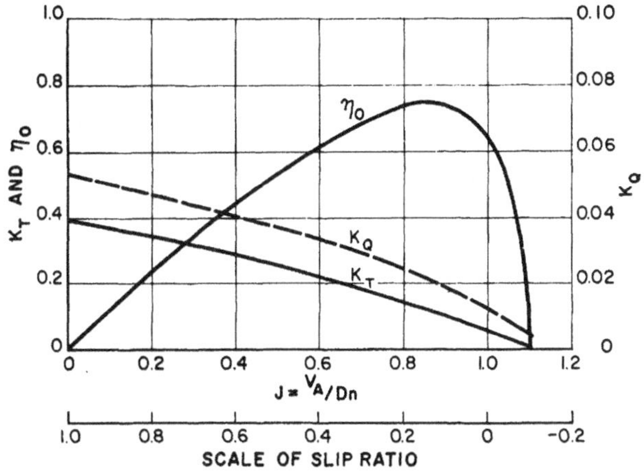

Fig. 16 Typical curves of thrust, torque and efficiency for propeller in open water

No. of blades = 4
Face pitch ratio = 1.00

sometimes occurring in practice, such as for a tug pulling at a stationary bollard or ship.

Since J or V_A/nD is the same for model and ship, we can replace V_A by nD and obtain new coefficients which do not have this disadvantage:

$$\text{Advance ratio, } J = \frac{V_A}{nD}$$

$$\text{Thrust coefficient, } K_T = \frac{T}{\rho n^2 D^4}$$

$$\text{Torque coefficient, } K_Q = \frac{Q}{\rho n^2 D^5}$$

$$\text{Propeller efficiency (in open), } \eta_0 = \frac{J}{2\pi} \times \frac{K_T}{K_Q}$$

where K_T, K_Q, and η_0 are functions of J. These coefficients are nondimensional in any consistent system of units.

3.2 Open water tests. Typical open-water curves are shown in Fig. 16. These show that this propeller reaches its maximum efficiency at a J-value of about 0.85. Since in this particular case the face pitch ratio is 1.00, $(1 - V_A/Pn) = (1 - V_A/Dn)$, and the scale of J is also a scale of (1-slip ratio). It will be noticed that the thrust does not vanish at $J = 1.00$, or zero slip ratio, but at some higher value nearer $J = 1.10$. This is due to the effect of the zero lift angles of the blade sections, so that the effective pitch is greater than the nominal or face pitch. The value of J at which the thrust does vanish can be used as a measure of the effective or analysis pitch ratio.

In practice, in order to obtain as high a value of Reynolds number as possible for the flow over the blade sections, the requirement to run at the correct Froude number is often ignored. Instead, the open-water tests are carried out at a higher speed of advance, the slip being varied to cover the necessary range by a variation in revolutions. In other cases, the slip range is covered by running at constant revolutions and different speeds of advance. Provided that the propeller is run with adequate immersion, so that there is no wave-making on the surface, the lack of Froude-number identity will not have any important effect.

Section 4
Interaction Between Hull and Propeller

4.1 General. The preceding discussion has related to a propeller working in *open water*, in which condition it is advancing into undisturbed water. When it is in its correct location behind the model or ship hull, the conditions are considerably modified. The propeller is now working in water which has been disturbed by the passage of the hull, and in general the water around the stern has acquired a forward motion in the same direction as the ship. This forward-moving water is called the wake, and one of the results is that the propeller is no longer advancing relatively to the water at the same speed as the ship, V, but at some lower speed V_A, called the speed of advance.

As we have seen, the propeller when developing thrust accelerates the water ahead of it, and this has the effect of lowering the pressure around the stern and also increasing the velocity there, both of which effects augment the resistance of the ship above that measured by towing the hull.

Also, the relations between thrust, torque, and revolutions in open water where the inflow is uniform, cannot be expected to remain the same behind the hull in the variable flow conditions experienced there. This leads to the possibility of differing propeller efficiencies in open water and behind the hull.

4.2 Wake. The difference between the ship speed V and the speed of advance V_A may be called the *wake speed*. Froude expressed the wake speed as a fraction of the speed of advance V_A, calling this ratio the *wake fraction* w_F, so that

$$w_F = \frac{V - V_A}{V_A} \tag{25}$$

and

$$V_A = \frac{V}{1 + w_F}$$

The expression $(1 + w_F)$ is called the *Froude wake factor*.

For a forward wake w_F is positive, but it may have small negative values in high-speed ships such as destroyers.

Taylor introduced a different definition of wake fraction by expressing the wake speed as a fraction of the ship speed, so that

$$w = \frac{V - V_A}{V} \quad (26)$$

and

$$V_A = V(1 - w)$$

This definition has much to recommend it, since a wake of 50 percent then means that the wake speed is 50 percent of the ship's speed, whereas in the Froude notation a 50 percent wake implies that the wake speed is 33 percent of the ship's speed. The Taylor definition is becoming more or less universal in current literature, but the difference must be remembered when using older published data, particularly British, much of which gives Froude-wake values. The two are related by the equations

$$w = \frac{w_F}{1 + w_F}$$

$$w_F = \frac{w}{1 - w}$$

The wake is due to three principal causes:

(*a*) The frictional drag of the hull causes a following current which increases in velocity and volume towards the stern, and produces there a wake having a considerable forward velocity relative to the surrounding water.

(*b*) The streamline flow past the hull causes an increased pressure around the stern, where the streamlines are closing in. This means that in this region the relative velocity of the water past the hull will be less than the ship's speed and will appear as a forward or positive wake augmenting that due to friction.

(*c*) The ship forms a wave pattern on the surface of the water, and the water particles in the crests have a forward velocity due to their orbital motion, while in the troughs the orbital velocity is sternward. This orbital velocity will give rise to a wake component which will be positive or negative according to whether there is a crest or a trough of the wave system in the vicinity of the propeller.

The total wake is made up of these three components, and in the great majority of cases is positive. Exceptions arise in very high-speed craft such as destroyers and high-speed motor boats. At a speed of 34 knots, the wave length of the system created by the ship will be some 200 m (656 ft), so that a destroyer 100 m (328 ft) in length would have a trough in the vicinity of the propellers, and the wave wake will be negative. With such a fine hull the potential or streamline wake would be small, and with large-diameter propellers much of the disk will be outside the frictional wake. Under these conditions the total wake over the propeller may be zero or slightly negative. Measurements of wake on destroyers have indeed shown the variation of wake with speed quite clearly as successive wave crests and troughs occurred at the stern (Newton, 1960).

The wake fraction can be measured in a number of ways. If we are interested in details of the wake pattern, the wake velocity can be measured by pitot tubes, the axial, radial and tangential components being obtained in the neighborhood of the intended propeller position. Curves of equal wake velocity can then be drawn. Examples for the axial components are shown in Figs. 17 and 18(*a*) and (*b*), for a single-screw ship and two twin-screw ships, respectively. For the former the wake is most intense over the upper part of the disk, rather less so down the vertical centerline, and much smaller over the outer lower quadrants. In a twin-screw ship the average wake over the propeller disk will, as a rule, be less than in a single-screw ship of the same fullness, because of the different propeller location, but there will be a considerable concentration immediately behind the ends of the bossings or behind the struts in the case of open shafts.

In both cases, the water flow has a general upward and inward direction, and this gives rise to further asymmetry so far as the propeller is concerned.

As a propeller blade rotates, a section at any given radius passes through regions of very different wake concentrations. We can make the propeller with a pitch which varies from hub to tip in such a way as to suit the *average* circumferential wake at any particular radius. These average wakes can be found from wake diagrams such as those shown in Figs. 17 and 18 or can be measured by using vane wheels of different radii which integrate the wake around particular radial annuli.

Wakes measured by such methods give the flow existing in the absence of the propeller, and are usually referred to as *nominal wakes*. They are modified when the propeller is present and developing thrust, and it is possible to deduce from the propeller performance behind the hull and in open water a wake factor which represents the *effective wake* as felt by the propeller.

Suppose that a propeller driving a hull at V-knots develops a thrust T when turning at n-revolutions per unit time. Reference to the open-water curves for the propeller will show that at the same revolutions n the propeller will develop the thrust T at some lower speed V_A. The latter is the effective speed of advance, and

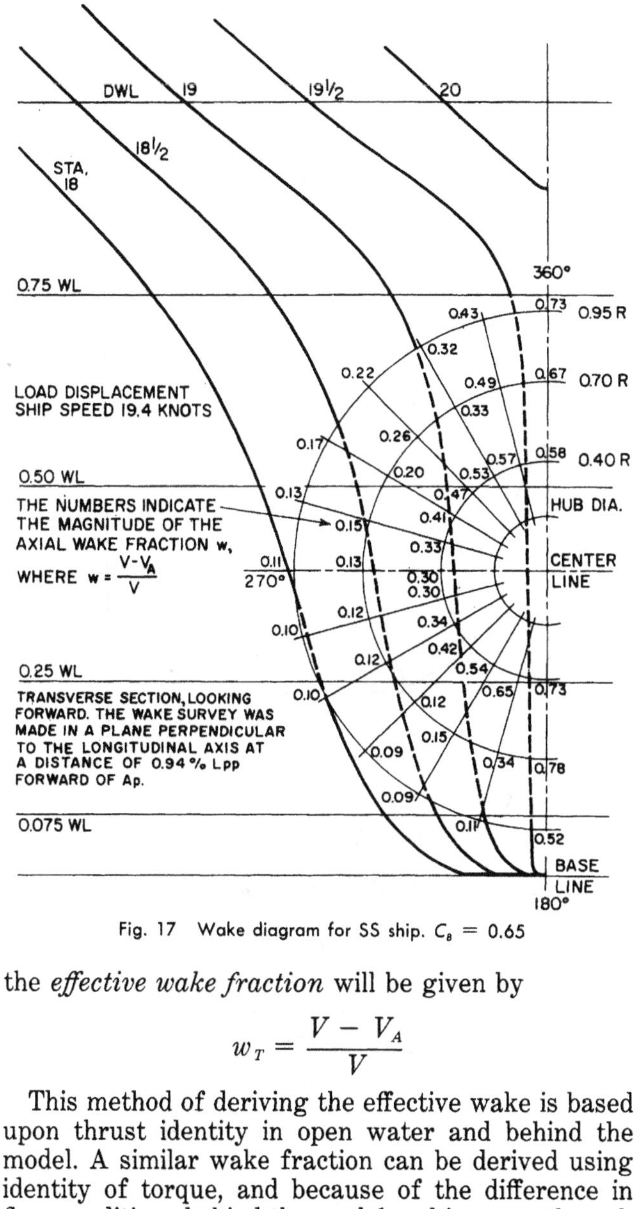

Fig. 17 Wake diagram for SS ship. $C_B = 0.65$

the *effective wake fraction* will be given by

$$w_T = \frac{V - V_A}{V}$$

This method of deriving the effective wake is based upon thrust identity in open water and behind the model. A similar wake fraction can be derived using identity of torque, and because of the difference in flow conditions behind the model and in open, the values of thrust and torque-derived wakes are usually somewhat different. For model work thrust-identity wakes are to be preferred, because thrust can be measured more accurately than torque. However, for comparison with ship trial results, it is often necessary to use a torque-identity wake, since in most cases thrust-meters are not fitted to ships and the only basis of comparison is on the measured torques.

The effect of the propeller in inducing an inflow velocity reduces the forward wake to some extent, the effective wake usually being three or four points lower than the nominal wake.

The nonuniformity of the wake has other highly undesirable consequences. As the blades rotate, periodic forces and couples are created which are transmitted through the water and the shaft bearings to the ship and are one of the principal sources of hull vibration. The variation in inflow velocity also results in a periodic change in angle of attack on the blades, and consequently is conducive to the onset of cavitation with its resultant vibration, noise and blade erosion. For all these reasons great attention should be paid to the shape of the stern lines and of appendages such as bossings, together with propeller clearances, to ensure that the wake inequalities over the propeller are kept as small as possible. Assistance in this problem can be obtained from model experiments in a circulating water channel, where the flow is made visible by tufts, dye, or other means, and by pitot-tube surveys in the towing tank. The effect of such forces on hull and machinery vibration are dealt with in Chapter VII, but the time to prevent or minimize them is in the early design stages, not when the ship is built.

To obtain a better picture of the wake non-uniformity a Fourier analysis may be carried out. In that case the axial wake at a specific point in the propeller disk is written as:

$$w_x(r) = \sum_{n=0}^{N} [a_n (r) \cos n\phi + b_n (r) \sin n\phi]$$

where a_n and b_n are the amplitudes of the Fourier components, r is the radius and ϕ is the angle of the point in the propeller plane. Hadler, et al (1965) present a large collection of data on wake distributions measured for a number of ships. They also show results of wake analysis using the above Fourier decomposition and provide information on the fluctuations of the hydrodynamic pitch angle as the blade makes one rotation.

To obtain an idea of the fluctuating forces on the propeller axis generated by the non-uniform velocity field, a simple analysis will be carried out, following a method proposed by Lewis (1935). The lift of a profile per unit span is given by:

$$L = \pi \rho c \beta U^2 \qquad (27)$$

because the lift coefficient for small hydrodynamic pitch angles equals $2\pi\beta$. Neglecting the fact that the lift direction does not coincide with the thrust direction, Equation (27) also holds for the thrust T of a propeller blade with unit span. The velocity U is made up of the rotational speed and the advance speed,

$$\begin{aligned} U^2 &= [\omega r]^2 + [V(1 - w_x)]^2 \\ &= \omega^2 r^2 + V^2 (1 - 2w_x + w_x^2) \qquad (28) \\ &= \omega^2 r^2 + V^2 + V^2 - 2w_x V^2 + w_x^2 V^2 \end{aligned}$$

with w_x the axial wake. If the axial wake is symmetric with respect to the vertical through the propeller axis the decomposition contains only cosine-terms:

$$w_x = \sum_{n=0}^{N} a_n \cos (n\phi) \qquad (29)$$

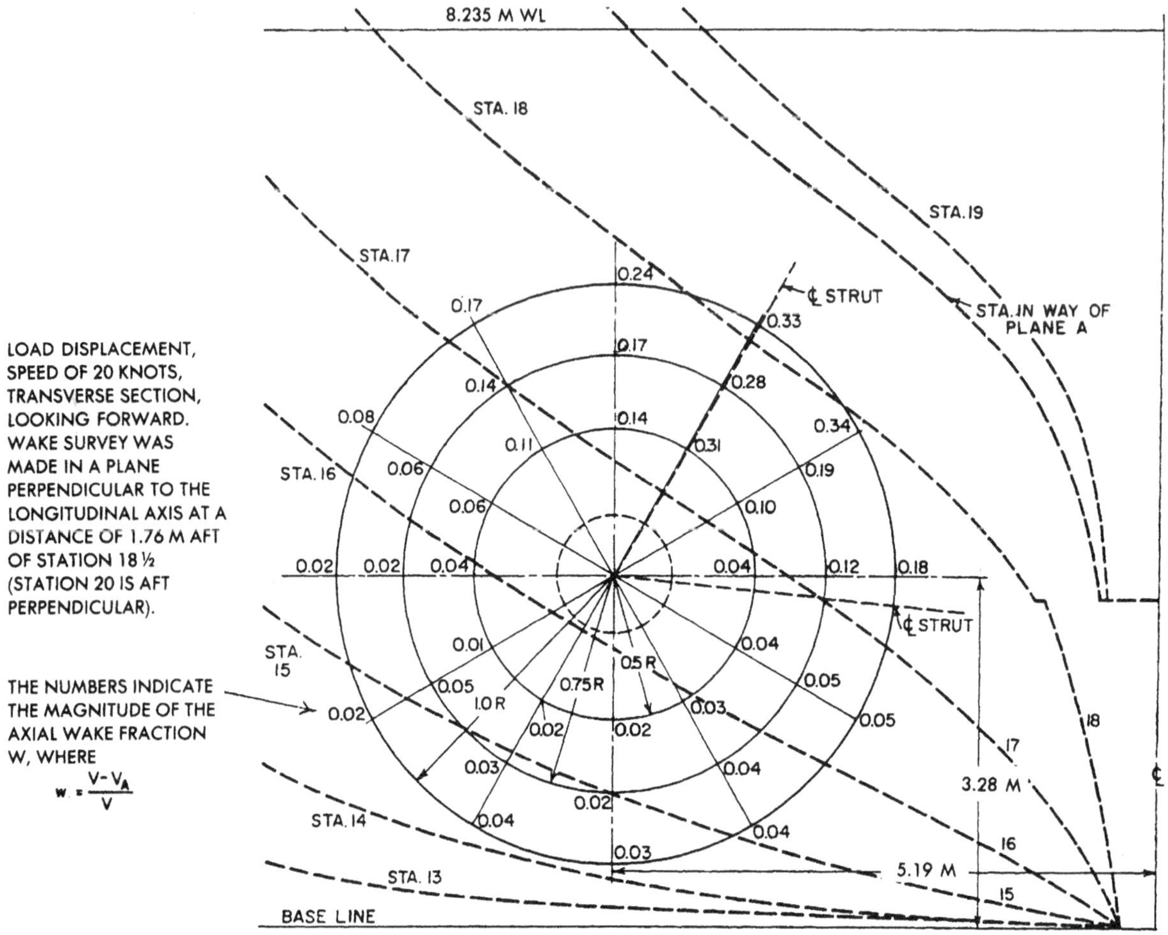

Fig. 18(a) Wake diagram for TS ship fitted with shaft-struts

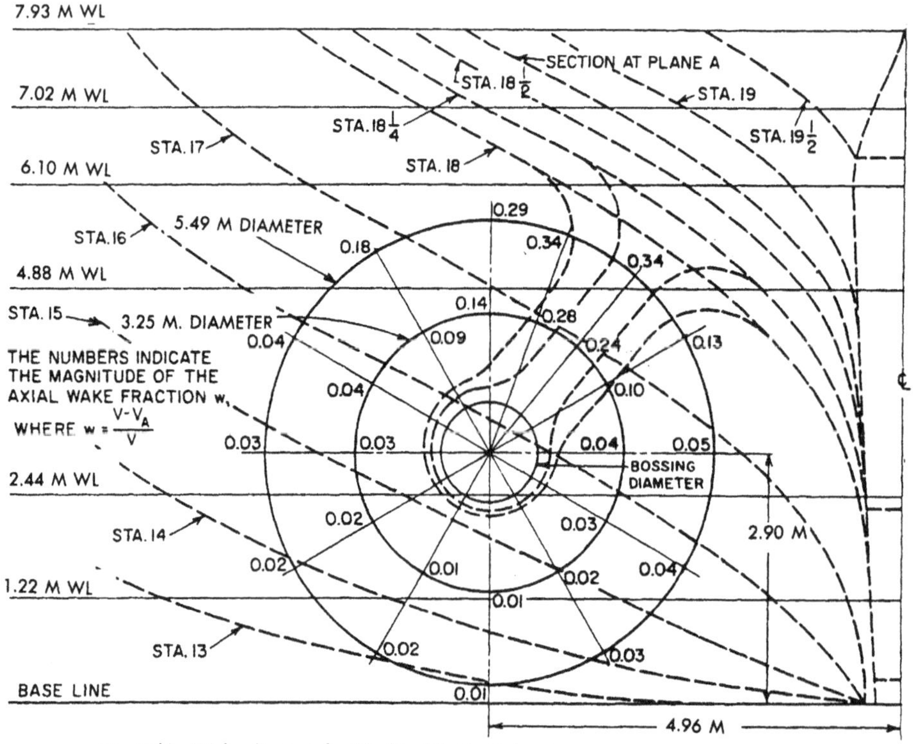

Fig. 18(b) Wake diagram for TS ship fitted with bossings

Inserting (27) into (28) and combining the result with Equation (27), the following formula for the thrust per unit span may be obtained.

$$T = \sum_{p=0}^{N} A_p \cos\,(p\theta + \phi_p) \qquad (30)$$

with ϕ_p the phase angle and θ the instantaneous position of the blade, which may be given by $\theta = \omega t - \frac{2\pi}{Z}(m - 1)$. A similar relation may be obtained for the torque per unit span. Summarizing over all blades, one obtains the following expression for the time-dependent thrust and torque:

$$T(t) = Z \sum_{k=0}^{N} f_{kz} \cos\,(kz\omega t - \phi_{kz}) \qquad (31)$$

$$Q(t) = Z \sum_{k=0}^{N} g_{kz} \cos\,(kz\omega t - \zeta_{kz}) \qquad (32)$$

Similar relations can be found for the bending moment (Hadler, et al, 1965). This analysis leads to the following conclusions:

(*a*) The unsteady thrust and torque are only determined by the terms of the Fourier decomposition that are integer multiples of the number of blades.

(*b*) The unsteady bending moments, M_x, M_y, vertical and horizontal side forces, F_x, F_y, are determined by the $(kz + 1)$th and $(kz - 1)$th terms of the Fourier series at integer multiples of blade frequency.

Table 2 summarizes these results for the number of blades ranging from 3 to 6. Note that this table gives the harmonic components of the axial wake field that contribute to the unsteady forces. It does not give the frequencies of the unsteady forces.

In this way one may obtain a judicious choice of the number of blades, Z. Of course, the above analysis is only a simple approximation, but Hadler (1965) showed that this approach worked well for a few cases.

Apart from the number of blades, the harmonic wake content determines the unsteady forces. This content and the amplitude of Fourier components depend on the hull form and possibly on speed. Hadler, et al (1965) drew the following conclusions on the basis of their results:

(*a*) The wake pattern is mainly affected by aft body shape. The variations in forebody have negligible effect.

(*b*) The amplitudes of the harmonics decrease with increasing order both for the longitudinal and tangential velocity component. The decrease is monotonic for open-stern ships but the even order amplitudes are relatively stronger for single-screw ships with conventional sterns.

(*c*) The radial distributions of mean longitudinal velocity and volumetric mean velocity have characteristics shapes. The open-type stern tends to provide a near uniform distribution with a magnitude comparable to the model speed. The single-screw ship with conventional stern provides a distribution with values lower at the inner radii and higher at the outer radii. The magnitudes depend on the fineness and shape of the afterbody; in general, the finer the stern, the higher the values.

(*d*) In most of the models analyzed, it has been found that the tangential velocity variation is large and sinusoidual in nature due to the upward flow in the propeller plane, which in turn shows up predominantly in the 1st harmonic. This, in conjunction with a large 1st harmonic amplitude of the longitudinal velocity, will result in a shaft-frequency vibration when some imperfections exist in the blade geometry which will cause hydrodynamic unbalances.

(*e*) In general, for minimizing the cavitation and vibration problems, the open-type stern, i.e., transom stern with struts supporting the shaft, is superior to the conventional type for the single-screw ship.

(*f*) The rudder may have an effect on the mean longitudinal velocity but does not have an appreciable effect on the derived quantities.

(*g*) The effect of variations in speed appears to be small.

(*h*) The effect of propeller location in the aperture can be important.

(*i*) Changes in displacement and trim may result in large variations in wake patterns and, consequently, in the cavitation and vibration characteristics.

In this respect the conclusions of Van Manen (1965) may also be mentioned. He notes (based on tests with some 40 different ship models) that for prismatic coef-

Table 2—Influence of Harmonic Wake Components on Unsteady Forces and Moments.

Harmonic Components

	Z	0	1	2	3	4	5	6	7	8	9	10	11	12	13	14	15	16	17	18	19	20
T, G	3	x			x			x			x			x			x			x		
	4	x				x				x				x				x				x
	5	x					x					x					x					x
	6	x						x						x						x		
M_x, F_x	3		x	x		x	x		x	x		x	x		x	x		x	x		x	x
	4		x		x		x		x		x		x		x		x		x		x	
M_y, F_y	5		x			x		x			x		x			x		x			x	
	6		x				x		x				x		x				x		x	

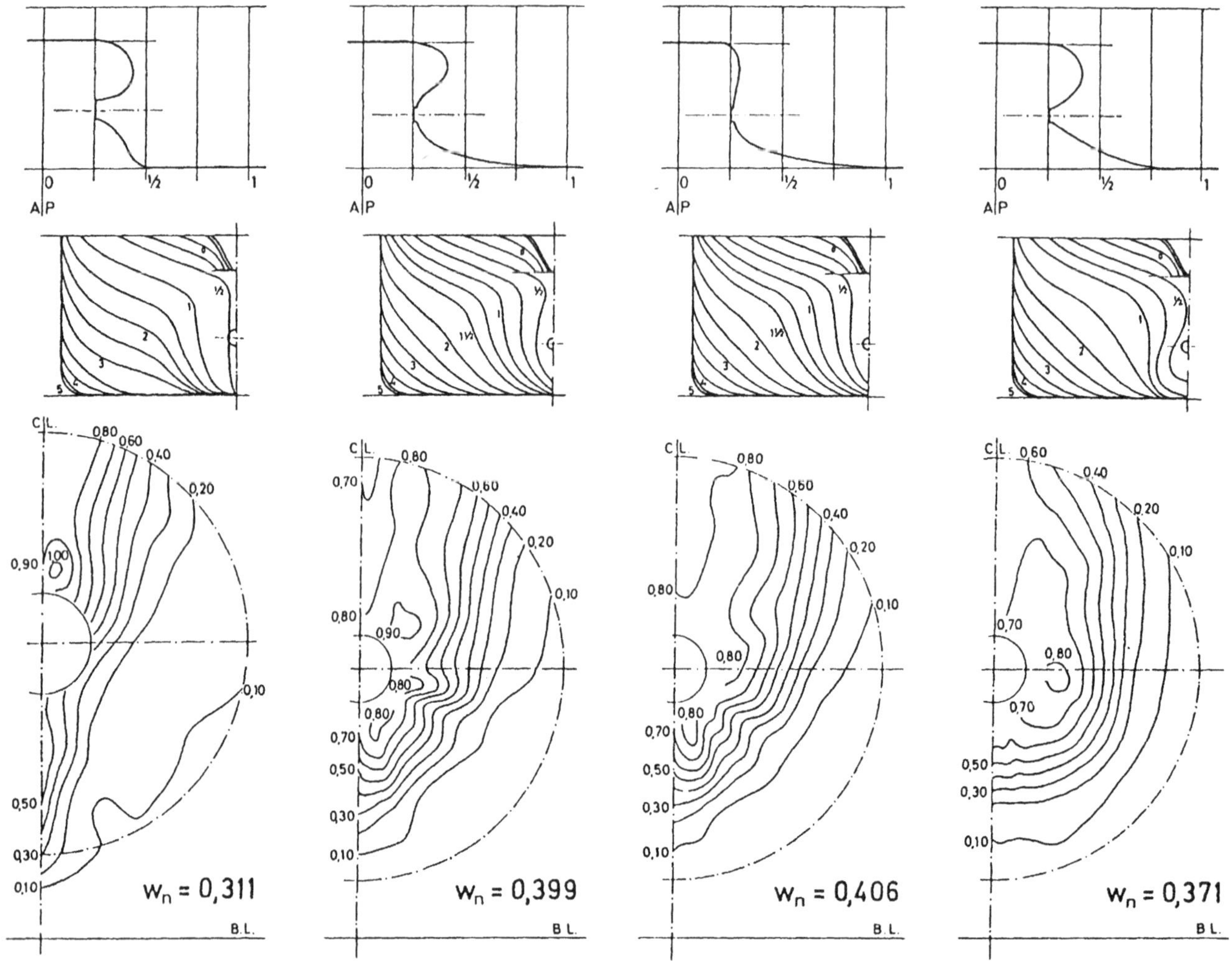

Fig. 19 Dependence of wake distribution on form of afterbody (Harvald, 1980)

ficients of the aft body between 0.73 and 0.79 in conjunction with four-bladed propellers there is an 80 percent probability that the first harmonic of the torque fluctuation will be 6.5 percent of the mean torque. Also the first harmonic of the thrust fluctuations is 10 percent of the mean thrust. For a five-bladed propeller these percentages are 1.5 and 2 percent resepectively. Deviations larger than 2 percent did not occur on any tested model. Van Manen also noted that fine-ended vessels such as warships can have substantially greater fluctuations.

The unsteady forces on the propeller blades and the possible occurrence of cavitation will also lead to fluctuating pressures on the ship hull. This problem may be more severe than the fluctuating forces on the propeller axis and bearings. For a detailed investigation the work of Van Oossanen, et al (1972) may be mentioned.

Naturally one would want to know the detailed wake field once the lines plan has been established or at least the amplitude of the different harmonic once the main particulars of the ship have been determined. Also with this knowledge designers could make alterations in hull form to accommodate certain requirements.

Large efforts have been made towards the solution of the equations of motion of a turbulent viscous fluid to determine theoretically the expected wake distribution. The numerical problems encountered and the computer time required have until now prohibited the adequate solution of the problem. For a summary of present activities and for a summary of relevant literature reference is made to the report of the Resistance Committee of the ITTC (1984).

Experimentally a number of authors have tried to correlate hull form and wake distribution. Van Gent, et al (1973) undertook an analysis, similar to the abovementioned one of Hadler, of the wake field of large tankers with block coefficients between 0.82 and 0.85. Results for the lower harmonics for five tankers with closed screw aperture showed a better mutual agreement than the remaining six models with open aperture. However, no specific tendencies with hull form parameters could be observed.

Van Gent, et al (1973) carried out a similar analysis for six third-generation twin-screw containerships. Scatter of the results were larger in this case because of the different shaft and support arrangements.

Harvald (1980) carried out wake measurements for a single-screw merchant ship with different aft body shapes for a few loading conditions. In line with previous results he found only a slight influence of speed. The influence of trim was found to be small at equal draft at the aft perpendicular. Changing the aft draft had a pronounced effect on the wake distribution, especially in the upper region of the propeller disk.

Harvald showed that changing the aft body by introducing a stern bulb, removing the deadwood and increasing the aperture above the propeller shaft, can lead to a much better wake field; Fig. 19. He also mentioned the possible use of shields above the propeller, a concept that has also been considered by Hylarides (1978). The application of a partial stern tunnel proved to be favorable for a containership and a dredger although application to a Great Lakes carrier did not lead to significant improvements, for the effective wake field the tunnel also proved to be somewhat beneficial.

An interesting comparison of wake fields is given in the publication of Jonk, et al (1980) who have carried out investigations into the wake distribution of simplified hull forms. The basic model represented a 455,000-ton deadweight product carrier. Simplifications of the hull form were made step by step, and the resulting changes in propulsive performance and wake distribution were measured.

Holden, et al (1980) presented on early design-stage approach to reducing pressure forces on the hull caused by a cavitating propeller. To this end they carried out a statistical analysis of the wake field distribution of 20 slender V-shaped aftbody ships and 49 full-form vessels. Both main particulars and local form characteristics were used as independent parameters.

By systematically varying the hull form parameters Holden, et al (1980) were able to indicate requirements for wakes with low peak values:

(*a*) Large waterline angles and blunt waterline endings should be avoided.

(*b*) The stern post half-breadth should be less than 0.05 D for waterlines situated in the range from 0.2 to 0.6 D above the propeller shaft.

(*c*) The maximum angle of the waterline with respect to the longitudinal ship axis should be kept below 30 degrees.

For full ships the following indications were also given:

(*d*) The area coefficient A/BT of the frame situated at 0.1 L_{pp} forward of AP should be in the range from 0.3 to 0.6.

(*e*) The angle with respect to the vertical of the frame situated at 0.1 L_{pp} forward of AP at the waterline through the propeller shaft should be less than 17 degrees. This means that the frames should be U-shaped in this region.

(*f*) The vertical gradient of the tangent of the waterline angles at the frame 0.1 L_{pp} forward of AP, should be between 0.25 and 0.46. The vertical gradient should be determined at the waterline through the propeller shaft.

(*g*) The value of the above-mentioned vertical gradient should be greater than 0.05. This means that the waterline angle should increase with increasing height above the shaft center.

For V-shaped ships, Holden, et al (1980) give some additional general recommendations.

Concerning measurements of the effective wake field the work of Hoekstra (1977) may be mentioned. He applied a diffusor to simulate the action of the propeller. By varying the diffusor length at constant diffusor angle a variation of the simulated propeller loading was obtained. The measurements for the axial and transverse velocities for a bulbous stern tanker showed that with increasing propeller loading:

- The bilge vortex shifts radially towards the propeller axis and downwards;
- the strength of the bilge vortex increases;
- the wake peak associated with the center of the bilge vortex is reduced;
- the wake fraction decreases.

4.3 Real and Apparent Slip Ratio. The real slip ratio has been defined (Section 2.5) as

$$s_R = 1 - \frac{V_A}{Pn}$$

For the screw working behind the hull, another slip ratio can be calculated using the ship speed V instead of the speed of advance of the propeller. This is called the *apparent slip ratio*, s_A, given by

$$s_A = 1 - \frac{V}{Pn}$$

The real slip ratio is the only real guide to the ship's performance and requires a knowledge of the effective wake fraction. However, the apparent slip ratio, which needs only the values of ship speed, revolutions, and propeller pitch for its calculation, is often recorded in ships' log books.

4.4 Relative Rotative Efficiency. The propeller in open water, with a uniform inflow velocity, at a speed of advance V_A, has an *open water efficiency* given by

$$\eta_0 = \frac{TV_A}{2\pi n Q_0} \qquad (33)$$

where Q_0 is the torque measured in open water when the propeller is delivering thrust T at n revolutions.

Behind the hull, at the same effective speed of advance V_A, the thrust T and revolutions n will be associated with some different torque Q, and the *efficiency behind the hull* will be

$$\eta_B = \frac{TV_A}{2\pi n Q} \quad (34)$$

The ratio of behind to open efficiencies under these conditions is called the *relative rotative efficiency*, being given by

$$\eta_R = \frac{\eta_B}{\eta_0} = \frac{Q_0}{Q} \quad (35)$$

The difference in torque found *behind* and *in open* is due to two main reasons—because of the heterogeneous wake behind the model, the flow conditions over a given blade section as it rotates differ greatly from those in open, so that the efficiency of any particular blade element will not necessarily be the same, and the relative amounts of laminar and turbulent flow on the propeller blades may be different in the two cases, the turbulence in the water behind the hull being greater than that in open water.

The value of the relative rotative efficiency does not in general depart materially from unity, being in the region of from 0.95 to 1.0 for most twin-screw ships and between 1.0 and 1.1 for single-screw.

4.5 Augment of Resistance and Thrust Deduction. When a hull is towed, there is an area of high pressure over the stern which has a resultant forward component reducing the total resistance. With a self-propelled hull, however, the pressure over some of this area is reduced by the action of the propeller in accelerating the water flowing into it, the forward component is reduced, the resistance is increased and so also the thrust necessary to propel the model or ship.

It is found in model work, where the necessary measurements can be made, that if the resistance of a hull when towed is R_T the thrust necessary to propel the model at the same speed V is greater than R_T and the increase is called the *augment of resistance*. It is expressed as the ratio of the increase in thrust to the resistance, so that

$$a = \frac{T - R_T}{R_T} = \frac{T}{R_T} - 1 \quad (36)$$

or

$$T = (1 + a)R_T$$

a is called the *resistance augment fraction* and $(1 + a)$ the *resistance augment factor*.

Although viewing the problem from the resistance point of view is the more logical one, the common practice is to look upon this increase in resistance as a deduction from the thrust available at the propeller, so that although the screw provides a thrust of T tons, say, only R_T tons are available to overcome resistance. This "loss of thrust" $(T - R_T)$ expressed as a fraction of the thrust T is called the *thrust-deduction fraction*, t, where

$$t = \frac{T - R_T}{T} = 1 - \frac{R_T}{T} \quad (37)$$

or

$$R_T = (1 - t)T$$

The expression $(1 - t)$ is the *thrust-deduction factor*.

It is common practice to fit rudders and other stern appendages to the model for self-propelled tests, and this has introduced some problems into the interpretation of t. It is usual to consider R_T as being the bare-hull resistance, with no appendages, but T has to overcome not only the augmented resistance $R_T(1 + a)$ but also the resistance of rudder and other appendages. Thus the value of t found from the experiments will depend not only on the shape of hull and the propeller characteristics and arrangement, as reflected in the augment a, but also on the type of rudder, and so on.

As an example, a model representing a 121.9 m (400 ft) cargo ship was run self-propelled without rudder, sternpost or other appendages (Todd, 1934). The value of t was found to be 0.2, and this was essentially due to the augment of resistance effect on the hull. With a plate rudder and square sternpost behind the hull, t increased to 0.29, this representing a considerable loss of propulsive efficiency. When the fore side of the post was faired into a fin, t dropped to 0.24, showing that most of the gain in efficiency was due to the reduction in head resistance of the post. Lastly the rudder and post were carried separately and their resistance measured. Using as the resistance in calculating t the sum of the bare hull and this separately measured appendage resistance, the value of t came out in all cases as 0.20, and the changes in propulsive efficiency were then exactly reflected in the changes in total resistance. When using published figures for thrust-deduction fraction for the purpose of design, therefore, it is most important to know the exact model conditions for which they were obtained.

4.6 Hull Efficiency. The work done in moving a ship at a speed V against a resistance R_T is proportional to the product $R_T V$ or the effective power P_E.

The work done by the propeller in delivering a thrust T at a speed of advance V_A is proportional to the product TV_A or the thrust power P_T.

The ratio of the work done on the ship to that done by the screw is called the *hull efficiency*, η_H, so that

$$\eta_H = \frac{P_E}{P_T} = \frac{R_T V}{TV_A}$$

or

$$\eta_H = \frac{1 - t}{1 - w} \text{ in Taylor notation} \quad (38)$$

and

$$\eta_H = \frac{1 + w_F}{1 + a}$$

$$= (1 + w_F)(1 - t) \text{ in Froude notation,}$$

from Equations (25, 26, 36, 37).

4.7 Propulsive Efficiency. In Equation (1) the quasi-propulsive coefficient is defined as

$$\eta_D = \frac{\text{effective power}}{\text{delivered power}} = \frac{P_E}{P_D}$$

With R_T and T in kN and speed in m per sec

$$P_E = R_T V$$

$$P_D = \frac{TV_A}{\eta_B}$$

from Equation (34). Hence

$$\eta_D = \frac{R_T V}{TV_A} \times \eta_B$$

From the relationships already developed, this can be expanded into the form

$$\begin{aligned} \eta_D &= \frac{(1 - t)}{1 - w)} \eta_B \\ &= \frac{(1 - t)}{1 - w)} \frac{\eta_B}{\eta_0} \eta_0 \\ &= \frac{(1 - t)}{1 - w)} \eta_R \eta_0 \\ &= \eta_H \eta_R \eta_0 \end{aligned} \qquad (39)$$

or quasi-propulsive efficiency equals hull efficiency times relative rotative efficiency times open propeller efficiency.

In Froude notation:

$$\eta_D = \frac{(1 + w_F)}{(1 + a)} \eta_R \eta_0$$

$$= (1 + w_F)(1 - t)\eta_R \eta_0$$

The division of the quasi-propulsive coefficient into factors in this way is of great assistance both in understanding the propulsion problem and in making estimates of propulsive efficiency for design purposes.

Section 5
Model Self-Propulsion Tests

5.1 Methods of Conducting Experiments. Information on all the components of propulsive efficiency can only be derived from model experiments, since only in this way can all the necessary measurements be made and the performance of the propeller in both open and behind conditions be determined.

For such self-propulsion experiments the model is fitted with stern tube or tubes, shafting, and such external fittings as stern post, rudder and, in multi-screw ships, bossings or open shafts and struts. The propellers are driven from inboard by an electric motor, with dynamometers in the shaft line which record thrust, torque and revolutions. The model is attached to the resistance dynamometer on the towing carriage in the same way as for resistance experiments. The carriage is run at any desired steady speed and the difference between the model resistance and the propeller thrust is measured on the dynamometer. This difference may be either positive or negative, depending on the relation of propeller revolutions to model speed, the model being under or overpropelled, respectively.

The model propellers should be sufficiently large for accuracy of manufacture, to enable accurate measurements of thrust and torque to be made during the tests and to avoid serious scale effect as compared with the ship screws. It is not yet possible to set down any absolute rule as to the minimum size of propellers necessary to avoid serious scale effect. Laminar flow has been detected on model propellers operating both in water tunnels and in the relatively turbulent wake of a ship model, while the frictional part of the turbulent resistance of the blade section and the lift are subject to Reynolds number effect. If possible, the diameter should not be less than 15 cm (8 in.) and preferably 22.5 to 30 cm (16 in.) which, in the case of large, multi-screw designs, may call for a model length of 10 m (32.8 ft) or more.

The Propeller Committee of the ITTC in 1975 concluded that in open water the minimum Reynolds number, based on chord length, should be 4×10^6. In the turbulent flow behind the model hull the minimum Reynolds number is appreciably smaller. If turbulence stimulation is adopted near the leading edge of the blades a minimum Reynolds number of 3×10^5 is advised. Since a Rn of 4×10^6 cannot be realized in most test facilities, some form of turbulence stimulation on the blades should therefore be adopted. In calculating the value of Rn, the length dimension L is the width of the blade at 0.7 radius and the speed V

is the resultant velocity at 0.7 radius found by compounding the circumferential speed of the blade with the speed of advance, ignoring any inflow effect, i.e.

$$V = [(0.7\ \pi n D)^2 + (V_A)^2]^{1/2}$$

where

n is revolutions per second

D is diameter of propeller

V_A is speed of advance

Research in this field is continuing. From work carried out by Meyne (1972), Kuiper (1981) and others, it would seem that laminar flow is present on propeller blades during most model tests and that the use of turbulent boundary layer stimulation devices at the leading edges of the blades is therefore recommended. Kuiper adopted artificial roughness at the leading edge of about 0.06 mm in size.

By running such experiments at a number of different model speeds and propeller revolutions, a complete picture may be established of the propulsive coefficient under a variety of propeller-loading conditions, and in association with the resistance experiments on the hull and the open-water characteristics of the propeller, the wake, thrust deduction, and propeller efficiencies may be determined. The relative efficiencies of different designs of propellers, the merits of various types of rudders or bossings and the effect of different choices of propeller revolutions can be obtained quickly and easily.

The procedure for carrying out self-propelled model experiments differs in matters of detail in different model basins. In the three-dimensional extrapolation procedure from model to ship (Sections 3.6 and 6.4 of Chapter V) for a given value of the Froude number $\text{Fn} = V/\sqrt{gL}$ we write, in terms of specific resistance coefficients

$$C_{TS} = C_{TM} - (1 + k)(C_{FOM} - C_{FOS}) + C_A \quad (40)$$

Ignoring for the moment the model-ship correlation allowance C_A, we see that the resistance coefficient of the ship is smaller than that of the model by the amount of the skin-friction correction $(C_{FOM} - C_{FOS})$ times $(1 + k)$, where k is the form factor, accounting for the difference of the three-dimensional hull on the flat-plate friction.

If the model propeller must overcome the full model resistance by actually self-propelling the model it will be working at a higher value of the specific thrust-loading coefficient $T/\frac{1}{2}\rho D^2(V_A)^2$ than will the ship propeller, and due to this extra loading the efficiency will be lower. To take account of this difference there are two common procedures in use in towing tanks.

In the first of these, the difference in specific resistance as measured by the factor $(1 + k)(C_{FOM} - C_{FOS})$ or, when a two-dimensional extrapolation procedure is adopted, by the factor $(C_{FOM} - C_{FOS})$ is converted to model scale. A weight is then placed on the appropriate scale pan of the resistance dynamometer of the amount required to provide such a towing force on the model, or the tow force is adjusted to the desired value on a load cell. If on a particular run the propeller revolutions are adjusted to give a zero reading on the resistance dynamometer, the propeller will then be operating at the thrust coefficient $T/\frac{1}{2}\rho D^2(V_A)^2$ appropriate to the full-scale conditions. This is known as the ship self-propulsion point. Values of $(1 + k)(C_{FOM} - C_{FOS})$ or of $(C_{FOM} - C_{FOS})$ and the corresponding pan weight for different speeds can be prepared in advance, and the correct weight chosen before each experiment. If the form factor value k is not known approximately before the experiment, it is common to adopt a correction based only on the difference $(C_{FOM} - C_{FOS})$.

This procedure will give propeller loadings corresponding to the case when C_A is zero. Usually C_A is not zero so that the weight to be added will be less than before, corresponding to $(1 + k)(C_{FOM} - C_{FOS}) - C_A$.

In the second method a series of experiments is made at a particular carriage speed, the propeller revolutions being set on successive runs to cover a range of thrust above and below that required for self-propulsion of the model. The resistance dynamometer is used to supply any defect or excess towing force required to maintain the model speed, and the values of (thrust minus augmented resistance) are measured. By plotting these against a base of revolutions per minute the self-propulsion point of the model at that speed can be determined accurately. From the corresponding curves of thrust and torque, together with those for open water, the various propulsion factors can be found. For other propeller loadings, corresponding to the smooth ship or the ship with any desired value of C_A, the same information can be obtained by moving along the curve of (thrust minus resistance) to the appropriate new crossing points. A number of sets of such experiments, each at a fixed speed and varying rpm, have to be run to obtain curves of delivered power to a base of ship speed. This second method is now frequently adopted by model testing organizations since in this way the influence of propeller loading on the results of propulsion tests can be determined, irrespective of the extrapolation technique or the C_A value adopted. This can best be done in the first method by running complete tests for each desired value, although correction methods can be applied.

The effective wake can be determined by calculating the coefficients K_T and K_Q, using the thrust, torque and rpm measured in the self-propelled model test, and entering the open-water characteristic curves of the propeller with these values, from which the real slip ratio s_R can be found. The apparent slip ratio s_A can be calculated from the self-propelled results using the speed of the ship V.

Since

$$s_R = 1 - V_A/Pn$$

and

$$s_A = 1 - V/Pn$$

we can find w from the relation

$$(1 - w) = V_A/V = \frac{1 - s_R}{1 - s_A} \tag{41}$$

As stated in Section 4.2, two different wake values can be determined in this way, one based on thrust identity, the other on torque identity, because the values of K_T and K_Q measured behind the model will not cut the K_T and K_Q curves for the open-water condition at the same value of slip ratio.

The thrust-deduction coefficient is determined when the resistance and thrust are known for the self-propulsion point, since $(1 - t) = K_T/T$. The resistance is that for the towed model, and to obtain a true thrust-deduction coefficient the value of T should be that required to propel the model in the same condition. Otherwise the value of t obtained will depend on the type and nature of the appendages, as already demonstrated.

5.2 Standard Procedures for Performance Predictions. The conduct of resistance tests, propeller open-water tests, and self-propulsion tests is a specialized task, and to ensure accuracy great care has to be taken in preparing for and running the tests and in the analysis and interpretation of the results. In the past there were many differences in detail between the practices of the various tanks which made comparisons of research results difficult and led to many problems in the interpretation of performance predictions.

In 1969 the ITTC requested its Performance Committee to attempt to formulate a common method of predicting performance with a sound physical basis to facilitate future ship-model correlation studies. This led to the adoptation by the ITTC of a performance prediction method for single screw ships. The method predicts rate of revolution and delivered power of a ship from model resistance and propulsion test results. The procedure used can be described as follows:

The viscous and the residuary resistance of the ship are calculated from the model resistance tests assuming the form factor to be independent of scale and speed.

(*a*) The total resistance coefficient of a ship without bilge keels is

$$C_{TS} = (1 + k)\, C_{FS} + C_R + C_A + C_{AA} \tag{42}$$

where

- k is the form factor determined from the resistance test
- C_{FS} is the frictional coefficient of the ship according to the ITTC-1957 ship-model correlation line
- C_R is the residual resistance calculated from the total and frictional coefficients of the model in the resistance tests: $C_R = C_{TM} - (1 + k)\, C_{FM}$
- C_A is the correlation allowance obtained from Equation (58) (Section 6.4, Chapter V):

$$C_A = \left[105 \left(\frac{k_s}{L}\right)^{1/3} - 0.64\right] \times 10^{-3}$$

where the roughness of the hull k_s is taken as $k_s = 150 \times 10^{-6}$ m.

- C_{AA} is the air resistance coefficient. For still air conditions the ITTC (1978) suggests the following relation:

$$C_{AA} = 0.001 \frac{A_T}{S}$$

where A_T is the transverse projected area of the ship above the waterline. This is equivalent to setting the air resistance coefficient to 1.0. More details concerning wind resistance have been given in Section 5.2, Chapter V.

(*b*) If the ship is fitted with bilge keels the total resistance coefficient is as follows:

$$C_{TS} = \frac{S + S_{BK}}{S}\left[(1 + k)\, C_{FS} + \delta C_F\right] + C_R + C_{AA} \tag{43}$$

where S_{BK} is the wetted surface of the bilge keels, assuming that eddy resistance of the bilge keels is negligible, and δC_F replaces C_A under the assumption that the added resistance is primarily frictional.

The ITTC standard predictions of rate of revolutions and delivered power are obtained from the full-scale propeller characteristics. These characteristics shall be determined by correcting the model values for drag scale effects according to a simple formula. Individual corrections then give the final predictions.

Thrust, T, and torque, Q, found from the self-propulsion tests are expressed in non-dimensional form

$$K_{TM} = \frac{T}{\rho D^4 n^2} \text{ and } K_{QM} = \frac{Q}{\rho D^5 n^2}$$

With K_{TM} known, J_{TM} and K_{QTM} are read off the model propeller characteristics, and calculations are made of the model wake fraction, which is by definition

$$w_{TM} = 1 - \frac{J_{TM} Dn}{V}$$

and the relative rotative efficiency

$$\eta_R = \frac{K_{QTM}}{K_{QM}}$$

The thrust deduction (assuming no scale effect) is obtained from

$$t = \frac{T + F_D - R_C}{T}$$

where R_C is the resistance corrected for differences in temperature between resistance and self-propulsion tests:

$$R_C = \frac{(1 + k) \times C_{FMC} + C_R}{(1 + k) \times C_{FM} + C_R}$$

where C_{FMC} is the frictional coefficient at the temperature of the self-propulsion test. In some model basins a *blockage* correction also is required.

The characteristics of the full-scale propeller are calculated from the model characteristics as follows,

$$K_{TS} = K_{TM} - \delta K_T$$

$$K_{QS} = K_{QM} - \delta K_Q$$

where

$$\delta K_T = -\delta C_D \, 0.3 \frac{P}{D} \frac{cZ}{D}$$

$$\delta K_Q = \delta C_D \, 0.25 \frac{cZ}{D}$$

The difference in drag coefficient δC_D is

$$\delta C_D = C_{DM} - C_{DS}$$

where

$$C_{DM} = 2\left(1 + 2\frac{t}{c}\right)\left[\frac{0.044}{(R_{nco})^{1/6}} - \frac{5}{(R_{nco})^{2/3}}\right] \text{ and}$$

$$C_{DS} = 2\left(1 + 2\frac{t}{c}\right)\left(1.89 + 1.62 \log \frac{c}{k_P}\right)^{-2.5}$$

In the formulas mentioned above, c is the chord length, t is the maximum thickness, P/D is the pitch ratio and R_{nco} is the local Reynolds number at $x = 0.75$. The blade roughness k is put $k = 30 \cdot 10^{-6}$ m. R_{nco} must not be lower than 2×10^5 at the open-water test.

Assume the thrust deduction to be independent of scale, which is supported by some theoretical investigations (Dyne, 1973). Then the full-scale wake factor can be assumed to be represented by,

$$w_{TS} = (t + 0.04) + (w_{TM} - t - 0.04)\frac{C_{VS}}{C_{VM}} \quad (44)$$

where the thrust deduction t and the wake fraction w_{TM} are determined from the resistance, self-propulsion and open-water tests, as discussed previously. In many cases the value of w_{TS} thus obtained exceeds the model value w_{TM}. In that case w_{TS} should be set equal to w_{TM} (ITTC, 1984). The factor 0.04 takes account of the effect of the rudder. Therefore, this factor may be deleted for twin-screw ships having a single rudder in the ship's center plane.

In the above formula for w_{TS}, C_V is the viscous resistance coefficient,

$$C_{VS} = (1 + k)\, C_{FS} + C_A,$$

$$C_{VM} = (1 + k)\, C_{FM}$$

with C_F the frictional resistance coefficient according to the ITTC 1957-line.

The load on the full-scale propeller is obtained from

$$\frac{K_T}{J^2} = \frac{S}{2D^2} \cdot \frac{C_{TS}}{(1 - t)(1 - w_{TS})^2} \quad (45)$$

With this K_T/J^2 as input value the full-scale advance coefficient J_{TS} and the torque coefficient K_{QTS} are read off the full-scale propeller characteristics and the following quantities are calculated:

- the rate of revolutions,

$$n_S = \frac{(1 - w_{TS}) V_S}{J_{TS} D}$$

- the delivered power,

$$P_{DS} = 2\pi \rho D^5 n_S^3 \times \frac{K_{QTS}}{\eta_R} \times 10^{-3}$$

- the thrust of the propeller,

$$T_S = \frac{K_T}{J^2} \times J_{TS}^2 \rho D^4 n_S^2$$

- the torque of the propeller,

$$Q_S = \frac{K_{QTS}}{\eta_R} \times \rho D^5 n_S^2$$

- the effective power,

$$P_E = C_{TS} \tfrac{1}{2}\rho V_S^3 S \times 10^{-3}$$

- the total efficiency,

$$\eta_D = \frac{P_{DS}}{P_E}$$

- the hull efficiency,

$$\eta_H = \frac{1 - t}{1 - w_{TS}}$$

At this stage there are two options for improving correlation between prediction and full-scale trials given in the ITTC method. The first is to adopt the so-called C_P-C_N correction factors in arriving at the final trial predictions according to

$$n_T = C_N n_S$$

for the rate of revolution and

$$P_{DT} = C_P P_{DS}$$

for the delivered power.

Table 3—Model-Ship Correlation Analysis of Fine Single-Screw Ships Performed for the 17th ITTC (1984)

Inst.	1	2		3	4	5		6	7		8	9
	Fine	Fine	Full	Fine	Fine	Fine	Full	Fine	Fine	Full	Fine	Fine
Data Pt.	86	12	28	17	18	61	59	18	49	86	31	21
$\overline{C}_P$	0.994	1.009	0.986	1.005	0.986	0.984	1.028	0.911	1.039	1.008	1.014	1.067
$\sigma C_P/\overline{C}_P$	0.094	0.082	0.094	0.034	0.079	0.043	0.080	0.058	0.081	0.059	0.078	0.033
$\overline{C}_N$	1.024	1.019	0.978	1.004	1.011	0.996	0.985	1.000	1.025	0.994	1.001	1.010
$\sigma C_N/\overline{C}_N$	0.036	0.040	0.039	0.018	0.026	0.013	0.017	0.029	0.022	0.023	0.039	0.019
$\overline{C}_{NP}$	1.014			1.002	1.014	1.002	0.979	1.023	1.016	0.992	0.996	0.990
$\sigma C_{NP}/\overline{C}_{NP}$	0.036			0.013	0.016	0.010	0.012	0.023	0.016	0.016	0.039	0.014

Values of C_N and C_P are determined by the towing tanks on the basis of correlation studies. Values of C_N and C_P are typically 0.98 and 1.01, respectively. In the absence of appropriate correlation data C_N and C_P should be taken as unity.

The second option is to adopt δC_{FC}-δW_C corrections according to, ITTC (1978):

$$\frac{K_T}{J^2} = \frac{S}{2D^2} \frac{\delta C_{TS} + \delta C_{FC}}{(1 - t)(1 - w_{TS} + \delta w_C)^2}$$

With this K_T/J^2 as input value, J_{TS} and K_{QTS} are read off from the full-scale propeller characteristics and

$$n_T = \frac{(1 - w_{TS} + \delta w_C) V_S}{J_{TS} D} \ (r/s)$$

$$P_{DT} = 2\pi \rho D^5 n_T^3 \frac{K_{QTS}}{\eta_{RM}} \times 10^{-3} \text{ in kW}$$

Values of δC_{FC} and δw_C are again determined by the towing tanks on the basis of correlation studies and should be taken as zero if correlation data are lacking.

The different ITTC Performance Committees have carried out an analysis of C_P, C_N, δC_{FC}, δw_C factors. Table 3 shows the results of this analysis (ITTC, 1984). The results are shown individually for a number of institutes. Fig. 20 summarizes these results. There seems to be no significant differences between full and fine ships (the border put at $C_B = 0.7$). The only exception appears to be for hull forms having open transom sterns. A direct application of the above method results in underprediction of P_D and number of revolutions. The report of ITTC (1984) also discusses C_P, C_N values for twin-screw vessels.

Values of δC_{FC} and δw_C may be derived from the C_P, C_N values by an analysis of trial results. They are interrelated. In Table 3 one also finds values for C_{PN}. This factor represents the same quantity as C_N in case the full-scale number of revolutions is calculated on the basis of torque identity instead of the thrust identity given above (ITTC, 1981).

Consequently, the best extrapolation technique available to date still leads to deviations in speed and number of revolutions of approximately 2 percent. For this the following causes may be mentioned (ITTC, 1978),

(*a*) Model tests
- errors in measurements, instrumentation
- errors in procedure such as turbulence stimulation

(*b*) Full-scale trials
- errors in measurements, instrumentation
- errors in procedure: differences in trim, displacement and influence of wind, waves and fouling

(*c*) Extrapolation and prediction
- errors in correlation allowance for roughness effects
- scale effects in wake factor
- no correction is applied for scale effects in lift
- dependence of propulsion factor in load

The U.S. Navy's policy regarding correlation allowances for naval vessels is given in Design Data Sheet 051-1 (NAVSEA, 1984), "Prediction of Smooth-Water Powering Performance for Surface Displacement Ships." Procedures are similar to ITTC practice, as outlined above, but differ in detail. In addition, it is customary to add a design margin on estimated power, by means of a *Power Margin Factor (PMF)*, which varies from 1.10 during feasibility and preliminary design (no lines drawing or model tests) to 1.04 during final contract design (final lines, propeller design and self-propelled model tests available). The PMF is applied to the estimated effective power, P_E, or to C_{TS}. It must be noted that this design margin is intended to allow for design uncertainties under trial conditions; it is entirely separate from the Service Power Allowance discussed in Section 8.5, Chapter V.

Hagen, et al (1986) present a survey of available data on correlation allowances for naval vessels of different lengths and types of hull coating. Results support the use of Equation (58) of Chapter V for, "Surface ships in the new-ship condition and painted with MilSpec vinyl paint" in the range of 58–293 m length (190–960 ft) Above this range it is assumed that $C_A = 0.0002$ and below it $C_A = 0.0008$. Different values would apply if other hull coatings are to be used.

5.3 Values of Wake, Thrust Deduction Fractions and Relative Rotative Efficiency. When designing a propeller for a new ship, it is necessary to estimate the probable values of the wake fraction, thrust-deduction fraction, and relative rotative efficiency, since all these enter into the calculations and the final assessment of power.

By far the most reliable values of these factors will be found from preliminary self-propelled model tests, in which a model of the new ship is propelled by a stock propeller chosen to have its principal characteristics as near as possible to the probable final design. These values will form a reliable basis for the design of the propeller, and if in the final model tests slight differences are found these can be allowed for in the manufacture of the ship propeller. This practice is quite common in many towing tanks.

In the absence of such information, estimates must be made from other sources, and many model investigations have been carried out to supply the designer with such data.

Systematic experiments to determine wake and thrust deduction made by Luke many years ago are still of use, especially for twin-screw ships. Self-propelled models of single and twin-screw ships with various block coefficients were used in the tests. The twin-screw models were run without bossings and with four designs of bossing having angles to the horizontal varying from 0 to 67½ deg. The results are shown in Fig. 20.

The wake in all cases increased with block coefficient, but the variation with speed was small. The propeller dimensions, except diameter, had little or no effect on either the wake or the thrust deduction. On the single-screw models a smaller diameter screw experienced a higher wake and smaller thrust deduction than a larger screw. This of course is due to the fact that the outer parts of the blades of the larger screw are working in areas of smaller wake velocity, so that the average wake is reduced.

On the twin-screw models the bossing angle of least resistance was 45 deg; smaller or greater angles increased the resistance but did not influence thrust deduction appreciably. High wakes were obtained when horizontal bossings were combined with outboard propeller rotation and vertical bossings with inboard rotation.[4] The best propulsive efficiency was obtained with outboard-turning screws when the bossing angle was somewhat smaller than the angle of least resistance and with inboard-turning screws when the bossing angle was somewhat greater than the angle of least resistance.

Another important investigation of wake was carried out by Bragg (1922) at Michigan with two groups of single-screw models. The models were towed at constant speed and the wakes were measured by means of current meters placed somewhat abaft the usual propeller position. The measured values were therefore nominal wakes freed from the influence of propeller action. Qualitatively the test results were consistent with the results obtained from self-propelled tests; quantitatively, they were consistently higher, for the reason already mentioned (Section 4.2). The results showed that the wake fraction increased with the vertical prismatic coefficients and the beam-draft ratios of the models, and with the elevation of the propeller shaft, but decreased with an increase in propeller diameter.

A third investigation of wake by systematic experiments with models was made by Admiral Taylor. The tests were made in the U.S. Experimental Model Basin with a single-screw model equipped with various propellers. It was found that the wake increased as the propeller diameter was reduced and as the propeller was raised towards the surface, but that it decreased as the clearance between the hull and the propeller was increased. The thrust deduction had in general the same trend as the wake fraction, but did not vary in exactly the same proportion.

For estimating the wake fraction in preliminary propeller design, Taylor gave average values derived from an analysis of the trial results of more than 150 ships in conjunction with open-water characteristics derived from model tests. These values are listed in Table 4 taken from the 1933 edition of *Speed and Power of Ships*.

It should be noted that these wake fractions are of a hybrid nature, the self-propulsion data being taken from ship trials and the open-water characteristics from model tests. They are thus subject to the possibility of some scale effect being present.

Schoenherr in 1934 made an analysis of experimental wake and thrust-deduction values determined in routine tests with self-propelled models in the EMB. The work was extended and reported in the first edition of

Table 4—Values of Wake Fraction from Taylor

Block coefficient C_B	Wake fraction (Taylor): Twin-screw ships	Wake fraction (Taylor): Single-screw ships
0.50	−0.038	0.230
0.55	−0.021	0.234
0.60	+0.007	0.243
0.65	0.045	0.260
0.70	0.091	0.283
0.75	0.143	0.314
0.80	—	0.354
0.85	—	0.400
0.90	—	0.477

[4] A propeller is said to be *outboard* or *outward* turning when, for ahead propulsion, the tips at the top of the disk are moving away from the hull.

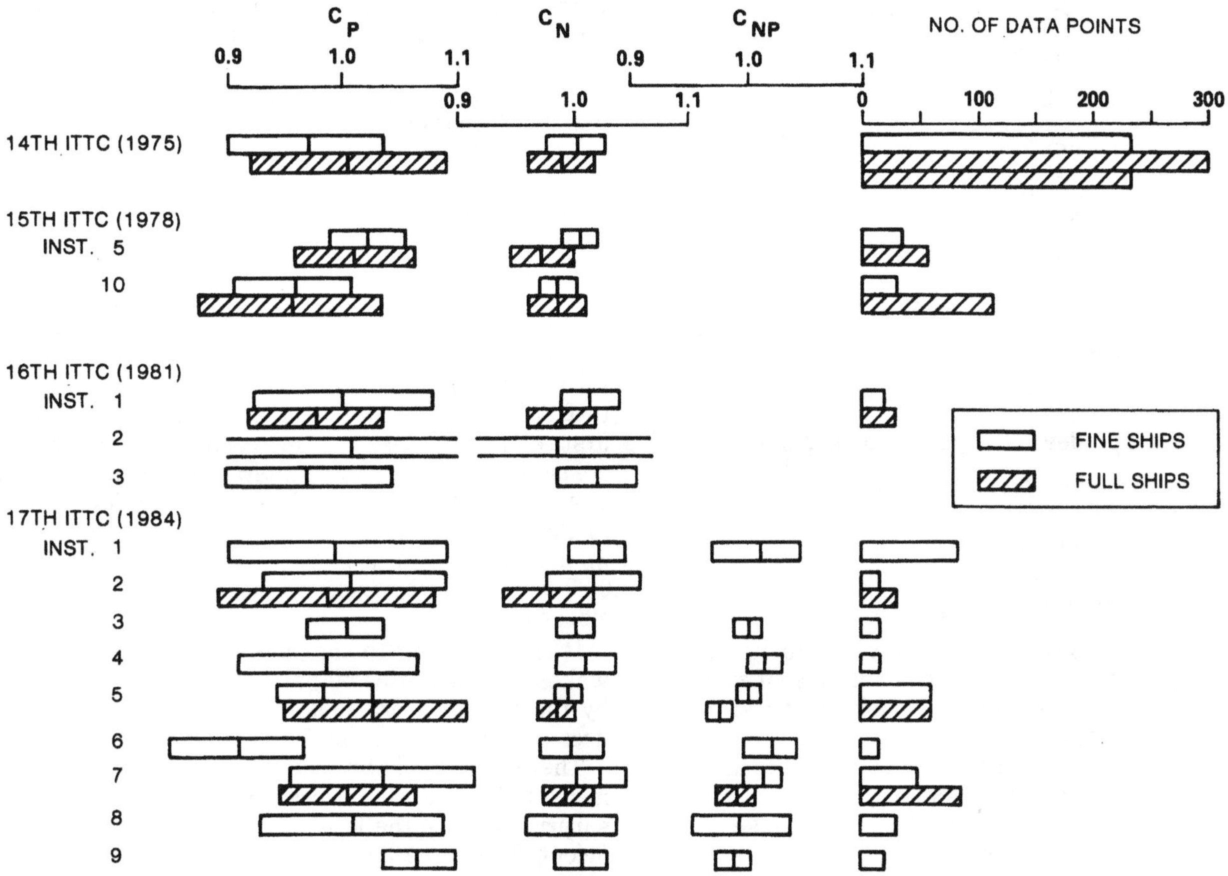

Fig. 20 Comparison of model-ship correlation factors between full and fine single-screw ships

Principles of Naval Architecture in 1939 (page 149). At that time data from 65 tests with 61 single-screw models had been analyzed, and certain empirical formulas derived.

For the wake fraction of single-screw ships

$$w = 0.10 + 4.5 \frac{C_{VP}\, C_P\, B/L}{(7 - 6\, C_{VP})\,(2.8 - 1.8C_P)} + \tfrac{1}{2}\left(\frac{E}{T} - \frac{D}{B} - k'\, K\right) \qquad (46)$$

where

C_{VP} is the vertical prismatic coefficient
C_P is the prismatic coefficient
B is the breadth of ship
L is the length of ship
T is the draft of ship
E is the height of propeller shaft above base-line
D is the propeller diameter
K is the rake angle of propeller blades, radians
k^1 is the a coefficient which has the value 0.3 for a normal type stern and 0.5 to 0.6 for a stern with cutaway deadwood

The precision measure of this formula for 65 tests with 61 models was found to be ±0.027.

For the thrust-deduction fraction of single-screw ships

$$t = kw \qquad (47)$$

where

k = 0.50 to 0.70 for vessels equipped with streamlined or contrarudders
k = 0.70 to 0.90 for vessels equipped with double-plate rudders attached to square rudder posts
k = 0.90 to 1.05 for vessels equipped with old-style single-plate rudders

For the wake fraction of twin-screw vessels, the following formulas were derived:

(*a*) With bossings and outboard turning propellers

$$w = 2C_B^{\,5}(1 - C_B) + 0.2\,(\cos)^2 \times \tfrac{3}{2}\,\psi - 0.02$$

(average deviation for 38 spots ±0.023)

(*b*) With bossings and inboard turning propellers

$$w = 2C_B^{\,5}(1 - C_B) + 0.2\,(\cos)^2 \times \tfrac{3}{2}(90 - \psi) + 0.02$$

(average deviation for 7 spots ±0.012)

(*c*) Propellers supported by struts

$$w = 2C_B^5(1 - C_B) + 0.04$$

(average deviation for 15 spots ±0.024)

For the thrust deduction of twin-screw ships

(*a*) With bossings

$$t = 0.25w + 0.14$$

(average deviation for 45 spots ±0.018)

(*b*) With struts

$$t = 0.70w + 0.06$$

(average deviation for 15 spots ±0.014)

In all the foregoing formulas

w is Taylor wake fraction
C_B is block coefficient of hull
ψ is angle of bossing to horizontal

For the models considered in the foregoing analysis the average value of the relative rotative efficiency η_R was 1.02 for the single-screw models and 0.985 for the twin-screw models.

The choice between open shafts with struts and complete bossings having different angles to the horizontal will affect the propulsive coefficient because of the different wake and thrust deductions which result.

For block coefficients of 0.5 and 0.6 the foregoing equations give the values of w and t shown in Table 5.

The wake fraction without bossings is intermediate between those for bossings having slopes of 30 and 60 deg to the horizontal. This is generally in keeping with the results in Fig. 21. The principal point of interest is the lower value of t with open shafts, which results in the hull efficiency η_H, an important component of the propulsive efficiency, having a higher value than either of those with bossings.

It would appear, therefore, that open shafts have some small advantage over complete bossings as regards propulsive coefficient, but the overall effect on shaft power must take into account the different resistance qualities, as discussed in Section 5 of Chapter V.

Apart from power considerations, the open shafts will give a more uniform wake and so tend to reduce the likelihood of vibration and cavitation arising from the propellers.

The foregoing formulas are valid only for merchant ships of normal form operating at Froude numbers below 0.3. For high-speed ships of the destroyer type, the wake fraction usually lies between −0.02 and +0.02 when the ship is equipped with struts and between 0.04 and 0.08 when equipped with bossings. The thrust-deduction fraction for ships of this type is correspondingly small and as a first approximation may be assumed equal to the wake fraction.

Other valuable sources of data on wake, thrust deduction, and propulsion factors will be found in the results of methodical series published by various towing tanks. For example, Todd (1963) gives such information for the single-screw merchant ship forms comprising Series 60, covering a range of block coefficients from 0.60 to 0.80. Table 6 shows the wake and thrust-deduction fractions and other propulsion factors for the five parent models of Series 60, the values being those at the service speed in each case.

It will be seen that with increasing fullness of hull form, the wake fraction increases steadily from 0.249 to 0.352. The thrust-deduction fraction decreases at first, and then begins to increase again, but this increase is insufficient to counteract the continuing rise in wake, so that the hull efficiency η_H continues to rise to the fullest model. The quasi-propulsive coefficient η_D increases up to a block coefficient of 0.75, but at 0.80 is reduced because of the lower open propeller efficiency.

It is interesting to compare the measured effective wakes for the Series 60 parents with those estimated by the formulas previously quoted. The relevant figures are shown in Table 7.

There is remarkably close agreement between the measured values and those given by Taylor, which is surprising since the latter depend only on block coefficient. The Luke figures, which also depend only on block coefficient, are consistently low, while those based on the formula given by Schoenherr, although showing the correct trend, are always too large, the excess increasing with block coefficient.

The last line in Table 7 shows the value of $k = t/w$, Equation (35), for the Series 60 measured results. From Schoenherr's analysis k varied from 0.5 to 0.7, and this is seen to cover the range of values found for Series 60. These results suggest that the value of $k = 0.7$ will apply to fine ships around $C_B = 0.60$, and that k will decrease with increasing block coefficient.

Some of the Series 60 models were run with propellers of different diameters, and the effect upon w and t can be found from the results given in Todd (1963). Typical figures are quoted in Table 8. These show the expected reduction in wake with increasing diameter. The value of t remains unaffected by changes in diameter within the limits covered in these experiments. As the block coefficient is decreased, the wake

Table 5—Effect of Shaft Arrangements on Hull Efficiency

C_B	0.5			0.6		
ψ	30 deg	Open Shafts	60 deg	30 deg	Open Shafts	60 deg
w	0.111	0.071	0.011	0.143	0.103	0.043
t	0.168	0.110	0.143	0.176	0.132	0.151
$\eta_H = \frac{1-t}{1-w}$	0.936	0.958	0.867	0.961	0.968	0.887

Table 6—Propulsion Factors for Parent Models of Series 60 (Todd, 1963)

C_B	0.60	0.65	0.70	0.75	0.80
F_n	0.229	0.236	0.212	0.190	0.167
L/B	7.50	7.25	7.00	6.75	6.50
B/T	2.50	2.50	2.50	2.50	2.50
$L/\nabla^{1/3}$	6.166	5.869	5.593	5.335	5.092
LCB % L_{PP} from ⊗	1.50A	0.50A	0.50F	1.50F	2.50F
w	0.249	0.268	0.277	0.307	0.352
t	0.176	0.167	0.161	0.171	0.200
η_H	1.097	1.138	1.160	1.196	1.235
η_0	0.669	0.659	0.666	0.653	0.624
η_R	1.035	1.026	1.010	1.014	1.014
η_D	0.759	0.769	0.781	0.792	0.783

Table 7—Comparison of Measured and Estimated Wakes for Series 60

C_B	0.60	0.65	0.70	0.75	0.80
w measured	0.249	0.268	0.277	0.307	0.352
w Taylor[a]	0.243	0.260	0.283	0.314	0.354
w Schoenherr[b]	0.257	0.285	0.322	0.365	0.416
w Luke[c]	0.202	0.226	0.250	0.275	0.300
$k = t/w$ for measured values	0.707	0.624	0.582	0.556	0.568

[a] From Table 4. [b] From Equation (46). [c] From Fig. 21.

decreases. We might therefore anticipate a similar decrease, though of less amount, as the LCB is moved forward for a given C_B because of the progressive fining of the stern. This is confirmed by the Series 60 results as shown in Table 9. Thus, for example, for a block coefficient of 0.65, as the LCB is moved forward from a position 2.46 percent L aft of midships to 1.37 percent L forward, w decreases from 0.310 to 0.229 and t from 0.206 to 0.136. Since these two have opposite effects upon the hull efficiency, the latter remains more or less constant.

It will be noticed that none of the empirical formulas already quoted takes account of the effect of the position of LCB on w and t, and since the Series 60 data on this point are not included in Todd (1963), they are given here in full in Table 9.

The results of the research on Series 60 contained in Todd (1963) also give details of the effect upon w and t of changes in speed, in proportions such as variations in L/B and B/T, and of changes in displacement and trim. These enable estimates to be made of the results of departures from the conditions used in the basic parent series.

Holtrop, et al (1982) and Holtrop (1984) derived expressions for the wake fraction, thrust-deduction fraction and the relative rotative efficiency by statis-

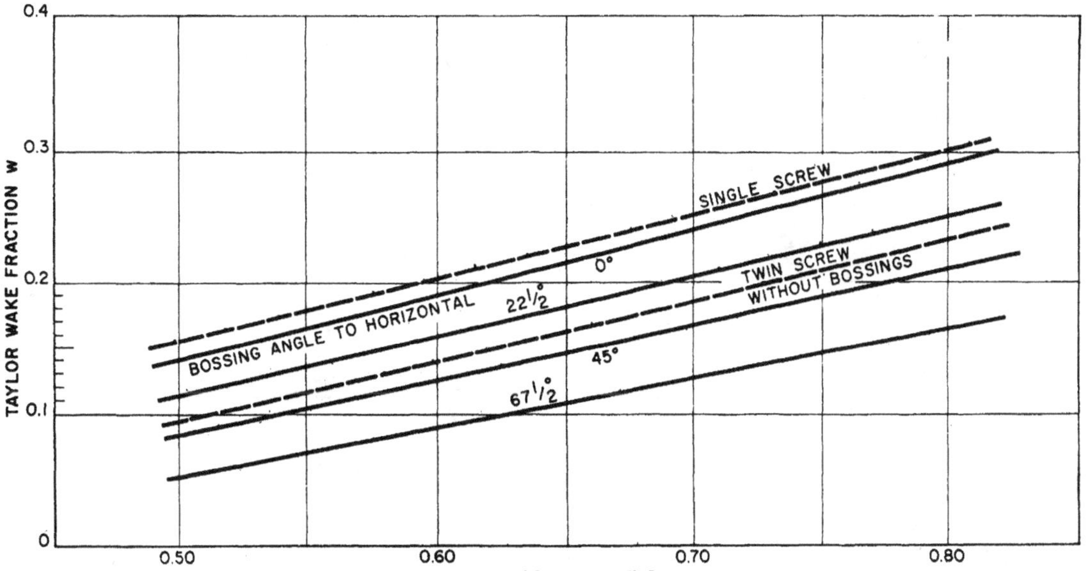

Fig. 21 Wake fractions for single and twin-screw models; w is in Taylor's notation: $V_A = (1 - w)V$

tically analyzing the results of propulsion tests of more than 200 models of various types (see also Section 8.12, Chapter V). The equations obtained for single screw ships with a conventional stern arrangement are as follows:

$$w = c_9 c_{20} C_V \frac{L}{T_A}\left(0.050776 + 0.93405\, c_{11}\, \frac{C_V}{(1 - C_{P1})}\right)$$

$$+ 0.27915\, c_{20} \sqrt{\frac{B}{L(1 - C_{P1})}} + c_{19} c_{20} \qquad (48)$$

The coefficient c_9 depends on the coefficient c_8 defined as:

$c_8 = BS/(L\, D\, T_A)$
when $B/T_A < 5$

or

$c_8 = S(7B/T_A - 25)/(LD(B/T_A - 3))$
when $B/T_A > 5$

$c_9 = c_8$
when $c_8 < 28$

or

$c_9 = 32 - 16/(c_8 - 24)$
when $c_8 > 28$

$c_{11} = T_A/D$
when $T_A/D < 2$

or

$c_{11} = 0.0833333(T_A/D)^3 + 1.33333$
when $T_A/D > 2$

$c_{19} = 0.12997/(0.95 - C_B) - 0.11056/(0.95 - C_P)$
when $C_P < 0.7$

or

$c_{19} = 0.18567/(1.3571 - C_M) - 0.71276$
$+ 0.38648\, C_P$ when $C_P > 0.7$

$c_{20} = 1 + 0.015\, C_{stern}$

$C_{P1} = 1.45\, C_P - 0.315 - 0.0225\, lcb$

The coefficient C_V is the viscous resistance coefficient with

$$C_V = (1 + k)\, C_F + C_A$$

The following formula was obtained for the thrust deduction fraction of single-screw ships:

$$t = 0.25014(B/L)^{0.28956}\,(\sqrt{BT}/D)^{0.2624}/$$

$$(1 - C_P + 0.0225\, lcb)^{0.01762} + 0.0015\, C_{stern} \qquad (49)$$

Table 8—Effect Upon Wake and Thrust Deduction of Change in Propeller Diameter

Block coefficient C_B	0.60		0.80	
Propeller diameter / Draft = D^*/T	0.666	0.800	0.521	0.730
w	0.26	0.23	0.43	0.36
t	0.17	0.17	0.20	0.20

* Standard value of D/T for basic series is 0.70.

Table 9—Effect of LCB Position on Wake, Etc. Series 60 Models.

C_B	0.60				0.65					0.70					0.75				0.80			
L/B	7.50				7.25					7.00					6.75				6.50			
B/T	2.50				2.50					2.50					2.50				2.50			
Fn	0.259				0.236					0.212					0.190				0.167			
V_k, knots,	17.38				15.84					14.28					12.74				11.24			
LCB (% L)	2.48A	1.50A	0.51A	0.52F	2.46A	1.54A	0.50A	0.38F	1.37F	2.05A	0.55A	0.50F	1.54F	2.55F	0.48F	1.50F	2.57F	3.46F	0.76F	1.45F	2.50F	3.51F
w	0.285	0.246	0.230	0.218	0.310	0.263	0.252	0.247	0.229	0.316	0.274	0.271	0.255	0.259	0.367	0.311	0.297	0.288	0.363	0.362	0.355	0.355
t	0.179	0.177	0.154	0.140	0.206	0.167	0.155	0.160	0.133	0.204	0.161	0.180	0.156	0.164	0.227	0.171	0.170	0.176	0.330	0.238	0.200	0.170
η_H	1.148	1.093	1.098	1.100	1.151	1.130	1.129	1.116	1.121	1.164	1.156	1.125	1.133	1.128	1.221	1.203	1.181	1.157	1.053	1.195	1.240	1.287
η_o	0.659	0.682	0.683	0.686	0.634	0.665	0.671	0.669	0.669	0.641	0.670	0.669	0.675	0.667	0.612	0.651	0.654	0.657	0.583	0.613	0.627	0.637
η_R	1.011	1.030	1.020	1.001	0.994	1.023	1.018	1.021	1.025	1.015	1.003	1.014	1.030	1.007	1.086	1.011	1.049	1.041	1.053	1.065	1.012	1.048
η_D	0.765	0.764	0.762	0.756	0.725	0.769	0.771	0.762	0.769	0.757	0.781	0.763	0.788	0.758	0.811	0.792	0.810	0.791	0.647	0.780	0.786	0.859

Note: All Figures Ships 121.9 m Lpp.

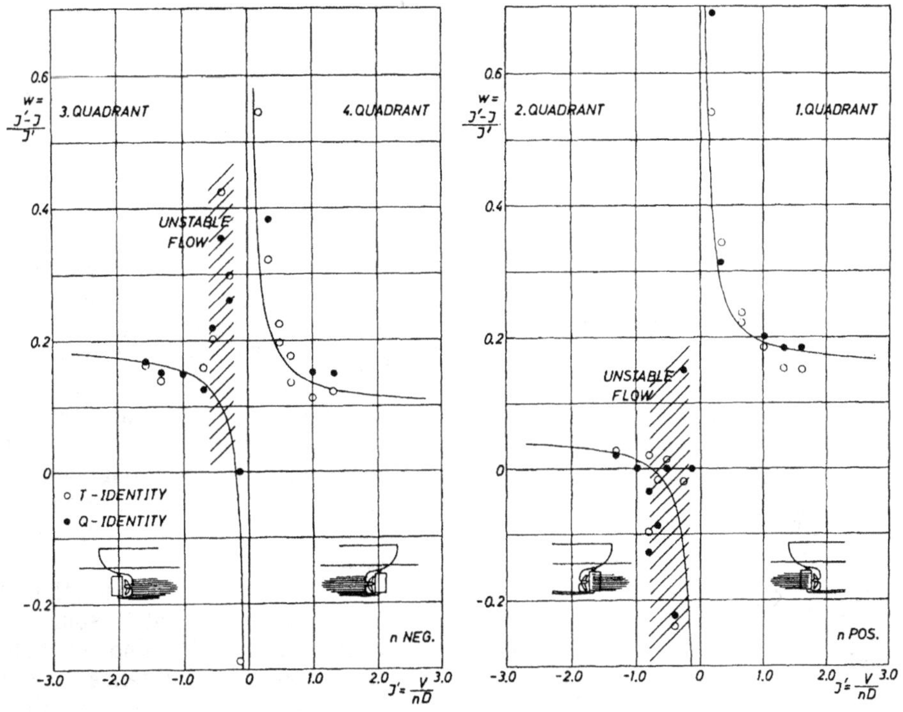

Fig. 22 w — J′-diagram, constant rate of revolution (Harvald, 1967)

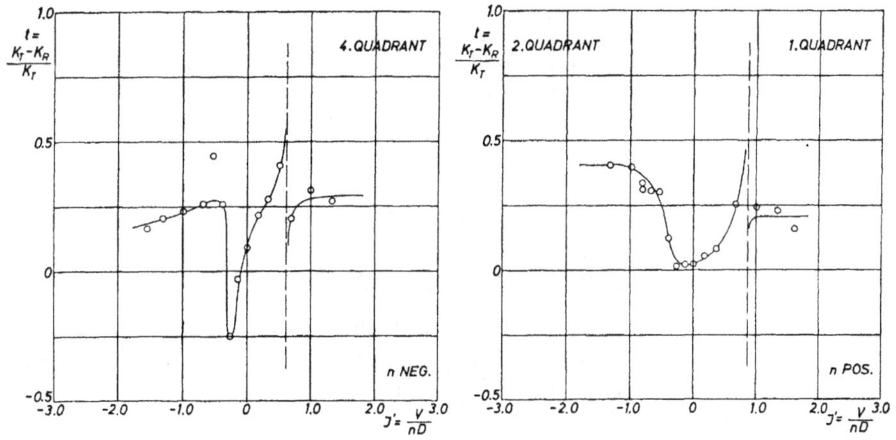

Fig. 23 t — J′ diagram, constant rate of revolution (Harvald, 1967)

The factor of C_{stern} has been given in Section 8.12, Chapter V.

The relative-rotative efficiency can be predicted well by the formula:

$$\eta_R = 0.9922 - 0.05908\ A_E/A_O + 0.07424(C_P - 0.0225\ lcb) \quad (50)$$

For single-screw ships with open sterns, such as transom-stern types, the following equations were obtained,

$$\begin{aligned} w &= 0.3\ C_B + 10\ C_V C_B - 0.1 \\ t &= 0.10 \text{ and } \eta_R = 0.98 \end{aligned} \quad (51)$$

The coefficients in these equations are based on a limited number of model test results.

For twin-screw ships the following equations were derived,

$$\begin{aligned} w &= 0.3095\ C_B + 10\ C_V C_B - 0.23\ D/\sqrt{BT} \\ t &= 0.325\ C_B - 0.1885\ D/\sqrt{BT} \\ \eta_R &= 0.9737 + 0.111(C_P - 0.0225\ lcb) + \\ &\quad - 0.06325\ P/D \end{aligned} \quad (52)$$

The previously mentioned research concerns free-running ships. However, a large number of ships operate at overload conditions such as cable-laying vessels, trawlers and tugs. Also vessels in dynamic positioning mode operate in such conditions, continuously accelerating and decelerating to keep station.

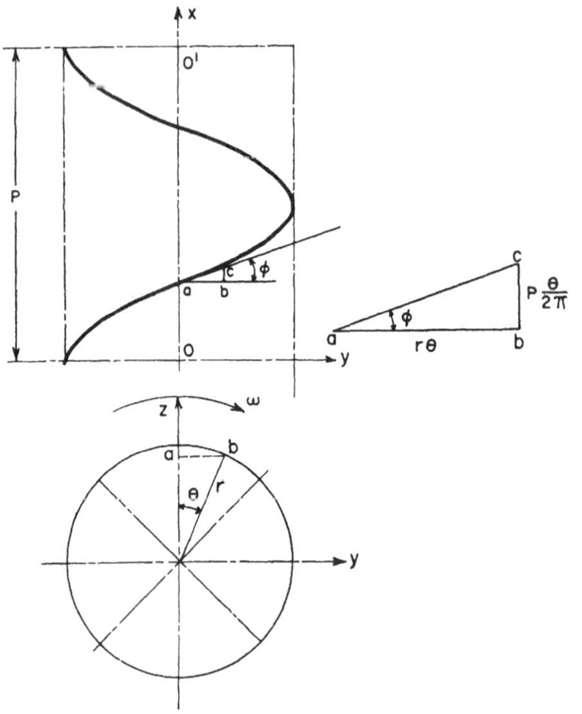

Fig. 24 Geometry of helix

For these cases the thrust deduction and wake factors may differ considerably from those valid for free-running. Troost (1957) already pointed to this fact. For dynamically positioned vessels or vessels in course-tracking mode these factors are also dependent on the relative current angle, i.e. the angle between the speed through the water and the course of the ship.

Harvald (1967) has presented data on both w and t for overload conditions. His work considered the ahead and astern conditions with the propeller working either ahead or astern. This way all four quadrants were covered for a range of J-values. These experiments were carried out for a trawler and a bulk carrier and showed a considerable influence of the propeller load and the advance on the propulsive coefficients w and t; Figs. 22 and 23. Part of these variations, however, have to be attributed to the definition of w and t. For instance, for $V = 0$ the wake fraction $w = (J' - J)/J$ will approach infinite values. Here J' is the advance of the propeller derived from open-water diagram and J is the advance based on the ship speed. Similarly t will not be defined for values of J' for which the thrust approaches zero.

The results of Harvald show that the variation of t and w and J is dependent to some extent on whether V or n are changed. He also concluded that the stern wave has very little influence on the variation in w. For bollard pull conditions the value of t was found to be considerably larger for astern pull which obviously is caused by propeller slip-stream being directed towards the ship. For forward bollard pull a thrust deduction factor of 0.04 is common for single-screw ships.

Section 6
Geometry of the Screw Propeller

6.1 General Characteristics. The general characteristics of screw propellers have been discussed briefly in Section 2.5, and a typical design with definitions of certain terms is shown in Fig. 8. To be able to design a propeller it is necessary to go somewhat further into the geometry involved.

The design of a screw propeller almost invariably start with a helicoidal surface, which either forms the face of the blade or serves as a reference frame from which offsets are measured to describe the blade. This helicoidal surface may be a true one or more generally a warped helicoidal surface having the properties to be described.

As stated in Section 2.5, a helical surface is that surface swept by a straight line AB, Fig. 7, one end of which, A, advances at uniform speed along an axis OO' while the line itself rotates about the point A with uniform angular speed ω radians in unit time. The space curves that are traced by the various points of the generating straight line are called helices. These helices lie on the surfaces of circular cylinders coaxial with the line OO', and all have the same advance per revolution, that is, the same pitch P. Therefore, the true helicoidal surface can be defined as a surface of double curvature, each line element of which is a helix of a constant pitch.

If the helical-line elements have different pitches, or if the radial line is curved, a more general surface is obtained which, while it cannot be described mathematically, is fully and definitely described by giving the shape of the radial reference line and the pitches of a number of helices at various distances from the axis of rotation OO'. This general surface, when used as a reference frame, enables us to describe any type of screw propeller likely to be used in practice. It is called the pitch surface of the propeller and the line elements, which are true helices, the pitch lines.

6.2 Geometry of Helix. The motion of the point on the cylindrical surface may be expressed in mathematical form, using rectangular coordinates x, y, and z, where the axis of x coincides with OO', the axis of revolution, Fig. 24.

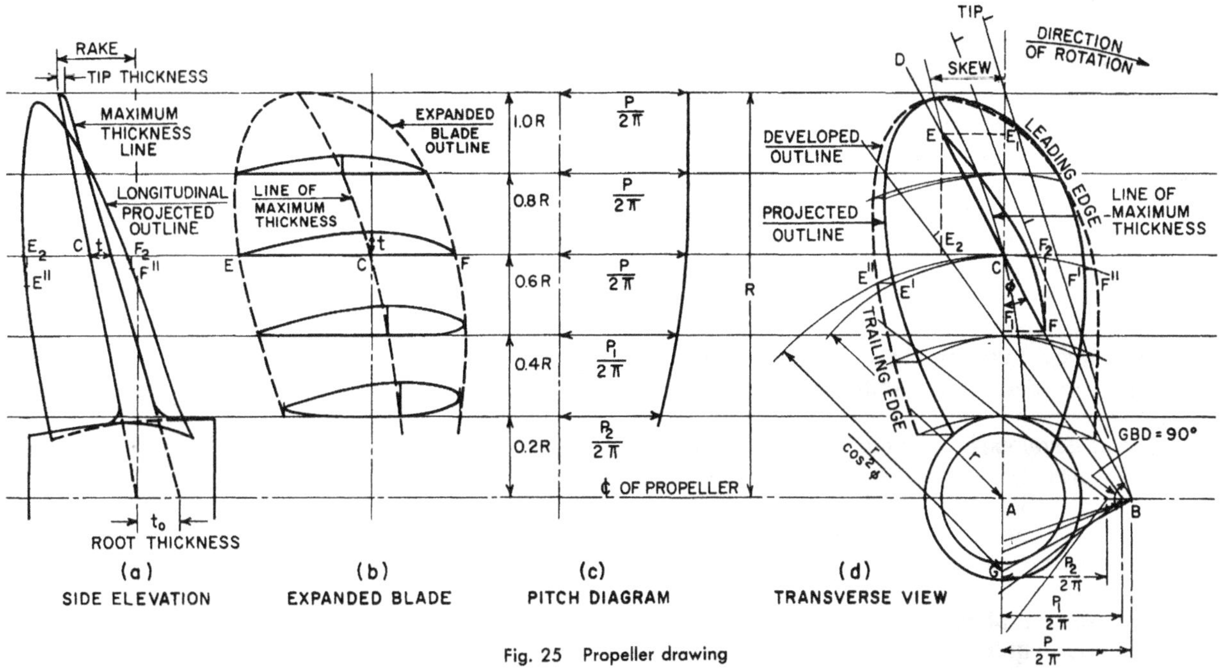

Fig. 25 Propeller drawing

If a cylinder of radius r is unrolled to form a flat surface, the helix will develop into a straight line, the pitch angle ϕ being given by

$$\tan \phi = \frac{P}{2\pi r} \qquad \text{(see Fig. 9).}$$

If the angle θ and the time t are measured from the instant when the generating line r is in the vertical position, then $\theta = \omega t$. The pitch P is the distance which r advances while it makes a complete revolution, i.e., when θ changes by 2π. For any other value of θ, r will advance a distance equal to $P\theta/2\pi$.

The ordinates of a point on the helix are therefore

$$x = P\theta/2\pi$$

$$y = r \sin \theta$$

$$z = r \cos \theta$$

The expanded length of a portion of the helix such as ac in Fig. 24 can be found by developing the cylinder to a flat surface, as in Fig. 9. The triangle abc will then have sides

$$ab = r\theta$$

$$bc = P\theta/2\pi$$

the angle bac being equal to the pitch angle ϕ.

The side ac will be given by

$$(ac)^2 = (ab)^2 + (bc)^2$$

$$= r^2\theta^2 + \frac{P^2\theta^2}{4\pi^2}$$

$$= \frac{\theta^2}{4\pi^2}(4\pi^2r^2 + P^2)$$

or

$$ac = \frac{\theta}{2\pi}(4\pi^2r^2 + P^2)^{1/2}$$

It also can be proved that the radius of curvature of the helix is given by

$$R_n = \frac{r}{\cos^2\phi}$$

6.3 Propeller Drawing. The design drawing for a propeller usually consists of four parts, which are illustrated in Fig. 25. These show, respectively, a side elevation of the propeller (*a*), an expanded blade outline with details of the section shapes (*b*), the pitch distribution if it is not uniform (*c*), and a transverse view (*d*).

For simplicity, we will assume in the first instance that the propeller has sections with flat faces. The choice of the blade outline will depend on a number of design features, and we will assume that for the present this has already been chosen.

The side elevation shows the rake of the propeller (the fore-and-aft slope of the generating line) and a hypothetical section showing the variation of maximum blade thickness from tip to root. It also shows the projected outline of the blade shape on a centerline longitudinal plane.

The section shapes are shown in view (*b*) with their

pitch faces all drawn parallel to the base line and at their correct radii from the axis. These are cylindrical sections, with the thickness measured parallel to the axis, and not normal to the face. The shapes of the sections are shown by the necessary dimensions giving the thickness at various distances along the chord, edge thicknesses, edge radii, and other details. The outline of the blade drawn as in view (*b*) of Fig. 25 is called the expanded blade outline.

The pitch line shown in view (*c*) gives the variation of pitch with radius from the axis. In the example shown, the pitch is constant over the outer part of the blade, and reduced towards the root, a common practice in single-screw merchant ship propellers.

In the transverse view (*d*) are shown the transverse projection and developed outline of the blade, the line of maximum thickness of the sections and the amount of skew. (See also Section 2.5). The developed outline is drawn through the edges of the blade when the blade widths are set out around helical arcs as shown in view (*d*).

Consider a section at radius r where the pitch is P. If AB is set out equal to $P/2\pi$, and the line BCD drawn, the angle ACB is the pitch angle ϕ and BCD is the pitch face line for that section. The transverse projection of the section EF will be E_1CF_1, and if an arc of a circle of radius r is drawn with center A and CE^1 ($= CE_1$) and CF^1 ($= CF_1$) are measured around this arc from C, E^1 and F^1 will be points on the transverse projected blade outline.

In the same way, the longitudinal projection of the section on the centerline plane will be E_2CF_2. If the offsets CE_2 and CF_2 are set off in view (*a*) as shown, and then the points E_2 and F_2 are dropped down to the same levels as E^1 and F^1 in view (*d*), E^{11} and F^{11} will be points on the longitudinal projected outline.

In this way the whole transverse and longitudinal projected outlines can be drawn.

The position of maximum thickness of the sections can be set off along each radial line and a locus of maximum thickness drawn in view (*d*). Since this line will not lie in the longitudinal centerline plane, there is no true plane section through the blade at this position, which is why the thickness line in view (*a*) was referred to as a hypothetical one.

The developed outline is not so easy to draw, but for most purposes it is sufficient to expand the section along an arc of a circle which has a radius equal to the radius of curvature of the helix at the point C. If BG is drawn perpendicular to BC, then $CG = r/\cos^2 \phi$, which is the radius of curvature of the helix at the point C having a radius r and pitch P.

Hence if CE and CF are set off from C around the arc of a circle struck from G with radius GC, the resultant points E^{11} and F^{11} will be points on the developed blade outline. This is nearly a correct construction for narrow and medium-width blades, but is not so accurate in the case of wide blades.

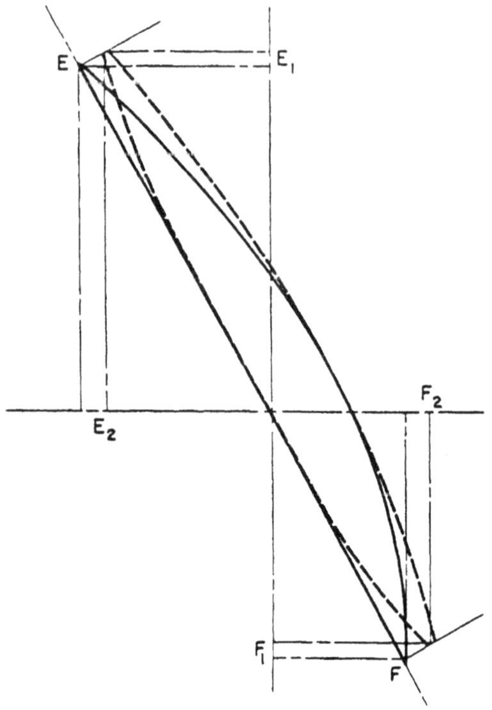

Fig. 26 Effect of blade section shape on projected blade outlines

When the sections are set out as in view (*b*) with the pitch lines all parallel to the axis, the ends of the sections can be joined to give still another outline, called the expanded blade outline.

If the sections are not of the type with a flat face and circular back, but are of airfoil shape with nose and tail lifted from the pitch face, somewhat different developed and projected outlines will be obtained for the same total chord width, as shown in Fig. 26.

The characteristics of propellers are customarily expressed in the form of nondimensional ratios, the most commonly used being

$$\text{Pitch ratio } PR = \frac{\text{propeller pitch}}{\text{propeller diameter}} = \frac{P}{D}$$

$$\text{Disk area } A_0 = \frac{\pi}{4} D^2$$

$$\text{Expanded-area ratio} = \frac{A_E}{A_0}$$

$$= \frac{\text{expanded area of all blades, outside hub}}{\text{disk area}}$$

$$\text{Developed-area ratio} = \frac{A_D}{A_0}$$

$$= \frac{\text{developed area of all blades, outside hub}}{\text{disk area}}$$

Projected-area ratio $= \frac{A_P}{A_0}$

$= \frac{\text{projected area of all blades, outside hub}}{\text{disk area}}$

Mean width ratio

$= \frac{\text{mean developed or expanded width of one blade}}{\text{diameter}}$

$= \frac{\text{area of one blade outside hub} / \text{length of blade outside hub}}{\text{diameter}}$

Blade-width ratio $= \frac{\text{maximum width of blade}}{\text{diameter}}$

Blade-thickness fraction

$= \frac{\text{maximum blade thickness produced to shaft axis}}{\text{diameter}}$

$= \frac{t_0}{D}$ in Fig. 25.

6.4 Constructional Details of Marine Propellers. For many years marine propellers were usually made with either three or four blades, the latter being almost universal in single-screw ships. Model experiments had indicated that such propellers were likely to be the most efficient, any increase in the number of blades giving rise to interference effects between the blades near the hub with a consequent decrease in efficiency. Also keeping the expanded area constant, an increased number of blades will lead to blades having higher aspect ratio and smaller chord length. The section lift-to-drag ratio will decrease, also because the blades will have relatively larger thickness to satisfy the requirements imposed by strength. The higher aspect ratio will have virtually no beneficial effect if the camber and pitch distribution may be optimized. On the other hand, the increase of the number of blades will result in better efficiency if the isolated blades have the same lift-drag ratio. This effect, however, is more than balanced by the reduced blade efficiency. Comparing results for the B-4-70 and B-7-70 propellers confirms this result. (See Section 8.3.)

With the continual increase in the speeds and powers of ships, particularly those with a single propeller, it became necessary to increase the blade area in order to delay the onset of cavitation. To avoid excessively wide blades, an increase in their number was a possible alternative. This same increase in power, and so in the thrust developed per blade, also increased the periodic forces transmitted from the propeller to the hull, both through the water by pressure effects and through the shaft bearings. As a result hull-vibration problems were intensified. These propeller forces are predominantly of blade frequency. An increase in the number of blades not only reduces the thrust per blade, thereby reducing the intensity of the disturbing forces, but also increases their frequency. This fact, in certain cases, may be used to avoid resonant conditions, and the forced vibration will be of lesser amount because of the increase in the internal damping of the hull structure at higher frequencies. Because of such reasons many ships are now fitted with propellers having 5, 6 or more blades, and it has been found that by careful design such screws need pay only a small penalty, if any, in efficiency.

The blades are either cast integral with the hub or cast separately and bolted to it; the two types are referred to as solid and built-up propellers, respectively. Built-up propellers have the advantages that damaged blades are easily replaced and that small adjustments in pitch can be made by turning the blades on the hub. Their disadvantages as compared with solid propellers are higher first cost, greater weight and somewhat smaller efficiencies because of the larger hub.

The hub is usually cylindrical or conical in outline, the diameter ranging from $0.15D$ to $0.25D$. It is bored to fit the taper of the tail shaft and is usually secured to the shaft by one or more keys and a nut, the latter being covered by a streamlined fairwater or cap.

The pitch ratios used for marine propellers range from about 0.6 for highly loaded propellers such as those on tugboats, up to 2.0 or more on high-speed motor boats. The radial distribution of pitch in twin-screw ships is sometimes constant, but in single-screw ships, where the wake variation over the disk is much greater, with a high concentration over the inner radii, the pitch is often reduced towards the hub. In highly loaded propellers, the pitch is often reduced towards the tip also, in order to decrease the thrust loading there and so delay the onset or reduce the severity of tip-vortex cavitation.

Propeller blades are generally given some rake aft, which increases the clearance from the hull, bossings or shaft brackets, and is generally beneficial to efficiency and in reducing the periodic propeller forces which induce hull vibration. The latter are also reduced by using skewed blade outlines, so that the leading edges of the blades enter the wake concentrations behind the hull and appendages more smoothly.

The blade area depends very much upon the thrust loading, and developed blade-area ratios vary from 0.35 to values over unity in very high-speed ships. Thrust loading is further discussed in Section 7.7.

The choice of blade area and blade-outline shape having been made for a given design, the thickness of the sections must be such as to provide adequate strength. The hydrodynamic efficiency of a propeller-blade section depends upon its thickness ratio, and from this point of view it is desirable to keep the thickness as small as possible. From the viewpoint of delaying the onset of certain types of cavitation, however, a larger blade thickness is beneficial. These conflicting requirements have resulted in blade designs which are not as thin as some years ago.

Propeller materials need to be light in weight, have a smooth surface and have a high resistance to erosion. Those commonly used include cast iron, cast steel, manganese bronze and different varieties of manganese-nickel-aluminum alloys. The manganese bronzes and alloys make tough blades, which take a high polish and are resistant to erosion, thus achieving and maintaining high efficiencies. Cast-iron propellers are relatively cheap but have little tensile strength, have to be relatively much thicker and moreover corrode badly in salt water and have a low resistance to cavitation erosion, which latter features reduce their efficiency. They are chiefly used on harbor tugs, ice-breakers and similar craft, because they tend to break off cleanly when striking an obstacle without causing damage to hull or machinery.

The working stress allowed for cast iron is only about one half that for manganese bronze, so that cast-iron propellers will be much thicker and heavier. On the other hand, the nickel-aluminum-bronze is considerably lower in density, has a higher allowable working stress, and will give thinner and lighter propellers of high efficiency. This material also gives a smooth surface finish and has a high resistance to cavitation erosion.

Since both the chord length and the thickness of the blade sections are important parameters in propeller design, and since the strength of the blades must satisfy the requirements of the classification societies concerned, it is necessary to be able to calculate the minimum thickness of the blades at an early stage in the design. Many methods have been developed for making such calculations. In the simplest ones, the blade is treated as a cantilever, with the total thrust and torque forces acting at some estimated points on the blade, and the stress in a typical section near the root is calculated. A refinement of this method is to replace the concentrated thrust and torque loads by a loading pattern varying from root to tip in accordance with that found from circulation theory. This still ignores the variation of load across a chord, which can only be taken into account by more sophisticated theory.

In the case of the average merchant ship, one of the simpler methods is generally adequate, and for incorporation in the classification society rules the calculation must be reduced to a relatively simple formula with numerical coefficients.

Schoenherr (1963) dealt in considerable detail with the problem of propeller-blade strength, taking into account the bending moments due to the hydrodynamic loading, centrifugal force, rake and skew. His stated objective was "to derive a formula for the minimum blade thickness required to ensure a blade of adequate strength for normal service conditions." He obtained such a formula for the required thickness at any desired section, which can be evaluated when the operating conditions, blade parameters and section coefficients are known. Tables of functions were provided to assist in carrying out the computation.

This formula was further simplified to suit the designer's need for a simple working rule:

(*a*) The thickness t was assumed to decrease linearly from root to tip, since the analysis had shown that when t is calculated for a section near the hub and such a linear variation assumed, the thickness elsewhere is greater than necessary to maintain constant stress. It is therefore only necessary to compute t for the most heavily stressed section.

(*b*) Average values were assumed for the blade shape and section parameters.

(*c*) One representative point was taken on the chosen section at which to calculate the stress—this was the point on the face of the blade at the position of maximum thickness, where the stress is a maximum and is tensile for ahead operation.

These simplifications were considered justifiable in view of other unknowns in the problem—our ignorance of locked-in stresses and the dynamic effects of blade vibration, ship motion in a seaway, and crash reversals.

The American Bureau of Shipping adopted the simplified Schoenherr formula, choosing as the typical strength section that at 0.25 radius. The slightly modified requirements as they appear in the current rules (American Bureau of Shipping, 1987) are given below.

(*a*) *Blade thickness.* Where the propeller blades are of standard design, the thickness of the blades shall not be less than determined by

$$t = S\left(289\sqrt{\frac{AH}{C_n\,CRN}} \pm \frac{C_s}{C_n}\frac{BK}{4C}\right) \qquad (53)$$

where t = minimum thickness of blade at 0.25 radius in mm
H = shaft power in kW
R = rpm at maximum continuous rating
N = number of blades
K = rake of propeller blade, where
$K = 500D \tan \epsilon$,
D = propeller diameter in m
ϵ = rake angle
R = propeller radius
S = a constant dependent on the value of D, viz.:

$D \leq 6.1$ m	$6.1 \leq D \leq 7.624$ m	$D \geq 7.624$ m
$S = 1.00$	$S = \sqrt{(D + 24)/30.1}$	$S = 1.025$

A is a constant determined by the pitch distribution, viz.:

$$A = 1.0 + \frac{6}{P_{0.7}} + 4.3P_{0.25}$$

Table 10—Propeller Material Constants

Materials	English units f	English units w	SI units f	SI units w
Type 1 Manganese bronze	68	0.30	2.10	8.3
Type 2 Nickel-manganese bronze	73	0.29	2.13	8.0
Type 3 Nickel-aluminum bronze	85	0.27	2.62	7.5
Type 4 Manganese-nickel-aluminum bronze	85	0.27	2.37	7.5
Type 5 Cast Iron	25	0.26	0.77	7.2

B is a constant determined by the following formula

$$B = \frac{4300\ Wa}{N} \times \left(\frac{R}{100}\right)^2 \times \left(\frac{D}{20}\right)^3$$

C is a constant determined by the formula

$$C = (1 + 1.5\ P_{0.25}) \times (Wf - B)$$

where $P_{0.25}$ is pitch at 0.25 radius divided by propeller diameter

$P_{0.7}$ is pitch at 0.7 radius divided by propeller diameter

w and f are material constants from Table 10.

W = width of blade at 0.25 radius in mm

C_n and C_s are determined from

$$C_n = \frac{I_o}{U_f\ W\ T^2}$$

$$C_s = \frac{a_s}{W\ T}$$

in which, a_s is area of cross-section of propeller blade at 0.25 radius in mm^2

I_o is moment of inertia of cross-section of propeller blade with respect to a straight line through the center of area of the cross-section and parallel to the pitch line or the nose-tail line of the section, in mm^4

U_f is maximum distance of the axis about which I_o is to be calculated to points on the pressure side (face) of the blade section, in mm

T is maximum thickness of the cross-section at 0.25 radius in mm

a is expanded blade area divided by disk area

If C_n exceeds 0.1, the value 0.1 has to be used in the formula for t. If the propeller has forward rake the minus sign in the formula is to be used, else the plus sign.

(*b*) *Fillets at the root of the blades* are not to be considered in the determination of blade thickness.

A number of notes are attached to the use of the formula under certain conditions of operation.

In the course of the analysis, data are available for a detailed calculation of the weight and polar moment of inertia of a blade. The former is important from a cost point of view, and both are necessary when considering shaft-vibration problems. Schoenherr gives approximate formulas for both items for rapid calculations or checks:

$$W = 1.982\tau\zeta\gamma R^3$$
$$I_P = 0.2745\ WR^2$$

where

W is weight of all blades

I_P is polar moment of inertia of all blades

τ is blade-thickness fraction

ζ is blade-area ratio (for whole propeller)

γ is specific weight of blade material

R is propeller-tip radius

The weight and moment of inertia of the hub are not included. The units used must be consistent.

For controllable pitch propellers the American Bureau of Shipping has adopted the following formula (American Bureau of Shipping, 1987):

$$t_{0.35} = 232\sqrt{\frac{A\ H}{C_n\ C\ R\ N}} \pm \left(\frac{C_s}{C_n}\right)\left(\frac{B\ K}{6.3\ C}\right) \quad (54)$$

Where $t_{0.35}$ is minimum thickness of blade at 0.35 radius, in mm

C_s is $a_s/(W\ T)$, for the cross-section of the propeller blade at 0.35 radius

C_n is $I_o/(U_f\ W\ T^2)$, for the cross-section at 0.35 radius

and

$$A = 1.0 + \frac{6}{P_{0.7}} + 3P_{0.35}$$

$$B = \frac{490c\ w\ a}{N}\left(\frac{R}{100}\right)^2\left(\frac{D}{20}\right)^3$$

$$C = \left(1 + 0.6\ P_{0.35}\right)\left(Wf - B\right)$$

in which $P_{0.35}$ is pitch at 0.35 radius divided by the propeller diameter

W is width of blade at 0.35 radius, in mm

Other classification societies have adopted other formulas. Usually the results for the minimum blade section thickness of the various formulas differ by only a small amount.

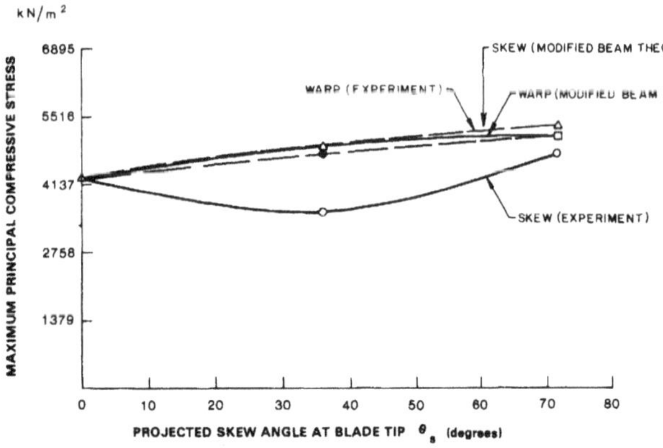

Fig. 27(a) Compressive Stresses (Boswell et al, 1976)

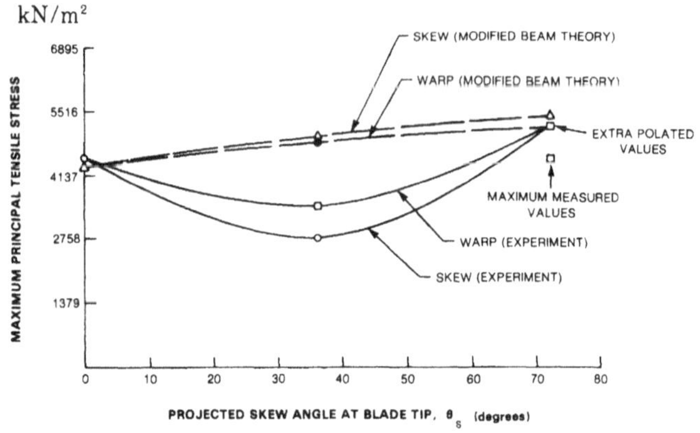

Fig. 27(b) Maximum principal stresses under uniform loading of 6.89 kN/m²

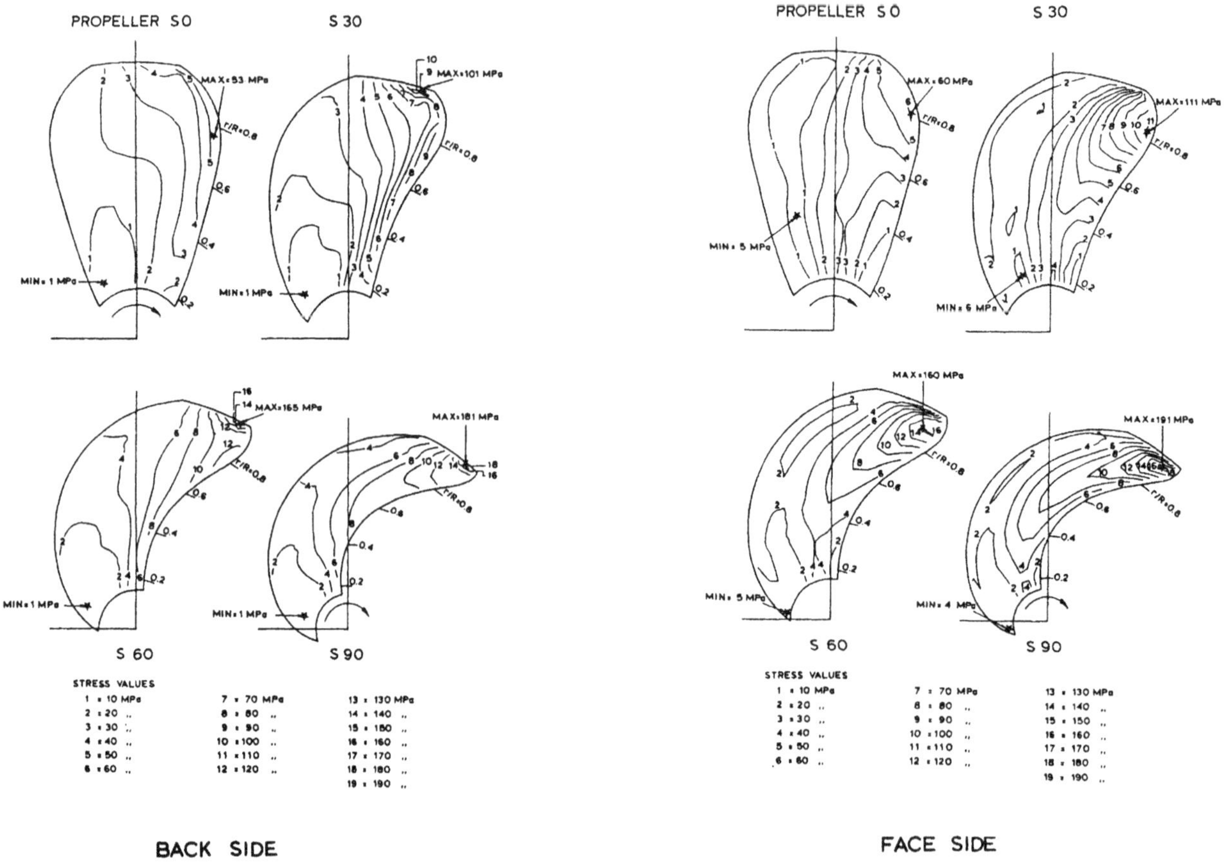

Fig. 28 Iso-curves of the Huber-Hencky-Von Mises equivalent stress, calculated for the backing bollard condition (full-scale values) (Luttmer et al, 1984)

Propeller manufacturers usually adopt thicknesses in excess of those which can be derived from these classification society formulas. Besides wanting to reduce the sensitivity of the pressure distribution on the blade sections to angle of attack, important for obtaining improved cavitation characteristics, propeller manufacturers generally want to incorporate a greater safety margin against structural failure.

For highly skewed propellers (skew angles exceeding some 40 degrees) stress calculations with cantilever beam methods may result in propellers having insufficient strength. Also the above-mentioned formulas may not be valid for these propellers. Cumming, et al (1972) discussed the major results of the investigations into the stresses associated with highly-skewed propellers to that date. Boswell (1969) found that predictions of beam theory yielded neither the correct value for the maximum principal stress nor the correct chordwise distribution. His measurements concerned a propeller with 120 degrees skew. To obtain more information Boswell, et al (1976) carried out strain measurements for one-bladed propellers with different amounts of skew. For all propellers he found the maximum stress to occur in the region from the hub to the 50 percent radius. At the 30 percent radius the maximum principal stress is near the half-chord for the unskewed propeller and was found to move toward the trailing edge with increased skew. Fig. 27 shows the dependence of the maximum stress on the skew angle. These results are for uniform air pressure and only show the effect of skew on the stresses. Centrifugal forces are not included. This figure also includes results for propellers with skew and rake (warped propellers).

Apart from the average loading and stress, the unsteady forces and the associated unsteady stresses may also be of importance. These are generated by the operation of the propeller in the wake field. The unsteady forces depend on the harmonic content of the wake field and the amount of skew. This is discussed in some more detail in Sections 4.2 and 8.2. Other contributions to unsteady forces may be caused by propeller vibrations. For conventional propellers the fundamental natural frequency of the blade usually is higher than the primary excitation frequencies. For high skew the natural frequency decreases significantly (for constant thickness and chord length) (Cumming, et al 1972). However, for constant stress this effect would be less.

Boswell, et al (1973) report on strain measurements on a two-bladed propeller with 60 degrees skew. The maximum stress was found to occur near the trailing edge. They found considerably lower stresses for the backing condition assuming equal maximum principal stresses on the face and back of the propeller. However, they note that conditions related to crash stop maneuvers may change this picture. The increased stress levels in that case may be enhanced by elastic deflection: the blade will bend such that the effective pitch is increased leading to still higher loading. This may lead to a phenomenon called static divergence, which instability leads to damage. The experiments of Boswell, et al (1973) showed that a propeller is more susceptible to this instability for increased skew. However, the instability itself has not yet been observed to occur for marine propellers.

Luttmer, et al (1984) carried out an analysis of the influence of skew on the stresses in backing propellers. They calculated the open-water performance for four propellers having skew angles of 0, 30, 60 and 90 degrees. This was compared with measurements. Thereby they obtained satisfactory agreement with realistic hydrodynamic loading levels. Next, a finite element analysis of the propeller was carried out for the astern backing condition, which was considered to be representative for conditions to be expected at crash stop maneuvers. Luttmer found the stress isocurves depicted in Fig. 28. These curves show, contrary to the previously mentioned research, maximum stresses at the tip (trailing edge for normal ahead condition). These stresses increase drastically with skew.

Special strength requirements have been adopted by the classification societies for propellers which have to operate in ice-covered waters.

Section 7
Cavitation

7.1 The Nature of Cavitation. Cavitation is a phenomenon met with in highly loaded propellers in which, beyond certain critical revolutions, there is a progressive breakdown in the flow and a consequent loss of thrust. In its extreme form, it may prevent the ship from reaching the desired speed. Before this stage is reached, however, it manifests itself by noise, vibration and erosion of the propeller blades, struts and rudders.

In early days these problems were confined to high-speed ships, but as speeds and powers have increased the erosion aspect of cavitation has become more and more important, particularly in very high-powered, single-screw ships. In such ships, there is a large wake variation over the propeller disk, which encourages cavitation, and it is necessary to give special consideration to propeller clearances from the hull and to the effects of overloading of the propeller in bad weather or when the ship's bottom is fouled. Avoidance of cavitation and erosion has become an important requirement in the design of nearly all propellers, and it is necessary to consider the problem of cavitation before going on to methods of propeller design.

One of the earliest references to cavitation on marine propellers was made by Osborne Reynolds, who in 1875 referred to the effect of racing of propellers (Newton, 1961). The first fully recorded case of its occurrence on a ship is that of the British destroyer *Daring* in 1894 (Barnaby, 1897). With the original twin three-bladed propellers the ship on trial only reached a speed of 24 knots instead of the desired 27. When these screws were replaced by another pair with 45 percent more blade area, not only was 24 knots achieved with

17 percent less power, but a top speed of 29.25 knots was reached, with the elimination of much of the vibration previously experienced.

At about the same time Sir Charles Parsons was building the *Turbinia* with which to demonstrate his invention of the marine steam turbine. The ship was first run in 1894, having a single shaft carrying one two-bladed propeller, 0.752 m in diameter, and gave very disappointing results (Burrill, 1951). Parsons then learned about the *Daring* trials, suggesting that the limiting thrust which that ship's propellers could deliver (owing to the formation of cavities in the water) was equivalent to a pressure of 76.7 kN / m^2 (10.8 psi).

He then replaced the single propeller by three in tandem on the one shaft, well spaced fore and aft, which gave a considerable improvement, the speed being 19¾ knots. Finally, Parsons fitted the *Turbinia* with three shafts, with three tandem propellers on each shaft, having a diameter of 0.457 m (18 in.) and a developed blade-area ratio of 0.60. These gave a speed on trial of 32.75 knots at 1475 kW (1976 hp) although speeds in excess of 34 knots were claimed on later occasions.

Since those early days much has been learned about the causes of cavitation and many criteria for estimating its probable occurrence have been proposed. We return to this point in Section 7.7.

To understand the mechanism of cavitation, consider a blade section or airfoil set at a small angle of attack in a two-dimensional, steady, nonviscous flow, Fig. 29. Let the uniform steady velocity far ahead of the section be V_0 and the corresponding total pressure p_0.

For a particular streamline such as AB, Bernoulli's theorem gives the relation

$$\frac{p_0}{w} + \frac{(V_0)^2}{2g} = \text{constant}$$

or

$$p_0 + \tfrac{1}{2}\rho\,(V_0)^2 = \text{constant}$$

At any point P on the streamline where the pressure and velocity are p_1 and V_1 we have

$$p_1 + \tfrac{1}{2}\rho\,(V_1)^2 = p_0 + \tfrac{1}{2}\rho\,(V_0)^2$$

and the change in pressure will be

$$\delta p = p_1 - p_0 = \tfrac{1}{2}\rho\,[(V_0)^2 - (V_1)^2]$$

If V_1 is greater than V_0, i.e., if the flow is accelerating, then p_1 is less than p_0, the pressure is decreased and δp is negative, and vice-versa.

At some point S near the nose of the section the flow divides, and the fluid following the dividing streamline is turned through 90 deg, losing all its velocity and momentum in the direction of motion along the streamline. Hence at the point S the velocity V_1 is zero, and

$$\delta p = p_1 - p_0 = \tfrac{1}{2}\rho(V_0)^2$$

The increase in pressure at S over the ambient pressure p_0 is therefore $\frac{1}{2}\rho V_0^2$; S is called a stagnation point, and the dynamic, stagnation or ram pressure of the flow, given the symbol q, is

$$q = \tfrac{1}{2}\rho\,(V_0)^2$$

The fluid above the dividing streamline passes over the upper surface or back of the blade with increased velocity, resulting in a decrease in pressure, while that below is slowed down, giving increased pressure over the face, Fig. 29. It is this differential-pressure effect which gives rise to the lift on the section.

At a point on the back of the blade where the pressure and velocity are p_1 and V_1

$$\begin{aligned} p_1 &= p_0 + \tfrac{1}{2}\rho\,[(V_0)^2 - (V_1)^2] \\ &= p_0 + \delta p \end{aligned}$$

Hence p_1 will become zero if

$$\delta p = -\,p_0$$

Since water cannot support tension, the flow will break down at this point with the formation of bubbles and cavities, and cavitation will result.

In practice, this situation will come about somewhat earlier—when p_1 has fallen not to zero but to the vapor pressure of water, p_V, at which it begins to "boil" and form cavities. The criterion will then be that

$$p_V = p_0 + \delta p$$

or

$$\delta p = -\,(p_0 - p_V)$$

Dividing by the dynamic pressure $\frac{1}{2}\rho V_0^2$ or q, cavitation will begin when

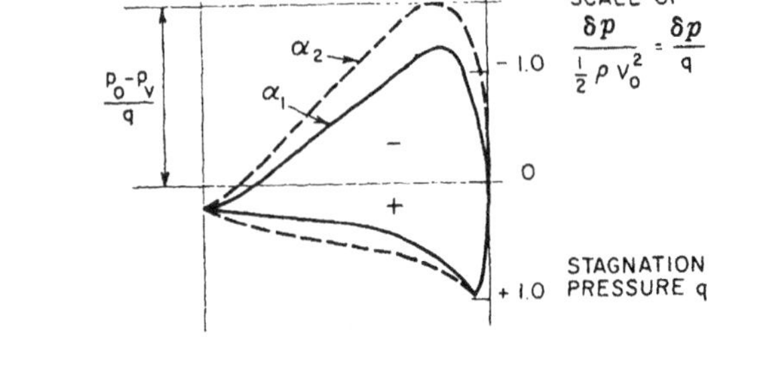

Fig. 29 Flow and pressure around an airfoil

$$-\frac{\delta p}{q} \geq \frac{p_0 - p_V}{q}$$

The expression

$$\frac{p_0 - p_V}{q} = \sigma$$

is called the *cavitation number.*

In any particular case, σ can be calculated. p_0 is the total static pressure, equal to the sum of the hydrostatic and the atmospheric pressures, p_V is dependent on the temperature and q upon the density of the fluid and speed of flow.

$\delta p/q$ is a function of the shape of the particular section under consideration and of the angle of attack to the flow. It can be determined for different positions around the section either by experiment or by calculation. A typical distribution is shown by the full line in the lower part of Fig. 29. If a line is drawn on this diagram to show the value of $(p_0 - p_V)/q$, one can see whether the reduction in pressure at any point on the back is sufficient to cause cavitation. For the conditions shown by the full line, this state has not yet been reached. If now the angle of attack is increased, the value of $\delta p/q$ will increase, and at some angle the maximum pressure reduction on the back will reach the value $(p_0 - p_V)$, as shown by the dotted pressure curve, and cavitation will result.

Although it is usual to assume that cavitation will occur when the pressure has fallen to the vapor pressure of water, this view is too optimistic. The vapor pressure for fresh distilled water is very small, at 14 deg C (57 deg F) only 1.70 kN/m^2 (0.2 psi). But sea water contains much dissolved and entrained air and many minute nuclei of other kinds which encourage earlier formation of cavities or bubbles, and cavitation may occur at local pressures as high as 17 kN/m^2 (2.5 psi). In assessing the probability of cavitation in sea water it is therefore wise to retain some margin.

In the case of a propeller with an immersion to the centerline of h, the total pressure available there will be

$$p_A + \rho g h - p_V$$

where p_A is the atmospheric pressure. If we consider a section at radius r from the center, the pressure there will be a minimum when the blade is in the upper vertical position:

$$p_A + \rho g h - \rho g r - p_V$$

As the propeller rotates the pressure on the section will change by an amount $\pm\ \rho g r$, and in a uniform flow we can expect transient cavitation to occur first in the upper part of the propeller disk. In considering cavitation on a propeller-blade section, this effect should be taken into account. The velocity also should be that of the water past the blade, which is compounded of the speed of advance V_A, the speed of rotation, and the inflow velocity. Neglecting the latter, which is not known unless a detailed screw calculation is made, the relative velocity V_R past the blade will be

$$(V_R)^2 = (V_A)^2 + \omega^2 r^2$$

where ω is the angular speed of rotation. We may thus write a local cavitation number as

$$\sigma_L = \frac{p_A + \rho g h - \rho g r - p_v}{\frac{1}{2}\rho\,[(V_A)^2 + \omega^2 r^2]}$$

The total lift force on a blade section is given by the sum of the areas under the pressure curves on face and back, which reinforce one another. It will be obvious from Fig. 29. That it is not the *average* reduction of pressure on the back of the section that initiates cavitation, but the *maximum* reduction. Hence, for a given lift force and therefore a given area under the pressure curves, the sections most resistant to cavitation will be those with the most uniform pressure distribution and the most subdued peak.

7.2 Types of Cavitation. Knapp, et al (1970) classify hydrodynamic cavitation in general into the following categories:

- Travelling,
- fixed,
- vortex,
- vibratory.

The travelling cavitation type occurs in the free stream and consists of individual bubbles moving with the flow. Fixed cavitation occurs on boundaries of immersed bodies. It is fixed in the sense that the cavitation is attached to the body, as such being inherent to the body and not to the flow. A vortex occurs in a high shear flow. To balance centrifugal forces the pressure in the flow near the vortex must decrease continuously as the vortex is approached. These low pressures cause cavitation. Vibratory cavitation is caused by pressure pulsations in the liquid. Usually these pressure pulsations are caused by a submerged object vibrating normal to its face, thus producing pressure waves. Cavitation is caused when these pressure variations are large enough to cause the pressure in the liquid to reach, and fall below, the vapor pressure.

The cavitation forms on marine propellers are mainly of the fixed and vortex type. These types can be subdivided either according to the position on the propeller where the cavitation occurs, or according to the physical nature of the cavitation. Classifying propeller cavitation according to the latter method, i.e., according to the physical nature or its appearance, it is possible to specify the following types:

- Sheet,
- bubble,
- cloud,
- tip vortex,
- hub vortex.

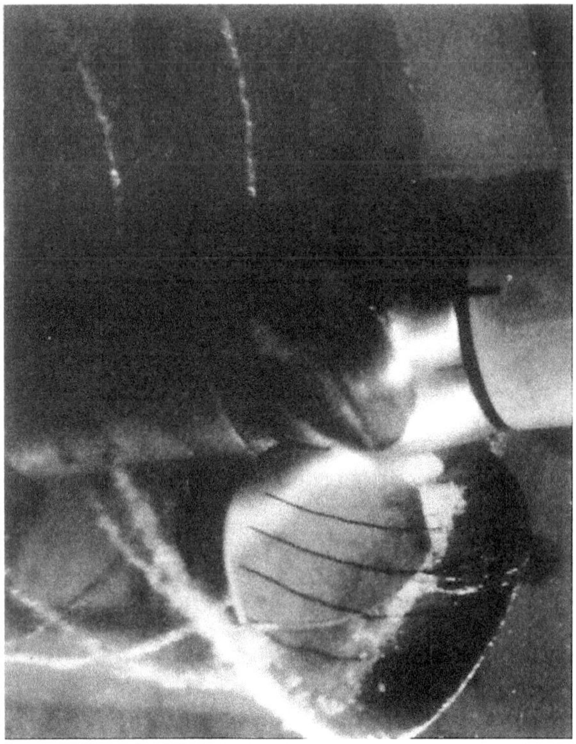

Fig. 30 Stable sheet cavitation

Fig. 31 Bubble cavitation

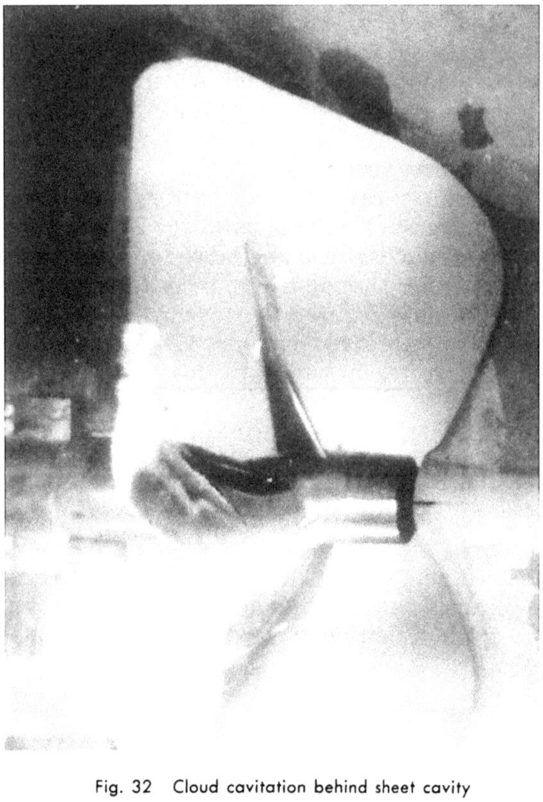

Fig. 32 Cloud cavitation behind sheet cavity

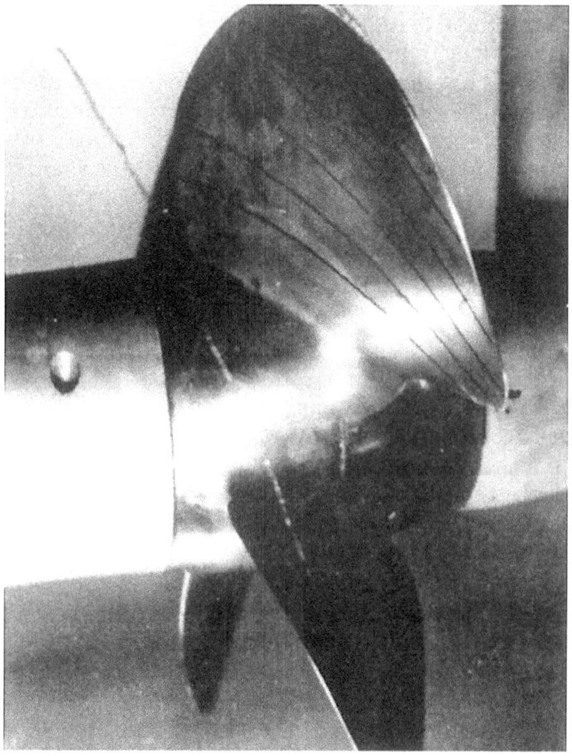

Fig. 33 Unattached tip vortex cavitation

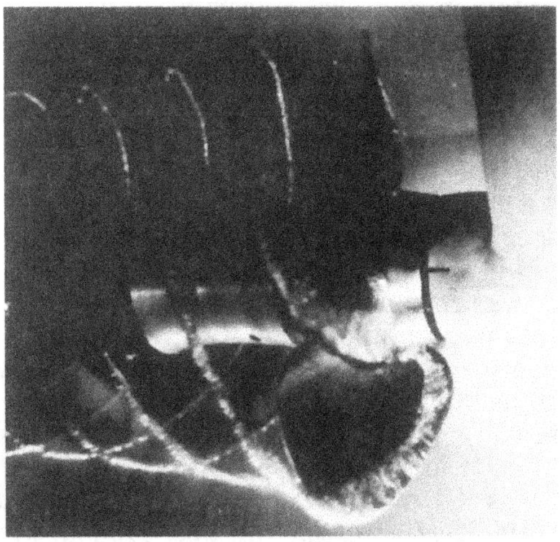

Fig. 34 Attached tip vortex cavitation

Sheet cavitation first occurs at the leading edges of propeller blades on the suction side (back) when the blade sections work under positive angles of attack, and on the pressure side (face) when the blade sections work under negative angles of attack. This is due to the fact that these non-shock free angles of attack (for which the lift is due not only to the camber of the sections) cause very low pressures in the leading edge region. This cavitation form may develop to cover the complete suction side of a blade, spreading inward from the leading edge in the form of a *sheet,* in which case it will often have a very stable character as shown in Fig. 30. When working in a wake, however, this cavitation type often has a very unstable character.

Bubble cavitation first occurs at the midchord or at the position of maximum thickness of the blade sections, at shock-free entry of the flow. As such it occurs in non-separated flows. This cavitation type appears as large individual *bubbles,* growing and contracting rapidly; see Fig. 31.

Cloud cavitation often occurs behind strongly developed stable sheet cavities and, in general, in moderately separated flows in which many small vortices form kernels form many small cavities. Cloud cavitation appears as a mist or a *cloud* of very small bubbles as shown in Fig. 32.

The vortex type of cavitation occurs at the tip and hub of the propeller. The flow around the blade tips from the pressure to the suction side causes an unstable *vortex* which is shed from the tip and the hub into the flow in the same way as an aerofoil of finite aspect ratio generates a vortex at each end. The pressure is least in the center of the vortex, and it is this vortex core which cavitates. Tip vortex cavitation usually starts somewhat behind the tip of the propeller blades. In this early stage, the cavitation is therefore unattached to the tip as shown in Fig. 33. When the vortex becomes stronger, or the liquid pressure is reduced, the cavitating tip vortex becomes attached as shown in Fig. 34. The hub vortex is formed by the combined vortices from the blades at the blade root, which by themselves are too weak to cavitate. With a converging hub form this hub vortex can, however, be very strong and cavitate readily. The resulting vortex cavitation is very stable and appears as a thick *rope* with strands corresponding to the number of blades as shown in Fig. 35.

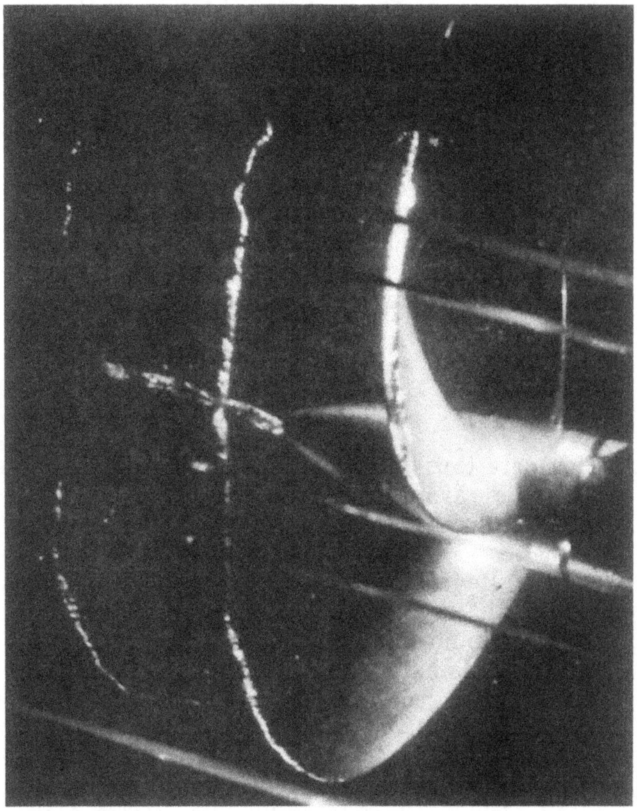

Fig. 35 Hub cavitation

7.3 Law of Similitude for Cavitating Propellers. In Section 3 dimensional analysis was used to determine the law of similitude for testing model propellers, and it was found that the thrust coefficient could be expressed functionally in the form

$$\frac{T}{\frac{1}{2}\rho D^2(V_A)^2} = f\left[\frac{gD}{(V_A)^2}, \frac{nD}{V_A}, \frac{p}{\rho(V_A)^2}, \frac{v}{V_A D}\right] \qquad (55)$$

If all the parameters on the right-hand side have the same values for two geometrically similar but different sized propellers, the flow patterns will be similar and the value of $T/\frac{1}{2}\rho D^2(V_A)^2$ will be the same for both.

In the discussion on this equation in Section 3, it was shown that the first, second, and last terms were equivalent to the statement that the propeller tests should be run at the same Froude number, at the same slip ratio or J-value, and at the same Reynolds number. It was also pointed out that the third term, $p/\rho(V_A)^2$, was not the same for model and ship in the towing-tank experiments, since the atmospheric pressure was not reduced to scale, but that since the forces on the propeller blades were due to *differences* in pressure, they would not be affected so long as the flow was not subject to cavitation.

Now, however, in dealing with experiments with cavitating propellers, we must take heed of this requirement and ensure that the expression $p/\rho(V_A)^2$ is the same in model and ship. This is, in fact, equivalent to saying that the cavitation number σ must be the same, since

$$\sigma = \frac{p_0 - p_V}{\frac{1}{2}\rho(V_A)^2}$$

There are thus four requirements for similitude in tests on the cavitating propellers:

(*a*) They should be run at speeds of advance which give the same Froude number.

(*b*) The Reynolds number on the blade sections should be the same.

(*c*) The slip must be the same.

(*d*) The tests must be run at the same value of σ.

The first two requirements, as for hull models and propellers in open water, are incompatible. To run at the Froude speed of advance would mean that the Reynolds numbers on the model propeller sections would be low and lead to problems of scale effect in blade friction. On the other hand, if the Froude law is not followed, the local distribution of pressure across the disk will be different in model and ship. This latter is generally taken to be the lesser of two evils, and although the Reynolds law is not attained exactly, by making the models as large as possible and running at high speeds the friction scale effect may be expected to be small. In order to meet condition (*d*), the tests must be run in a facility where the pressure can be reduced to the correct scale amount, and the third condition will then entail running the model propeller at the same slip as that of the ship, or at the same value of V_A/nD or J.

To attain true similitude between model and ship, some relationship should also exist between the air, gas, and nuclei content of the water in the two cases, but no final formulation of this aspect of testing has yet been reached.

7.4 Cavitation Tests With Model Propellers. The maximum pressure reduction on the back of the blade which can be tolerated before cavitation begins depends on the total pressure head ($p_A + \rho gh - \rho gr - p_V$) at the point where the section is working at a particular instant. In model experiments in a towing tank, h and r are reduced to scale but p_A, the atmospheric pressure, is not. The pressure on the model blade is therefore much too large to scale, and the experiments will not give a proper indication of the onset of cavitation.

To obtain information on cavitation performance of propellers, therefore, some other model technique is required, using a facility in which the total pressure can be varied at will.

In solving the problems connected with *Turbinia*, Parsons built the first such cavitation tunnel in 1897. It was a closed oval circuit, in the vertical plane, made of copper with a square cross section 0.152 m by 0.152 m (6 in.) in which he ran 5 cm (2 in.) diam propellers. The air was removed from above the water, to reduce the pressure, and the water was also heated to raise the vapor pressure and so achieve a low value of σ. In 1910 he built a much larger tunnel, with a working section 0.686 m by 0.762 m (27 x 30 in.) in which he ran 30 cm (12 in.) diam propellers. The maximum water speed was 4.4 m (15 ft) per sec. Parsons did some methodical series testing in this tunnel, which remained in use until his death in 1931. It is noteworthy that in both tunnels the propeller was illuminated once per revolution by a mirror rotating on the shaft, the light source in the larger tunnel being a searchlight. The time of illumination was 1/20,000 to 1/30,000 sec and excellent photographs were obtained (Godfrey, 1959).

In 1929 a cavitation tunnel was built at the U.S. Experimental Model Basin, and in the years that followed cavitation tunnels were built at what is now DTRC, the Hamburg Model Basin, the National Physical Laboratory in Feltham, England (now the National Maritime Institute), Massachusetts Institute of Technology, MARIN, and King's College, Newcastle. By this time many refinements and improvements over Parsons' tunnels had been developed.

Typical of more recent cavitation tunnels is the large cavitation tunnel of the DTRC. This water tunnel consists of a closed circuit in the vertical plane, the test section in the top horizontal arm having a diameter of 0.91 m (35.8 in.). The water is circulated by an impeller in the bottom horizontal arm, where the pressure head is high to inhibit cavitation. The maximum water speed through the measuring section is 25 m per sec (80 ft per sec). In order to reduce the pressure in the measuring section most tunnels have an air chamber at the highest part of the circuit, in which the air pressure can be varied as required. Since it is desirable to be able to control the air content of the water, this tunnel was designed without such an air chamber, the pressure in the tunnel being controlled by pumping water into or out of the circuit.

When the water passes the model propeller at high speed under cavitation conditions, the resulting low pressures bring air out of solution, the water becomes cloudy with bubbles and as these circulate around the tunnel, visual observation of the propeller becomes impossible. After a little time, this air is abstracted by the vacuum maintained in the air chamber, and the water is deaerated and cleared. Visual observation is now possible, but the water is no longer representative as regards air content. In an effort to overcome this a *resorber* was built into the circuit, consisting of a large cylinder, 7.6 m (25 ft) in diam, set 21 m (70 ft) into the ground. This is divided internally so that the water makes four vertical passes through it before going on to the measuring section. Because of the large diameter of the resorber the water spends considerable time there under high pressure, which ensures reabsorption of the air, which arrives at the measuring section without an increase in air content. Recent measurements of the nuclei content of sea water and research on how nuclei content influences the onset of cavitation have revealed that due to a lack of nuclei on model scale, significant scale effects can occur. Often the onset of cavitation occurs at higher speeds (i.e., lower pressures) indicating that the use of resorbers might not lead to an accurate simulation of cavitation characteristics. In some facilities additional nuclei have to be generated to arrive at a satisfactory correlation (Kuiper, 1981).

The appearance of the cavitation on the propeller is observed through windows, either visually or photographically, the propeller being illuminated stroboscopically. Measurements are made of thrust, torque, revolutions, water speed, and pressure. The tests are usually carried out at a series of values of σ, the slip ratio being varied at each one by varying the propeller revolutions or water speed.

The conditions in the tunnel differ in a number of ways from those behind the ship. The inflow is uniform instead of being irregular due to the variation of wake, the flow is parallel to the shaft whereas in practice this is seldom the case in twin-screw ships, the air content is different, and so on. To allow for these factors it has been fairly common practice to run the tests in the tunnel at a σ value 15 to 25 percent less than that calculated for the ship. Some tunnels are equipped to simulate shaft rake by moving the propeller drive shaft in the tunnel to angles up to 15 deg or so, and it is becoming fairly common practice to simulate the wake measured behind the hull model in the towing tank by inserting or installing a sufficiently large portion of a ship model in the test section, in front of the propeller model. Because of the limited dimensions of the tunnel it is necessary to change the ship model. Usually a standard shortened fore body is applied as the wake is almost completely determined by the aft body shape. The ship form in the vicinity of the pro-

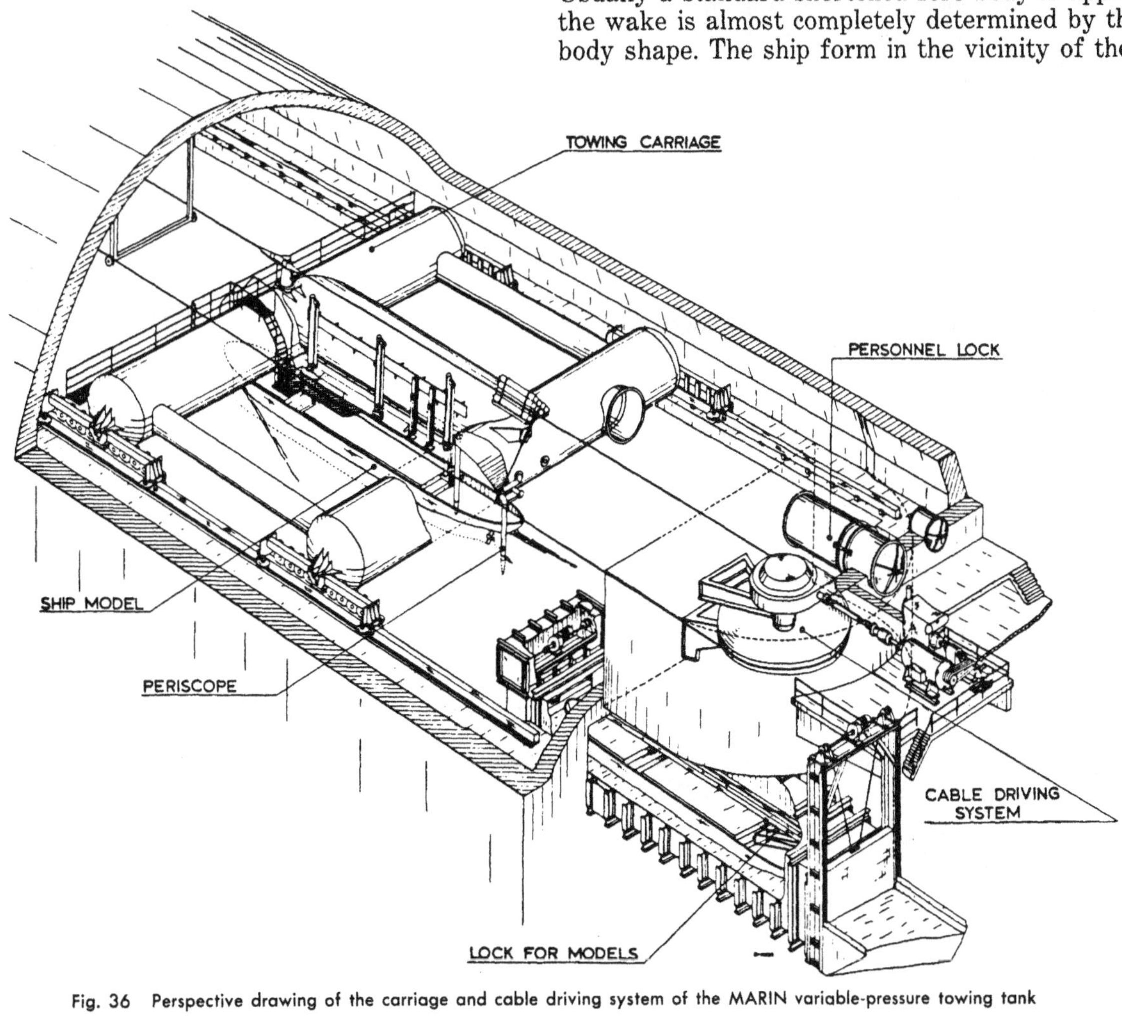

Fig. 36 Perspective drawing of the carriage and cable driving system of the MARIN variable-pressure towing tank

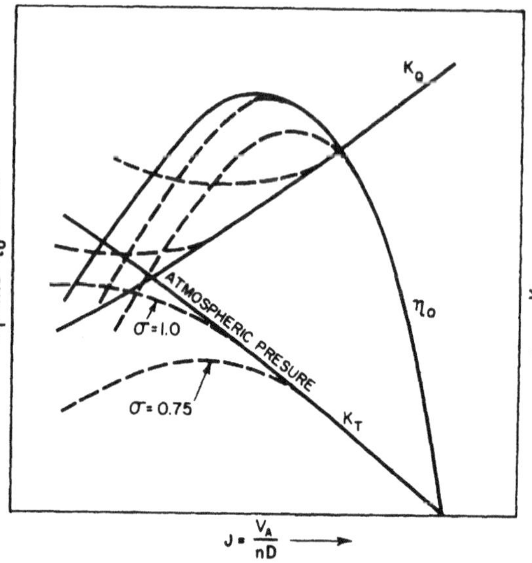

Fig. 37 Characteristic curves for propeller in cavitation tunnel

peller is retained as much as possible but the width of the model is usually reduced because of the tunnel dimensions. To compensate for the lack in breadth wire gauze can be attached to the flanks of the model. The resulting wake has to be checked and further adaptations may be necessary.

For hull pressure measurements the driving shaft, situated downstream of the propeller, has to be supported outside the ship model to avoid vibrations transmitted by the shaft. These would generate undesired extra pressure fluctuations.

Another way to simulate the ship conditions is to use a towing tank or circulating water channel in which the air pressure can be controlled. The model then can be run with the correct scale pressure and the cavitation and propulsion tests combined. In this respect the variable-pressure towing tank of MARIN is unique. This facility is 240 m long, 18 m wide and the water depth is 8 m (787 x 59 x 26 ft). The tank is constructed of special reinforced concrete. The upper section or roof is cantilevered from each side toward the center of the building. When the air is evacuated, the upper section must resist a pressure difference of nearly 100 kN/m^2 (14.1 psi). The air pressure in the tank can be lowered to about 4 kN/m^2 (0.56 psi).

7.5 Presentation of Data. The performance characteristics of a propeller in a cavitation tunnel are plotted in the same way as open-water results, except that there are separate sets of curves for each value of σ, Fig. 37.

Another type of diagram has been used by Newton to illustrate the development of cavitation (Newton, 1961). Experiments are run at a number of values of σ and the J-values noted at which different types of cavitation appear and disappear. These points can then be marked on a J-σ chart and boundaries drawn; Fig. 38. The area at the right marked by hatched lines is that which is free from all cavitation.

Since for every ship speed there are corresponding values of J and σ, a line can be drawn showing where the ship propulsion performance lies relative to the cavitation picture.

7.6 Detrimental effects of cavitation. (*a*) *Effects on propeller performance.* When the amount of cavitation on a propeller blade increases, the flow over the blade is changed. This in turn influences the pressure distribution on the blade, resulting in changes in propeller thrust and torque. Systematic experiments with airfoils have shown that at constant angle of attack, the pressure distribution for values of the cavitation number,

$$\sigma = \frac{p_0 - p_V}{\frac{1}{2}\rho V_0^2} \tag{56}$$

is as shown in Fig. 39. It is seen that cavitation reduces the under-pressure peak at the leading edge, spreading it out over the chord length. In most cases this change results in a decrease of the lift of the airfoil after an initial increase. The effect on the drag of the airfoil is quite analogous, but lags behind somewhat and is not so extreme. As such, the effectiveness, or the lift-drag ratio, of propeller blade sections decreases with increasing cavitation, after an initial increase, as shown in Fig. 40. In this figure sectional lift and drag coefficients are used:

$$C'_L = \frac{L}{\frac{1}{2}\rho V_0^2 c} \tag{57}$$

$$C'_D = \frac{D}{\frac{1}{2}\rho V_0^2 c} \tag{58}$$

and α is angle of attack. In Equations (56) to (58),

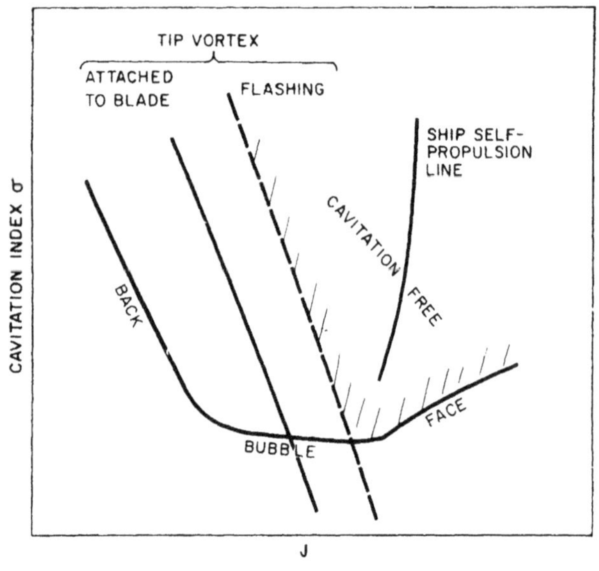

Fig. 38 Development of cavitation patterns (Newton, 1961)

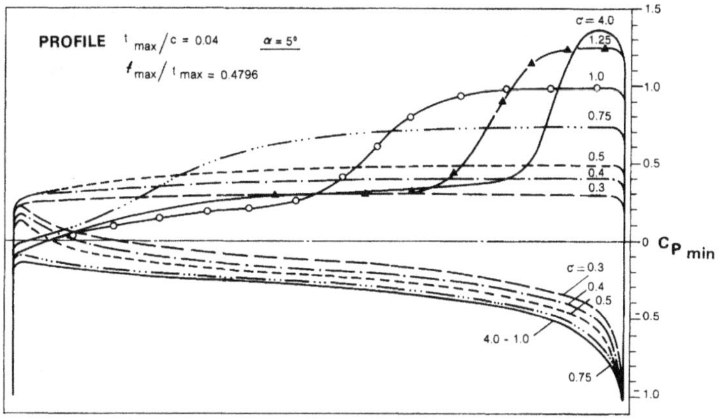

Fig. 39 Measured pressure distribution on a profile at different cavitation numbers

p_0 is static pressure
p_v is vapor pressure
V_0 is undisturbed flow velocity
c is profile chord length

The resultant drop in propeller thrust and torque influences the propeller efficiency a great deal. This is shown in Fig. 41, in which:

$$\sigma = \frac{p_0 - p_V}{\frac{1}{2}\rho n^2 D^2}$$

To obtain a definite ship speed, a cavitating propeller suffering from thrust breakdown requires a larger power and thus a larger number of revolutions. As will be obvious, the described changes in propeller performance, resulting from the presence of cavitation are dependent on propeller geometry and propeller inflow conditions. As a stepping-stone to determine propeller performance under cavitating conditions and to design propellers with optimum cavitation properties, it is important to know the cavitation characteristics of the applied propeller blade sections. This subject is discussed in Section 8.

(*b*) *Cavitation damage of propellers.* From the very first, erosion of marine propellers has been encountered. Originally it was thought that this form of cavitation damage was due to corrosion. Today it is realized that the mechanism of cavitation is responsible for erosion and other forms of damage occurring to screw blades such as bent trailing edges. Parsons, et al (1919) first found the connection between cavitation phenomena and erosion. Many theories and hypotheses have since been formulated to account for the various cavitation damage forms (Eisenberg, et al, 1965).

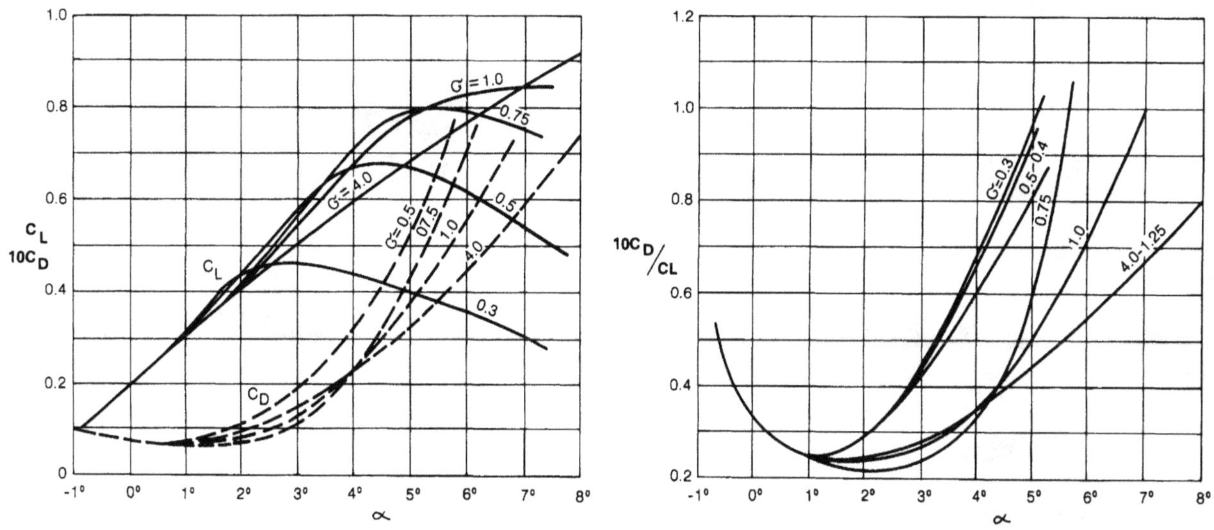

Fig. 40 Values of C_L, C_D and C_D/C_L as a function of α at different cavitation numbers for the airfoil in Fig. 39

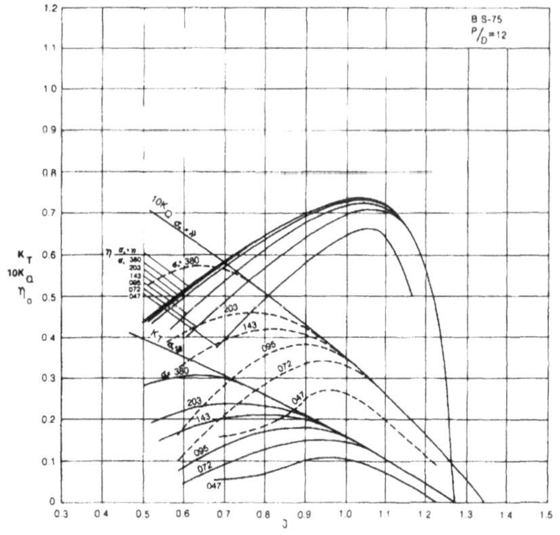

Fig. 41 Values of K_T, K_Q and η_o as a function of the advance ratio J at different cavitation number for the Wageningen B5-75 propeller

Nowadays it is assumed that cavitation damage is primarily caused by the process of cavitation bubble collapse on the propeller blade surface. The energy associated with cavitation bubble collapse has been calculated to be extremely high. This is particularly due to the fact that this bubble collapse apparently occurs in the form of shock waves on very small portions of the blade surface. This explains the pitted nature of the cavitation damage form termed erosion as shown in Fig. 42. Intense and continued erosion often leads to disastrous damage as shown in Fig. 43. Hydrofoil vessels and other high-speed craft experience erosion not only on the propellers, but also on the foils and other underwater parts such as struts and fins; see Section 9.7, Chapter V.

Cavitation erosion has been observed at places where cloud and bubble cavitation occurs. The individual bubbles associated with these cavitation forms collapse at the downstream end of the cavitation zone where the pressure increases. This type of cavitation collapse also occurs behind sheet cavitation when the sheet breaks up into individual bubbles. Recent theories on the mechanism of bubble collapse propose that the energy causing the damage is brought about by a re-entrant jet into the bubble on the boundary surface. Ellis (1965, 1966) has in fact photographed this.

Even though erosion and corrosion of propeller blades are different phenomena, interaction can occur. When the smooth surface of a propeller blade is eroded away by cavitation, corrosion starts, accelerating the damage. On the other hand, should corrosion start, thereby roughening the surface, the flow may cavitate more readily if the pressure is low enough.

Prevention of cavitation erosion may be realized by choosing suitable propeller materials or by applying protective, metallic or non-metallic coatings. The best method, however, is to ensure that harmful cavitation does not occur. If this is impossible then a supercavitating condition, i.e. a fully-developed cavitating condition should be designed for, ensuring that the cavitation bubbles collapse in the flow downstream of the boundary surfaces. Sometimes an intermediate propeller design, showing no individual cavitation bubbles on the blades, is possible.

Another form of cavitation damage is the bending of trailing edges of propeller blades of high-powered single-screw ships as shown in Fig. 44. Van Manen (1963) ascribes this damage form to the simultaneous collapse of a large number of bubbles on the trailing edge. Leaving the wake peak a blade section camber increase is induced when the angle of attack decreases again. Sheet cavitation at the leading edge disappears and bubbles are formed at the midchord position. These bubbles leave the propeller blade together when the induced blade section curvature decreases. The relatively long period available for bubble growth and the simultaneous arrival of these bubbles at the trailing edge with its associated high pressure causes a si-

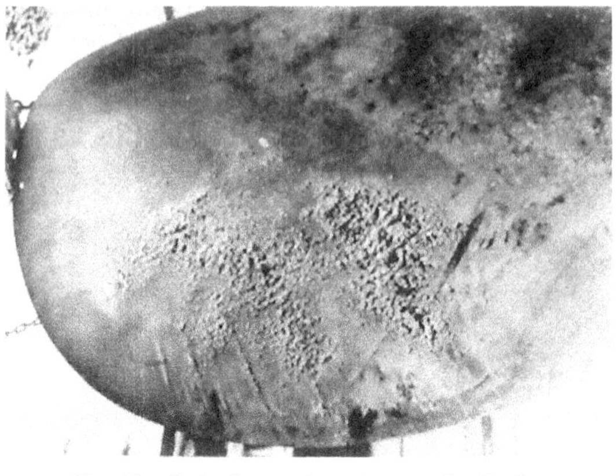

Fig. 42 Cavitation erosion of a propeller blade

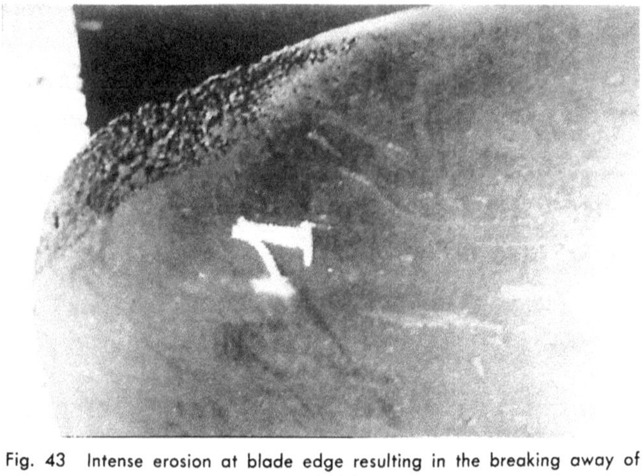

Fig. 43 Intense erosion at blade edge resulting in the breaking away of material

Fig. 44 Bent trailing edge of propeller blade caused by cavitation

multaneous collapse. This explains the bending of trailing edges towards the pressure side. For the prevention of this form of cavitation damage, a good compromise between the camber distribution of the blade section and the angle of attack at which the blade section works is necessary.

(c) *Cavitation-induced vibrations and noise.* Ship vibrations are determined by the response characteristics of the ship structure and by the excitation level. Propeller-induced vibratory forces on the afterbody of a ship form the largest part of these excitation forces. This aspect of the influence of cavitation has been the subject of significant experimental studies. It was found that cavitation considerably influences the whole problem of ship-propeller interaction. Cavitation was found both to influence propeller blade stresses and to modify the flow ahead of the propeller. The largest effect, however, was found in the pressure fluctuations induced on the ship's afterbody. Not only the amplitude but also the phase angle of the propeller-induced fluctuating pressures are affected. Propeller cavitation increases the amplitude of these vibratory pressures, depending on blade number and extent of cavitation, with a factor between 1 and 10 and sometimes even higher. This is primarily due to the variation in angle of attack of the flow causing large variations in the size of the cavities on the blades thereby causing large volume variations. When the cavities on the propeller blade do not fluctuate as much, such as occurs in a uniform flow, the pressure fluctuations on a nearby body are not increased as much; see Chapter VII.

Cavitation not only influences low frequency propeller-induced pressure fluctuations on the ship hull but also increases high frequency noise levels in ships. For warships this aspect is particularly disturbing. The increase of underwater self-noise with increasing cavitation (i.e., with increasing ship speed) reduces the ship's sonar-detection capabilities considerably. It is therefore important for a warship to have propellers with a maximum cavitation-free speed range. A high cavitation inception ship speed is nowadays considered to be essential in design of naval propellers.

7.7 Criteria for Prevention of Cavitation. Many criteria have been proposed for predicting the onset of cavitation. The earliest ones, using the average thrust per unit area of projected blade surface, are today not sufficient in many sophisticated designs, although still useful as a first guide. The criterion devised by Barnaby from the *Daring* trials was to limit the pressure to 76.7 kN/m^2 (10.8 psi) of projected area for a tip immersion of 0.28 m (11 in.), increasing this limit by 2.5 kN/m^2 (0.35 psi) for every additional 0.305 m (ft) of immersion.

For the same lift coefficient C_L on a section, the maximum reduction in pressure on the back depends on the shape of the section and on the conditions under which it is operating. Any proposed criteria must take account of these factors, and it is difficult to find one which is really satisfactory.

The modern approach is to calculate the pressure distributions around suitable sections, or to measure them in a wind or water tunnel. A knowledge of the real incidence angle obtained from circulation theory can then be used to determine the maximum reduction of pressure on the back of the section for comparison with the static pressure $(p_0 - p_V)$ available. The true angle of incidence depends on the wake pattern in which the propeller is working, and such calculations have to be made using the average circumferential wake at each particular radius. In practice the angle of incidence will vary both above and below the average, so that cavitation will occur at somewhat lower revolutions, and allowance must be made for this fact. By the same token, cavitation will be delayed or avoided by making the wake more even by attending to shape of hull, clearances, propeller rake and the alignment of bossings or shafts and struts with the average flow direction.

Many propellers are still designed from charts derived from methodical series tests, and in designing by the circulation theory it is necessary to begin with some chosen propeller diameter, also determined from a design chart. Some general criterion is therefore needed for the choice of blade area to avoid cavitation. A diagram designed to provide such guidance in order to avoid excessive cavitation and erosion under average service conditions at sea was given by Burrill (1943). He used a coefficient τ_c expressing the mean thrust loading on the blades, defined as

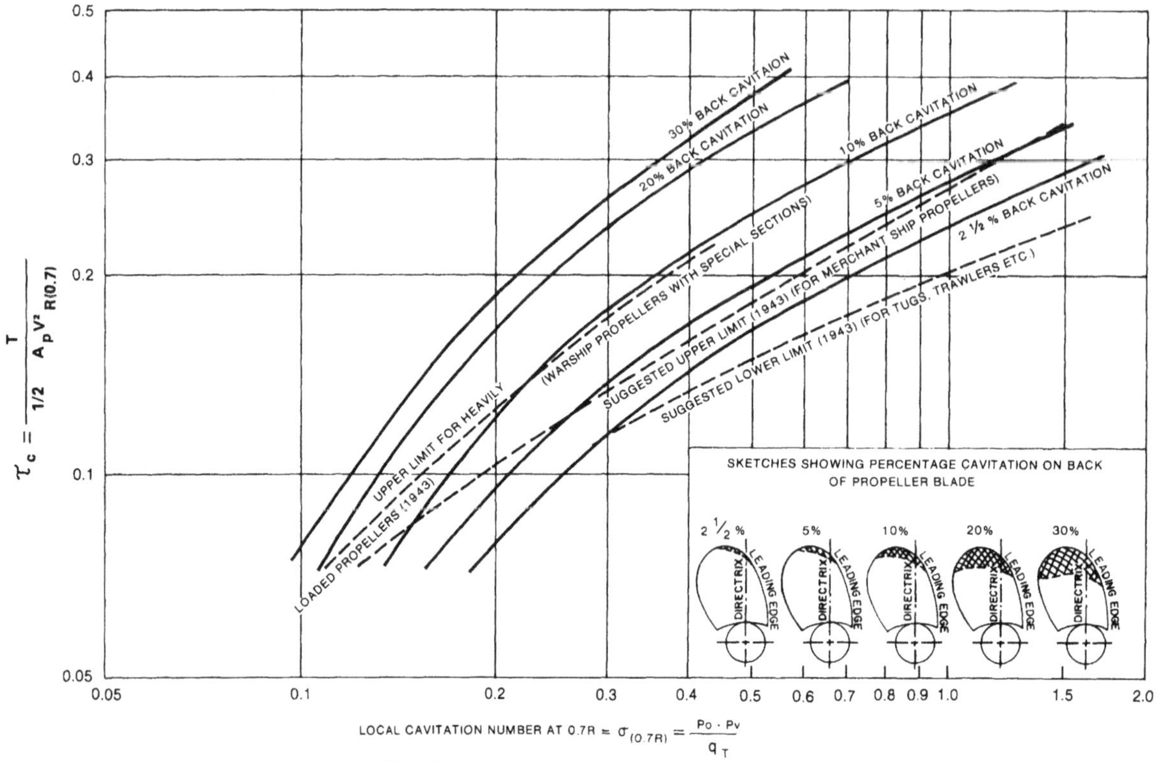

Fig. 45 Simple cavitation diagram (Burrill, et al, 1943, 1962-63)

$$\tau_c = \frac{T/A_P}{\frac{1}{2}\rho(V_R)^2}$$

plotted to a base of $\sigma_{0.7R}$, where

T is thrust in kN
A_p is projected blade area in m^2
V_R is relative velocity of water at 0.7 radius in m/sec.
$T_{0.7R}$ is local cavitation number at 0.7 radius
ρ is mass density, kg/L.

The projected blade area A_P can be found from the more usual developed area A_D by using Taylor's approximate formula

$$A_P/A_D = 1.067 - 0.229 \times \text{pitch ratio} \qquad (59)$$

The thrust can be calculated from P_E or P_D from the expressions

$$T = \frac{P_E}{(1-t)V}$$

or

$$T = \frac{P_D\,\eta_D}{(1-t)V}$$

The cavitation number $\sigma_{0.7R}$ is calculated using the relative velocity V_R at 0.7 radius and the pressure at the centerline of the propeller, viz.

$$\sigma_{0.7R} = \frac{p_0 - p_V + \rho g h}{\frac{1}{2}\rho(V_A 2 + (0.7\pi n \mathrm{D})^2} \qquad (60)$$

an approximate formula for which is

$$\sigma_{0.7R} = \frac{188.2 + 19.62\,h}{V_A\,2 + 4.836 n^2 D^2} \qquad (61)$$

where $(p_0 - p_V)$ is pressure at screw center line in N per m^2, h is head of water at screw center line, m

The chart, reproduced in Fig. 45, was based originally on experience with full-sized propellers and gave lines for suggested upper limits of T/A_P for heavily loaded (warship) propellers with special sections and for merchant ship propellers in order to avoid serious back cavitation. A third line indicated the lower limits of T/A_P to avoid face cavitation on tugs and trawlers. Systematic tests of a series of model propellers with circular back sections in the cavitation tunnel at King's College, Newcastle, confirmed in a general way that the model results were in reasonable agreement with the practical experience on which the chart was based (Gawn, et al, 1957). Later a series of four-bladed merchant ship propeller models was run in the same tunnel covering a range of pitch ratio and values of σ (Burrill, et al, 1962–63.) From these tests lines were added to the diagram, Fig. 45 indicating 2½, 10 and 30 percent back cavitation. It will be seen that the line for 5 percent back cavitation lies very near that given originally for merchant ships. The authors state that observations on many other propellers in the tunnel running at the average service condition have shown cavitation of this kind and extent, and have been found reasonably free from erosion after several years in service. They therefore concluded that the line indicating 5 percent back cavitation was a suitable criterion at which to aim in practical design calculations, and that "even recent experience with propellers of modern design would not suggest any material alteration in the positioning of this upper limiting line for aerofoil type propellers."

A useful formula for obtaining a first indication as to the required expanded blade area ratio was derived by Keller (1966),

$$\frac{A_E}{A_o} = \frac{(1.3 + 0.3Z)T}{(p_o - p_v)\,D^2} + k \tag{62}$$

where:

T is thrust in N (or kN)
Z is number of propeller blades
$p_o - p_v$ is pressure at centerline of propeller in N per m^2 (or kN per m^2)
k is a constant varying from 0 (for transom-stern naval vessels) to 0.20 (for high-powered single-screw vessels).

The subject of cavitation criteria in propeller design can really only be dealt with adequately by incorporating pressure distribution, angle-of-attack, and cavitation number information into a detailed design process, for every radius. Criteria such as the Burrill chart and the Keller formula do not reflect the influence of the wake or propeller blade geometry such as pitch, camber and thickness distribution. They should therefore be used with care.

Section 8
Propeller Design

8.1 Methods of Propeller Design. The design of a marine propeller is almost invariably carried out by one of two methods, although each method covers a number of procedures differing in detail.

In the first of these, the design is based upon charts giving the results of open-water tests on a series of model propellers. These cover variations in a number of the design parameters such as pitch ratio, blade area, number of blades, and section shapes. A propeller that conforms with the characteristics of any particular series can be rapidly designed and drawn to suit the required ship conditions.

The second method is used in cases where a propeller is heavily loaded and liable to cavitation, or has to work in a very uneven wake pattern, when it is desirable to carry out a detailed design using circulation theory. Basically this involves finding the chord width, section shape, pitch, and efficiency at a number of radii to suit the average circumferential wake values and give optimum efficiency and protection from cavitation. By integration of the resulting thrust and torque-loading curves over the blades, the thrust, torque, and efficiency for the whole propeller can be found. Before such detailed design can be started it is necessary to know preliminary dimensions and in general these are found from standard charts.

Also some choices as to the propeller characteristics have to be made such as number of blades, skew, etc. The next sub-section gives a general philosophy on propeller design with respect to these choices. After that two methods of design will be presented, one based on systematic series and another one based on theoretical calculations.

8.2 General Propeller Design Philosophy. Continuing with a qualitative discussion of the considerations that may influence the propeller design (Cumming, et al, 1974), a detailed account of an actual propeller design and its evaluation by calculations and model tests has been given by Boswell, et al (1973) concerning a highly skewed propeller.

(*a*) *Diameter*. The maximum diameter is usually limited by the geometry of the aperture although sometimes tunnels may be applied to allow larger propellers. Another limiting factor is imposed by propeller induced unsteady hull forces which decrease with increasing clearance. The propeller efficiency usually increases with increasing diameter. A larger diameter will change the radial distribution of the wake in which the propeller operates, however, which can lead to serious detrimental effects if the blades extend into a region of greater flow non-uniformity. Also, keeping the rate of revolution constant, a larger diameter will lead to higher tip velocities and hence to a reduced cavitation index. The latter effect usually more or less balances the delay of tip vortex cavitation caused by the smaller gradient of the bound circulation. This gradient is smaller since the same thrust load is spread over a larger distance. Finally, the hull efficiency may be reduced by applying a larger diameter.

As can be seen from the above discussion the choice of the diameter will be a compromise. Also, changing the diameter will lead to changes in other parameters. Therefore this discussion is necessarily qualitative. The final choice of all parameters involved will be interrelated with the speed to be attained and by the allowable levels of vibrations, cavitation, noise and installed power.

(*b*) *Number of revolutions*. In many cases this parameter is selected beforehand and may not be a part of the final design. A reduction of the RPM tends to be beneficial as the local section velocities due to section thickness become smaller. This in turn leads to a large margin between the cavitation index and the minimum pressure coefficient. This may be partly offset by the increased fluctuation of the angle of attack at lower rotational velocity. Further, vibration considerations may restrict the allowable range of RPM as the rotational speed and the number of blades together determine the frequency of unsteady forces. The rotational speed should be chosen, if possible, to be sufficiently different from resonant frequencies of the hull, shafting and propulsion machinery; Chapter VII.

Having discussed the diameter and RPM it is appropriate to refer to the recent developments in large-diameter, low-RPM propeller design, (Hadler, et al 1982). Muntjewerf (1983) mentions a possible increase of propulsive efficiency of 10 to over 15 percent.

(*c*) *Number of Blades*. In discussing the analysis of the wake field Section 4.2 it was shown that the blade number affects the unsteady force levels. Therefore the permissible levels of the exciting forces may influence the number of blades. Efficiency considerations also have their impact on the choice of the number of blades, the optimum open water efficiency (without any other restrictions) decreasing with increasing Z.

Increasing the number of blades and keeping the same blade area ratio and thickness-to-chord ratio will lead to a significantly reduced section modulus and an increased stress level. Near the blade tips an increased blade area may be necessary in that case.

(*d*) *Radial distribution of loading*. The optimum distribution for wake-adapted propellers can be derived from Lerbs induction method which will be discussed in Section 8.4. However, it may be of advantage to unload the tip to reduce the susceptibility to cavitation, to reduce the blade stress levels and to reduce the blade-frequency pressures on the hull (Boswell, et al, 1973).

A somewhat lower efficiency is the cost of these improvements. Extreme reductions in tip loading may aggravate the leading-edge sheet cavitation due to the variation in angle of attack.

(*e*) *Blade Outline*. Decreasing the blade area (or the chord) increases the efficiency because of the decreased frictional drag. This tendency holds up to the point where strength requirements cause the thickness to chord ratio to become too large with an associated increase in form drag.

A higher efficiency results for propellers with narrower blade tips. A lower limit to the width is imposed by cavitation considerations. The magnitude and radial distribution of blade area may have important effects on the unsteady shaft forces and bending moments.

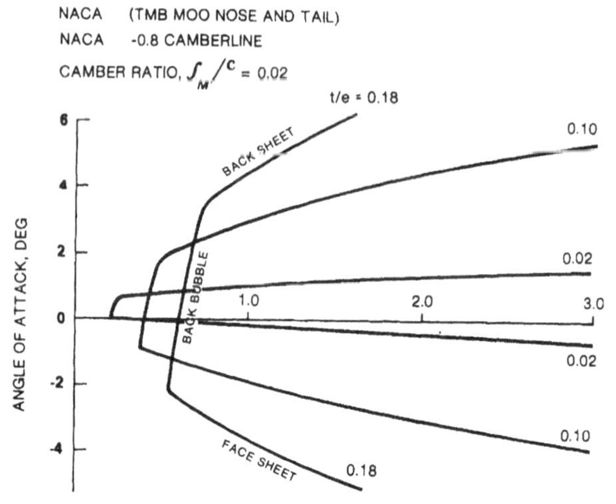

Fig. 46 Characteristic cavitation buckets

This depends on the harmonic content of the wake and on their phases in the wake. The chord distribution is coupled to the thickness. Apart from strength considerations thin foils will result in high efficiency, but also in reduced resistance against cavitation at differing angles of attack such as occur in the wake. Thicker foils have better cavitation performance in this respect but are more susceptible to bubble cavitation which is apt to result in erosion. These trends may be observed in a cavitation-bucket diagram, Fig. 46. In this figure σ_r denotes a local cavitation number; $\sigma_r = 2gH/V_r^2$ with H the local head.

(*f*) *Camber and angle of attack*. Once the design lift is determined and its chordwise distribution is chosen, the camber and angle of attack may be specified. This may be the ideal angle of attack for minimum drag. By selecting a larger angle of attack and the corresponding camber a section will be obtained which is less susceptible to pressure-side cavitation but more susceptible to suction side cavitation. The reverse tendency applies if the angle of attack is decreased. The effect on efficiency is small unless the angle of attack becomes very high. Also the effect on overall strength and vibration is small unless the volume and type of cavitation change significantly, and this only affects the unsteady pressure forces (Cumming, et al, 1974).

(*g*) *Skew*. With a view to the recent developments of highly skewed propellers, it is appropriate to discuss their properties in somewhat more detail. Cumming et al (1972) mention the following advantages of properly designed highly skewed propellers:

- Decrease in propeller-induced unsteady bearing forces and moments
- Decrease in propeller-induced unsteady pressure forces
- Decreased susceptibility to cavitation when operating in a wake.

These advantages are at the expense of:

- Decreased backing efficiency

- More difficulty in manufacture
- Strength-related problems for very high skew and for backing conditions.

The strength problems of highly skewed propellers have been dealt with in Section 6.4. Valuable research and development work on highly skewed propellers has been carried out by the Maritime Administration, including full-scale tests (Hammer, et al, 1978). Reference may also be made to a report discussing the design and evaluation of the highly skewed propeller of a roll-on/roll-off cargo vessel (Anonymous, 1979).

A proper design of a highly skewed propeller requires correcting the blade camber and pitch distribution to account for the blade distortion. Cumming et al (1972) derived lifting surface corrections for highly skewed propellers of four, five and six blades, for blade area ratio of 0.75, for two hydrodynamic pitch ratios of 0.8 and 1.2 and for 50 and 100 percent skew (100 percent skew equals a skew angle of $360/Z$ degrees). The calculations were carried out for a modified NACA 66 thickness distribution and $a = 0.8$ camberline. The computations showed that skew did not affect the chordwise camber distribution. Also the camber correction factor (i.e. the ratio between the maximum camber for the propeller section and the maximum camber for the NACA two-dimensional profile) was only slightly changed by skew for moderate skew angles (45 degrees for the four-bladed propeller). Effects of skew were more pronounced for higher skew angles and at higher hydrodynamic pitch. The dominant effect of skew is on the ideal angle correction K_α,

$$K_\alpha(r) = \frac{\alpha_i\,(r)}{\alpha_{i,\,1.0}\;C_L} \tag{63}$$

where $\alpha_i\,(r)$ is the angle of attack computed by Cumming. $\alpha_{i,\,1.0}$ is the two-dimensional ideal angle of attack for the NACA profile at $C_L = 1.0$. A positive angle was induced toward the blade root and a negative angle towards the tip.

Concerning the open-water performance, Boswell (1971) and Cumming found that skew has insignificant influence on efficiency for the ahead condition. Backing efficiency on the other hand shows large reductions with skew. Initial experiments indicated that high skew may be very effective in reducing unsteady bearing forces (Miller, 1969). He obtained a reduction of a factor 10 in unsteady thrust and torque and a factor 2 in the vertical and horizontal bearing forces. Tsakonas, et al (1967), (1969) correlated their calculations with unsteady lifting surface theory with the experimental results and found close agreement.

Cumming used the method of Tsakonas to carry out calculations for a highly skewed propeller in the wake of a Series 60 ship. He concluded that a significant reduction of unsteady shaft forces could be obtained by a judicious choice of the amount of skew. This was, however, strongly dependent on the way in which the phase angle of the different harmonics varied with the radius. If the phase angle also shows skewness, some skew angles of the propeller may lead to an increase of the unsteady forces.

The other contribution to unsteady phenomena, the hull pressure fluctuations, was first investigated by Denny (1967) for highly skewed propellers. A reduction of 50 percent was obtained for a 100 percent skewed propeller compared to an unskewed propeller. Cumming noted that rake has no influence on the reduction

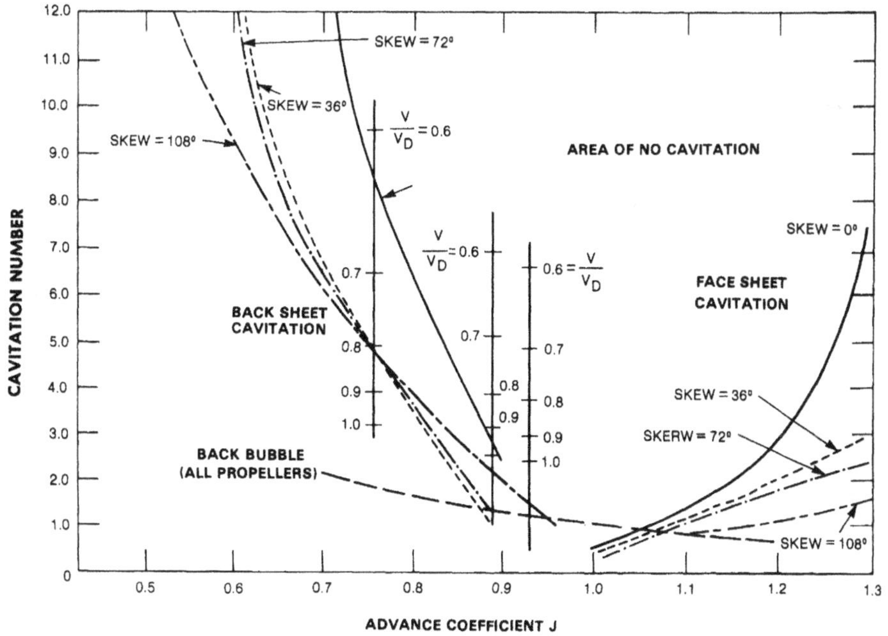

Fig. 47 Initial inception curves for skew series propellers

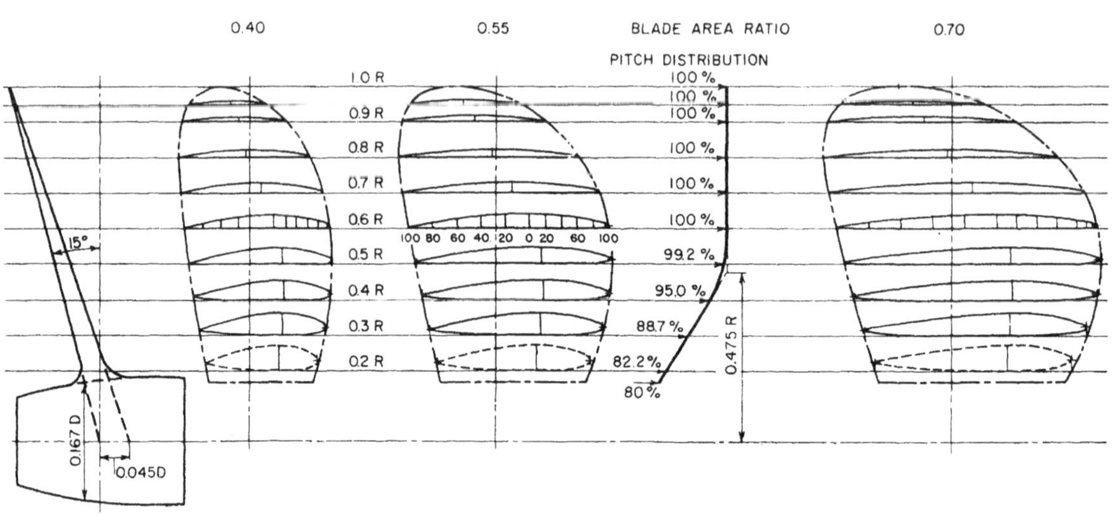

Fig. 48 Blade form and sections of B.4 Series

of the peak pressure amplitudes achieved by the application of highly skewed propellers.

A cavitation bucket diagram has been shown in Fig. 47 for a series of skewed propellers. This shows for a particular σ an increased width of the cavitation free zone: large fluctuations in the local advance and hence in the angle of attack are permitted. From this figure it is also possible to derive at what speed V (as a percentage of the design speed V_D) inception begins. Thrust and torque breakdown caused by cavitation were found to be independent of the amount of skew.

8.3 Propeller Design From Methodical Series Charts. In conducting model propeller series tests, a parent model is chosen of given blade area, number of blades, blade outline, section shape, and blade thickness, and a number of models are built, all having these same characteristics but different pitch ratios. These models are then run in open water in a towing tank over a range of slip from 0 to 100 percent. Additional groups can be run in which one of the other parameters is changed and the effect of the alteration on performance determined.

An important method of screw design is that based on the results of such open-water tests with systematically varied series of screw models. Among the well-known propeller series are those developed by Schaffran, Taylor, Gawn and others, and the B-screw series of MARIN.

The first tests with systematic series of screw propellers were performed in 1936. From model experiments carried out by Baker, et al (1934) and Baker (1934) it had become evident that screws with circular-back blade sections and elliptical blade outline such as the Taylor and Schaffran series, were inferior as regards efficiency to propellers having airfoil sections. These results led to the development of a series of model propellers having airfoil sections. This MARIN series was designated A4-40, as it was a series having 4 blades and a blade area ratio of 0.40. The results of the open-water tests with this series were given by Troost (1938).

Later it was found that the A4-40 screw series was only suitable for use in cases where no cavitation danger was present. This was found to be a result from the fact that the narrow blade tips combined with the airfoil-shaped blade sections gave rise to very unfavorable pressure distributions on the blades. In addition, the A4-40 screw series possessed undersirable backing characteristics. It was therefore decided to design a new screw series having wider blade tips, circular blade sections near the tips and airfoil blade sections near the hub. The first member of this new

Table 11—Table of Existing Wageningen B-screw Series.

Blade number Z	Blade area ratio A_E/A_O														
2	0.30														
3		0.35			0.50			0.65			0.80				
4			0.40			0.55			0.70			0.85		1.00	
5				0.45			0.60			0.75					1.05
6					0.50			0.65			0.80				
7						0.55			0.70			0.85			

Table 12—Dimensions of Four, Five, Six and Seven-bladed Wageningen B-screw Series.

r/R	$\frac{c_r Z}{D A_E/A_O}$	a_r/c_r	b_r/c_r	$S_r/D = a_r - b_r Z$	
				a_r	b_r
0.2	1.662	0.617	0.350	0.0526	0.0040
0.3	1.882	0.613	0.350	0.0464	0.0035
0.4	2.050	0.601	0.350	0.0402	0.0030
0.5	2.152	0.586	0.350	0.0340	0.0025
0.6	2.187	0.561	0.389	0.0278	0.0020
0.7	2.144	0.524	0.443	0.0216	0.0015
0.8	1.970	0.463	0.479	0.0154	0.0010
0.9	1.582	0.351	0.500	0.0092	0.0005
1.0	—	0	—	0.0030	0

Table 13—Dimensions of Three-bladed Wageningen B-screw Series.

r/R	$\frac{c_r \cdot Z}{D A_E/A_O}$	a_r/c_r	b_r/c_r	$S_r/D_r = a_r - b_r Z$	
				a_r	b_r
0.2	1.633	0.616	0.350	0.0526	0.0040
0.3	1.832	0.611	0.350	0.0464	0.0035
0.4	2.000	0.599	0.350	0.0402	0.0030
0.5	2.120	0.583	0.355	0.0340	0.0025
0.6	2.186	0.558	0.389	0.0278	0.0020
0.7	2.168	0.526	0.442	0.0216	0.0015
0.8	2.127	0.481	0.478	0.0154	0.0010
0.9	1.657	0.400	0.500	0.0092	0.0005
1.0	—	—	—	0.0030	0

B-series was designated B-4-40 and because of its great popularity this screw series was gradually extended to other blade numbers and blade area ratios. The results of the first open-water tests were given by Troost (1938), (1940), (1951) and Van Lammeren, et al (1948). Fig. 48 shows the geometry of the four-bladed B-4 series.

Any particular group of Series B is described by combinations such as B.5.60, where the 5 indicates the number of blades and 0.60 the expanded blade-area ratio.

The original 4-bladed series had a constant pitch from the tip to 0.5 radius, from which point it was reduced to 80 percent of that value at the hub, Fig. 48. This was done to accommodate to some extent the known wake concentration on the inner radii on single-screw ships. The 3-bladed series, run at the same time, was intended primarily for twin-screw ships, and in view of the more uniform wake in such ships no pitch reduction was used.

Subsequent experiments showed that the effective pitch for the B.4.40 propellers was only about 1 percent less than the constant pitch over the outer half of the blades, and, in view of this small effect, the pitch on the 5, 6 and 7-bladed screws was made constant.

Table 11 lists the available members of the B-screw series, and Tables 12, 13 give their geometrical properties. About 120 model propellers of the B-series have been manufactured and tested to date at MARIN.

In Tables 12, 13, c_r is chord length of blade section at radius r/R,

- a_r is distance between leading edge and generator line at radius r/R
- b_r is distance between leading edge and maximum thickness of blade section at r/R,
- S_r is maximum blade section thickness at radius r/R.

The required coordinates of the profiles can be calculated by means of formulas, analogous to the formulas given by Van Gent, et al (1973) and Van Oossanen (1974), viz:

$$\left.\begin{aligned} y_{\text{face}} &= V_1(t_{\max} - t_{l.e.}) \\ y_{\text{back}} &= (V_1 + V_2)(t_{\max} - t_{l.e.}) \end{aligned}\right\} \text{for } P > 0$$
$$\left.\begin{aligned} y_{\text{face}} &= V_1(t_{\max} - t_{t.e.}) \\ y_{\text{back}} &= (V_1 + V_2)(t_{\max} - t_{t.e.}) \end{aligned}\right\} \text{for } P \leq 0 \qquad (64)$$

where, as shown in Fig. 49,

- y_{face}, y_{back} are vertical ordinates of a point on a blade section on the face and on the back with respect to the pitch line,
- $t_{\max}$ is maximum thickness of blade section,
- t_{te}, t_{le} are extrapolated blade section thickness at the trailing and leading edges, respectively
- V_1, V_2 are tabulated functions dependent on r/R and P,
- P is non-dimensional coordinate along pitch line from position of maximum thickness to leading edge (where $P = 1$), and from position of maximum thickness to trailing edge (where $P = -1$).

Values of V_1 and V_2 are given in Tables 14 and 15. The values of $t_{l.e.}$ and $t_{t.e.}$ are usually chosen in accordance with rules laid down by classification societies or in accordance with manufacturing requirements.

In conjunction with the geometry of this propeller series, it is noted that at MARIN modified B-series propellers are now used and designed which have a slightly wider blade contour near the blade tip. These propellers are denoted as BB propellers. For the sake of completeness, Table 16 is included which gives the

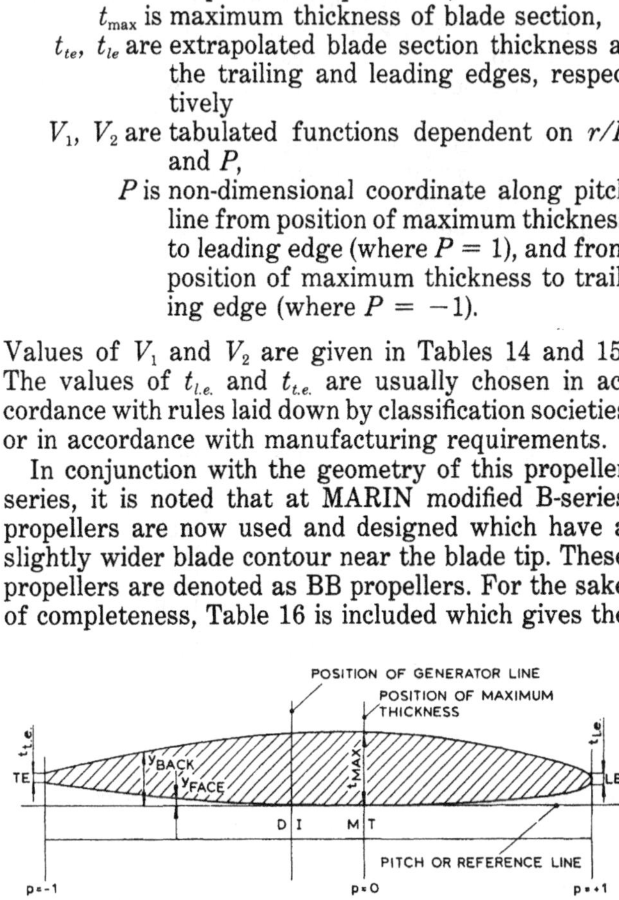

LE = LEADING EDGE
TE = TRAILING EDGE
MT = LOCATION OF MAXIMUM THICKNESS
DI = LOCATION OF DIRECTRIX

Fig. 49 Definition of geometric blade section parameters of Wageningen B- and BB-series propellers

Table 14—Values of V_1 for use in Equation (64)

r/R \ P	−1.0	−.95	−.9	−.8	−.7	−.6	−.5	−.4	−.2	0
.7-1.0	0	0	0	0	0	0	0	0	0	0
.6	0	0	0	0	0	0	0	0	0	0
.5	.0522	.0420	.0330	.0190	.0100	.0040	.0012	.0	0	0
.4	.1467	.1200	.0972	.0630	.0395	.0214	.0116	.0044	0	0
.3	.2306	.2040	.1790	.1333	.0943	.0623	.0376	.0202	.0033	0
.25	.2598	.2372	.2115	.1651	.1246	.0899	.0579	.0350	.0084	0
.2	.2826	.2630	.2400	.1967	.1570	.1207	.0880	.0592	.0172	0
.15	.3000	.2824	.2650	.2300	.1950	.1610	.1280	.0955	.0365	0

r/R \ P	+1.0	+.95	+.9	+.85	+.8	+.7	+.6	+.5	+.4	+.2	0
.7-1.0	0	0	0	0	0	0	0	0	0	0	0
.6	.0382	.0169	.0067	.0022	.0006	0	0	0	0	0	0
.5	.1278	.0778	.0500	.0328	.0211	.0085	.0034	.0008	0	0	0
.4	.2181	.1467	.1088	.0833	.0637	.0357	.0189	.0090	.0033	0	0
.3	.2923	.2186	.1760	.1445	.1191	.0790	.0503	.0300	.0148	.0027	0
.25	.3256	.2513	.2068	.1747	.1465	.1008	.0669	.0417	.0224	.0031	0
.2	.3560	.2821	.2353	.2000	.1685	.1180	.0804	.0520	.0304	.0049	0
.15	.3860	.3150	.2642	.2230	.1870	.1320	.0920	.0615	.0384	.0096	0

particulars of this series. The performance characteristics of these BB-series propellers may be considered identical with the original B-series propellers.

Some years ago it was decided to cross-fair the B-screw series open-water test results by means of a regression analysis. In this way the existing small errors in the diagrams would be eliminated and the resulting analytical expressions for the thrust and torque would be very welcome for use in preliminary design calculations by means of high speed computers. One reason for the small errors in the diagrams, in addition to experimental errors, was the inconsistency of the Reynolds number during open-water tests. The early open-water tests were carried out at a lower rotational propeller speed than the more recent tests. For the correction of the test results for Reynolds number effects the method developed by Lerbs (1951) was applied. This method is a so-called "equivalent profile method," consisting of replacing the propeller by one of its profiles, the equivalent profile, and de-

Table 15—Values of V_2 for use in Equation (64)

r/R \ P	−1.0	−.95	−.9	−.8	−.7	−.6	−.5	−.4	−.2	0
.9-1.0	0	.0975	.19	.36	.51	.64	.75	.84	.96	1
.85	0	.0975	.19	.36	.51	.64	.75	.84	.96	1
.8	0	.0975	.19	.36	.51	.64	.75	.84	.96	1
.7	0	.0975	.19	.36	.51	.64	.75	.84	.96	1
.6	0	.0965	.1885	.3585	.5110	.6415	.7530	.8426	.9613	1
.5	0	.0950	.1865	.3569	.5140	.6439	.7580	.8456	.9639	1
.4	0	.0905	.1810	.3500	.5040	.6353	.7525	.8415	.9645	1
.3	0	0.800	.1670	.3360	.4885	.6195	.7335	.8265	.9583	1
.25	0	.0725	.1567	.3228	.4740	.6050	.7184	.8139	.9519	1
.2	0	.0640	.1455	.3060	.4535	.5842	.6995	.7984	.9446	1
.15	0	.0540	.1325	.2870	.4280	.5585	.6770	.7805	.9360	1

r/R \ P	+1.0	+.95	+.9	+.85	+.8	+.7	+.6	+.5	+.4	+.2	0
.9-1.0	0	.0975	.1900	.2775	.3600	.51	.6400	.75	.8400	.9600	1
.85	0	.1000	.1950	.2830	.3660	.5160	.6455	.7550	.8450	.9615	1
.8	0	.1050	.2028	.2925	.3765	.5265	.6545	.7635	.8520	.9635	1
.7	0	.1240	.2337	.3300	.4140	.5615	.6840	.7850	.8660	.9675	1
.6	0	.1485	.2720	.3775	.4620	.6060	.7200	.8090	.8790	.9690	1
.5	0	.1750	.3056	.4135	.5039	.6430	.7478	.8275	.8880	.9710	1
.4	0	.1935	.3235	.4335	.5220	.6590	.7593	.8345	.8933	.9725	1
.3	0	.1890	.3197	.4265	.5130	.6505	.7520	.8315	.8920	.9750	1
.25	0	.1758	.3042	.4108	.4982	.6359	.7415	.8259	.8899	.9751	1
.2	0	.1560	.2840	.3905	.4777	.6190	.7277	.8170	.8875	.9750	1
.15	0	.1300	.2600	.3665	.4520	.5995	.7105	.8055	.8825	.9760	1

Table 16—Particulars of BB-series Propellers.

r/R	$\frac{c_r}{D}\frac{Z}{A_E A_O}$	a_r/c_r	b_r/c_r
0.200	1.600	0.581	0.350
0.300	1.832	0.584	0.350
0.400	2.023	0.580	0.351
0.500	2.163	0.570	0.355
0.600	2.243	0.552	0.389
0.700	2.247	0.524	0.443
0.800	2.132	0.480	0.486
0.850	2.005	0.448	0.498
0.900	1.798	0.402	0.500
0.950	1.434	0.318	0.500
0.975	1.122	0.227	0.500

a_r = distance between leading edge and generator line at r
b_r = distance between leading edge and location of maximum thickness at r
c_r = chord length at r

ducing the properties of the propeller at other scale and roughness values from the known properties of this profile.

The cross-fairing of the B-series was first attempted for each blade number separately. The results of the investigations for the four and five bladed B-series were given by Van Lammeren, et al (1969). It was later decided to include the blade number as an independent variable in the cross-fairing and also to include the Reynolds number as an independent variable in the polynomials for K_T and K_Q.

In the Lerbs equivalent profile method it is shown that the blade section at $0.75R$ is representative of the whole blade. At a specific value of the advance coefficient the lift and drag coefficients and the corresponding profile angle of attack is deduced from the K_T and K_Q values from the open-water test. In this way the polar curves for C_D and C_L on a basis of α are calculated from the known propeller characteristics K_T and K_Q on a basis of J. Reynolds number effects are only

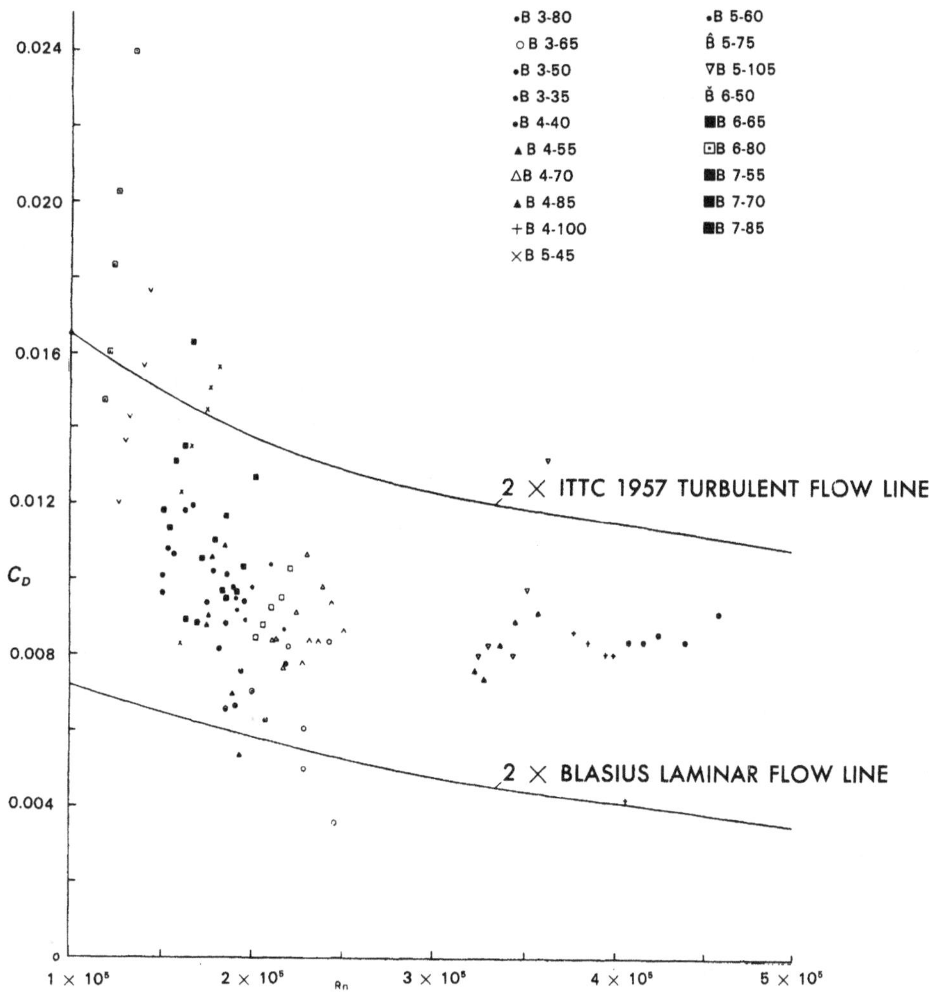

Fig. 50 Uncorrected value of the minimum drag coefficient of equivalent profile of B-series propellers

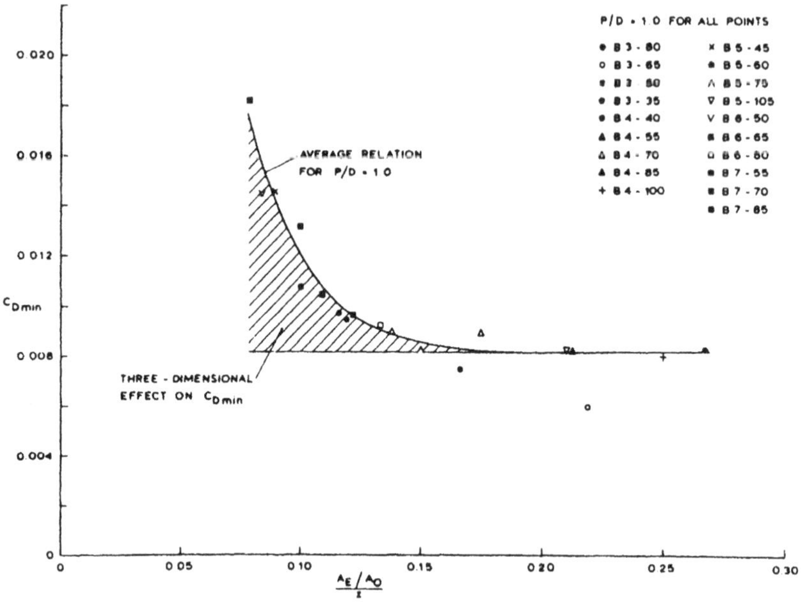

Fig. 51 Three-dimensional effect on minimum drag coefficient of equivalent profile of B-series propellers.

considered to influence the drag coefficient of the equivalent profile. It is furthermore assumed that the influence of Reynolds number on the drag coefficient is in accordance with a vertical shift of the C_D curve equal to the change in the minimum value of the drag coefficient. This minimum value is for thin profiles having mainly frictional resistance, the effect of the pressure gradient being small.

According to Hoerner (1965) the minimum drag coefficient of the profile is:

$$C_{D_{\min}} = 2C_f\left[1 + 2\left(\frac{t}{c}\right)_{0.75R}\right] \qquad (65)$$

in which

$$C_f = \frac{0.075}{[0.434291n(R_{n0.75R}) - 2]^2}$$

where

$$R_{n0.75R} = \frac{C_{0.75R}\sqrt{V_A^2 + (0.75\ \pi nD)^2}}{\mu}$$

C_f is the drag coefficient of a flat plate in a turbulent flow and the term $\left[1 + 2\left(\frac{t}{c}\right)_{0.75R}\right]$ represents the effect of the pressure gradient.

On setting out the minimum value of the drag coefficient as obtained from the lift and drag characteristics for each propeller on a base of Reynolds number, a large scatter was apparent as shown in Fig. 50. When this minimum value of the drag coefficient is set out against $\frac{A_E/A_O}{Z}$ for each pitch-diameter ratio, it is seen that below a specific value of the blade area/blade number ratio an increase in the $C_{D_{\min.}}$ value occurs. For a pitch-diameter ratio equal to 1.0, this is shown in Fig. 51. The existence of such a correlation of the $C_{D_{\min.}}$ value with propeller geometry points to the fact that the scatter in Fig. 50 is not entirely due to Reynolds number effects and experimental errors. Since the drag coefficient is influenced by three-dimensional effects, it is necessary, before correcting for Reynolds number according to the given equations, to subtract this three-dimensional effect from the $C_{D_{\min.}}$ value.

An estimation of this effect was obtained by applying regression analysis of which the results are given by Van Oossanen (1974). The resulting lift and drag coefficients were each expressed in polynomials as a function of blade number, blade area ratio, pitch-diameter ratio and angle of attack by means of a multiple regression analysis method. By applying this process in reverse, thrust and torque coefficient values were then calculated. The basis for this reverse process was formed by calculating C_L and C_D coefficients from the C_L and C_D polynomials for specific combinations of Z, A_E/A_O, P/D, α and Rn. The resulting values formed the input for the development of a thrust coefficient and a torque coefficient polynomial. The thrust and torque coefficients were then expressed as polynomials in the advance coefficient J, pitch ratio P/D, blade area ratio A_E/A_O and blade number Z and with the aid of a multiple regression analysis method the significant terms of the polynomials and the values of the corresponding coefficients were determined. For Rn $= 2 \times 10^6$ the polynomials obtained in this way are given in Table 17. The choice of a Reynolds number value of 2×10^6 for the characteristics on the model scale followed from the fact that the corresponding $C_{D_{min.}}$ values is an average of all model $C_{D_{min.}}$ values.

The original B-series results were presented in the form used by Taylor. The Taylor propeller coefficients are,

$$B_P = \frac{n(P_D)^{0.5}}{(V_A)^{2.5}}$$

and

$$\delta = \frac{nD}{V_A}$$

where

n is revolutions per minute
P_D is delivered power at propeller
V_A is speed of advance
D is propeller diameter

With the rotative speed of the propeller known, the optimum diameter can be found from specially-prepared $B_p - \delta$ charts. These charts consist of contours of constant propeller efficiency η_o and of constant advance coefficient δ, plotted on a grid of B_p and pitch ratio, P/D.

Oosterveld, et al (1975) published two new sets of charts, based on the thrust and torque polynomials given in Table 17. The first set gives curves of open water efficiency η_o and of constant $1/J$, plotted on a grid of $0.1739\sqrt{B_p}$ and pitch ratio, P/D.

The reasons for adopting the square root of B_p is to obtain a horizontal scale more suitable for interpolation, while the coefficient 0.1739 was introduced to make it possible to replace the B_p coefficient by the non-dimensional variable $K_Q^{1/4} \times J^{-5/4}$, viz

$$0.1739\sqrt{B_{p1}} = K_Q^{1/4} \times J^{-5/4} \tag{66}$$

This representation makes it possible to use these diagrams in either the conventional way (using knots, horsepower, feet and revolutions per minute) or in SI-units, adopting m/sec., kW, m, kN-m, revolutions per sec. and ρ in kg/L (t/m^3),

$$K_Q^{1/4} \times J^{-5/4} = \left[\frac{P_D\, n^2}{2\pi\rho V_A^{\,5}}\right]^{1/4} = \left[\frac{Qn^3}{\rho\, V_A^{\,5}}\right]^{1/4} \tag{67}$$

The subscript P_1 in Equation (66) is used to distinguish this Taylor propeller loading coefficient from another propeller loading coefficient which Oosterveld and Van Oossanen have termed B_{p2}, which can be used to find the optimum rotative speed of a propeller when the diameter is given, viz

$$B_{p2} = P_D^{1/2}\, D^{-1}\, V_A^{-3/2}$$

using horsepower, knots and ft.

In this case $\quad 1.75\sqrt{B_{p2}} = K_Q^{1/4}\, J^{-3/4} \quad (68)$

$$\text{with } K_Q^{1/4}\, J^{-3/4} = \left[\frac{P_D}{2\pi\rho D^2 V_A^{\,3}}\right]^{1/4} = \left[\frac{Qn}{\rho D^2 V_A^{\,3}}\right]^{1/4}$$

using either system of units. The Taylor advance coefficient δ has been replaced by the non-dimensional entity $1/J$,

$$0.009875\delta = 1/J = nD/V_A$$

In the $B_{p2} - 1/J$ charts also the square root of B_{p2} is adopted since then a linear scale on the horizontal axis can be used.

The B_{p1} − and $B_{p2} - 1/J$ charts for the 4-bladed B-series propellers are given in Figs. 52 through 61. For the B_{p1} − charts of the five-bladed B-series as well as the open-water diagrams of the four-bladed and the five-bladed B-series propellers the publication of Van Lammeren, et al (1969) may be mentioned. The charts for the 3-, 6-, and 7-bladed propellers have been published only prior to the inclusion of Reynolds-number effects. The new charts for these propellers can be acquired from MARIN. *(Continued on page 202)*

Table 17—Coefficients and Terms of the K_T and K_Q Polynomials for the Wageningen B-screw Series for Rn = 2 × 10⁶.

$$K_t = \sum_{s,t,u,v} \left[C_{s,t,u,v} \cdot (J)^s \cdot (P/D)^t \cdot (A_E/A_o)^u \cdot (z^v)\right]$$
$$K_q = \sum_{s,t,u,v} \left[C_{s,t,u,v} \cdot (J)^s \cdot (P/D)^t \cdot (A_E/A_o)^u \cdot (z^v)\right]$$

K_T = $C_{s,t,u,v}$	s (J)	t (P/D)	u (A_E/A_o)	v (z)	K_Q $C_{s,t,u,v}$	s (J)	t (P/D)	u (A_E/A_o)	v (z)
+0.00880496	0	0	0	0	+0.00379368	0	0	0	0
−0.204554	1	0	0	0	+0.00886523	2	0	0	0
+0.166351	0	1	0	0	−0.032241	1	1	0	0
+0.158114	0	2	0	0	+0.00344778	0	2	0	0
−0.147581	2	0	1	0	−0.0408811	0	1	1	0
−0.481497	1	1	1	0	−0.108009	1	1	1	0
+0.415437	0	2	1	0	−0.0885381	2	1	1	0
+0.0144043	0	0	0	1	+0.188561	0	2	1	0
−0.0530054	2	0	0	1	−0.00370871	1	0	0	1
+0.0143481	0	1	0	1	+0.00513696	0	1	0	1
+0.0606826	1	1	0	1	+0.0209449	1	1	0	1
−0.0125894	0	0	1	1	+0.00474319	2	1	0	1
+0.0109689	1	0	1	1	−0.00723408	2	0	1	1
−0.133698	0	3	0	0	+0.00438388	1	1	1	1
+0.00638407	0	6	0	0	−0.0269403	0	2	1	1
−0.00132718	2	6	0	0	+0.0558082	3	0	1	0
+0.168496	3	0	1	0	+0.0161886	0	3	1	0
−0.0507214	0	0	2	0	+0.00318086	1	3	1	0
+0.0854559	2	0	2	0	+0.015896	0	0	2	0
−0.0504475	3	0	2	0	+0.0471729	1	0	2	0
+0.010465	1	6	2	0	+0.0196283	3	0	2	0
−0.00648272	2	6	2	0	−0.0502782	0	1	2	0
−0.00841728	0	3	0	1	−0.030055	3	1	2	0
+0.0168424	1	3	0	1	+0.0417122	2	2	2	0
−0.00102296	3	3	0	1	−0.0397722	0	3	2	0
−0.0317791	0	3	1	1	−0.00350024	0	6	2	0
+0.018604	1	0	2	1	−0.0106854	3	0	0	1
−0.00410798	0	2	2	1	+0.00110903	3	3	0	1
−0.000606848	0	0	0	2	−0.000313912	0	6	0	1
−0.0049819	1	0	0	2	+0.0035985	3	0	1	1
+0.0025983	2	0	0	2	−0.00142121	0	6	1	1
−0.000560528	3	0	0	2	−0.00383637	1	0	2	1
−0.00163652	1	2	0	2	+0.0126803	0	2	2	1
−0.000328787	1	6	0	2	−0.00318278	2	3	2	1
+0.000116502	2	6	0	2	+0.00334268	0	6	2	1
+0.000690904	0	0	1	2	−0.00183491	1	1	0	2
+0.00421749	0	3	1	2	+0.000112451	3	2	0	2
+0.0000565229	3	6	1	2	−0.0000297228	3	6	0	2
−0.00146564	0	3	2	2	+0.000269551	1	0	1	2
					+0.00083265	2	0	1	2
					+0.00155334	0	2	1	2
					+0.000302683	0	6	1	2
					−0.0001843	0	0	2	2
					−0.000425399	0	3	2	2
					+0.0000869243	3	3	2	2
					−0.0004659	0	6	2	2
					+0.0000554194	1	6	2	2

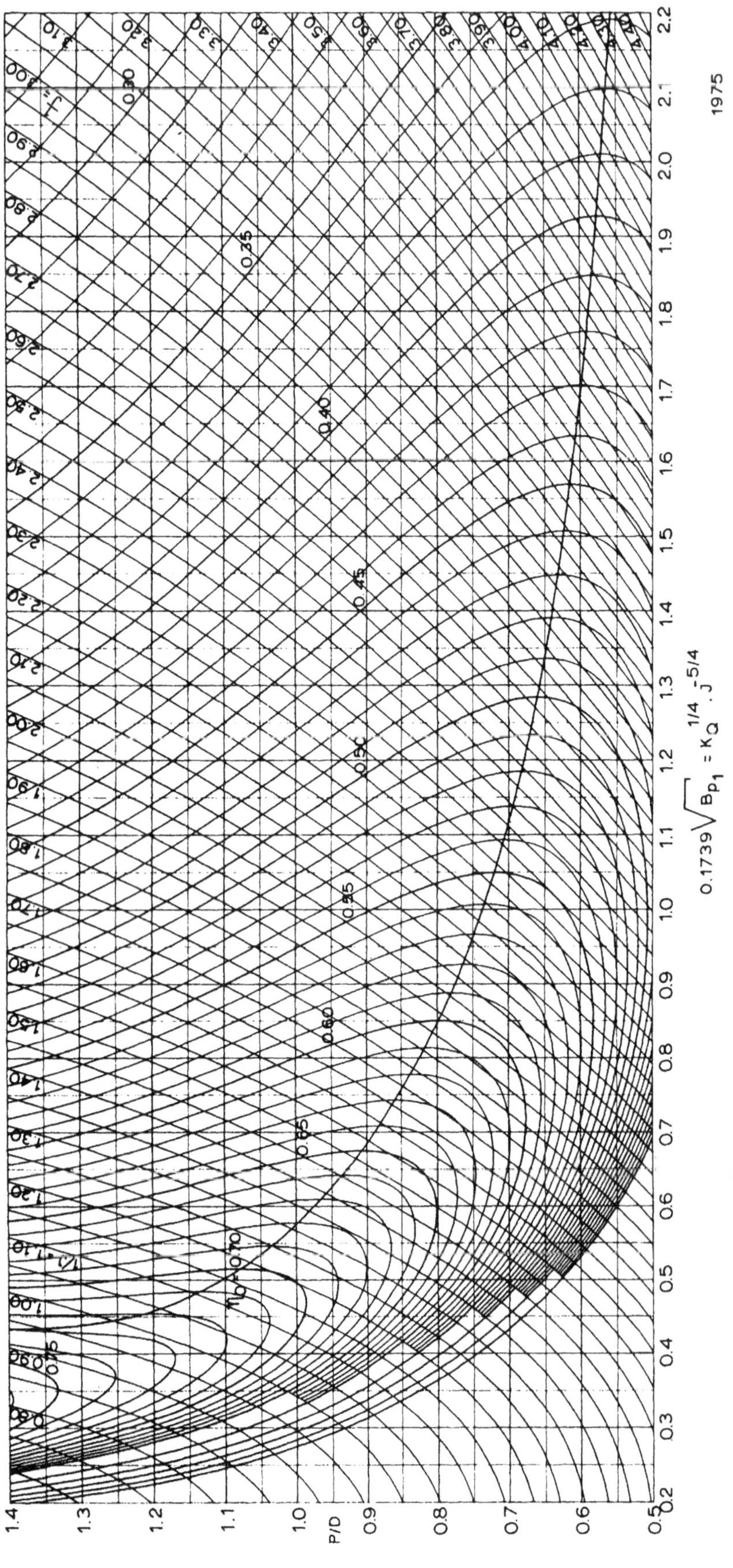

$B_{p_1} = NP^{1/2} \cdot V_A^{-5/2}$

N = PROPELLER RPM
V_A = V_S (1 - w)
V_S = SHIP SPEED IN KNOTS
w = WAKE FRACTION
P = SHAFT HORSEPOWER (BRITISH)

$K_Q^{1/4} \cdot J^{-5/4} = \left[\frac{Qn^3}{\rho V_A^5} \right]^{1/4}$

Q = PROPELLER TORQUE IN KGM
n = PROPELLER REVOLUTIONS PER SECOND
ρ = WATER DENSITY (TANK) = 101.94 $KGSEC^2 M^{-4}$
V_A = V_S (1 - w)
V_S = SHIP SPEED IN M/SEC.
w = WAKE FRACTION

Fig. 52

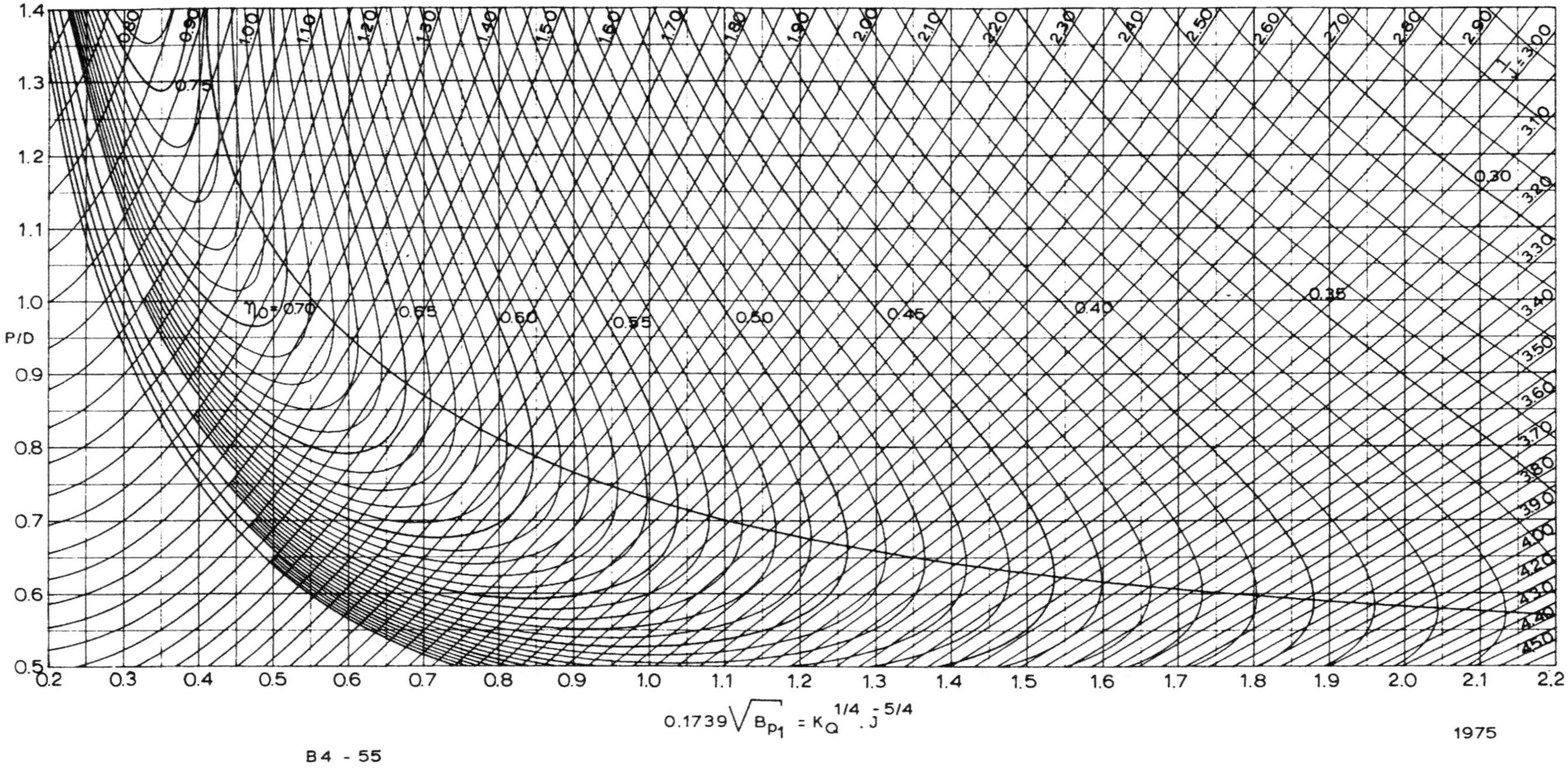

$B_{P_1} = NP^{1/2} . V_A^{-5/2}$	N = PROPELLER RPM V_A = V_S (1 - w) V_S = SHIP SPEED IN KNOTS w = WAKE FRACTION P = SHAFT HORSEPOWER (BRITISH)

$K_Q^{1/4} . J^{-5/4} = \left[\frac{Qn^3}{\rho V_A^5}\right]^{1/4}$	Q = PROPELLER TORQUE IN KGM n = PROPELLER REVOLUTIONS PER SECOND ρ = WATER DENSITY (TANK) = 101.94 $KGSEC^2 M^{-4}$ V_A = V_S (1 - w) V_S = SHIP SPEED IN M/SEC. w = WAKE FRACTION

Fig. 53

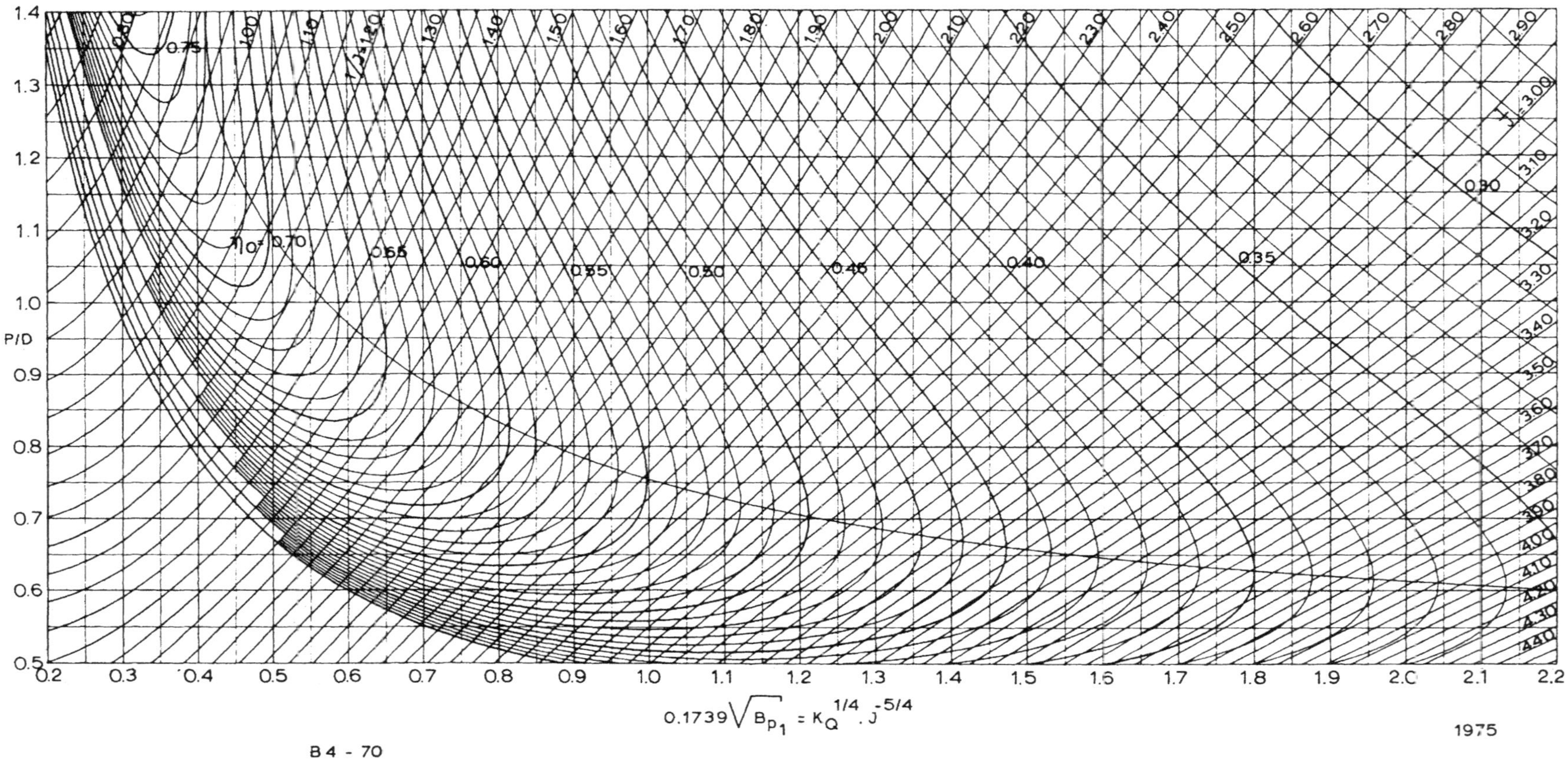

$B_{p_1} = NP^{1/2} . V_A^{-5/2}$

N = PROPELLER RPM
V_A = V_S (1 - w)
V_S = SHIP SPEED IN KNOTS
w = WAKE FRACTION
P = SHAFT HORSEPOWER (BRITISH)

$$K_Q^{1/4} . J^{-5/4} = \left[\frac{Qn^3}{\rho V_A^5} \right]^{1/4}$$

Q = PROPELLER TORQUE IN KGM
n = PROPELLER REVOLUTIONS PER SECOND
ρ = WATER DENSITY (TANK) = 101.94 $KGSEC^2 M^{-4}$
V_A = V_S (1 - w)
V_S = SHIP SPEED IN M/SEC.
w = WAKE FRACTION

Fig. 54

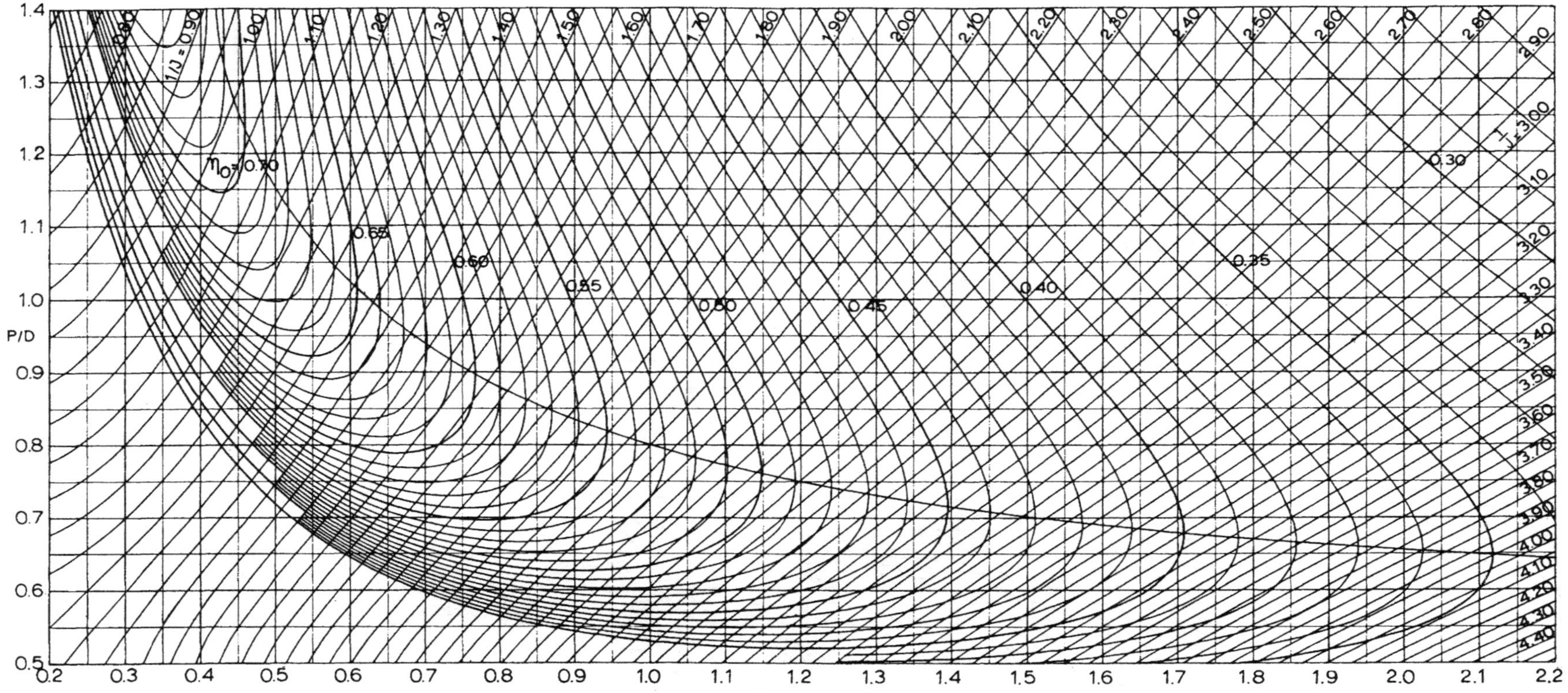

$0.1739\sqrt{B_{P_1}} = K_Q^{1/4} . J^{-5/4}$

B4 - 85

1975

$B_{P_1} = NP^{1/2} . V_A^{-5/2}$	N	=	PROPELLER RPM
	V_A	=	V_S (1-w)
	V_S	=	SHIP SPEED IN KNOTS
	w	=	WAKE FRACTION
	P	=	SHAFT HORSEPOWER (BRITISH)

$K_Q^{1/4} . J^{-5/4} = \left[\dfrac{Qn^3}{\rho V_A^5} \right]^{1/4}$	Q	=	PROPELLER TORQUE IN KGM
	n	=	PROPELLER REVOLUTIONS PER SECOND
	ρ	=	WATER DENSITY (TANK) = 101.94 $KGSEC^2 M^{-4}$
	V_A	=	V_S (1-w)
	V_S	=	SHIP SPEED IN M/SEC.
	w	=	WAKE FRACTION

Fig. 55

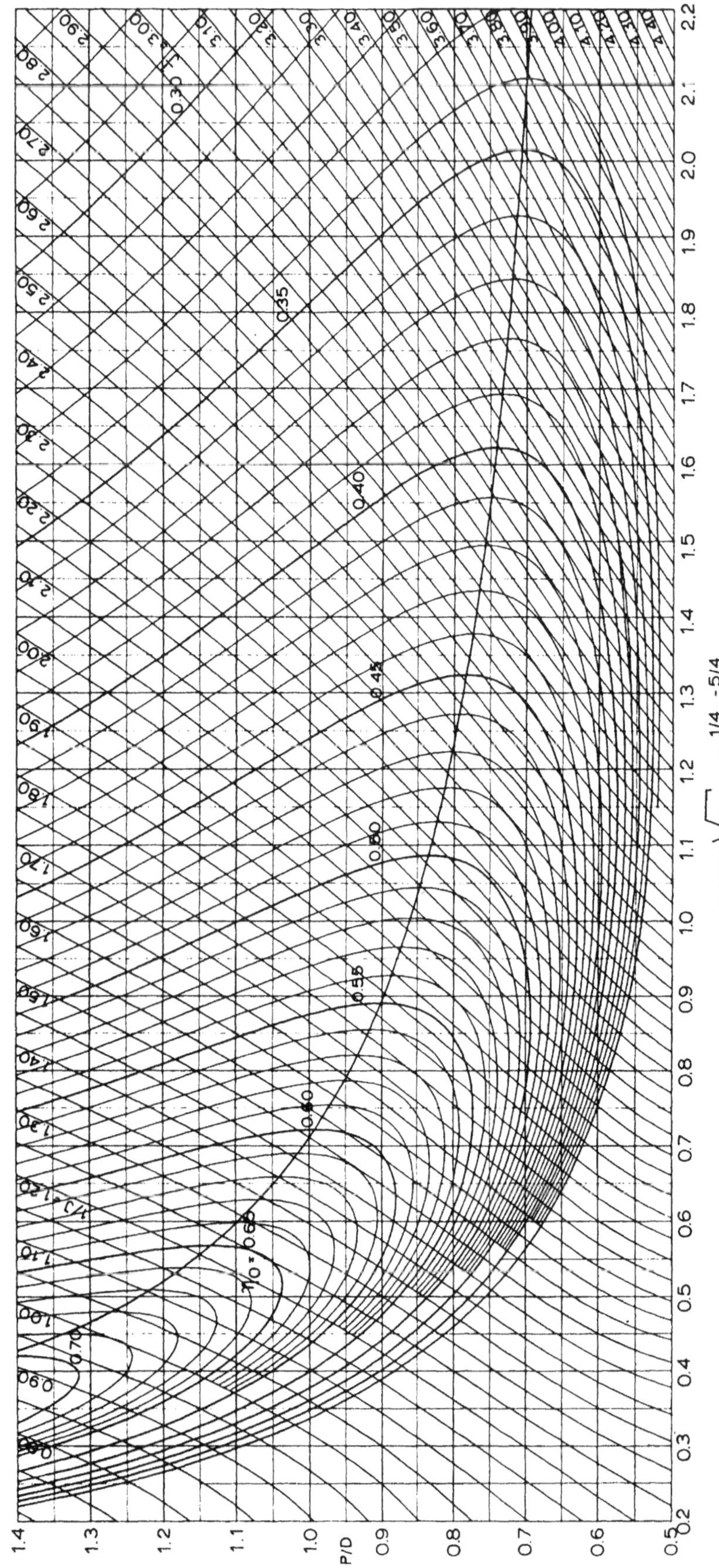

B 4 - 100

1975

$B_{p_1} = NP^{1/2} . V_A^{-5/2}$

N = PROPELLER RPM
$V_A = V_S\ (1 - w)$
V_S = SHIP SPEED IN KNOTS
w = WAKE FRACTION
P = SHAFT HORSEPOWER (BRITISH)

$K_Q^{1/4} . J^{-5/4} = \left[\frac{Qn^3}{\rho V_A^5} \right]^{1/4}$

Q = PROPELLER TORQUE IN KGM
n = PROPELLER REVOLUTIONS PER SECOND
ρ = WATER DENSITY (TANK) = 101.94 $KGSEC^2 M^{-4}$
$V_A = V_S\ (1 - w)$
V_S = SHIP SPEED IN M/SEC.
w = WAKE FRACTION

Fig. 56

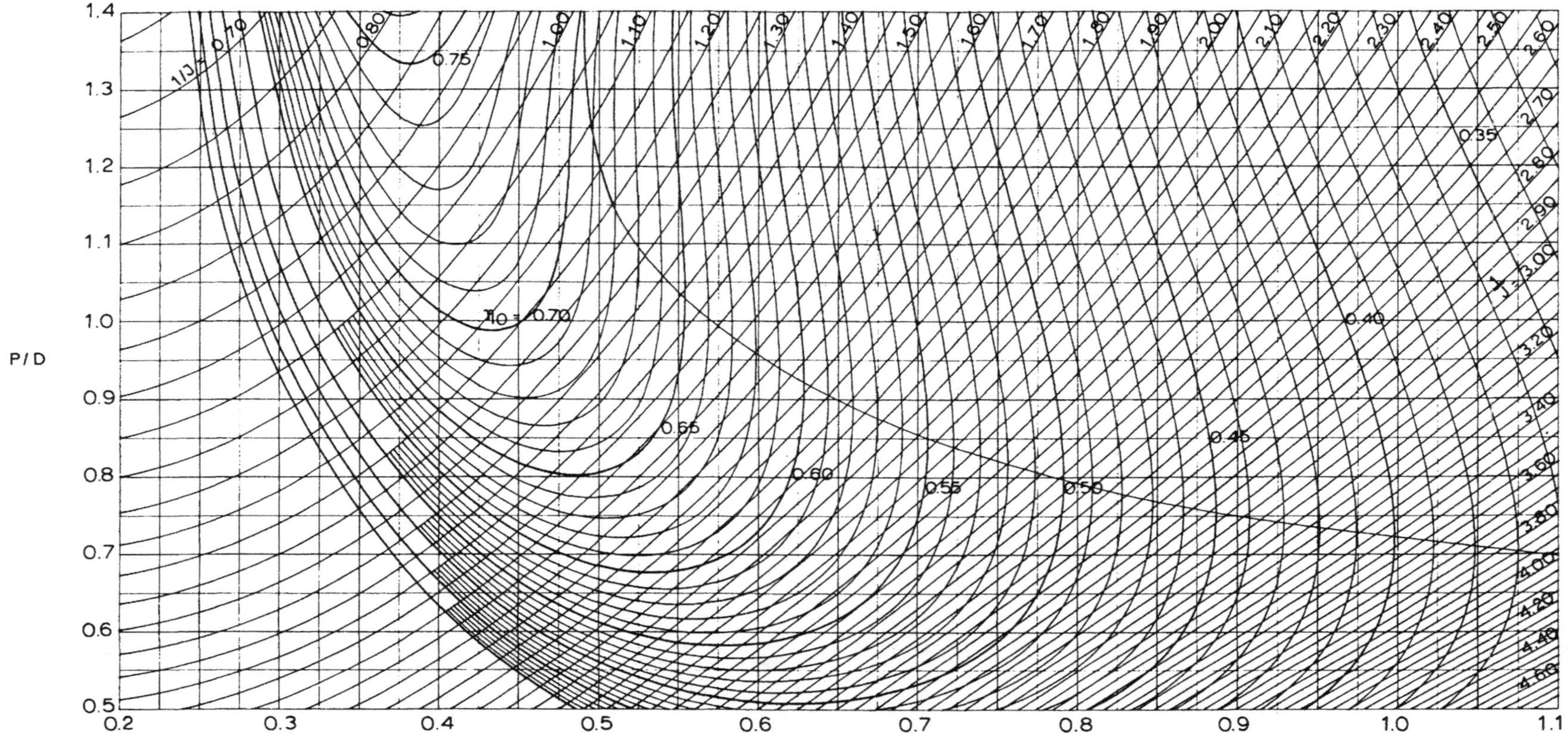

$1.75\sqrt{B_{P_2}} = K_Q^{1/4} . J^{-3/4}$

B4 - 40

1975

$B_{P_2} = P^{1/2} . D^{-1} . V_A^{-3/2}$

D = PROPELLER DIAMETER IN FEET
V_A = V_S (1 - w)
V_S = SHIP SPEED IN KNOTS
w = WAKE FRACTION
P = SHAFT HORSEPOWER (BRITISH)

$$K_Q^{1/4} . J^{-3/4} = \left[\frac{Qn}{\rho D^2 V_A^3}\right]^{1/4} = \left[\frac{P}{2\pi \rho D^2 V_A^3}\right]^{1/4}$$

Q = PROPELLER TORQUE IN KGM
n = PROPELLER REVOLUTIONS PER SECOND
ρ = WATER DENSITY (TANK) = 101.94 $KGSEC^2 M^{-4}$
D = PROPELLER DIAMETER IN METER
V_A = V_S (1 - w)
V_S = SHIP SPEED IN M/SEC.
w = WAKE FRACTION
P = SHAFT POWER IN KGM/SEC.

Fig. 57

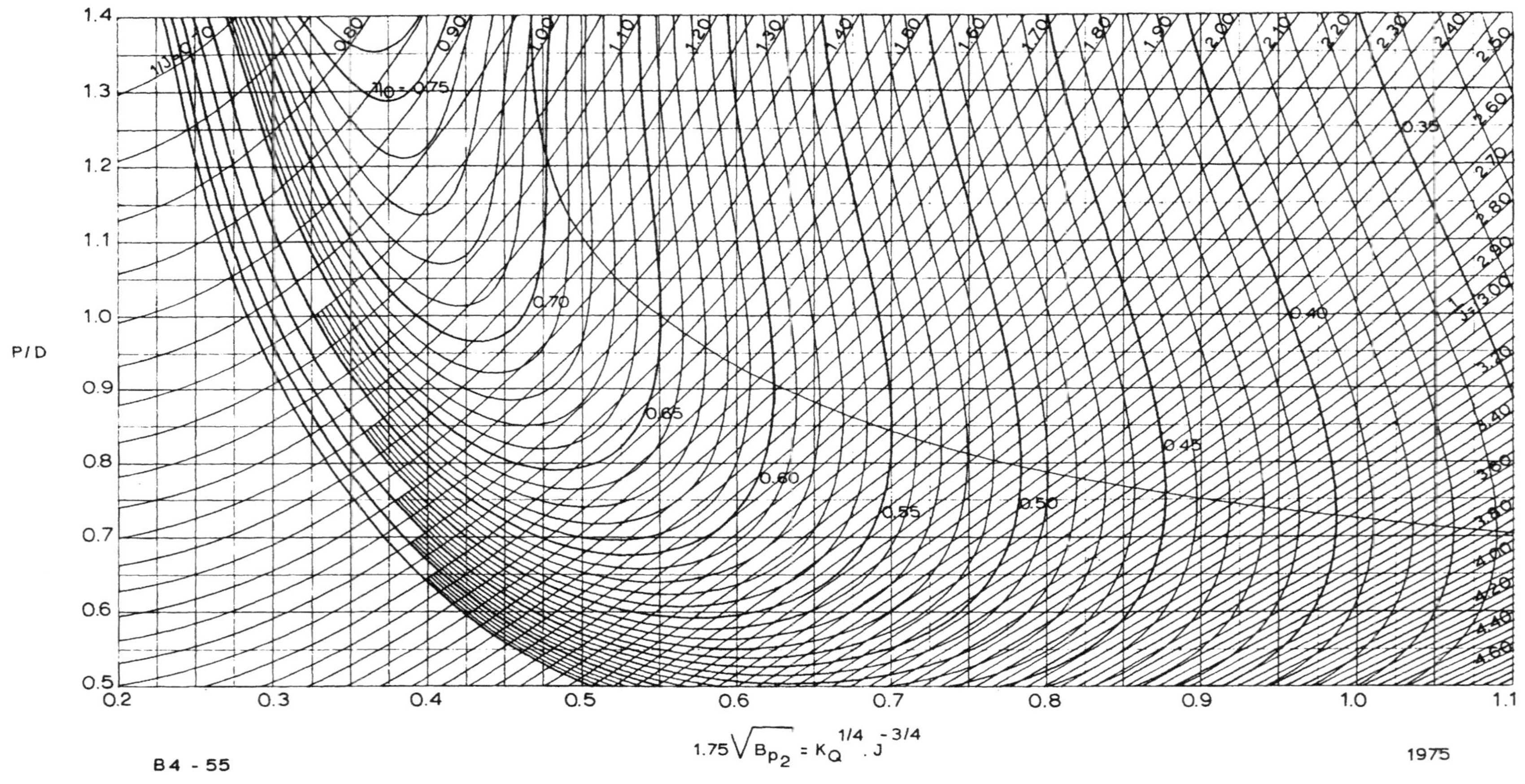

$B_{p_2} = P^{1/2} . D^{-1} . V_A^{-3/2}$	D	=	PROPELLER DIAMETER IN FEET
	V_A	=	V_S (1 - w)
	V_S	=	SHIP SPEED IN KNOTS
	w	=	WAKE FRACTION
	P	=	SHAFT HORSEPOWER (BRITISH)

$$K_Q^{1/4} . J^{-3/4} = \left[\frac{Qn}{\rho D^2 V_A^3}\right]^{1/4} = \left[\frac{P}{2\pi \rho D^2 V_A^3}\right]^{1/4}$$

Q = PROPELLER TORQUE IN KGM
n = PROPELLER REVOLUTIONS PER SECOND
ρ = WATER DENSITY (TANK) = 101.94 $KGSEC^2 M^{-4}$
D = PROPELLER DIAMETER IN METER
V_A = V_S (1 - w)
V_S = SHIP SPEED IN M/SEC.
w = WAKE FRACTION
P = SHAFT POWER IN KGM/SEC.

Fig. 58

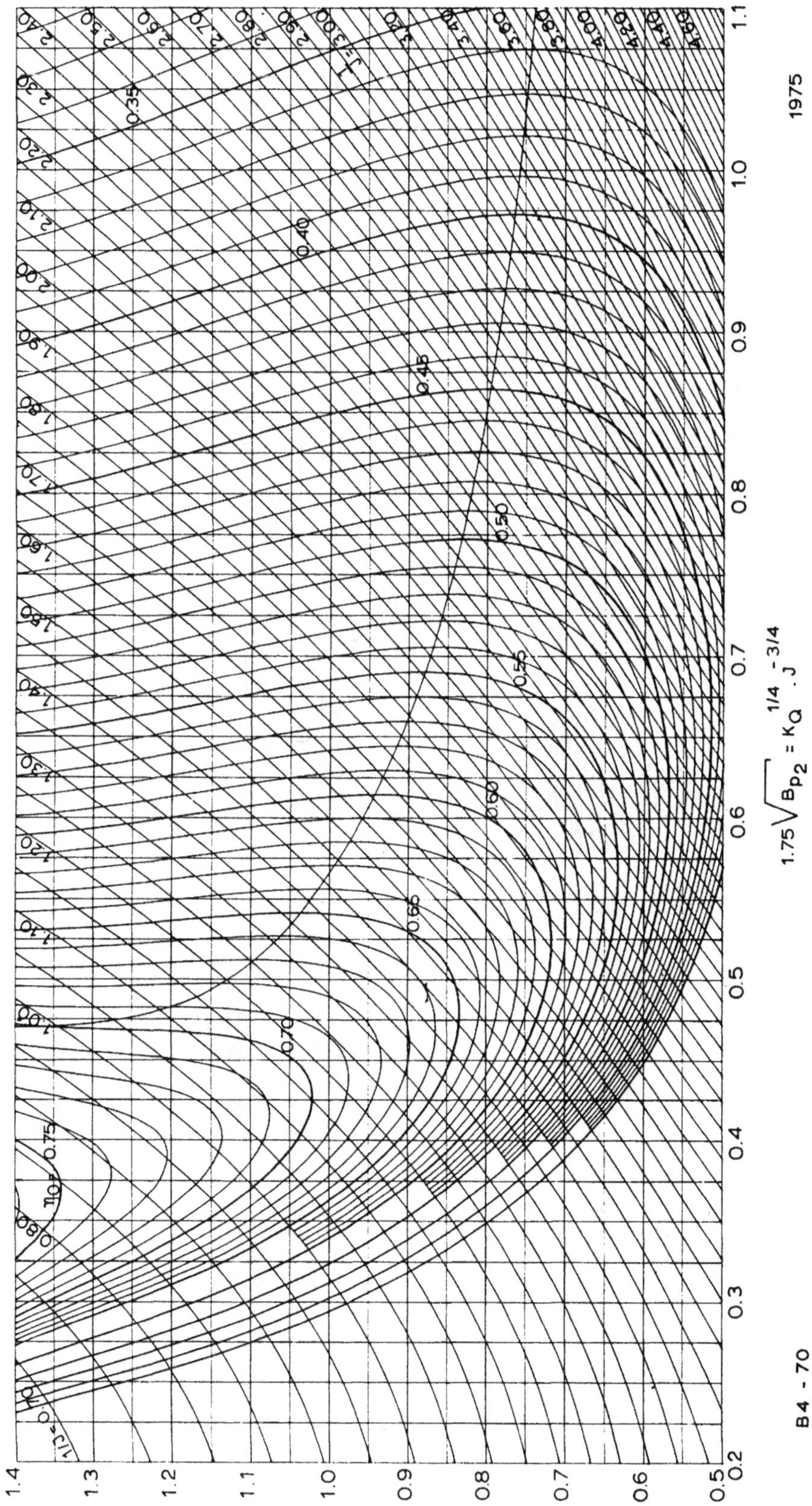

$B_{P_2} = P^{1/2} . D^{-1} . V_A^{-3/2}$

D = PROPELLER DIAMETER IN FEET
V_A = V_S (1 - w)
V_S = SHIP SPEED IN KNOTS
w = WAKE FRACTION
P = SHAFT HORSEPOWER (BRITISH)

$K_Q^{1/4} . J^{-3/4} = \left[\frac{Qn}{\rho D^2 V_A^3}\right]^{1/4} = \left[\frac{P}{2\pi\rho D^2 V_A^3}\right]^{1/4}$

Q = PROPELLER TORQUE IN KGM
n = PROPELLER REVOLUTIONS PER SECOND
ρ = WATER DENSITY (TANK) = 101.94 $KGSEC^2 M^{-4}$
D = PROPELLER DIAMETER IN METER
V_A = V_S (1 - w)
V_S = SHIP SPEED IN M/SEC.
w = WAKE FRACTION
P = SHAFT POWER IN KGM/SEC.

Fig. 59

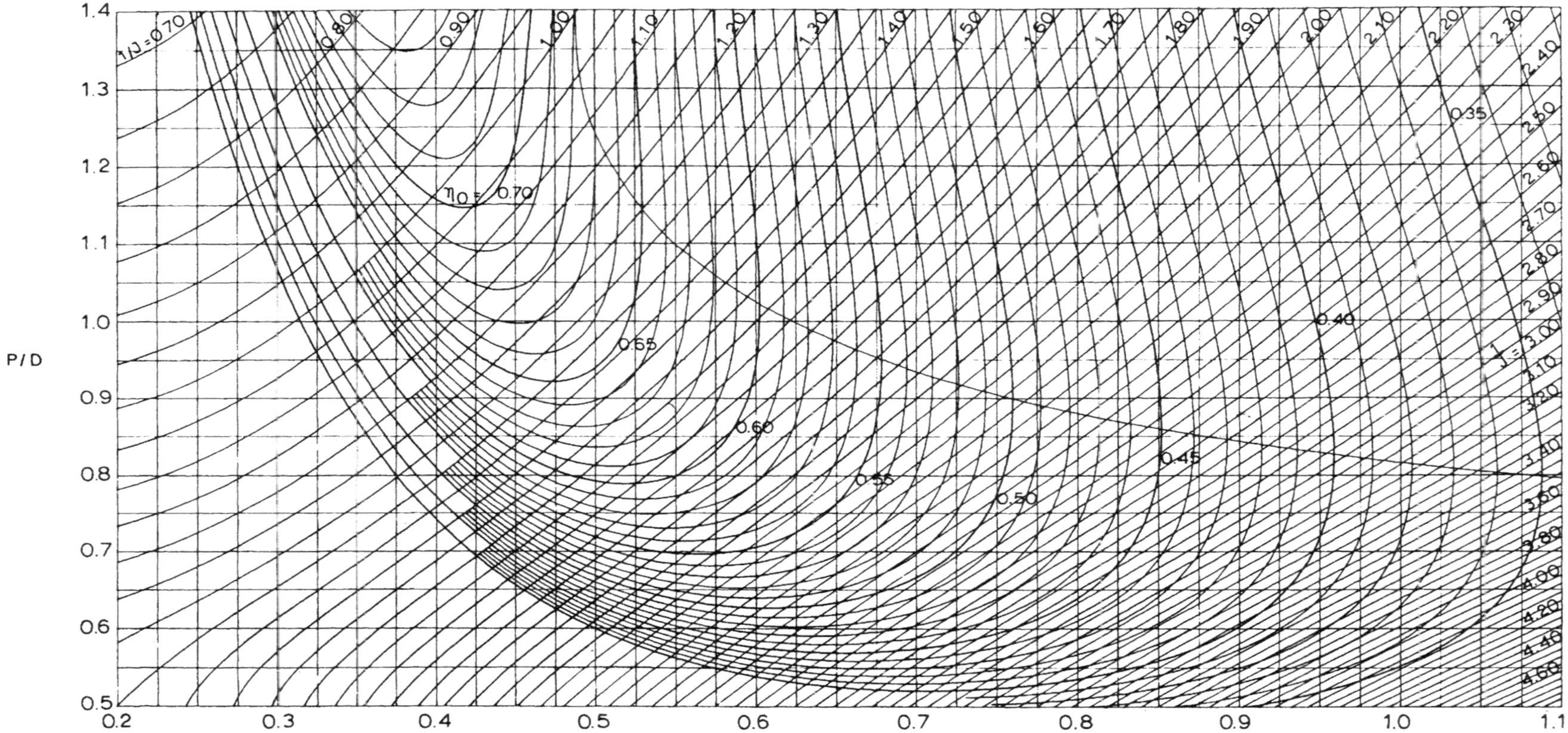

$$1.75\sqrt{B_{P_2}} = K_Q^{1/4} . J^{-3/4}$$

B4 - 85

1975

$$B_{P_2} = P^{1/2} . D^{-1} . V_A^{-3/2}$$

D = PROPELLER DIAMETER IN FEET
V_A = V_S (1 - w)
V_S = SHIP SPEED IN KNOTS
w = WAKE FRACTION
P = SHAFT HORSEPOWER (BRITISH)

$$K_Q^{1/4} . J^{-3/4} = \left[\frac{Qn}{\rho D^2 V_A^3}\right]^{1/4} = \left[\frac{P}{2\pi \rho D^2 V_A^3}\right]^{1/4}$$

Q = PROPELLER TORQUE IN KGM
n = PROPELLER REVOLUTIONS PER SECOND
ρ = WATER DENSITY (TANK) = 101.94 KGSEC^2 M^{-4}
D = PROPELLER DIAMETER IN METER
V_A = V_S (1 - w)
V_S = SHIP SPEED IN M/SEC.
w = WAKE FRACTION
P = SHAFT POWER IN KGM/SEC.

Fig. 60

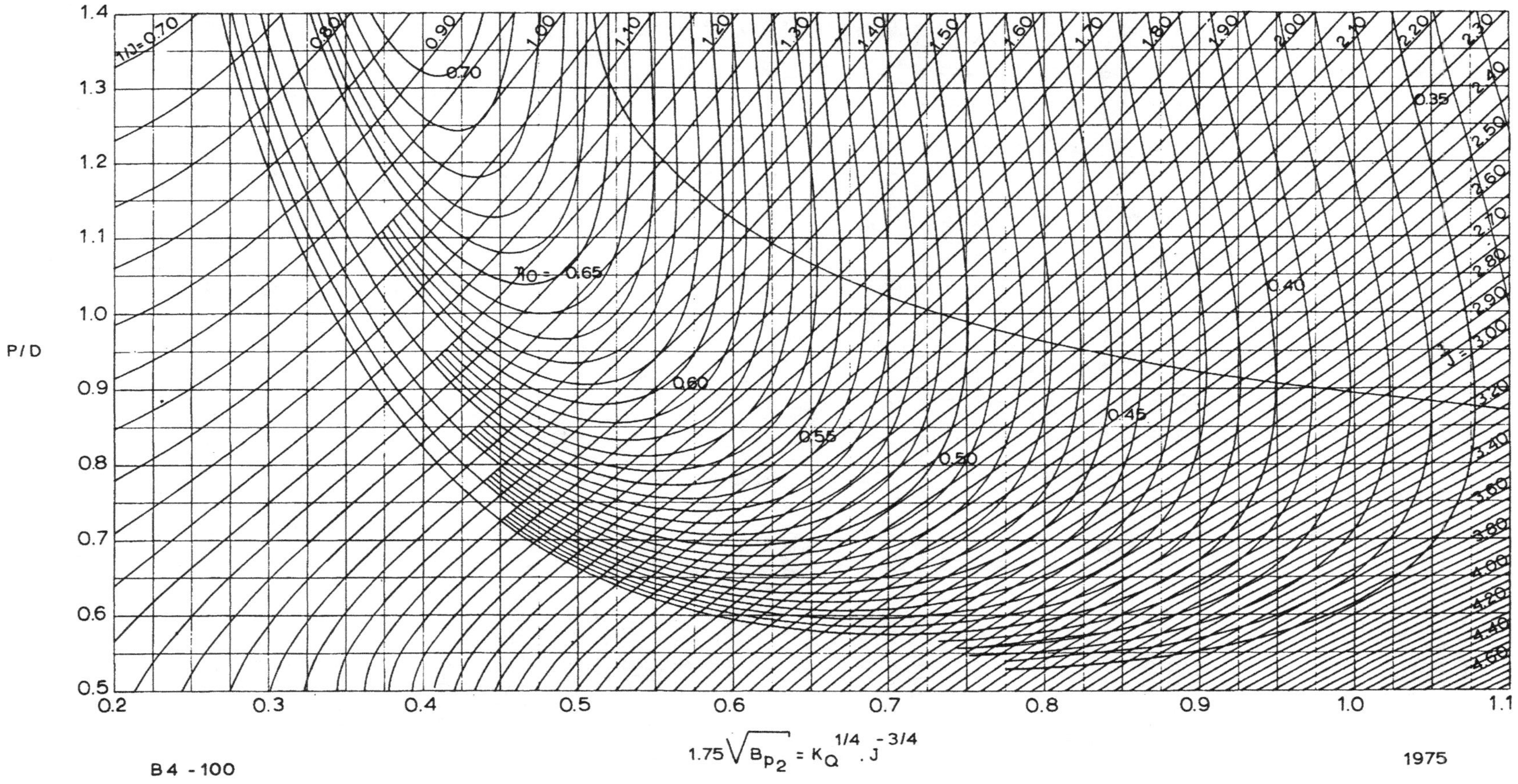

$B_{P_2} = P^{1/2} . D^{-1} . V_A^{-3/2}$

D = PROPELLER DIAMETER IN FEET
V_A = V_S (1 - w)
V_S = SHIP SPEED IN KNOTS
w = WAKE FRACTION
P = SHAFT HORSEPOWER (BRITISH)

$$K_Q^{1/4} . J^{-3/4} = \left[\frac{Qn}{\rho D^2 V_A^3}\right]^{1/4} = \left[\frac{P}{2\pi \rho D^2 V_A^3}\right]^{1/4}$$

Q = PROPELLER TORQUE IN KGM
n = PROPELLER REVOLUTIONS PER SECOND
ρ = WATER DENSITY (TANK) = 101.94 $KGSEC^2 M^{-4}$
D = PROPELLER DIAMETER IN METER
V_A = V_S (1 - w)
V_S = SHIP SPEED IN M/SEC.
w = WAKE FRACTION
P = SHAFT POWER IN KGM/SEC.

Fig. 61

(Continued from page 191)

The model results used to construct the charts are based upon experiments in fresh water and allowance for this must be made in the design calculations.

The information required for making a propeller design, and checking the estimated shaft power and shaft revolutions, from these charts may be summarized as follows:

(*a*) Principal dimensions, proportions and form coefficients for ship, to estimate wake and thrust-deduction factors and other propulsion information, if not available from self-propelled model tests.

(*b*) Effective power P_E from model tests or estimated from other available data, using information under (*a*).

(*c*) Engine power and rated rpm.

(*d*) Speed of ship.

(*e*) Any restrictions, such as a limit on the maximum diameter of the propeller.

The P_D at the propeller can be estimated from Equation (1).

$$P_D = P_E/\eta_D$$

where the quasi-propulsive coefficient η_D can be taken from such published data as Series 60, from the trial results of a similar ship or, in many cases, from experiments run with a model of the ship in question with a stock propeller of approximately the correct diameter and pitch. The estimated value will later be checked in the course of the design.

For a new ship in trial condition, P_E may be estimated either from specific model tests or by series data, in either case with the addition of a suitable model-ship correlation allowance C_A. This latter will depend on a number of factors such as the method of shell construction, quality of paint surface, time out of dock and size of ship. A discussion of appropriate values for C_A can be found in Section 6.4, Chapter V. In commercial work, clients may require estimates involving other values of C_A which they have derived from an analysis of the performance of their own ships. The P_E must also be increased to include the resistance of any appendages not fitted during model tests, such as bilge keels or bossings, as well as an addition for still air resistance (Section 5.2, Chapter V).

It is desirable in general to design the propeller to suit the expected average service conditions rather than those on trial, which are unlikely to be repeated. The service allowance should be chosen upon the basis of the average weather conditions on the ocean routes on which the ship is expected to trade. The North Atlantic, for example, makes more exacting demands in this respect than most oceans. In the absence of any such special requirements, it is usual to specify the service power as being 15, 20 or 25 percent above that required for the same speed on trial, as discussed in Section 8.5.

The shaft horsepower P_S at the torsionmeter will be, very nearly

$$P_S = \frac{P_D}{0.98} \text{ for ships with engines aft}$$

and

$$P_S = \frac{P_D}{0.97} \text{ for ships with engines amidships}$$

However, it is often assumed that $P_S = P_D$, with shaft losses being assumed to be negligibly small.

The speed of advance V_A is obtained from V by model tests or by using the data already described in Section 1.

The choice of number of blades is largely dependent on vibration considerations, but so far as efficiency is concerned, this can be explored by the use of the charts themselves. Blade area is governed by the need to avoid cavitation, and an estimate of the minimum area required can be made from one of the criteria described in Section 7.7. Where cavitation requirements indicate that only a small blade-area ratio is necessary, it is nevertheless unwise to reduce this too much. It is a matter of experience that ships with larger blade area maintain speed better in bad weather, and where cavitation requirements are not critical it is well to be guided to some extent by experience in the choice of area.

In the preliminary design stage only P_E and V are fixed, and it is then possible by the use of the charts to explore the best combination of diameter, RPM and pitch ratio to give the best efficiency. In the final design this choice may be affected by limitations on propeller diameter and by the characteristics of propelling machinery available. With n known, the value of $K_Q^{1/4} \cdot J^{-5/4}$ can be determined or, with D known, the value of $K_Q^{1/4} \cdot J^{-3/4}$ and from the charts the value of $1/J$ and pitch ratio can be found to give maximum efficiency, the latter being indicated on the charts. Then $D = (1/J)(V_A/n)$ or $n = (1/J)(V_A/D)$.

This is the propeller of maximum efficiency in open water, but for conditions behind the ship it is usual to reduce this diameter by 2 percent for full ships and by 1 percent for more slender ships. The ITTC (1984) mentions an average value of 5 percent. If the propeller is designed for the greatest loading expected, and the diameter is chosen to give the highest efficiency under those conditions, then at lighter loads the screw may work on the low-slip side of the efficiency curve, where the efficiency falls off rapidly. For this reason a value of $1/J$ (and therefore of D or n) should be chosen somewhat smaller than that given by the line for maximum efficiency. Moreover, because of the circumferential variation of wake in single-screw ships, it is desirable to reduce D still further so as to avoid the greatest wake inequalities. The foregoing figures are based on such reasoning and upon many experiments with model propellers of different diameters behind model hulls. In ships fitted with struts instead of bossings, the circumferential inequalities in the wake

Table 18—Data for Design Calculation

Speed in service, V_k	21 knots (10.8 m/sec.)
P_E from model tests (including C_A allowance)	9592 kW
Propulsion arrangement	single-screw, engines amidships
Estimated $\eta_D = \frac{P_E}{P_D}$	0.75
Immersion of propeller shaft	7.5 m
Estimated P_D at 21 knots (including service allowance)	12,789 kW
wake fraction, w	0.20
thrust deduction fraction, t	0.15
relative rotative efficiency, η_R	1.05

are smaller, and no reduction in diameter is necessary.

With the same $K_Q^{1/4} \cdot J^{-5/4}$ or $K_Q^{1/4} \cdot J^{-3/4}$ value, but a new value of $1/J$ corresponding to the new D or η_0, respectively, the chart is again used to determine η_0 and P/D.

In some cases it may not be possible to use the value of D so found, because of restrictions of draft or other reasons, such as the maintenance of adequate water over the screw to prevent air-drawing when in ballast or when pitching, the need for adequate tip clearance, or the desire to keep down the size of bossings, struts and shafting to avoid excessive appendage resistance. If for these or other reasons D is limited to some value less than that indicated on efficiency grounds, then both $K_Q^{1/4} \cdot J^{-5/4}$ and $1/J$ are known and the charts will give the open efficiency η_0 and the pitch ratio P/D. Also, if η is found to yield unacceptable values for some reason the appropriate correction to the $K_Q^{1/4} \cdot J^{-3/4}$ and $1/J$ values can be made and the corresponding η_0 and P/D values determined.

In the same way, the effect of different choices of diameter, blade area, RPM, and number of blades can be explored. Knowing η_0 the assumed propulsive coefficient can be checked from the expression in Equation (39),

$$\eta_D = \frac{1 - t}{1 - w} \cdot \eta_0 \cdot \eta_R$$

where t, w, and η_R are known from model tests with a stock propeller or have been estimated by methods given in Section 5.3.

If the value of η_D is the same as, or is close to, that assumed to estimate P_D, well and good—if not, we must use the new propulsive efficiency to estimate a new value of P_D and repeat the calculation—more than once, if necessary, until substantial agreement is reached.

Many other methodical series have been tested besides those already mentioned. Among these is a series of wide-bladed propellers with circular-back sections having blade-area ratios up to 1.10 run by Gawn (1937) at the Admiralty Experiment Works. The pitch ratio ranged from 0.8 to 1.4. The results were presented in $B_P - \delta$ form by deGroot and Hoffman (1949) of MARIN. The AEW series was extended in 1953 to cover pitch ratios from 0.4 to 2.0 (Gawn, 1953). These series form a valuable source of information for the design of warship-type screws with circular-back sections.

A number of series have been run at NPL, England, of medium blade-area ratios (0.4 to 0.6) with 4, 5 and 6 blades (Doust, et al, 1958-59) and the effect of blade thickness has been investigated on screws with circular-back sections and with airfoil sections (O'Brien, 1957).

Cavitation data on the Wageningen B-series have been published by Van Lammeren and Van Oossanen. Other information in this area has been obtained by a number of series run in the cavitation tunnel at King's College, Newcastle, England (Burrill, et al, 1953). Additional experiments have been made on propellers representative of modern merchant ship practice, and from all these data cavitation criteria have been derived, as already explained (Burrill, et al, 1962–63).

To follow the various steps in designing a propeller from methodical series charts, we will work through a typical example. Let us assume the basic design data given in Table 18.

In this case the propeller diameter is restricted to 6.4 m (21 ft) to give adequate clearance in aperture and coverage of water. A four-bladed propeller is desirable from considerations of vibration forces. Calculating in the SI System:

$$V_A = V(1 - w) = 10.80\,(1 - 0.2) = 8.64 \text{ m/sec}$$

Since Wageningen B-series charts are derived from results of experiments in fresh water, the corresponding P_D will be 12,789/1.025 = 12,477 kW in salt water.

On adopting the largest possible propeller diameter of 6.4 m, so as to obtain the highest open-water efficiency, the propeller RPM is calculated from the B-series charts as follows:

$$K_Q^{1/4} \cdot J^{-3/4} = \left[\frac{P_D}{2\pi\rho D^2 V_A^3}\right]^{1/4}$$

$$= \left[\frac{12{,}477}{2\pi \times 1.025 \times (6.4)^2 \times (8.64)^3}\right]^{1/4}$$

$$= 0.5190$$

For this propeller loading value the data given in Table 19 can be derived from the $B_{p2} - 1/J$ charts for the B4-40, B4-55 and B4-70 propellers

To choose the correct blade area, it is necessary to apply a cavitation criterion discussed in Section 7.7. The required thrust is

$$T = \frac{R}{1 - t} = \frac{R\ V}{(1 - t)V} = \frac{P_E}{(1 - t)V}$$

$$= \frac{9592}{(1 - 0.15) \times 10.80} = 1044.9 \text{ kN}$$

The Keller area criterion for a single-screw vessel gives

$$A_E/A_0 = \frac{(1.3 + 0.3Z)T}{(P_0 - P_V)D^2} + 0.20$$

with $P_0 - P_V = 98{,}100 - 1750 + 1025 \times 9.81 \times 7.5$

$$= 171{,}764 \text{ N/m}^2$$
$$= 171.8 \text{ kN/m}^2$$

$$A_E/A_0 = \frac{(1.3 + 0.3 \times 4)\ 1044.9}{171.8 \times (6.4)^2} + 0.2$$
$$= 0.571$$

By interpolation from Table 19, $P/D = 1.087$ and $\eta_0 = 0.669$. The η_D value becomes:

$$\eta_D = \frac{1 - t}{1 - w} \cdot \eta_0 \cdot \eta_R = \frac{0.85}{0.80} \times 0.669 \times 1.05 = 0.746$$

This compares with the value of 0.750 assumed. If a larger difference had been found, a new estimate of the power would have to be made, using $\eta_D = 0.746$ and the calculations repeated until the difference between the assumed and calculated η_D values are within 1 percent.

The same procedure should be followed if the diameter is to be determined (the RPM being given). If in this case the engine RPM is specified at maximum continuous rating (mcr), while for the propeller calculations the RPM at some reduced P_D has to be adopted, use can be made of the fact that P_D/n^3 will be constant. With the use of the K_T and K_Q polynomials given in Table 46 it is relatively simple to carry out the calculation of optimum diameter, RPM, blade area ratio, P/D and η_0 by computer.

With blade area ratio, P/D, diameter and number of blades known, the propeller can now be drawn, see Tables 14 and 15. The strength of the blades must then be checked to ensure that they comply with classification society requirements.

8.4 Application of Circulation Theory to Propeller Design. When using Series charts, a propeller is designed to suit the average flow conditions behind the ship. No account is taken of the variation of the wake

Table 19—Data From Charts

Expanded blade area ratio	0.40	0.55	0.70
Value of $1/J$ at optimum rpm	1.260	1.275	1.290
rpm = $1/J \times V_A/D \times 60$	102.1	103.3	104.5
Corresponding, P/D	1.110	1.090	1.070
Open water efficiency η_0	0.673	0.670	0.663

over the propeller disk, except that in single-screw ships a somewhat arbitrary reduction of pitch is sometimes made near the root of the blade where the wake is heaviest. To avoid cavitation, reference is made to one of the criteria already discussed to ensure that the propeller has sufficient blade area.

Two of the advantages of a detailed design method are that the pitch of the sections can be chosen to suit the mean circumferential wake at each radius and that the shape of the sections can be chosen to minimize cavitation. The simple blade-element theory described in Section 2.5 attempts to do this, but suffers from the defects that it ignores tip-loss effects and the interference to the flow through the propeller by the blades themselves, which induces curvature and velocity changes in the flow. The better understanding of the propeller action resulting from the use of circulation theory described in Section 2.6 enables correction factors to be derived which to a large extent take account of these deficiencies. It also leads to avoidance or reduction of vibration, cavitation and noise.

The design of a marine propeller working in its complicated environment is a difficult problem. The circulation method involves a combination of theoretical and experimental knowledge, and its practical application has been greatly extended by the use of high-speed computers. Programs have been written to suit variations of the method and to calculate the appropriate correction factors. Many variations of a design can be obtained from the computer very quickly, thus enabling the effects of wide ranges in parameters to be investigated.

The basis of the circulation theory as applied to propeller design has been described in Section 2.6. The theory accounted for the lift of an airplane wing due to the development of circulation around each section in the span direction. It was postulated that the vortex movement around such an airfoil is continued in the fluid in the form of vortices trailing from the ends, and in the case of propeller blades, passing downstream in approximately helical paths from the tips. This concept of the shedding of vortices from the tips of propeller blades was shown to be true by means of photographs of the wake of a propeller. Many scientists subsequently endeavored to calculate the induced velocity associated with this system of trailing vortices. Prandtl (1918) succeeded, and the concept of trailing vortices became fully accepted. Prandtl concluded to state that the behavior of an element of an airfoil of finite span can only be considered the same as in two-dimensional flow when proper allowance is made for the induced velocities caused by the shedding of vortices from such an airfoil. This first vortex theory is often referred to as a lifting line theory due to the fact that a wing of finite span is replaced by a vortex line. In this way it was possible to account for spanwise variations in the circulation distribution, the circulation of the vortex line at each section being put equal to the chordwise integrated circulation at that section. This theory is therefore not able to calculate subsequent effects of the fact that the circulation around a section of a propeller blade or a wing is not concentrated at the position of the line vortex, but distributed along the chord. For moderate aspect ratios it was found that the Prandtl lifting line theory was satisfactory, and this soon led to the standard procedure in airscrew design to use two-dimensional lift and drag characteristics (so-called profile characteristics) at an angle of incidence corrected for the induced velocities. For the broad-bladed propeller this theory was still not satisfactory, however.

On assuming that the trailing vortices behind a propeller blade follow helical paths with a constant angle of advance (implying a uniform propeller inflow and that the induced velocities are small, i.e. that the propeller is lightly loaded), and on neglecting the profile drag of the blade sections. Betz (1919) succeeded in establishing the best load or thrust distribution along the blades for minimum induced drag. Due to the involved mathematical difficulties, Betz had to assume that the propeller had an infinite number of blades. In an appendix to Betz's paper, Prandtl established an approximate correction to account for a finite number of blades. The difficulty inherent in the finite blade number case lies in the complexity of calculating the induced velocities caused by the system of trailing vortices constituting a finite number of vortex sheets. Particular credit must be paid to Betz's paper, not only for determining the optimum radial load distribution from the viewpoint of efficiency, but also for being the first to successfully apply the Prandtl vortex theory to propellers and to define the mathematical model concerned.

With these new vortex conceptions, which in fact constituted an important break-through in propeller theory, various important propeller theories were developed in the years that followed. Bienen, et al (1924) extended Betz's 1919 paper and performed the additional calculations for the case that the effects of profile drag are included. Later Betz extended his work to the heavily loaded, free-running case. In the case of a heavily loaded propeller the influence of the induced velocities on the shape of the helical vortex sheets is taken into account as well as the effects of centrifugal forces and of the contraction of the induced velocity components. Only in the case of the lightly loaded propeller are the vortex sheets true helical surfaces.

Goldstein (1929) successfully considered the flow past a finite number of true helical vortex sheets and obtained an expression for the ratio between the mean circulation taken around an annulus and the circulation at the helical surfaces for a 2 and a 4-bladed propeller. From these values the ratio between the mean inflow velocity taken around any annulus and the corresponding larger inflow velocity in way of the helical vortex sheets at the position of the propeller blades was derived. The values of this ratio for various values of the propeller radius, the hydrodynamic pitch angle and the number of blades have since been designated as Goldstein factors. A large number of these values were calculated very accurately by Tachmindji and Milam (1956). The Tachmindji and Milam values are valid for the case of zero circulation at the hub, which case is now considered as correct.

Today these Goldstein factors are still used, and often in cases where they are not applicable. They are strictly valid only for uniform propeller-inflow (so-called free running propellers), having a constant radial virtual pitch, i.e. the Betz optimum radial circulation distribution. Furthermore, none of the devised marine propeller design methods based on the Goldstein factors are suitable for heavy screw loading. Very few recognize slipstream contraction and the influence of the radial pressure gradient. The most used marine propeller design methods based on the lifting line procedure incorporating Goldstein factors are those due to Burrill, Lerbs, Hill, Van Manen and Eckhardt and Morgan. When Lerbs (1952) published his lifting line method based on induction factors as defined by Kawada, it was a timely introduction of a method suitable for a radially varying wake and a non-optimum circulation distribution. Van Manen (1958) showed that important differences occur between the results of the induction factor method and the Goldstein factor method when applied to so-called wake adapted propellers.

As mentioned above, lifting-line procedures were found to be very suitable for the design of propellers and wings of moderate aspect ratios as low as 3. This was contrary to the case with wide-bladed propellers having lower aspect ratios. Soon after lifting-line procedures were introduced it was found necessary to supplement design procedures based on the lifting-line concepts for wide-bladed marine propellers with empirical or theoretical correction factors. It is now known that this is due to the fact that in the case of wide propeller blades it is no longer correct to calculate the induced velocity at the position of the vortex line representing the blade and to neglect the variation of the induced velocity along the chord. A distribution in the induced velocity or downwash along the chord results in a specific curvature of the flow over the blade which, among other effects, changes the effective camber of the blade sections. Such effects remain unaccounted for in lifting-line procedures. In the case of moderate aspect ratios it is more or less correct to consider the induced velocity at the lifting line as a correction to be applied to the geometric angle of incidence in determining an effective angle of incidence to which the corresponding blade section reacts as if it were in a two-dimensional flow. In the case of low aspect ratios, however, the decrease in effective camber and, in general, the way in which the blade sections react to the curved flow, must also be accounted for. Changes in the ideal angle of attack and, in the case of non-symmetric chordwise loading of the meanline about the midchord position, in the hydrodynamic pitch angle are a result.

Ludwieg, et al (1944) were the first to recognize this influence and they actually calculated some correction factors with which the amount of blade section camber could be corrected for the induced curvature of the flow. These camber corrections were given in the form of a ratio between the effective camber and the geometric camber of the blade sections as a function of propeller radius, blade area, blade number and hydrodynamic pitch. These particular correction factors were, however, only strictly valid for the optimum radial circulation distribution, uniform chordwise loading (circular arc meanlines) and a uniform propeller inflow. Because of the large amount of work involved in calculating such correction factors, the Ludwieg and Ginzel camber factors were used for a long time following 1945. Very often they were applied in cases where they were unsuitable, resulting in fact in larger errors than caused by the correction factors based on experiments with profiles in cascades, which they replaced.

Prior to the work performed by Ludwieg and Ginzel, all lifting-line procedures for the design or analysis of marine propellers incorporated correction factors based on previous experience or theoretical or experimental work on cascade effects. The opinion, that the differences occurring between theory and the results of experiments with marine propellers are due to so-called cascade effects, was then a general one, and many efforts were made at developing reliable theories and at obtaining relative experimental information in the 50-year period following the first suggestion to do so by Drzewiecki in 1892. Drzewiecki first found that the lift and drag forces appeared to depend considerably on the ratio of length to breadth of the propeller blades. The experimental investigations eventually culminated in the work of Gutsche (1933), (1938) who tested a series of cascades of airfoils and propeller blade sections at different pitch angles and various gap ratios.

After the Ludwieg and Ginzel paper, the application of cascade corrections in marine propeller problems continued. The reason for this was pointed out by Burrill in the discussion of his paper on the optimum diameter of marine propellers. He stated: "One of the deficiencies of the Ludwieg-Ginzel corrections is that they have been worked out for wide-tipped outlines and another is that the basic aerofoil (sic) section characteristics obtained from wind tunnel work must be corrected by similar lifting surface curvature effects in order that they may be applied to propeller design work. The use of the Gutsche cascade corrections may seem to be out-of-date, but they have the merit of simplicity and they do seem to be of the right order and to give satisfactory integrated values of thrust, torque and efficiency." Hence many designers felt that it was better to make use of the simpler Gutsche cascade corrections instead.

With the advent of high speed computers, gradually proper and more accurate lifting surface calculations were made, and now all propeller design procedures based on lifting line theory incorporate lifting surface correction factors. Cox (1961) derived a set of camber corrections valid for four different types of blade shapes, with 3, 4 and 5 blades, applicable to the case of constant chordwise loading at shock-free entry of the flow. An extensive set of lifting surface correction factors were derived by Morgan, et al (1968) for a family of non-skewed and skewed propellers. These correction factors were derived from the lifting surface programs developed by Cheng for blade loading and Kerwin for blade thickness. They are valid for the NACA $a = 0.8$ meanline and the NACA 66 thickness distribution. The number of blades for which these corrections are given are 4, 5 and 6. For the respective 3-bladed propellers of this series, the correction factors were derived by Minsaas, et al (1971) by means of the same programs.

Lifting surface theory for marine propellers has developed basically along two different paths. The first calculations and theories were really extended lifting-line methods. These eventually developed into so-called vortex lattice methods, in which the lifting surface is represented by a discrete lattice of vortices. Then Sparenberg (1959) derived the three-dimensional integral lifting surface equation for a screw propeller in a steady flow. This theory incorporated a continuous vortex sheet representation of the lifting surface, i.e. without physical or mathematical assumptions and models for the arrangement of the lattice. Hanaoka (1962) extended Sparenberg's work to the case of unsteady flow. This theory was then further developed by Pien, Yamazaki and others. In obtaining numerical results with this theory, various different numerical procedures have evolved to solve the integral equation. In this connection the work done by Cheng, Shiori, Tsakonas, Brown, Greenberg, Verbrugh and Van Gent should be particularly mentioned.

Lifting surface theory is now widely used to determine the pitch and camber distribution required to generate a prescribed loading over the propeller blades. For all other aspects of propeller design lifting line theory is still commonly used.

In the design of a propeller it is first necessary to

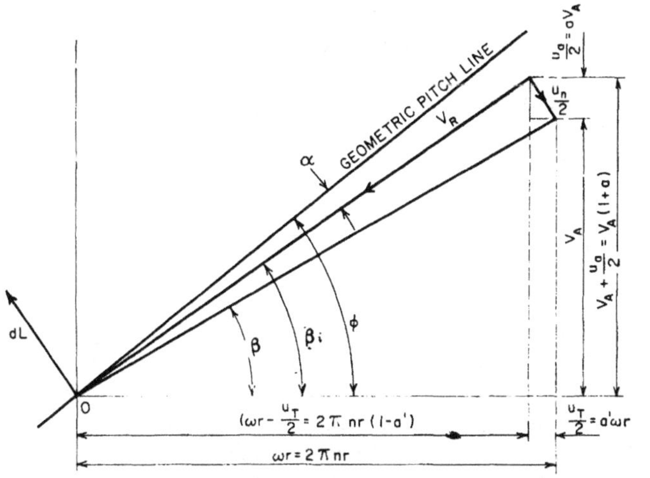

Fig. 62 Velocity diagram for "lifting-line" propeller in nonviscous flow

Blade section at radius r
β = advance angle
β_i = hydrodynamic pitch angle

$$\operatorname{Tan}\beta = V_A/\omega r = V_A/2\pi nr$$

$$\operatorname{Tan}\beta_i = \frac{V_A + Ua/2}{2\pi nr - ut/2} = \frac{V_A(1 + a)}{2\pi nr(1 - a')} = \frac{1 + a}{1 - a'}\tan\beta$$

carry out a preliminary calculation using series charts or other approximate means, to determine the diameter. The effect of changes in this preliminary choice of diameter can be determined subsequently by repeating the calculations. This is a simple matter when the method of design has been programmed for a high-speed computer. In some cases the draft or the stern arrangements may limit the diameter to less than the desired value.

As a propeller in open water develops thrust, it induces three inflow velocity components, radial, axial, and tangential. These are in general small compared with the speed of advance V_A, but, as we have seen in Section 2.5, the axial and tangential components have a large effect upon the angle of incidence, which is itself small and so very sensitive to changes in direction of the relative velocity, Fig. 62.

The radial component is due to the contraction of the slipstream in passing through the propeller, and this is small in all but heavily loaded propellers, so that the resulting radial inflow velocity is usually neglected.

From the circulation theory for the optimum, free-running propeller we have seen in Section 2.6 that, when the blades are replaced by lifting lines, an induced velocity u_n is generated far aft of the propeller, which is normal to the helicoidal vortex sheets, and that this induced velocity has the value $u_n/2$ at the lifting line itself. This is represented by the vector BA in Fig. 11, assuming the "condition of normality," i.e., that $u_n/2$ is normal to the resultant inflow velocity vector V_R. The axial and tangential inflow components, $a \cdot V_A$ and $a' \cdot \omega \cdot r$, are then the appropriate components of $u_n/2$, Fig. 62.

The assumption is made that the method also applies to a propeller under variable loading due to particular pitch or wake distributions, but we know that this assumption will not apply when the variations in pitch or wake are large.

Up to this point, the circulation theory has been applied to a propeller in a nonviscous fluid. When viscosity is introduced, an additional drag is experienced by the blade, and the general effect is to decrease the thrust and increase the torque; Fig. 63. To restore the thrust to the required value, it is necessary to increase the pitch at each section, the amount depending upon the particular blade-section shape being used.

In determining the pitch distribution over the blade, we can only make the blade section at any radius suit the *average* circumferential velocity of inflow around that particular radius, and so determine its *average* angle of attack. It will not work at maximum efficiency throughout, but we can get nearer to this condition over the whole blade than in chart designs.

The circumferential variation in the wake at different radii can be measured on a model by pitot tubes and the average at each radius determined. The average values at each radius can be measured also directly by the use of blade wheels. When such experiment results are not available for a particular design, the radial variation of the average wake can be estimated from published results.

Van Lammeren gave the distribution of average circumferential wake along the radius in the form of curves based upon a great number of tests made at MARIN in 1948 using blade wheels. Two diagrams are given, for single-screw and twin-screw models, respectively. They are plotted to a base of the vertical prismatic coefficient $C_{VP} = C_B/C_{WP}$, because this coefficient to some extent characterizes the shape of the

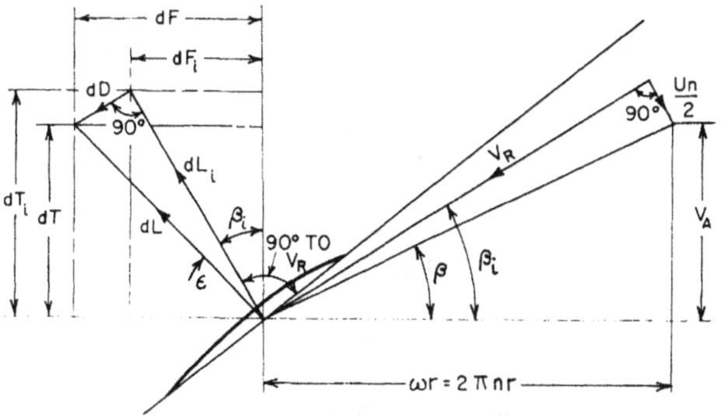

Fig. 63 Force diagram for viscous flow at a blade section at radius r

T = thrust
D = drag
L = lift
F = torque force

$\epsilon = \frac{D}{L}$ ratio
= angle in radians, approximately
i = subscript for ideal (nonviscous) flow

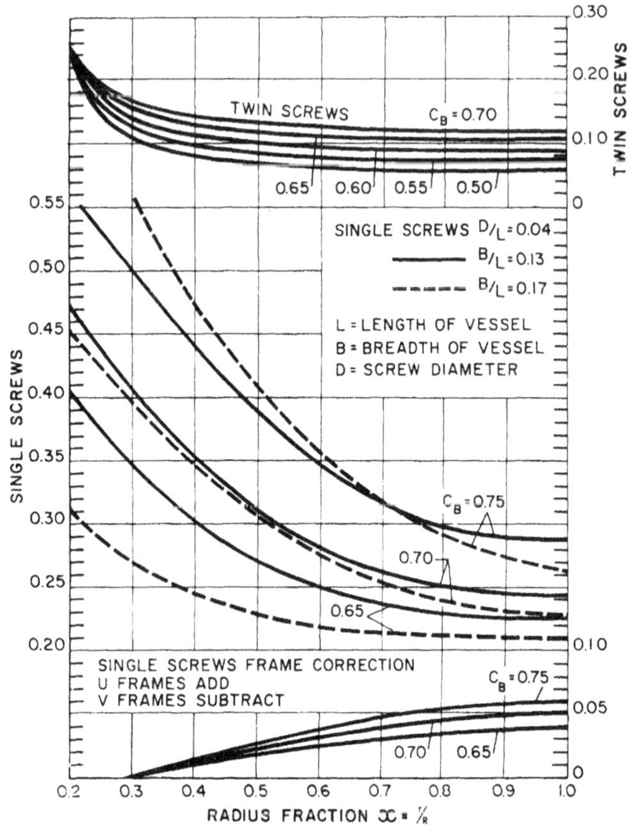

Fig. 64 Local wake fraction (Harvald)

transverse sections. Troost (1956) also has discussed the effect of radial-wake variation. Harvald (1950) has given curves of local wake fraction at different radii for both single and twin-screw models for a variety of block coefficients and length-beam ratios. His diagram is reproduced in Fig. 64.

The wakes determined by pitot tubes or blade wheels are *nominal* wakes, as discussed in Section 4.2. These are somewhat greater than the *effective* wakes measured by the propeller when operating behind the model. For a particular design the effective wake for the whole propeller can be found either by self-propelled tests with a stock propeller or from published results of self-propelled series data. If this effective wake is assumed to apply at 0.7 radius, the variation of the effective wake over the blade from hub to tip can be found by using one of the foregoing radial distribution curves, assuming that the ratio of effective to nominal wake at 0.7 radius applies at all other radii.

When detailed wake data are available it is preferable to calculate the volumetric mean wake[5] over the propeller disk and to apply the ratio of effective wake to volumetric wake to the measured mean circumferential wake velocity at each radius. A propeller designed in this way is referred to as *wake-adapted*.

In using the circulation theory, two approaches are possible. The optimum circulation distribution along the blade can be selected and the pitch distribution calculated, or the pitch distribution can be assumed and the resulting circulation computed. For the general case of a wake-adapted propeller, a relationship for the tangent of the hydrodynamic pitch angle β_i for an optimum wake-adapted propeller can be used. The resultant pitch variation depends upon the average effective wake for which each blade section is designed. Small errors in the assumed wake will not affect materially the resulting propeller efficiency, since model tests have shown that this efficiency is relatively insensitive to quite large variations in pitch distribution.

The first step is to estimate the propeller diameter from Series Chart data or other sources, remembering the reduction necessary below the optimum figure for the reasons discussed in Section 8.3. Ideally, the design should be made for the optimum combination of diameter and propeller revolutions per minute, but there are often many practical restrictions on the choice of these two factors. Diameter may be restricted because of draft limitations and the maintenance of adequate clearances, while the RPM may be limited by the available machinery. The calculation of diameter will follow the same general lines as described in Section 8.3. The number of blades will be chosen primarily to avoid vibration, taking into account the wake pattern and the natural frequencies of hull and shafting.

For an open-water propeller, i.e. one working in a uniform wake, the hydrodynamic pitch angle β_i, Fig. 11, is first calculated for each section. To do this, the thrust T is determined from the resistance and thrust-deduction fraction and the thrust coefficient obtained:

$$C_T = \frac{T}{\frac{1}{2}\rho A_0 (V_A)^2}$$

Since the propeller is first designed for nonviscous

---

[5] From the detailed wake data, the mean flow velocity $v(r, m)$ at any radius r can be determined.

The volume of water flowing through an annulus at radius r in unit time is

$$2\pi r\, dr\, v(r, m) = 2\pi r\, dr\, V[1 - w(r, m)]$$

The total volume flowing through disk is

$$\int_{r_1}^{R} 2\pi V[1 - w(r, m)] r\, dr$$

where R is propeller radius and r_1 is hub radius.

The total volume which would flow through the disk in the absence of any wake is

$$\pi(R^2 - r_1^2)V$$

and *volumetric mean wake factor* is

$$(1 - w_v) = \frac{\int_{r_1}^{R} [1 - w(r, m)] r\, dr}{\frac{1}{2}(R^2 - r_1^2)}$$

flow, C_T is changed to an ideal thrust coefficient C_{Ti} by the approximate formula

$$C_{Ti} = (1.02 \text{ to } 1.06)\, C_T \qquad (70)$$

Kramer (1939) has developed a set of curves for determining the ideal efficiency η_i of an optimum propeller in open water. Knowing the number of blades, the advance ratio[6] $\lambda = V_A/\pi nD$ or the whole propeller and the value of C_{Ti}, the Kramer curves give the appropriate value of η_i. This is then used to derive the hydrodynamic pitch angle at each radius, viz:

$$\tan\beta_i = (1/\eta_i)\tan\beta, \text{ where } \tan\beta = V_A/x\pi nD$$

and $x = r/R$ is the nondimensional radius. The induced axial and tangential velocities at the propeller, at radius x, are determined from

$$\frac{U_t}{V_A} = \left(\frac{\tan\beta_i}{\tan\beta} - 1\right)\left(\frac{\tan\beta_i}{\tan^2\beta_i + 1}\right)$$

$$\text{and } \frac{U_A}{V_A} = \frac{1}{\tan\beta_i}\frac{U_t}{V_a}$$

where U_t = induced tangential velocity at x
and U_a = induced axial velocity at x

The non-dimensional circulation G at x, is defined as:

$$G = \frac{\Gamma}{\pi D V_A} \qquad (71)$$

where $\Gamma = L/\sigma V$ = circulation around a propeller blade at x

in which V = resultant velocity at x.

The value of G follows from the equation:

$$G = \frac{2x\kappa}{Z}\frac{U_t}{2V_A} \qquad (72)$$

where κ is the appropriate Goldstein factor.

Curves of the Goldstein factors are given by Tachmindji for propellers with 3, 4, 5 and 6 blades. Alternatively the value for G can be found using the induction factors as adopted by Lerbs. The radial distribution of the thrust and power coefficients (without viscous effects) can then be found from

$$\frac{d\,C_{Ti}}{dx} = 4ZG\left(\frac{1}{\tan\beta} - \frac{U_t}{V_A}\right)$$
$$\text{and } \frac{d\,C_{p_i}}{dx} = 4ZG\,\frac{1}{\tan\beta}\left(1 + \frac{U_a}{V_A}\right) \qquad (73)$$

[6] It should be noted that this is not the advance ratio $J = V_A/nD$, used generally in this book in accordance with ITTC terminology.

After integration and iteration as necessary to obtain the required thrust (or power) values, the lift coefficient is calculated by means of

$$C_L\frac{c}{D} = \frac{2\pi\, G\cos\beta_i}{\left(\dfrac{1}{\tan\beta_i} - \dfrac{U_t}{V_A}\right)} \qquad (74)$$

where c is chord length at x. When a blade contour is chosen, C_L can be calculated. On adopting suitable values for the drag coefficient C_D for the various radii, the value of $\epsilon = C_D/C_L$ can be determined.

From the value of ϵ so derived, the assumption made in Equation (70) can be checked from the approximate formula

$$C_{Ti} = \frac{C_T}{(1 - 2\epsilon\lambda_i)} \qquad (75)$$

and if necessary the calculations repeated.

The pitch for the optimum propeller in open water can thus be obtained. When such a propeller is designed so that the sections operate under the condition of shock-free entry, where all the lift is due to the camber of the sections, the result is a constant-pitch propeller.

When the propeller is designed to suit a particular wake pattern, Kramer's curves no longer strictly apply, but they still work very well as a first approximation if applied to a typical section at 0.7 radius. To determine the pitch at different radii, the distribution of the mean circumferential wake must be assumed, as already discussed.

Lerbs and Van Manen have developed formulas for the hydrodynamic pitch angle of wake-adapted propellers in which the average circumferential wake varies from radius to radius. These formulas are as follows:

$$\tan\beta_i = \frac{\tan\beta}{\eta_i}\left(\frac{1-\omega}{1-\omega_x}\right)^{3/4} \qquad \text{(Van Manen)} \quad (76)$$

$$\tan\beta_i = \frac{\tan\beta}{\eta_i}\left(\frac{1-\omega}{1-\omega_x}\right)^{1/2} \qquad \text{(Lerbs)} \quad (77)$$

Either one of these relations, together with Kramer's curves, can be used to derive a first approximation to the radial distribution of β_i.

In the case of wake-adapted propellers it is recommended to use the Lerbs induction factor method for calculating the radial circulation distribution, since adoption of the Goldstein factors will lead to inaccuracies, as shown by Van Manen. This calculation procedure can be carried out by using the initial values for β_i, which allow induction factors to be calculated by means of formulas derived by Wrench (1957). The resulting axial and tangential induction factors, as well as the (unknown) circulation, are expanded in a Fourier

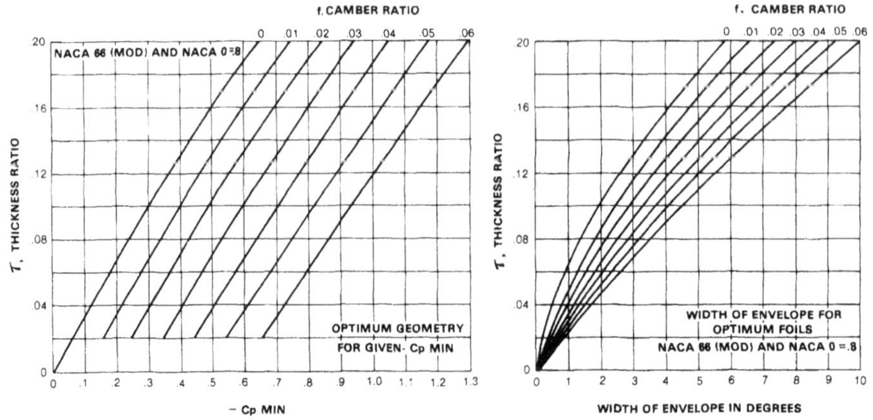

Fig. 65 Geometry of optimum NACA 66 a = 0.8 foils (Brockett, 1966)

series and on substituting these expressions in Biot-Savant's formula for a free helical vortex, one obtains expressions for the induced axial and tangential velocities, which still contain the unknown Fourier coefficients of the circulation. Therefore, it is necessary to have one more equation in the unknowns U_a, U_t and G. Such a relation is (see Fig. 62),

$$\tan \beta_i = \frac{V_s(1 - w_x) + U_a}{x\pi\, nD - U_t}$$

which can be written as:

$$\frac{U_a}{V_s(1 - w_x)} + \frac{U_t}{V_s(1 - w_x)} \tan \beta_i = \frac{\tan \beta_i}{\tan \beta} - 1 \quad (78)$$

With the help of this relation the values of U_a, U_t and G can be derived (Van Oossanen, 1974), and the ideal thrust and power loading coefficients can be determined,

$$C_{T_i} = 4Z \int_{x_h}^{1} G(1 - w_x)^2 \left(\frac{1}{\tan \beta} - \frac{U_t}{V_s(1 - w_x)}\right) dx \quad (79)$$

$$\text{and } C_{p_i} = 4Z \int_{x_h}^{1} \frac{G(1 - w_x)^3}{\tan \beta} \cdot \left(1 + \frac{U_a}{V_s(1 - w_x)}\right) dx \quad (80)$$

The required T_i and P_i values are determined from

$$C_{T_i} = \frac{8T_i}{\pi\rho\, V_A^2\, D^2} \quad (81)$$

$$\text{and } C_{p_i} = \frac{8P_i}{\pi\rho\, V_A^3\, D^2} \quad (82)$$

If the T_i and P_i values do not agree with the design requirements, the values for β_i are modified and the procedure is repeated. Once the correct value of tan β_i is known for each radius, it is possible to calculate the corresponding values of the lift parameter

$$\frac{C_L c}{D} \quad (83)$$

where

C_L = lift coefficient of blade section
c = length of section chord
D = propeller diameter

The values of c and C_L at any radius must be chosen so as to combine high efficiency, i.e., high lift/drag ratio, with the minimum amount of cavitation.

Many airfoil sections designed and tested by the National Advisory Committee for Aeronautics (NACA, now NASA) have characteristics suitable for marine propellers, i.e., high lift/drag ratios and chord-wise pressure distributions likely to delay surface cavitation (Abbot, et al, 1945). The sections are derived from a symmetrical streamlined shape disposed about a curved mean line.

The shapes considered by Eckhardt and Morgan were the NACA 16 and 66 sections, the latter with a thickened trailing edge suitable for construction requirements. In circumferentially varying flow, where the angle of incidence is changing, the elliptic nose of 16 is theoretically more favorable than the somewhat blunter nose of 66, but this superiority has not been confirmed by experiment. To delay cavitation at high loadings, it is desirable to have as uniform a pressure distribution as possible over the back of the section in order to avoid the occurrence of any local pressure-reduction peak. NACA has developed a number of curved mean lines, which are defined by the value of the symbol a, indicating the proportion of the chord from the leading edge over which the pressure is theoretically uniform. For $a = 1.0$, it would be uniform from leading to trailing edge. This mean line also gives the section of minimum chord to avoid incipient cavitation. With the mean line $a = 0.8$, for which the pressure distribution is uniform for 0.8 of the chord from the nose, dropping to zero at the trailing edge, the pressure reduction is slightly greater than on the $a = 1.0$ line at the same value of C_L, but the correction going from ideal to viscous flow is very close to zero.

When the type of thickness and camber lines have been chosen, the maximum thickness and the maximum camber (as yet uncorrected for lifting surface effects) is chosen by cavitation considerations. For the $a = 0.8$ mean line, the section maximum camber to chord ratio can be found from

$$f_M/c = 0.0679\ C_L \tag{84}$$

while the section ideal angle of attack in degrees of this mean line is

$$\alpha_i = 1.54\ C_L \tag{85}$$

Brockett in 1966 has prepared design charts for the NACA 66 thickness section and the $a = 0.8$ camber line, these are given in Figs. 65 and 66. These charts are based on "optimum" foil characteristics, defined as the foil allowing the greatest total angle change without occurrence of cavitation for a given σ. The procedure Brockett suggests relative to the use of these charts is as follows. In the design of cavitation-free foils, a design C_L is set, a minimum thickness from strength considerations is obtained, and a minimum operation σ is calculated. In some cases a variation in the operating angle of attack is known or can be estimated. It is now necessary to find a camber ratio, thickness ratio, and an average operating angle of attack such that the design C_L is met, the thickness is not less than the strength considerations permit, and such that $-C_{P_{min}}$ is less than σ over the range of angle-of-attack variations.

For situations when the angle-of-attack variation is

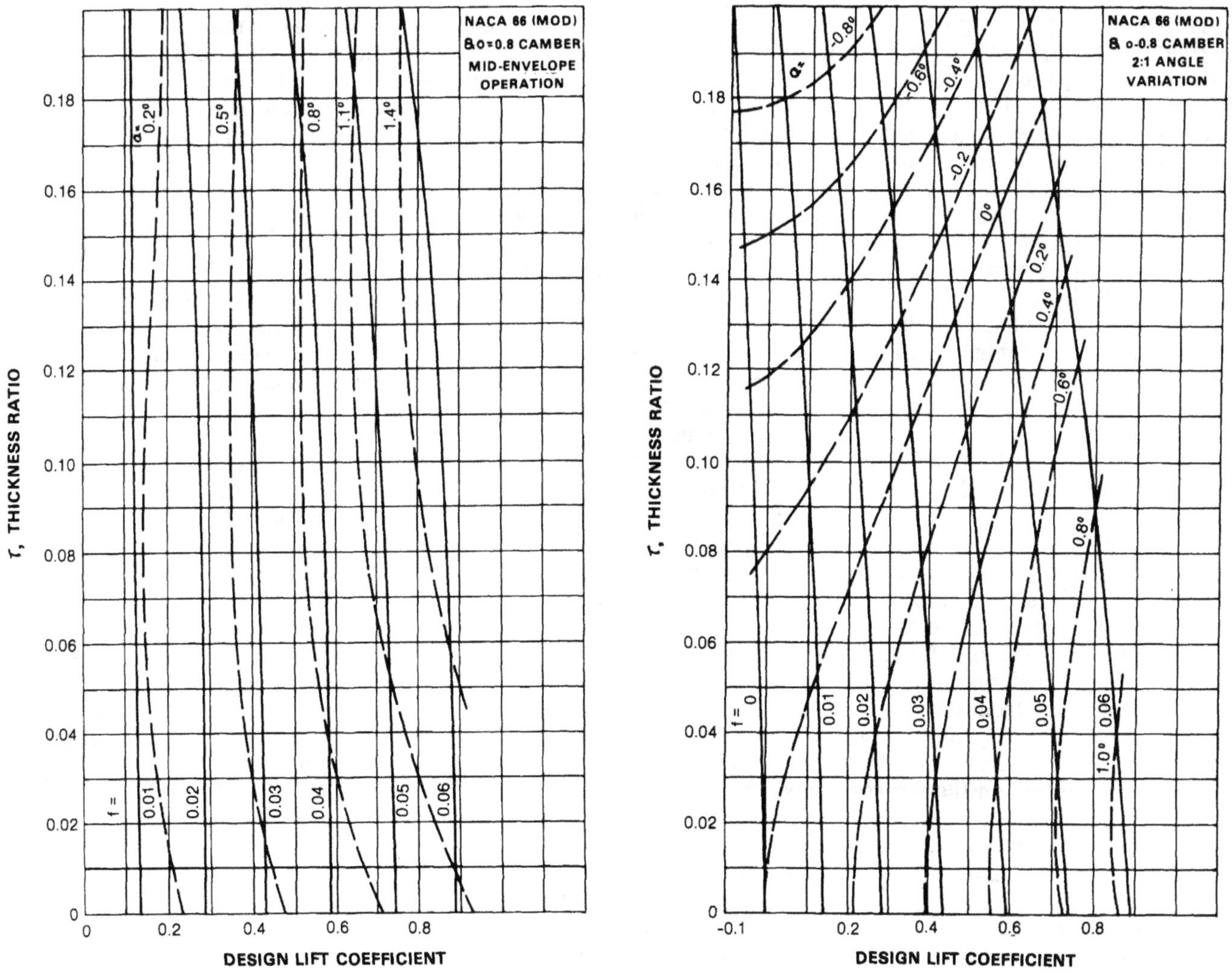

Fig. 66 Optimum characteristics for modified NACA 66 profile with $a = 0.8$ mean line. (Brockett, 1966).

not known or not critical, the following procedure is recommended: With the minimum thickness and known σ (i.e., $-C_{P_{\min}}$), enter Fig. 65 to obtain a camber ratio. Then enter Fig. 66 with a selected type of angle variation to obtain an operating incidence and C_L. In general, this C_L will not be the same as that required. Either the thickness may be increased or the chord lengthened—or both—and the process repeated until the required C_L is obtained for an "optimum" foil.

If the angle-of-attack variation is known and critical, then the known variation and known σ uniquely determine t/c and f/c from Fig. 65. Fig. 66 will give an operating incidence and C_L for the foil. Here too, it may be necessary to change the chord length to carry the necessary load, remembering that the thickness and camber ratio are fixed. If this section does not quite make the strength requirements, a judicious rereading of the charts is suggested since some latitude is permitted in the readings. For large disagreements, designing for a smaller angle variation is suggested if possible.

The above procedures are not rigid, of course, and are offered only as a guide. It is quite possible that other design approaches will be used. In some instances perhaps the camber, σ, and incidence are fixed. In this case, Fig. 65 will give an optimum thickness for the fixed σ and also the permissible angle variation. Fig. 66 will give the midpoint of the envelope. The endpoint incidences of the envelope width would be the mid-point plus or minus one-half the width. These endpoints permit a check that the operating angle of incidence is within their limits.

To illustrate and extend the remarks made in the previous paragraphs, a specific design problem will be presented. The problem is to determine a foil shape and incidence for a given C_L for the foil considered in this section (i.e. NACA 66 (mod)) and such that the minimum pressure envelopes extend approximately equal distances on both sides of the design angle of attack. For each foil, the average lift coefficient was taken to be 0.3 and σ (or $-C_{P_{\min}}$) was taken to be 0.6. For the 66 foil, Fig. 66 is entered with σ and C_L, respectively, and a common thickness and camber ratio found. This gives thickness ratio of 0.126, a camber ratio of 0.0225, and an operating incidence of 0.41 degrees. The second part of Fig. 65 gives a total permissible angle variation of 3.9 degrees.

In foil selection from a cavitation standpoint, several points are worth keeping in mind: First, for constant angle of attack in the favorable operating range (the nearly vertical line on the figures for which $-C_{P_{\min}}$ is low), the value of $-C_{P_{\min}}$ increases with both t/c and f/c. Second, the extent, with respect to α, of the favorable range increases with increasing t/c and also with increasing f/c. Third, in this favorable range $-C_{P_{\min}}$ increases more rapidly with f/c than with angle of attack for equal changes in C_L. Fourth, the thin-wing ideal angle of attack may be of use when designing cavitation-free foils to meet a given variation in angle of attack. Fifth, often it will not be possible to avoid cavitation for given σ and angle-of-attack variation. Estimations of the variation in angle of attack at a particular radius can be made from the results of a wake survey. Alternatively, use can be made of the diagrams given by Lerbs, et al (1962).

A strength calculation must now be carried out and, if necessary, the blade thicknesses or the chord lengths adjusted. In order to retain the value of $C_L \cdot c/D$ at any section, the value of C_L must be adjusted to suit the new chord length c at each section.

The camber ratios determined from cavitation charts must be corrected to take into account the curvature of the flow. This is done through use of lifting-surface theory. Lifting-surface theory is necessary to account for finite blade width effects since in lifting-line theory only three-dimensional spanwise effects are accounted for. The camber correction factor K_C accounts for the loss in lift due to the curvature of the distribution of the induced velocity along the chord. In the same way the ideal angle-of-attack correction factor K_α accounts for the change in ideal angle-of-incidence from the two-dimensional value. The angle of attack correction factor for thickness K_t accounts for the main effect of blade thickness. Morgan, et al (1968), Minsaas, et al (1971) and Cumming, et al (1972) published accurate values for these correction factors in tabulated form as a function of the number of propeller blades Z, the expanded blade area ratio A_E/A_O, the induced advance coefficient $\lambda_I = x \tan \beta_I$, the radial coordinate x and the skew angle θ_S.

Current trends in propeller design have led to complex blade shapes involving high skew and rake and extreme radial pitch distributions. Such designs require checking by means of lifting surface programs and model tests, particularly when off-design performance is important.

8.5 Service Power Allowances. After completing the final design of the propeller it is possible to confirm or modify the tentative selection of required machinery power, P_s. To do this the *service power allowance* must be added to the power required under ideal trial conditions (Section 5.2)—or the trial power must be multiplied by the corresponding factor—in order to obtain the service power needed under the influence of storm seas and wind, including the effects of fouling and corrosion. The *service power* is the total power required to enable the ship to maintain its intended sailing schedule or number of voyages per year in service.

The standard U.S. Maritime Administration service power factor for cargo ships has been 1.25 (i.e. service speed is that obtainable on trial with 1/1.25 = 80 percent of maximum designed power.). The U.S. Navy's *sustained* speed is determined in the same way. The factor for large, full super-tankers is often taken to be only 1.15.

On the other hand, fast containerships and LNG carriers on rough weather services, with rigid schedules to meet, may need a more detailed estimate of service power requirements, as noted by Giblon (1975). In such ships the attainable speed often depends on the severity of the ship motions and the resulting accelerations, shipping water or slamming, rather than on available power (Aertssen, et al, 1972). Thus it is not always possible to provide a desired rough weather speed by simply increasing power. Instead the power must be adequate to permit the ship to make up lost time when conditions lessen, if possible.

Such detailed analyses have been outlined by Lewis (1958) and given in detail by Levine, et al (1969). They involve direct estimates of the service power requirement, considering wind and sea conditions expected on the ship's intended service and all sources of added resistance (Section 6.4, Chapter V) of reduced propulsive efficiency and of environmental delays such as currents, fog, ice, etc. (It may be assumed that delays in ports and restricted waters can be allowed for in the sailing schedule.) The service power factor is then simply the ratio of total required service power to the power required on trial.

Methods of calculating added power in realistic irregular waves of different levels of severity and determining the overall power required for a particular ship in a specific service are discussed in Chapter VIII, Vol. III along with worldwide ocean wave data that have been gathered.

Section 9
Ducted Propellers

9.1 General. Although the idea of surrounding a propeller by a nozzle is very old, it was not until the early 1930's that the ducted propeller came into practical use. Luisa Stipa and later Kort (1934) experimentally showed the advantages that can be obtained by application of the nozzle. These investigations clearly showed that an increase in efficiency can be obtained when applied in the case of heavy screw loads. Primarily due to the work done by Kort, the application of ducted propellers behind certain ships (tugs, pushboats, supply vessels, trawlers) has become common practice. This may be the reason that the ducted propeller is frequently referred to as the *Kort Nozzle.*

Insight into the working principle of a ducted propeller can be gained by the application of fundamental momentum relationships. Fig. 67 shows the simplified system by which the ducted propeller can be replaced. Here the screw propeller is represented by an actuator disk rotating at infinite angular velocity. The tangential induced velocities and consequently the losses due to the rotation of the fluid are then zero. The influence of friction is neglected. With momentum theory the following expressions for the ideal efficiency η_i and the ratio between the velocity V_p at the impeller plane and the undisturbed stream velocity V_A can be derived:

$$\eta_i = \frac{2}{1 + \sqrt{1 + \tau C_T}} \tag{86}$$

$$V_P/V_A = \frac{C_r}{2[-1 + \sqrt{1 + \tau C_T}]} \tag{87}$$

where

$$C_r = \frac{T}{\frac{1}{2}\rho V_A^2 \frac{\pi}{4} D^2} \text{ and } \tau = \frac{T_P}{T}$$

T and T_p denote the total thrust and the impeller thrust respectively: D is the propeller diameter. These formulas are graphically represented in Fig. 68. From this diagram it can be seen that due to the nozzle action the inflow velocity of the impeller can be either less or greater than the inflow velocity of an open propeller under equal conditions. For a thrust ratio τ equal to 1.0, no force acts on the nozzle and the flow pattern is comparable with that of an open screw. With decreasing valves of τ, the nozzle produces a positive thrust, the inflow velocity of the impeller is increased, and an improvement in ideal efficiency is found. For thrust ratios greater than 1.0, a negative thrust or drag force acts on the nozzle, the inflow velocity of the impeller decreases and the ideal efficiency is lower.

Insight into the shape of the nozzle profile of a ducted propeller can be gained by means of Fig. 69.

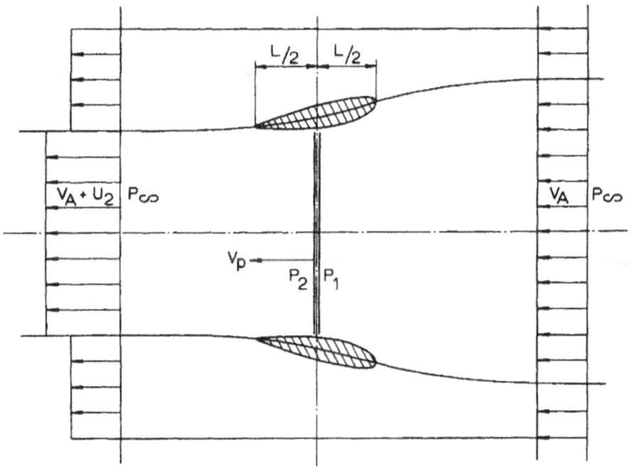

Fig. 67 Control for volume momentum considerations of ducted propeller.

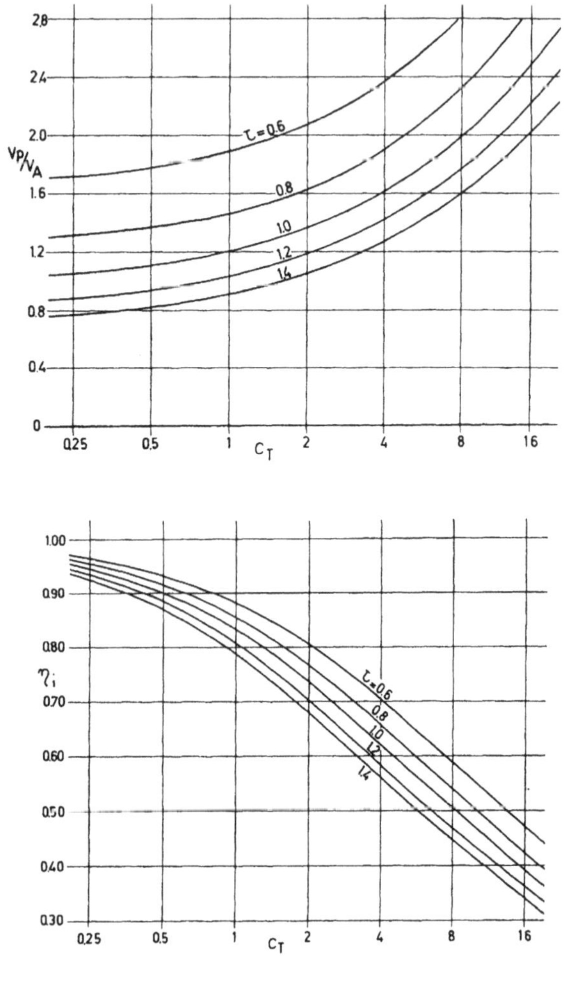

Fig. 68 Efficiency and mean axial velocity of a ducted propeller.

Here the flow through different types of ducted propellers is superimposed on the flow through an open propeller. Both the open- and the ducted propeller are designed for the same mass flow rate and velocity in the ultimate wake. Consequently the thrust and ideal efficiency of these systems are equal.

The ducted propeller with the accelerating flow type of nozzle is now used extensively in cases where the ship screw is heavily loaded or where the screw is limited in diameter. The accelerating nozzle offers a means of increasing the efficiency of heavily loaded propellers. The nozzle itself produces a positive thrust. In the case of the decelerating flow type of nozzle, the nozzle is used to increase the static pressure at the impeller. The duct will produce a negative thrust. This nozzle may be used if retardation of propeller cavitation is desired. For naval ships a reduction in noise level can be obtained, which may be of importance for tactical reasons.

Many studies on ducted propellers have been made during the last 40 years. An extensive summary of this work was made by Sacks, et al (1962). A general review of the more recent theoretical studies on ducted propellers has been given by Weissinger et al; (1968). Among the theoretical studies on ducted propellers the investigations of Horn, et al (1950), Kücheman, et al (1953) and Dickmann, et al (1955) may be mentioned in particular. Especially, the work of Dickmann and Weissinger was a first step to develop a more refined theory for ducted propellers. This paper was the basis for the work which has been performed at Karlsruhe by Dickmann, Weissinger, Wiedemer, Bollheimer, Brakhage, Maass and Rautmann. Some of the basic ideas used at Karlsruhe were also used by other investigators such as Ordway, Ritter, Greenberg, Hough, Kaskel, Lo, Sluyter, Sonnerup, Morgan, Caster, Chaplin, Voight, Nielsen, Krievel, Mendenhall, Sacks, Spangler, etc.

Most of the theoretical investigations on ducted propellers were concentrated to a large extent on linearized theory and on axisymmetrical nozzles in a uniform flow. These theories do not give data about the danger of flow separation on the nozzle. If flow separation occurs, which may happen if the nozzle is very heavily loaded, the drag of the nozzle will increase sharply. The efficiency of the system will decrease and the propeller will operate in a highly irregular flow. Hence flow separation on the nozzle surface should be avoided. For the design of a ducted propeller it is therefore necessary to have available a sound theoretical method supported by the results of carefully selected systematic experiments. A comparison of theory and experiments on ducted propellers has been made by Morgan, et al (1968).

Extensive systematic experiments on ducted pro-

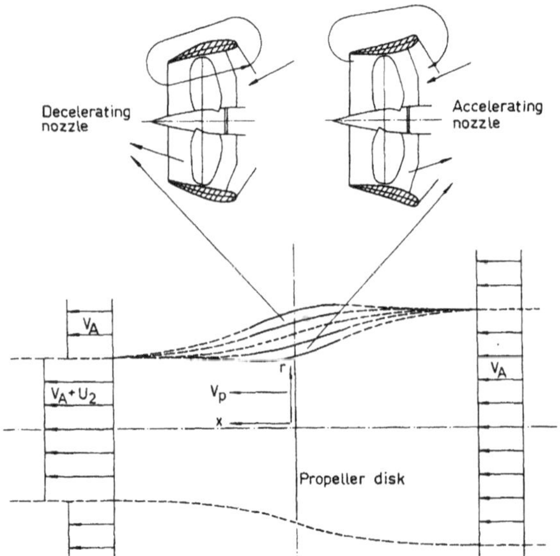

Fig. 69 Streamline forms induced by different nozzle types.

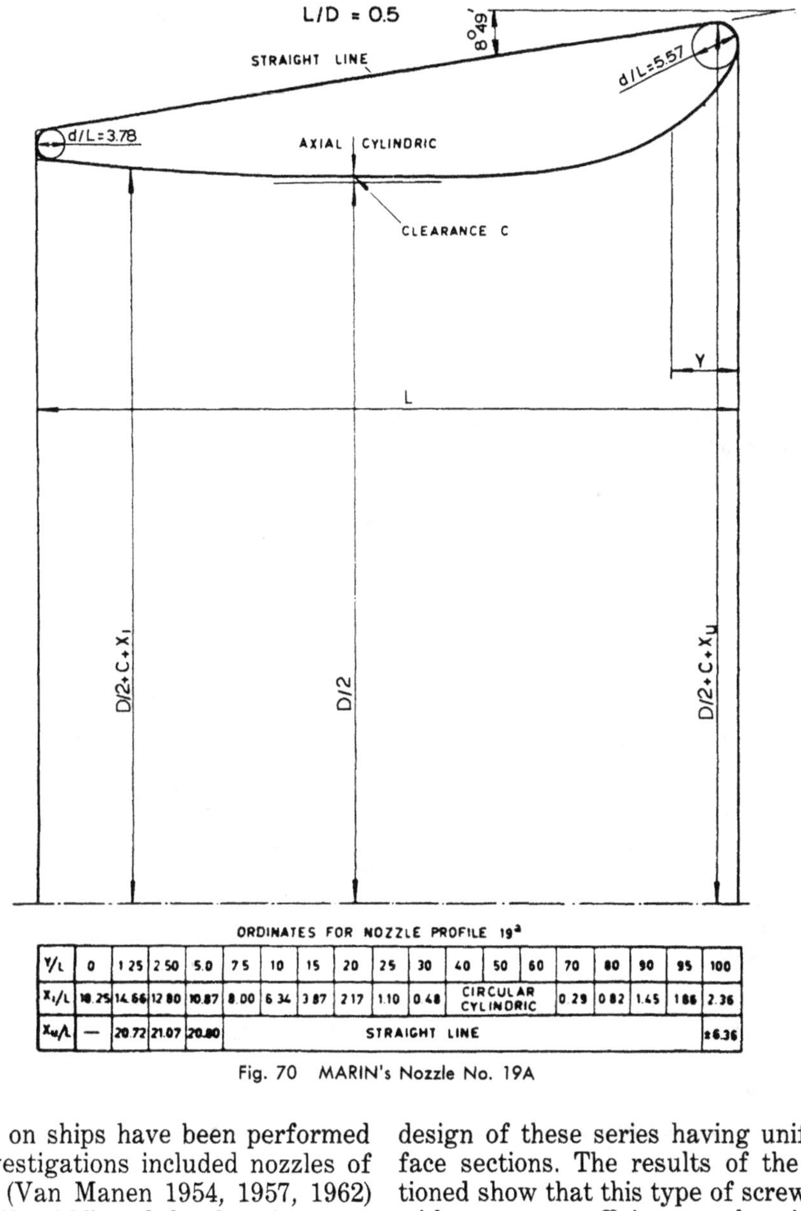

ORDINATES FOR NOZZLE PROFILE 19[a]

Y/L	0	1.25	2.50	5.0	7.5	10	15	20	25	30	40	50	60	70	80	90	95	100
X_l/L	18.25	14.66	12.80	10.87	8.00	6.34	3.87	2.17	1.10	0.48	CIRCULAR CYLINDRIC			0.29	0.82	1.45	1.86	2.36
X_u/L	—	20.72	21.07	20.80	STRAIGHT LINE													±6.36

Fig. 70 MARIN's Nozzle No. 19A

pellers for application on ships have been performed at MARIN. These investigations included nozzles of both the accelerating (Van Manen 1954, 1957, 1962) (Van Manen, et al 1959, 1966) and decelerating type (Oosterveld 1968, 1970).

9.2 Accelerating nozzles. The investigations on accelerating nozzles have led to the development of a nozzle designated MARIN 19A for application in the case of heavy screw loads. This nozzle has, from the structural point of view, a simple shape. The inner side of the nozzle at the location of the screw has an axial cylindrical form. The outside of the nozzle profile is straight and the trailing edge of the nozzle is relatively thick. The profile of Nozzle No. 19A is shown in Fig. 70.

For use in Nozzle No. 19A, special screw series (the so-called Ka-screw series) were designed. Screws of these series have relatively wide blade tips which make them less susceptable to blade tip cavitation Extensive investigations performed at MARIN have led to the design of these series having uniform pitch and flat face sections. The results of the experiments mentioned show that this type of screw has no drawbacks with respect to efficiency and cavitation. The particulars of these screw models are given in Table 20 and Fig. 71. Nozzle No. 19A has a length-diameter ratio L/D equal to 0.5. For application on pushboats and tugs, nozzles with larger length-diameter ratios may be attractive. Therefore two other nozzles were designed of which the basic form is equal to the shape of Nozzle No. 19A, possessing length-diameter ratios L/D of 0.8 and 1.0. These nozzles were designated as No. 22 and 24. The backing characteristics of these nozzle types are rather poor. For towing vessels (especially pushboats), the thrust which can be developed at bollard pull condition either with the propeller running ahead or astern is of the utmost importance. In such cases it is attractive to use a nozzle with a relative thick trailing edge. Therefore, a new type of nozzle, especially suited for astern operation was developed.

Table 20—Characteristics of the Ka-screw Series

Dimensions	r/R	0.2	0.3	0.4	0.5	0.6	0.7	0.8	0.9	1.0	
Length of the blade sections in percentages of the maximum length of the blade section at 0.6 R	from center lane to trailing edge	30.21	36.17	41.45	45.99	49.87	52.93	55.04	56.33	56.44	Length of blade section at 0.6 R
	from center line to leading edge	36.94	40.42	43.74	47.02	50.13	52.93	55.04	56.33	56.44	$1.969 \frac{1}{Z} \cdot \frac{A_E}{A_O}$
	total length	67.15	76.59	85.19	93.01	100.00	105.86	110.08	112.66	122.88	
Max. blade thickness in percentage of the diam.		4.00	3.52	3.00	2.45	1.90	1.38	0.92	0.61	0.50	Maximum thickness at center of shaft = 0.049 D
Distance of maximum thickness from leading edge in percentages of the length of the sections		34.98	39.76	46.02	49.13	49.98	—	—	—	—	

Ordinates

Distance of the ordinates from the maximum thickness

	Trailing edge					Leading edge						
r/R, percent	100	30	60	40	20	20	40	60	80	90	95	100
			Ordinates for the back									
0.2	—	38.23	63.65	82.40	95.00	97.92	90.83	77.19	55.00	38.75	27.40	—
0.3	—	39.05	66.63	84.14	95.86	97.63	90.06	75.62	53.02	37.87	27.57	—
0.4	—	40.56	66.94	85.69	96.25	97.22	88.89	73.61	50.00	34.72	25.83	—
0.5	—	41.77	68.59	86.42	96.60	96.77	87.10	70.46	45.84	30.22	22.24	—
0.6	—	43.58	68.26	85.89	96.47	96.47	85.89	68.26	43.58	28.59	20.44	—
0.7	—	45.31	69.24	86.33	96.58	96.58	86.33	69.24	45.31	30.79	22.88	—
0.8	—	48.16	70.84	87.04	96.76	96.76	87.04	70.84	48.16	34.39	26.90	—
0.9	—	51.75	72.94	88.09	97.17	97.17	88.09	72.94	51.75	38.87	31.87	—
1.0	—	52.00	73.00	88.00	97.00	97.00	88.00	73.00	52.00	39.25	32.31	—
			Ordinates for the face									
0.2	20.21	7.29	1.77	0.1	—	0.21	1.46	4.37	10.52	16.04	20.62	33.33
0.3	13.85	4.62	1.07	—	—	0.12	0.83	2.72	6.15	8.28	10.30	21.18
0.4	9.17	2.36	0.56	—	—	—	0.42	1.39	2.92	3.89	4.44	13.47
0.5	6.62	0.68	0.17	—	—	—	0.17	0.51	1.02	1.36	1.53	7.81

Note: The percentages of the ordinates relate to the maximum thickness of the corresponding section.

The profile of this nozzle is given in Fig. 72. In comparison with Nozzle No. 19A, this nozzle (designated as Nozzle No. 37) has a well-rounded and relatively thick trailing-edge. This prevents flow separation in reversed condition. Open-water tests were performed with all these nozzles in combination with the Ka 4-70 screw series. The fairing of the open-water test results was performed by means of regression analysis and the polynomials for the Ka-4-70 propeller in the Nozzles 19A and 37 are given in Tables 21 and 22. The coefficients for other propeller-nozzle combinations may be found in Van Gent, et al (1983). The coefficients given in these tables are those of the following polynomials:

$$K_T = \sum_{x,y} C_T(x,y)\,(P/D)^x\,(J)^y$$

$$K_Q = \sum_{x,y} C_Q(x,y)\,(P/D)^x(J)^y$$

$$K_{TN} = \sum_{x,y} C_{TN}(x,y)\,(P/D)^x(J)^y$$

where

K_T is total thrust coefficient
K_{TN} is nozzle thrust coefficient
K_Q is torque coefficient

Important factors in the selection of a nozzle propeller for tugs, pushboats etc. are:

- The forward static bollard pull,
- the astern static bollard pull,
- the free-running speed.

A comparison between the forward static bollard pull of Nozzles nos. 24, 22, 19A and 37 and a conven-

Table 21—Coefficients for Ka 4-70 in 19A-Nozzle

PROPELLER		Ka 4-70	NOZZLE 19A	
x	y	C_T	C_{TN}	C_Q
0	0	.030550	.076594	.006735
	1	−.148687	.075223	0.0
	2	0.0	−.061881	−.016306
	3	−.391137	−.138094	0.0
	4	0.0	0.0	−.007244
	5	0.0	−.370620	0.0
	6	0.0	.323447	0.0
1	0	0.0	−.271337	0.0
	1	−.432612	−.687921	0.0
	2	0.0	.225189	−.024012
	3	0.0	0.0	0.0
	4	0.0	0.0	0
	5	0.0	0.0	0.0
	6	0.0	−.081101	0.0
2	0	.667657	.666028	0.0
	1	0.0	0.0	0.0
	2	.285076	.734285	.005193
	3	0.0	0.0	0.0
	4	0.0	0.0	0.0
	5	0.0	0.0	0.0
	6	0.0	0.0	0.0
3	0	−.172529	−.202467	.046605
	1	0.0	0.0	0.0
	2	0.0	−.542490	0.0
	3	0.0	0.0	0.0
	4	0.0	0.0	0.0
	5	0.0	0.0	0.0
	6	0.0	−.016149	0.0
4	0	0.0	0.0	−.007366
	1	0.0	0.0	0.0
	2	0.0	0.0	0.0
	3	0.0	.099819	0.0
	4	0.0	0.0	0.0
	5	0.0	0.0	0.0
	6	0.0	0.0	0.0
5	0	0.0	0.0	0.0
	1	0.0	.030084	0.0
	2	0.0	0.0	0.0
	3	0.0	0.0	0.0
	4	0.0	0.0	0.0
	5	0.0	0.0	0.0
	6	0.0	0.0	0.0
6	0	0.0	0.0	−.001730
	1	−.017293	0.0	−.000337
	2	0.0	−.001876	.000861
	3	0.0	0.0	0.0
	4	0.0	0.0	0.0
	5	0.0	0.0	0.0
	6	0.0	0.0	0.0
0	7	0.0	0.0	0.0

Table 22—Coefficients for Ka 4-70 in 37-Nozzle

PROPELLER		Ka 4-70	NOZZLE 37	
x	y	C_T	C_{TN}	C_Q
0	0	−.162557	−.016806	.016729
	1	0.0	0.0	0.0
	2	0.0	0.0	0.0
	3	0.0	0.0	0.0
	4	−.077387	0.0	0.0
	5	0.0	0.0	0.0
	6	0.0	−.099544	.030559
1	0	.598107	0.0	−.048424
	1	−1.009030	−.548253	−.011118
	2	0.0	.230675	−.056199
	3	0.0	0.0	0.0
	4	0.0	0.0	0.0
	5	0.0	0.0	0.0
	6	0.0	0.0	0.0
2	0	.085087	.460206	.084376
	1	.425585	0.0	0.0
	2	0.0	0.0	.045637
	3	0.0	0.0	−.042003
	4	0.0	0.0	0.0
	5	0.0	0.0	0.0
	6	0.0	0.0	0.0
3	0	0.0	−.215246	−.008652
	1	0.0	0.0	0.0
	2	0.0	0.0	0.0
	3	0.0	0.0	0.0
	4	0.0	0.0	0.0
	5	−.021044	0.0	0.0
	6	0.0	0.0	0.0
4	0	0.042997	0.0	
	1	0.0	0.0	0.0
	2	0.0	0.0	0.0
	3	0.0	0.0	0.0
	4	0.0	0.0	0.0
	5	0.0	0.0	0.0
	6	0.0	0.0	0.0
5	0	0.0	0.0	0.0
	1	−.038383	0.0	0.0
	2	0.0	0.0	0.0
	3	0.0	0.0	0.0
	4	0.0	0.0	0.0
	5	0.0	0.0	0.0
	6	0.0	0.0	0.0
6	0	0.0	0.0	0.0
	1	0.0	0.0	−.001176
	2	.014992	0.0	.002441
	3	0.0	0.0	0.0
	4	0.0	0.0	0.0
	5	0.0	0.0	0.0
	6	0.0	0.0	0.0
0	7	0.036998	.051753	−.012160

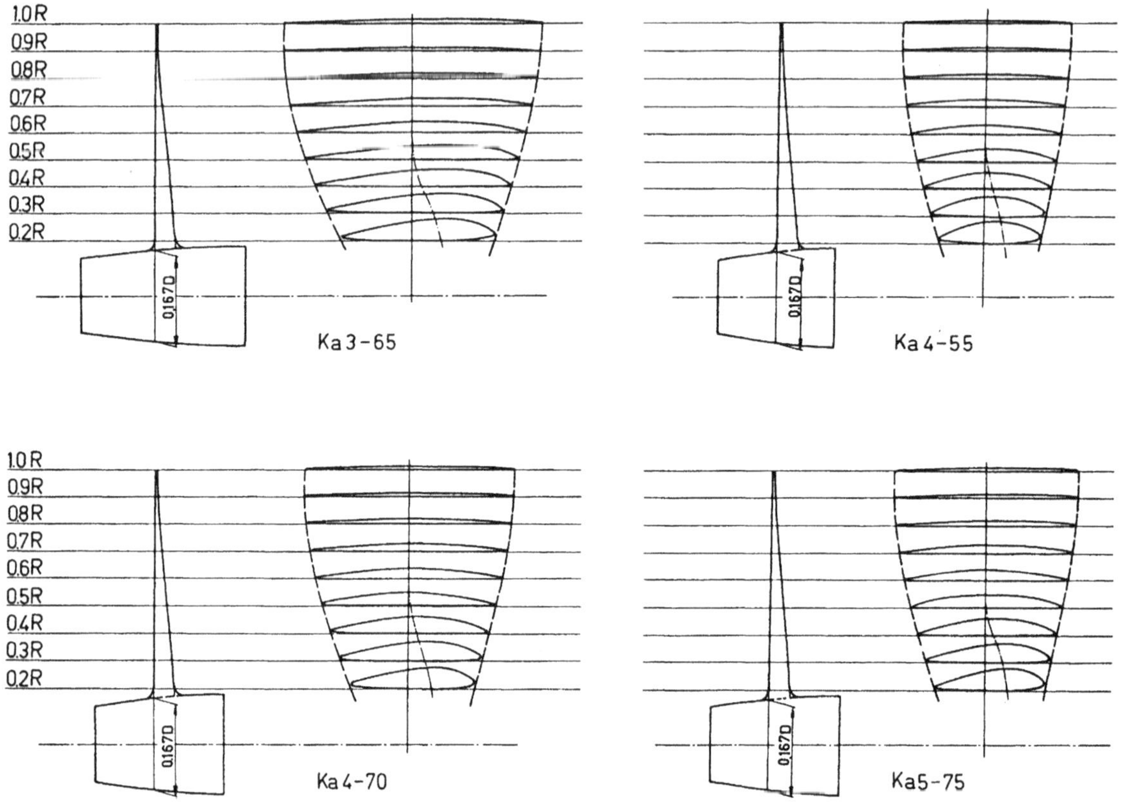

Fig. 71 Blade planform of the Ka-series propellers.

tional screw series (the B4-70 series) can be made with the aid of Fig. 73. In this diagram the thrust coefficient K_T, the torque coefficient K_Q, the impeller thrust-total thrust ratio τ and the efficiency coefficient η_d of the different propulsion devices are given on a base of the pitch ratio P/D. The efficiency coefficient η_d is defined as:

$$\eta_d = \frac{(K_T/\pi)^{3/2}}{K_Q}$$

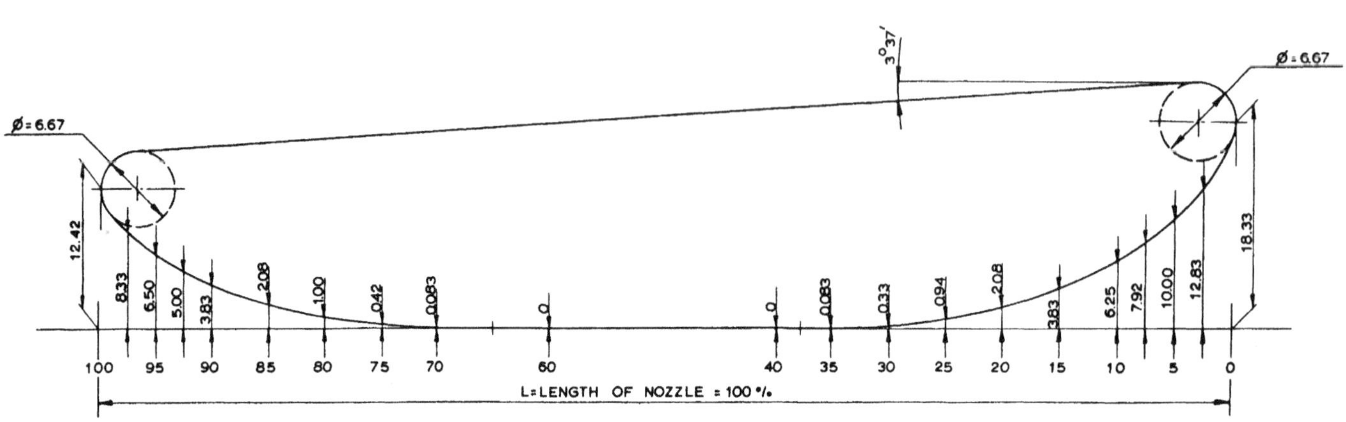

Fig. 72 MARIN's Nozzle No. 37, L/D = 0.5

This efficiency can be used as a direct measure for the effectiveness of different propulsion devices at the static condition, if systems with the same diameter and power are considered. However, this coefficient must not be used if there are restrictions in choosing the RPM of the different propulsion devices.

From Fig. 73 it can be seen that for ducted propellers the efficiency coefficient η_d is much higher than for conventional screws. Further, it can be seen that the effect of nozzle length on η_d is small. With increasing length-diameter ratio of the system a slight increase in the efficiency factor η_d has been found.

A comparison between the astern static bollard pull of Nozzles Nos. 19A and 37 and the B4-70 screw series can be made with the aid of Fig. 74. From this diagram it can be seen that the efficiency factor η_d for Nozzle No. 37 is much higher than for Nozzle No. 19A. This can be explained by the fact that Nozzle No. 19A suffers from flow separation when operating astern. The efficiency factor η_d of Nozzle No. 19A is still higher than the value of this factor for the B4-70 screw series.

For the method of selecting the propulsion device based on the free-running speed of the vessel, more practical information can be derived from the $K_T = K_Q$-J diagram. The most widely encountered design problem for the screws of cargo ships is that where the speed of advance of the screw V_A, the power to be absorbed by the screw P and the number of revolutions n are given. The diameter D is to be chosen such that the greatest efficiency can be obtained. Push-boats often operate in restricted water. This means that the ship must be limited in draft and have small propellers or tunnel sterns. In such cases it is more convenient to start the propeller design from given V_A, T and D or V_A, P_D and D. The problem of determining the optimum diameter or the optimum number of revolutions can be solved by plotting η_d and J as functions of:

K_Q/J^5 if V_A, P_D and n are given,
K_T/J^4 if V_A, T and n are given,
K_Q/J^3 if V_A, P_D and D are given,
K_T/J^2 if V_A, T and D are given.

For the Ka4-70 screw series in Nozzles Nos. 24, 22, 19A and 37, the curves for optimum diameter (on base of $(K_\Omega/J^5)^{1/2}$ and $(K_T/J^4)^{1/2}$) and for optimum number of revolutions (on base of (K_T/J^2) and (K_Ω/J^3)) are given in Figs. 75-78.

On comparing the characteristics of ducted propellers with the B4-70 screw series, it may be concluded that ducted propellers give a higher efficiency than conventional screws at larger screw loads. The differences between the different ducted propellers are small. Nozzles Nos. 19A, 22 and 24 have about a 2 percent higher efficiency than Nozzle No. 37. Nozzle No. 37 in comparison with Nozzles Nos. 19A, 22 and

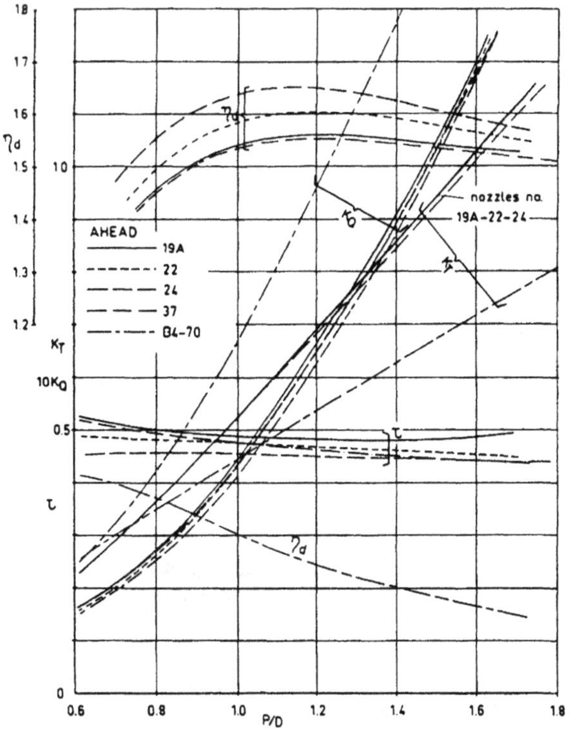

Fig. 73 Characteristics of different propulsion devices at forward static bollard condition.

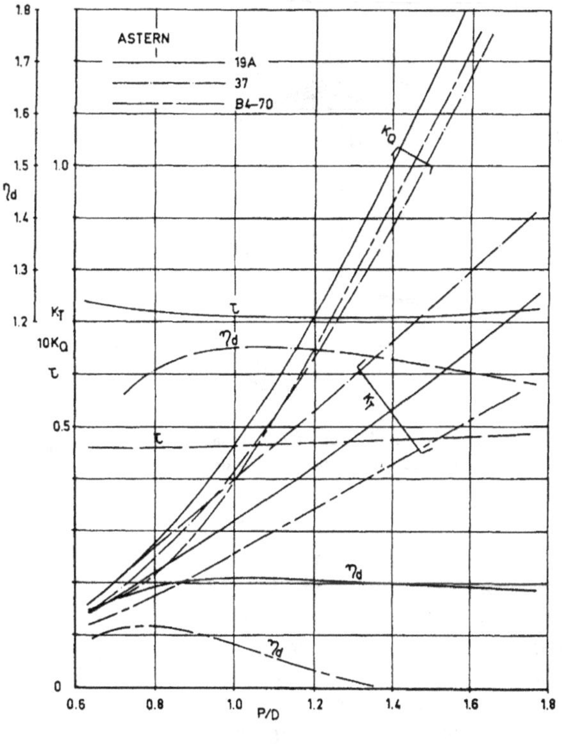

Fig. 74 Characteristics of different propulsion devices at astern static bollard condition.

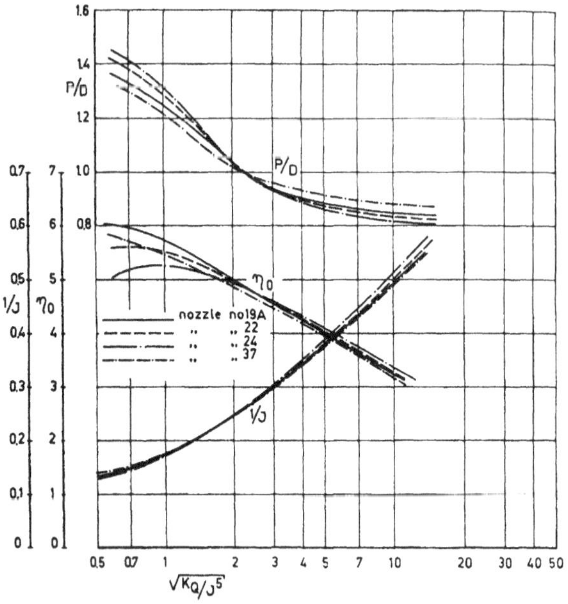

Fig. 75 Curves for the determination of the optimum diameter of various ducted propellers if V_A, P_D and n are given.

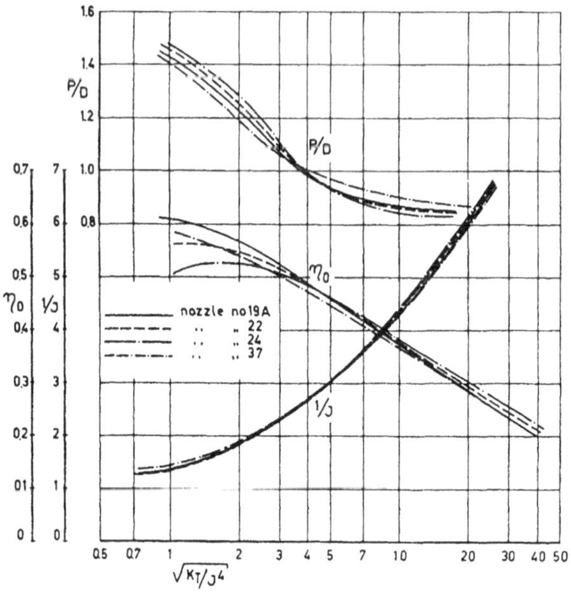

Fig. 76 Curves for the determination of the optimum diameter of various ducted propellers if V_A, T and n are given.

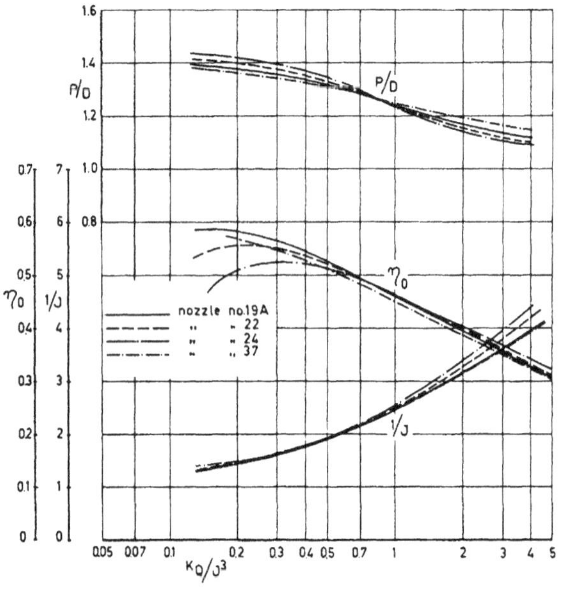

Fig. 77 Curves for the determination of the optimum number of revolutions of various ducted propellers if V_A, P_D and D are given.

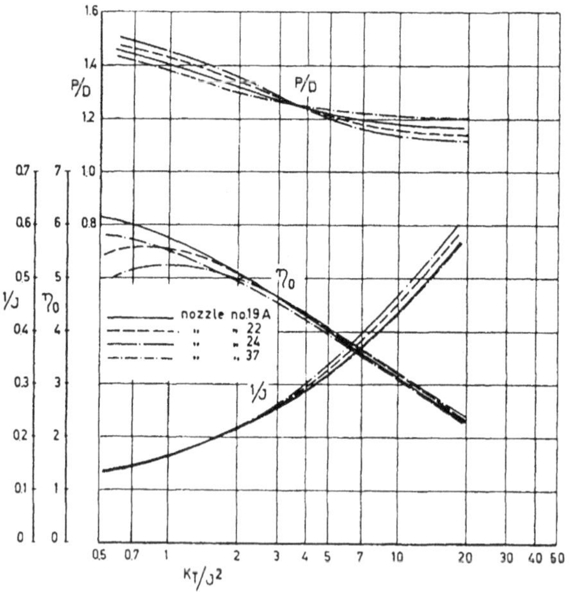

Fig. 78 Curves for the determination of the optimum number of revolutions of various ducted propellers if V_A, T and D are given.

24 presents, however, a very attractive solution when both ahead astern operation are of interest.

From these results it can be concluded that ducted propellers are to be recommended for heavy screw loads such as occur on towing vessels, trawlers etc. In addition, the ducted propeller lends itself very well to restricted water applications as the nozzle protects the propeller blades from striking fixed objects such as the bottom or banks. Furthermore the nozzle may protect the propeller from damage by ice.

Finally, it must be noted that the flow velocity at the impeller disk of a ducted propeller is far less sensitive to variations in ship speed than in the case of a conventional screw propeller. Consequently the power absorption of a ducted propeller is relatively less sensitive to variations in ship speed. This in turn leads to smaller variations in the required number of revolutions at the same maximum engine torque. These fea-

Fig. 79 Photo model pushboat showing the ducted propeller configuration.

tures have also been important in the choice of ducted propellers behind tugs, pushboats and trawlers. All these ships must operate satisfactorily at different loadings (towing and free-running) of the screw.

For single screw tugs spending an appreciable percentage of their service life to maneuvering, the ducted propeller configuration must include backing rudders forward of the propeller besides the traditional steering rudder located aft of the propeller. Without backing or flanking rudders a fixed nozzle does not provide directional control when backing. A twin-screw tug has satisfactory maneuvering qualities.

River pushboats can not operate safely without flanking rudders. Even on multi-screw pushboats, flanking rudders are to be used to provide satisfactory directional control when backing. In Fig. 79 a typical ducted propeller configuration for a pushboat is shown. In this figure the flanking rudders forward of the propellers can be seen.

Investigations have also been carried out to determine the effect on performance of various ducted propeller-rudder systems when applied to pushboats and tugs. For instance the effect of the flanking rudder angle θ on performance was investigated for a ducted propeller arrangement as shown in Fig. 80. The results of these tests are given in Fig. 81. From this diagram it can be seen that the flanking rudder angle has a marked effect on the characteristics of the system. With increased loading of the ducted propeller system a larger flanking rudder θ becomes attractive. This is due to the relatively larger slipstream contraction at larger screw loads. A flanking rudder angle θ of 15 to 20 degrees gives the best characteristics.

9.3 Decelerating nozzles. As already mentioned, the application of the flow decelerating nozzle may be attractive if retardation of propeller cavitation phenomena is desired. The reduction of the flow rate inside the decelerating type of nozzle results in an increase of the static pressure at the impeller. However, the duct itself will produce a negative thrust ($\tau > 1$). In order to compensate for this thrust loss (induced nozzle drag), the impeller loading must be increased. An improvement of the cavitation properties of the impeller will therefore only be obtained if the gain in static pressure at least compensates the unfavorable effect of the increased screw loading. The result of an analysis of the minimum pressures which may occur at the blades of a ducted propeller is given in Fig. 82. This diagram shows that improved cavitation properties for the particular screw considered (blade-area ratio $A_E/A_O = 1.0$ and number of blades $z = 5$) only occur for low values of the thrust coefficient C_T. If ducted propellers with larger blade-area ratios of the impeller or with more rotor- (and eventually more stator-) rows are considered, the decelerating nozzle may favorably affect the cavitation properties of the screw for larger values of C_T.

Application of the decelerating nozzle results, furthermore, in a reduction of the pressure at the exterior surface of the nozzle. From a comparison between the minimum pressures which occur at the exterior surface of the nozzle and at the impeller blades, it can be concluded that except in the case of very short nozzles or very low loaded systems, impeller cavitation is more critical. Systematic series of model tests with decelerating ducted propellers have also been performed. The design of the nozzles was based on the vortex theory as described in Oosterveld (1970). The variation of the design parameters considered is shown in Table 23. The experiments with these nozzles were all carried out with a series of five bladed Kaplan type screws (Kd 5-100 series). The Kd 5-100 series screws were designed in combination with MARIN Nozzle No. 33. The pitch distribution of the screws depend on the velocities induced by the nozzle at the impeller plane and on the radial load distribution. For the design, use was made of the method described by Van Manen in 1966. Particulars of the screw models are given in Table 24 and in Fig. 83. The screws were located in the nozzle with a uniform tip clearance of 1 mm (1/8 in.). The fairing of the open-water test results was performed by means of regression analysis. The poly-

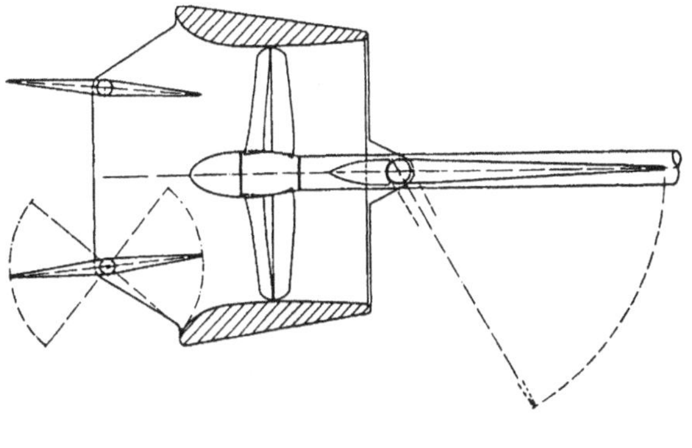

Fig. 80 Ducted propeller configuration with rudder arrangement for a pushboat.

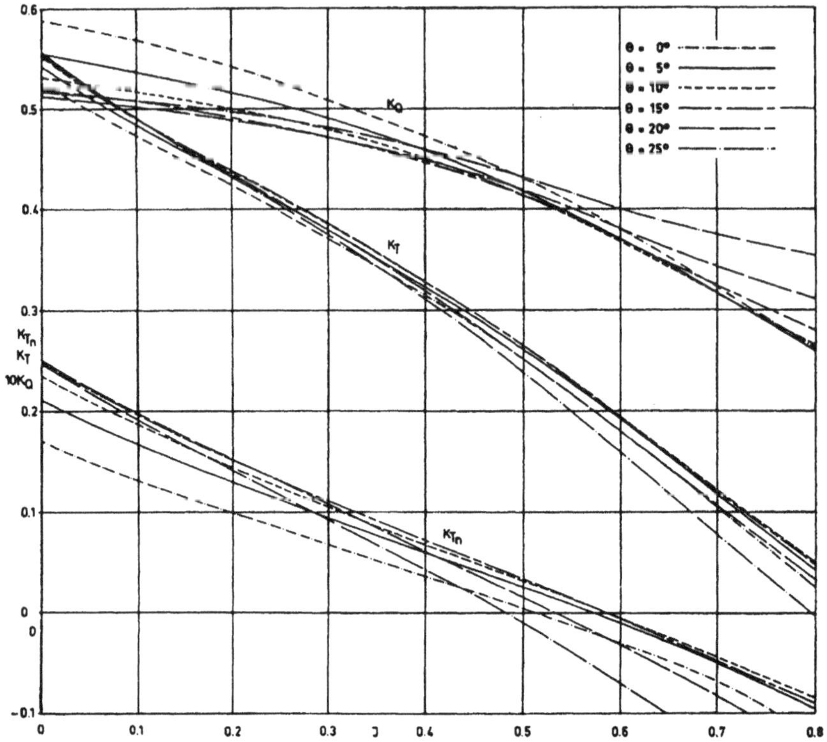

Fig. 81 Effect of flanking rudder angle θ on open-water characteristics of ducted propeller configuration for a pushboat.

nomials together with the coefficients are given in Table 25.

9.4 Further Remarks on Accelerating and Decelerating Nozzles. From all the model tests performed with accelerating and decelerating nozzles at MARIN it can be deduced that the relation between the thrust coefficient C_T and the thrust ratio τ of a ducted propeller system is approximately independent of the pitch ratio P/D of the impeller. Furthermore, it was found that for the considered nozzles there exists a fixed relation between the impeller disk area-nozzle exit area ratio A_O/A_{EX}. This result is shown in Fig. 84. In Fig. 85 the optimum efficiency which can be obtained with the different ducted propeller systems is also given. Furthermore it can be seen that with decreasing value

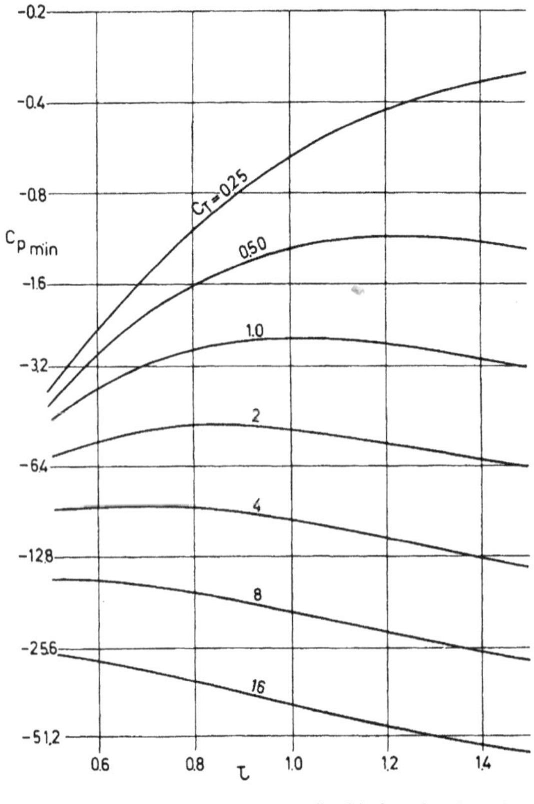

Fig. 82 Minimum static pressure at impeller blades of a ducted propeller.

Table 23—Design Parameters of Decelerating Nozzle Series

Nozzle number	C_{T_0}	r_o	L/D	a/L	S/L	d/D
30	0.95	1.00	0.6	0.5	0.15	0.20
31	0.95	1.15	0.6	0.5	0.15	0.20
32	0.95	1.30	0.6	0.5	0.15	0.20
33	1.0	1.2	0.6	0.5	0.15	0.20
34	1.0	1.2	0.6	0.5	0.09	0.20
35	1.0	1.2	0.9	0.5	0.10	0.20
36	1.0	1.2	1.2	0.5	0.075	0.20

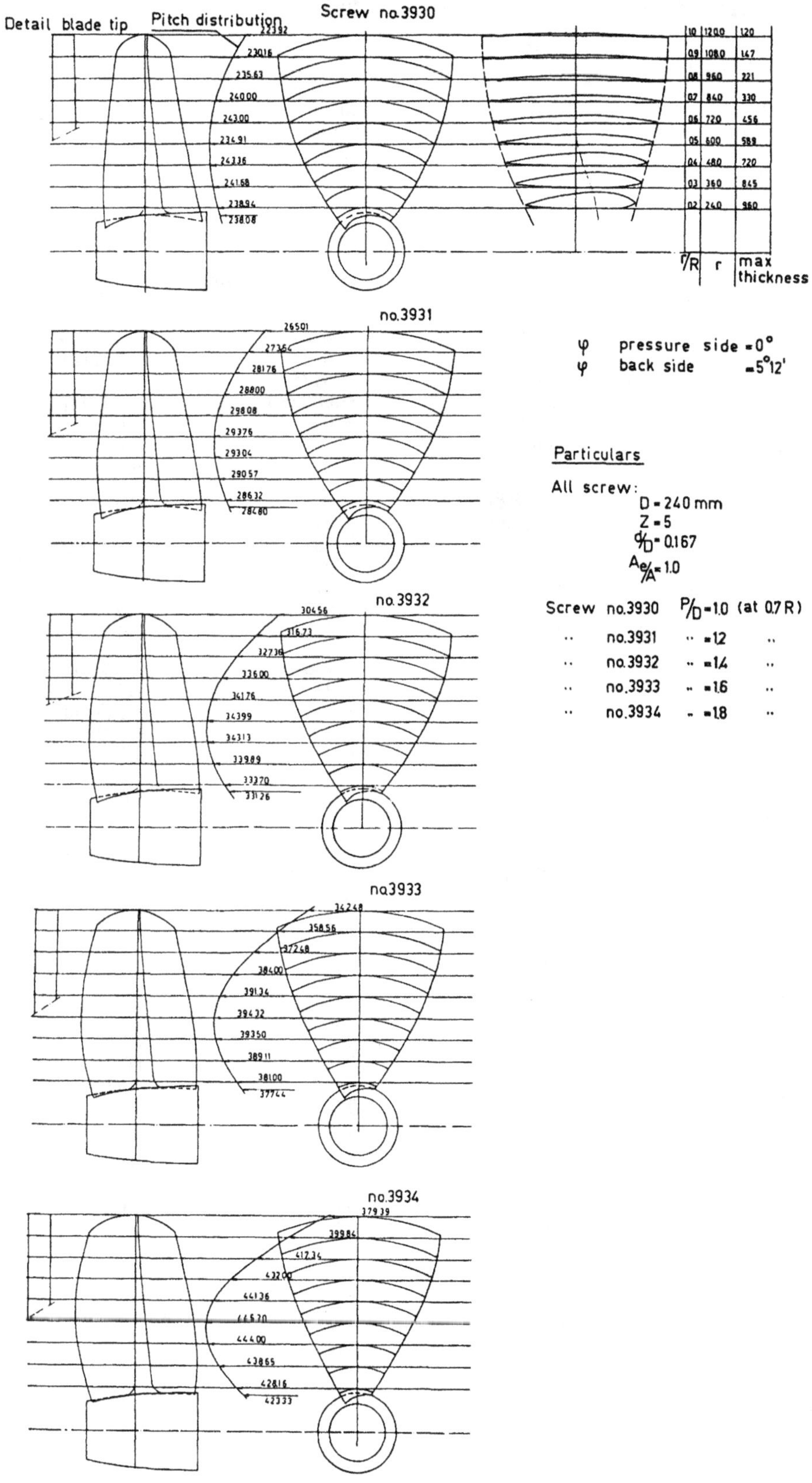

Fig. 83 Particulars of Kd 5-100 model propellers.

Table 24—Particulars of Screw Models of the Kd 5-100 Series

Diameter	D	240 mm (9.5 in.)
Number of blades	Z	5
Pitch ratio (at 0.7 D)	P/D	1.0−1.2−1.4−1.6−1.8
Blade area Ratio	A_E/A_O	1.00
Blade outline		Kaplan type
Blade section		NASA 16-parabolic camberline
Hub diameter rates	d/D	0.167

Table 25—Coefficients for Kd 5-100 in 33-Nozzle

PROPELLER		Kd 5-100	NOZZLE 33	
x	y	C_T	C_{TN}	C_Q
0	0	−.347562	.025149	−.007789
	1	−.321224	0.0	−.022424
	2	.075277	.317808	0.0
	3	0.0	−.083296	−.009087
	4	−.009560	−.070735	0.0
	5	0.0	.050083	0.0
	6	0.0	0.0	0.0
1	0	.963261	0.0	0.0
	1	−.215803	−.371072	0.0
	2	0.0	−.561715	−.010492
	3	0.0	.921327	0.0
	4	0.0	−.410495	0.0
	5	0.0	.067465	0.0
	6	0.0	0.0	0.0
2	0	0.0	.138501	.082463
	1	0.0	0.0	0.0
	2	0.0	−.315179	.026193
	3	0.0	0.0	−.009585
	4	0.0	0.0	0.0
	5	0.0	0.0	.001029
	6	0.0	0.0	0.0
3	0	0.0	0.0	0.0
	1	0.0	.235429	−.007692
	2	.013401	.077988	0.0
	3	0.0	0.0	0.0
	4	0.0	0.0	0.0
	5	0.0	0.0	0.0
	6	0.0	0.0	−.000094
4	0	−.016882	−.015350	−.003196
	1	0.0	−.073049	0.0
	2	0.0	0.0	0.0
	3	0.0	0.0	0.0
	4	0.0	0.0	−.000117
	5	0.0	0.0	0.0
	6	0.0	0.0	0.0
5	0	0.0	0.0	0.0
	1	0.0	0.0	0.0
	2	0.0	0.0	0.0
	3	0.0	0.0	0.0
	4	0.0	0.0	0.0
	5	0.0	0.0	0.0
	6	0.0	0.0	0.0
6	0	0.0	0.0	0.0
	1	0.0	0.0	0.0
	2	0.0	0.0	0.0
	3	0.0	0.0	.000152
	4	0.0	0.0	0.0
	5	0.0	0.0	0.0
	6	0.0	0.0	0.0
0	7	0.0	−.003473	0.0

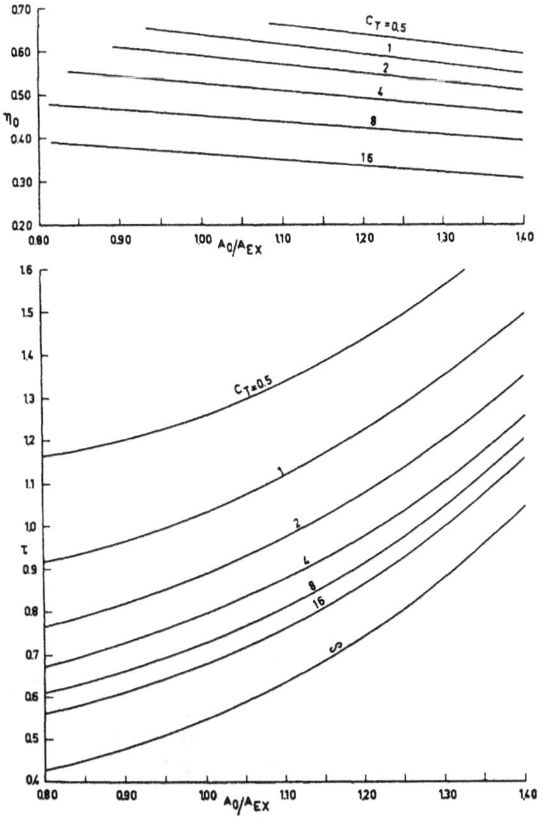

Fig. 84 Relations between impeller disc area-nozzle exit area ratio, thrust coefficient C_T and efficiency η_o of a ducted propeller system.

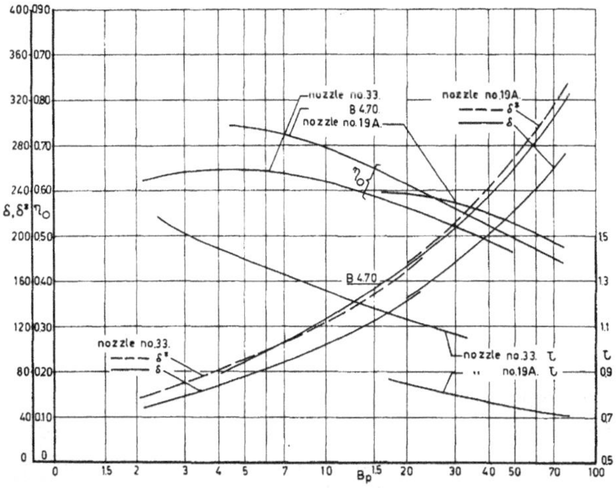

Fig. 85 Optimum relationship between η_o, δ and βp of the Ka 4-70 screw series in Nozzle no. 19A, the B4-70 screw series and the Kd 5-100 screw series in Nozzle no. 33.

Table 26—Typical B_p and C_T Values for Different Ship Types.

	B_p	C_T
Torpedo's	<10	<0.5
Twin-screw ships	10–15	0.5–1.0
Fast warships (frigates, destroyers)	10–25	0.5–1.5
Single screw cargo ships	15–35	1.0–2.5
Coasters	35–60	2.5–4.0
Tankers	35–70	2.5–5.0
Trawlers	60–100	4.0–8.0
Towing vessels (tugs, pushboats)	>80	>6.0

of the area ratio A_O/A_{EX}, the thrust ratio τ decreases and the efficiency η_o of the ducted propeller system increases. Thus with increasing positive loading of the nozzle, the efficiency of the ducted propeller system increases. The loading of the nozzle, however, is limited by the risk of flow separation on the nozzle.

Optimum curves for open-water efficiency η_O, diameter coefficient δ and thrust ratio τ of the Ka 4-70 screw series in Nozzle No. 19A, the Kd 5-100 screw series in Nozzle No. 33 and the B 4-70 screw series are presented in Fig. 85. Typical B_P values for different ship types are indicated in Table 26. The lightly loaded screws of fast ships are on the left side of Fig. 85, while the heavily loaded propellers of towing vessels are on the right. It can be seen from Fig. 85 that the accelerating nozzle (Nozzle No. 19A), when compared with a conventional screw (B 4-70 series), gives rise to an improvement in open-water efficiency η_o in the case of heavy screw loads. The decelerating nozzle (Nozzle No. 33) has a low open-water efficiency η_o. The curves of the diameter coefficient δ of the accelerating and the decelerating nozzle almost coincide; the B 4-70 series has a larger optimum screw diameter. It is interesting to note that the curves for the diameter coefficient δ^* based on the maximum diameter of the system of both the accelerating and the decelerating nozzle and the B 4-70 screw series almost coincide.

Section 10
Other Propulsion Devices

10.1 General. Although marine screw propellers are used to propel the vast majority of ships, there are other propulsion devices which have advantages in special circumstances. It is not possible to deal with these in full detail here, but a review of some of their merits and demerits may be useful. Further information on the various devices can be found in the references cited in the various paragraphs.

10.2 Jet Propulsion. This is the oldest known type of mechanical propulsion for ships. Patents were

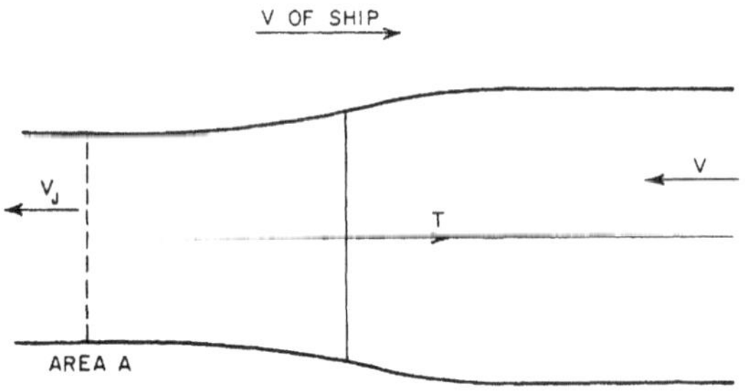

Fig. 86 Jet propulsion

granted to Toogood and Hayes in England in 1661, it was reproposed by Benjamin Franklin in 1775 and actually applied by James Rumsey in 1782 to propel a 24 m (80 ft) passenger boat on the Potomac between Washington and Alexandria.

The method usually consists of an impeller or pump inside the hull, which draws water from outside, imposes on it an acceleration, and discharges it astern as a jet at a higher velocity. It is therefore a reaction device exactly like the ordinary propeller, but in which the moving parts are contained inside the hull.

Many claims have been made for jet propulsion, but these cannot be substantiated on grounds of efficiency. Referring to Fig. 86, and for an ideal fluid:

Mass of water discharged in unit time $= \rho V_J A$

Change in velocity $= (V_J - V)$

The thrust is given by

$$T = \text{change in momentum in unit time} = \rho V_J A (V_J - V) \tag{88}$$

Useful work done on ship $= TV = \rho V_J A (V_J - V) V$

Kinetic energy lost in discharge $= \frac{1}{2}\rho V_J A (V_J - V)^2$

The efficiency is

$$\eta = \frac{\text{useful work}}{\text{useful work + lost jet energy}} = \frac{\rho V_J A (V_J - V) V}{\rho V_J A (V_J - V) V + \frac{1}{2}\rho V_J A (V_J - V)^2}$$

or

$$\eta = \frac{2V}{V_J + V} \tag{89}$$

The efficiency approaches unity as V_J approaches the velocity of the ship, but from Equation (88) the thrust at the same time approaches zero unless A is made extremely large. This is the same as with the ordinary propeller, where the efficiency increases as the amount of water used increases and the acceleration imparted to it decreases. To obtain such large values of A, it is obviously better to have the propeller outside the hull where the penalty on size is much less.

The foregoing analysis is on the assumption that the

Table 27—Examples of Jet Propulsive Efficiencies (Kim, 1964)

D = nozzle diameter; η_j = jet efficiency; η_P = pump efficiency; PC = propulsive coefficient

Type of ship	Nozzle diameter (m)	Loss factor 0.50 η_j	Loss factor 0.50 PC	Loss factor 1.00 η_j	Loss factor 1.00 PC
5.1-m power boat, 30 knots, $\eta_P = 0.90$	0.05	0.59	0.53		
	0.10	0.60	0.54 max		
	0.15	0.53	0.48		
	0.20	0.45	0.41		
	0.25	0.30	0.27		
161.0-m cargo ship, 25 knots, $\eta_P = 0.85$	1.83	0.55	0.47	0.49	0.42
	2.44	0.59	0.50	0.50	0.43 max
	3.05	0.60	0.51 max	0.49	0.42
	3.66	0.57	0.49	0.43	0.37
Destroyer, 33 knots, $\eta_P = 0.85$	1.22	0.58	0.50	0.50	0.43 max
	1.83	0.60	0.51 max	0.49	0.42
	2.44	0.57	0.49	0.43	0.37
	3.05	0.52	0.44	0.38	0.32
	3.66	0.46	0.39	0.32	0.27
Cargo ship, 16 knots, $\eta_P = 0.85$	1.22	0.38	0.32	0.36	0.31
	1.83	0.48	0.41	0.46	0.39
	2.44	0.55	0.47	0.49	0.42
	3.05	0.59	0.50	0.50	0.43 max
	3.66	0.60	0.51 max	0.49	0.42
Hydrofoil boat, 60 knots, $\eta_P = 0.90$	0.15	0.51	0.46	0.47	0.42
	0.31	0.59	0.53 max	0.48	0.43 max
	0.61	0.42	0.38	0.30	0.27
	0.92	0.28	0.25	0.18	0.16

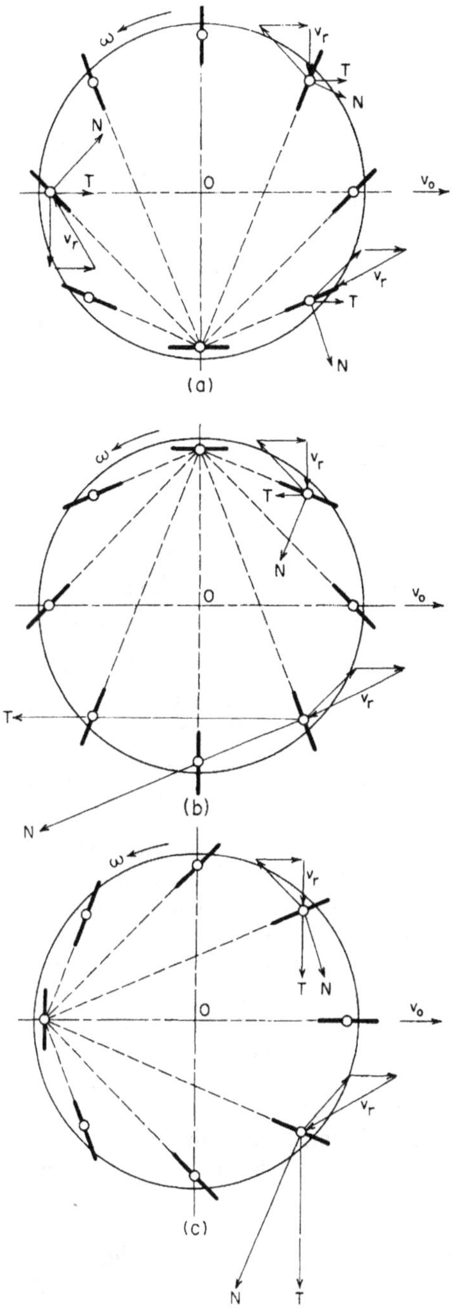

Fig. 87 Kirsten-Boeing propeller

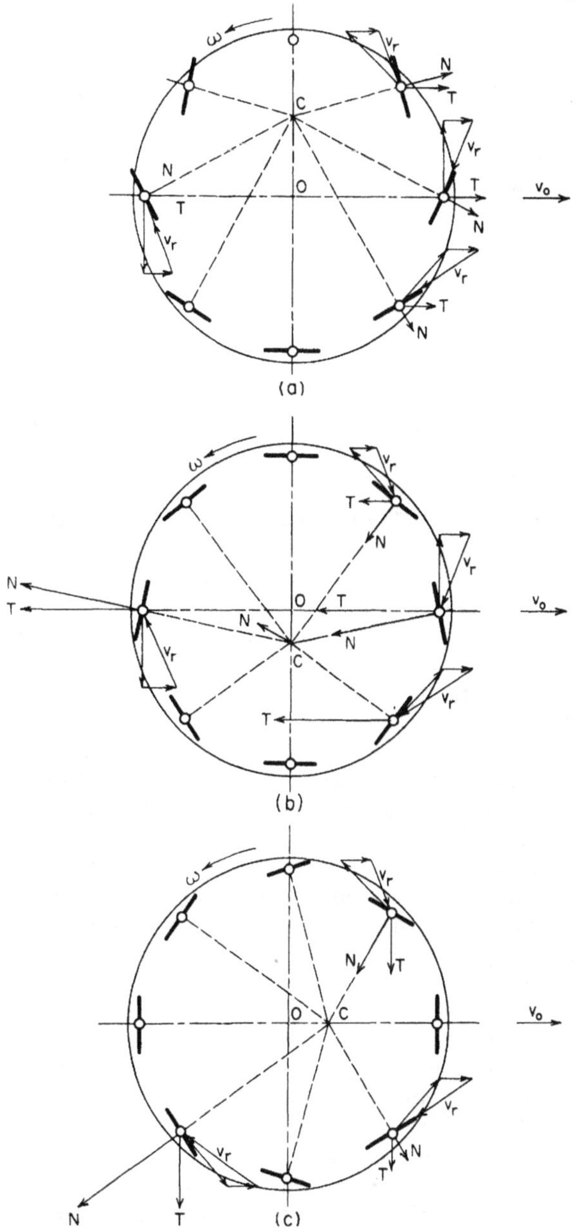

Fig. 88 Voith-Schneider propeller

water flowing into the impeller has a velocity equal to that of the ship, i.e., it has been taken in at the bow and suffered no intake losses. If these losses in fact are such as to bring the water almost to rest relative to the ship before reaching the impeller, or if the intake is in the bottom or side of the ship without any scoop, then the impeller has to supply the full acceleration from zero to V_J, and the additional kinetic energy necessary is $\frac{1}{2}\rho A V_J \cdot V^2$.

The efficiency is now

$$\eta = \frac{2(V_J - V)V}{(V_J)^2} \tag{90}$$

which has a maximum value of only 0.5 when $V_J = 2V$.

In these calculations, no allowance has been made for the efficiency of the pumping mechanism or for the inevitable frictional and other losses in the intake and ducting. The overall efficiency is therefore likely to be quite low in practice, a point brought out in some examples quoted by Kim (1964), (Table 27).

The loss factor is an estimate of the inlet and frictional losses, and Kim states that it will vary in actual cases between 0.25 and 0.50 for well-designed systems and 0.50 to 1.00 for poorly designed ones.

The maximum efficiencies are inferior to those attainable by normal propellers and it is also doubtful whether some of the values of pump efficiency η_P could be realized in practice.

Among other disadvantages of jet propulsion are

the loss of volume inside the ship due to ducting and impeller, or conversely the increase in size of the ship to restore the displacement to its original value, the danger of fouling of the impeller by debris or weeds unless resistful gratings are placed over the intakes, and the difficulty of obtaining uniform flow into the impeller when the water comes from the boundary layer around bends in the ducts.

The principal advantage is possibly in maneuvering ability when the discharge is made to operate as a movable rudder. If it can be made to deliver astern thrust as well, it eliminates the need for reversing gear on the engine. The fact that the impeller is inside the hull may also make for quietness, and external appendage drag is absent.

Chemicals or explosives may be used to accelerate and discharge the water, or a jet engine or rocket may be used above water. Aircraft jet engines have been used in racing boats. Rockets are not very efficient unless the speed of the craft is a reasonable fraction of the velocity of the gas in the jet, and there are problems arising from the heat and noise of the exhaust.

A particularly good overview of waterjet and gas-augmented waterjet propulsion systems and ejectors and rotary jets is given in *Marine Propulsion,* edited by Sladky (1976).

10.3 Pump Jets. This is the name given to an arrangement in which the impeller is external to the hull, taking the form of a rotating impeller with fixed guide vanes either ahead of it or astern, or both, the whole unit being enclosed in a duct or long shroud ring. The first device of this kind, called a *screw turbine,* was fitted by Thornycroft as early as 1881 to a great many shallow-draft steamers; a ship 42.2 m (138.5 ft) long with two such screws attained a speed of 15.25 knots (Barnaby, 1900).

In the modern pump jet, the duct diameter increases from the entrance up to the impeller, so that the velocity falls and the pressure increases. In this way the diameter of the impeller is larger and the thrust loading coefficient C_T is less, leading to a higher efficiency. However, there is some penalty to be paid for the resistance of the duct itself, and the final outcome depends very much on individual circumstances.

In the pump jet the velocity at the propeller is less and the pressure is higher than for the *open* propeller and so the incidence of cavitation and noise is delayed. If the tip clearance between impeller and ring is small, the formation of tip vortices is also inhibited.

Much general information on jet propulsion may be found in Wald (1970) and McCormick, et al (1962). The only known design method available in open literature can be found in Henderson, et al (1963, 1964).

10.4 Paddle Wheels. The general characteristics of paddle wheels are discussed in Section 1.1. The early wheels had fixed, flat blades, but by 1830 curved, feathering blades had been introduced which, for the same angle of entering the water, reduced the diameter by a factor of two or more (Saunders, 1957). With feathering wheels the efficiency of propulsion may approach or equal that of the screw propeller where draft and depth restrict the diameter of the latter.

It is important to locate the wheel over a crest in the wave profile, so as to take advantage of the forward motion of the water in the crest. This takes care of itself in the case of stern wheels, but requires careful attention, and usually should be checked by model tests, for side wheels. This is easy to ensure in vessels of fixed speed, such as river passenger vessels, but may be difficult in other classes. Tugs when towing have no serious wave system, and so side paddles can be placed to suit the free-running speed.

Much information on the design of paddle wheels has been given as the result of systematic series of models run at the Denny tank on the Clyde (Volpich et al, 1954–57).

10.5 Vertical-Axis Propellers. As mentioned in Section 1.1 the vertical-axis propeller had its origin as long ago as 1870. There are two types of such a propeller. One, the Kirsten-Boeing, has its blades so interlocked by gears that each blade is constrained to make a half-revolution about its axis for each revolution of the whole propeller, the action being as shown in Fig. 87 (Kirsten, 1928). The propeller is assumed to be advancing from left to right with uniform velocity V_0 and rotating with uniform angular velocity ω in the counterclockwise direction.

(*a*) With the blades set as in (*a*), the resultant velocity is as shown, and the normal forces N furnish a thrust T for going ahead.

(*b*) With the blades set as in (*b*), the blade forces oppose the motion of the ship.

(*c*) With the blades set as in (*c*), the blade forces are such that the resultant thrust acts at right angles to the motion of the ship.

A mechanism by which the blades can be set at will to any of the dispositions shown, as well as intermediate dispositions, is incorporated in the design, and it is therefore possible not only to propel the ship but to steer it also.

The second type, the Voith-Schneider, introduced in 1931, differs in that the blades describe a complete revolution about their own axes for each revolution of the disk, Fig. 88. The point C is a point to which the blades are connected by linkages and which can be moved to different positions in the plane of the disk.

(*a*) With C located as in (*a*), the blades have the disposition shown, and the resultant force is such as to propel the ship in the direction of motion.

(*b*) With C located as in (*b*), the resultant force is directed astern, and opposes the motion.

(*c*) With C located as in (*c*), the resultant force acts at right angles to the original motion of the ship.

A mechanism to alter the position of C while the propeller is in operation is incorporated in the design,

Fig. 89 Photo of model ship fitted with vertical-axis propellers.

so that again the ship can be propelled and steered at the same time. The distance of C to the center O, made dimensionless with the radius R, is called the eccentricity. It is always smaller than unity for Voith-Schneider propellers.

Propellers of this type have been fitted to a considerable number of ships and have proved entirely practical; details of some installations will be found in a paper by Mueller (1955). They have also been used at the bow to assist in maneuvering. A photograph of a vertical-axis propeller arrangement on a model is shown in Fig. 89.

The advantage of vertical-axis propellers lies in the fact that the propeller thrust can be used for steering and stopping the ship without stopping or changing the direction of rotation of the main engine. This makes it eminently suitable for the propulsion of ships that operate in crowded and restricted waters, requiring large steering power at low speeds.

For cycloidal propellers in general, the following three kinematically characteristic situations result:

- The rotational velocity ωR is grater than the translational velocity, V_o. The advance coefficient $J = \dfrac{V_o}{\omega R} < 1$. Here the blade centers describe an epicycloid.
- The speed of rotation ωR equals the translation velocity V_o. The advance coefficient $J = \dfrac{V_o}{\omega R} = 1$. The blade centers describe a normal cycloid.
- The speed of rotation is less than the translation velocity V_o. The advance coefficient $J = \dfrac{V_o}{\omega R} > 1$. The blade centers describe an elongated cycloid or trochoid.

The course of the effective angle of attack along the blade circle determines the unsteady load distribution along the blade circle and hence the system of free vortices shed in the propeller stream. From the consideration of this free vorticity system in the propeller slipstream, Sparenberg (1960) derived a condition for minimal energy loss; this indicated that the blade circulation along the foremost and the rearmost part of the blade circle must be held constant, so that only for blades near the mid-position, fore and aft, are the free vortices shed. The condition can be described in terms of the relationship for the effective angle of attack ($\beta - \beta_o = 0$)

$$(\beta - \beta_o) = -\frac{V_n}{V} + \frac{\Gamma}{V\pi C} \qquad (91)$$

in which V_n is the induced velocity at the blade.
V is the resultant blade velocity relative to the water
β is the geometric pitch angle
β_o is the hydrodynamic pitch angle, without induced velocity effects

This relation for the blade movement is different from the movement for the Voith-Schneider system, which is recommended from a constructional point of view.

Pitch for vertical axis propellers is again defined as the traversed path in the translation direction per revolution at zero slip, $\beta - \beta_o = 0$. Hence, $P/D = \pi J = \pi V_o/(\omega R)$. The pitch at zero slip is also related to the eccentricity, i.e., $e = J$.

A number of models of vertical-axis propellers have been tested at MARIN for the U.S. Office of Naval Research (Van Manen, 1966). The highest efficiencies were obtained with propellers having six blades. The propellers covered a number of pitch ratios, and the maximum efficiencies plotted to a base of B_P are shown in Fig. 90. The open-water test results are shown in Fig. 91. These hold for blades with aspect ratio 3.33. On the same diagram are shown the optimum open efficiencies for two Troost propellers having 4 blades and blade-area ratios of 0.40 and 0.85. The vertical-axis propeller is considerably less efficient, by amounts varying from 30 to 40 percent.

Van Manen also reports on tests with the blade movement according to the optimum efficiency criterion of Sparenberg. These tests show some 6 percent higher efficiency values for eccentricities exceeding 1.0.

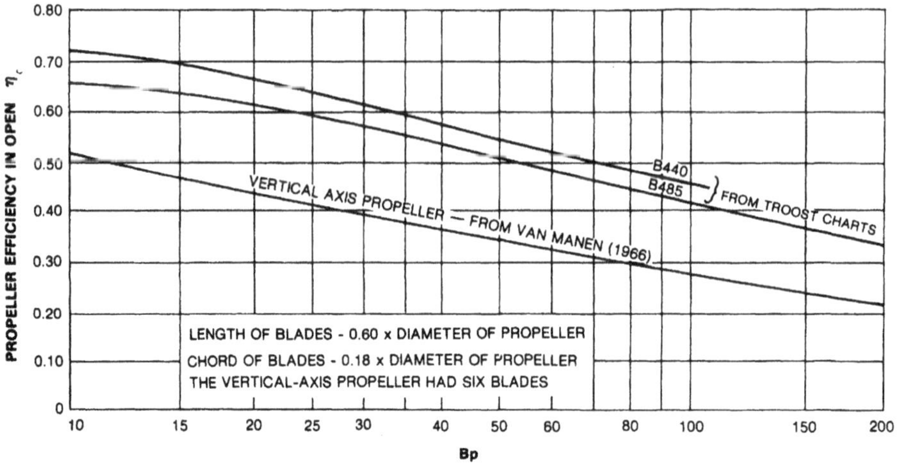

Fig. 90 Comparison of open water propeller efficiencies

For very high velocities ($J > 1$) Van Manen found high efficiency values for a four-bladed cycloidal propeller at eccentricities of about 1.6 (as high as 0.7). The trochoidal blade motion in that case is quite reminiscent of the movement of a fish. The lack of systematic data for the trochoidal case did not allow one to infer general conclusions from this test. Also cavitation may set a bound to such speeds (Isay 1957 and Van Manen 1966).

Ficken (1966) carried out tests for six pitch ratios ranging from 0.4π to 0.9π (J-values of 0.4 to 0.9). He found higher efficiencies compared to Van Manen for similar test conditions. Also the thrust and torque coefficients for bollard pull were significantly lower compared to Van Manen's results.

Ficken compared results for rectangular and elliptical blade form and found that the rectangular blades achieved some 2 percent higher peak efficiency.

A great number of researchers have developed theoretical models for vertical-axis propellers, e.g. Taniguchi (1960 and 1962), Isay (1958), Sparenberg (1960) and Sparenberg, et al (1967). Only the method of Taniguchi yielded quantitatively reasonable results at some advance values (Haberman, et al 1961, 1962). Taniguchi's method was found to be adequate for medium advance values and for lower eccentricities. It is based on the assumption that a quasi-steady state exists. The thrust and torque of the propeller are evaluated by integrating the lift and drag on each blade section. As such it is similar to the blade element theory covered in Section 2.5.

Taniguchi neglected the camber induced by the curved orbit and the rotation of the blade around its own axis. He also assumed the induced velocities to be constant over the entire blade span. In his method only the longitudinal induced velocities due to the trailing vortex system were taken into account, and the induced velocity was considered independent of the orbital blade position.

Zhu (1981) included the effects of induced camber and also obtained a better representation of section lift and drag coefficients. Comparison between his method, Taniguchi's method and experiments performed at the DTRC indicates a much better agreement with measurements.

Other notable contributions to the performance of vertical-axis propellers are the theoretical analysis carried out by Mendenhall, and the investigations into the steering characteristics of cycloidal propellers of Ficken. Jobst, et al (1972) obtained flow visualizations for a high-pitch cycloidal propeller.

10.6 Controllable-Pitch Propellers. Controllable-pitch propellers are propellers in which the blades are separately mounted on the hub, each on an axis, and in which the pitch of the blades can be changed, and even reversed, while the propeller is running, by means of an internal mechanism in the hub.

Although many designs of controllable-pitch propellers have been evolved over a long period, the modern usage goes back to about 1935, the mechanism consisting essentially of hydraulic pistons in the hub acting on crossheads.

From the ship point of view, the chief advantages of this type of propeller are in ships which have to meet very different operating conditions, such as tugs and trawlers or in ships that have non-reversing prime movers, usually naval ships powered by gas turbines. By reducing the propeller pitch when towing or trawling, for example, the engine can still run at its full revolutions, and so develop full power without increasing the mean effective pressure in the cylinders, an important matter in diesel engines. Ferries which have to stop, start, and reverse repeatedly also benefit by the fact that full-astern power is available by reversing the pitch while the engines continue to run in the same direction.

From the engine point of view, such propellers obviate the necessity of reversing mechanism in reciprocating engines and of astern turbines in turbine-powered ships, thus saving both weight and cost, and making the reversing of thrust in an emergency very much quicker.

The controllable-pitch propeller can be made almost as efficient as the solid, fixed-blade propeller at any particular chosen condition—say that in normal service—the only difference being the somewhat larger hub needed to house the pitch-changing mechanism. When the pitch is changed, all sections turn through the same angle, so that the pitch face is no longer a true helical surface.

Further information on controllable-pitch propellers can be found in Strom-Tejsen et al (1972), Stephens (1974), Boswell et al (1975) Müller-Graf (1978) and Tani, et al (1979).

10.7 Tandem and Contrarotating Propellers. When the diameter of a propeller is restricted on account of draft or other reasons, the loading factor is greater, the efficiency suffers and the liability to cavitation is increased. This situation can be relieved by dividing the load between two or more propellers, and we have already seen (Section 7.1), how this idea was used to solve the problem on the *Turbinia* by fitting three propellers to one of the three shafts.

Propellers arranged in this way, on the same shaft and trailing in the same direction, are called tandem propellers. Such screws were fitted to the 14 m (46 ft) long *Francis B. Ogden* by John Ericsson in England in 1837. After the *Turbinia* experience they were also used on a number of British destroyers around 1900, which had 4 shafts with 2 tandem propellers on each (Saunders, 1957). In these later applications it was realized that the after propeller, working in the race of the forward one, required a higher pitch to give the same power absorption.

With both propellers rotating in the same direction, the rotational energy in the race from the forward one is augmented by the working of the after one. The idea of regaining the rotational energy from the forward one is attractive, and has led to the development of contrarotating propellers working on coaxial, contrary-turning shafts. The after propeller of the pair must be of smaller diameter to suit the contracting race column of the forward one and must have the pitch designed to suit the required power absorption. Ideally there would be no rotation in the race behind the second propeller.

The first recorded use of contrarotating propellers seems to have been by John Ericsson in 1837 on the 21 m (69 ft) steam-driven ship *Robert F. Stockton* built in Birkenhead, England (Saunders, 1957). Rota in Italy found a gain in propulsive efficiency, and attributed this to an improvement in the propulsive factors of

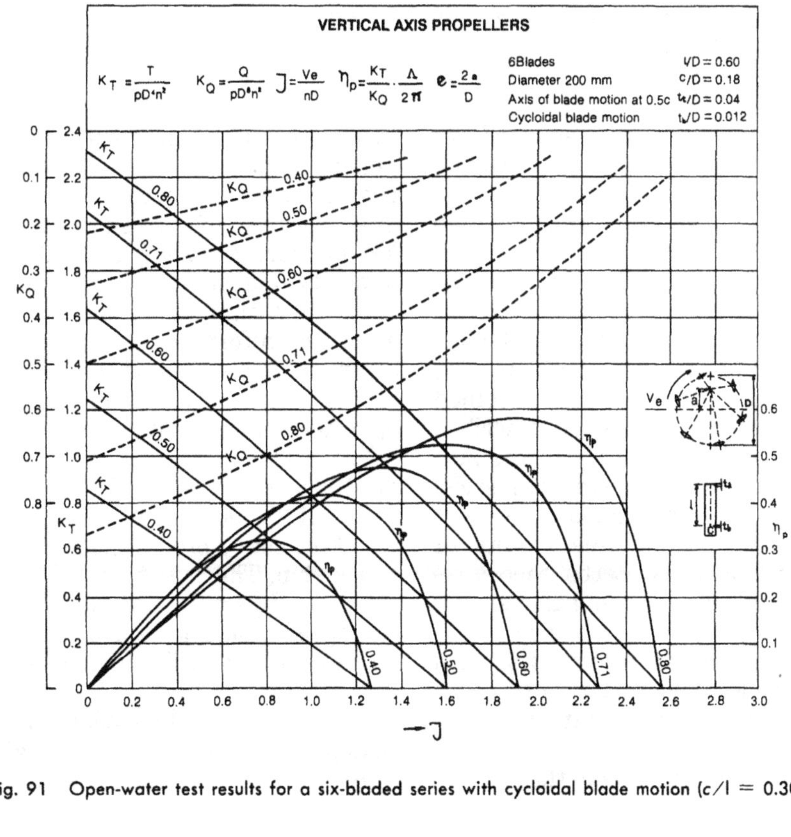

Fig. 91 Open-water test results for a six-bladed series with cycloidal blade motion ($c/l = 0.30$)

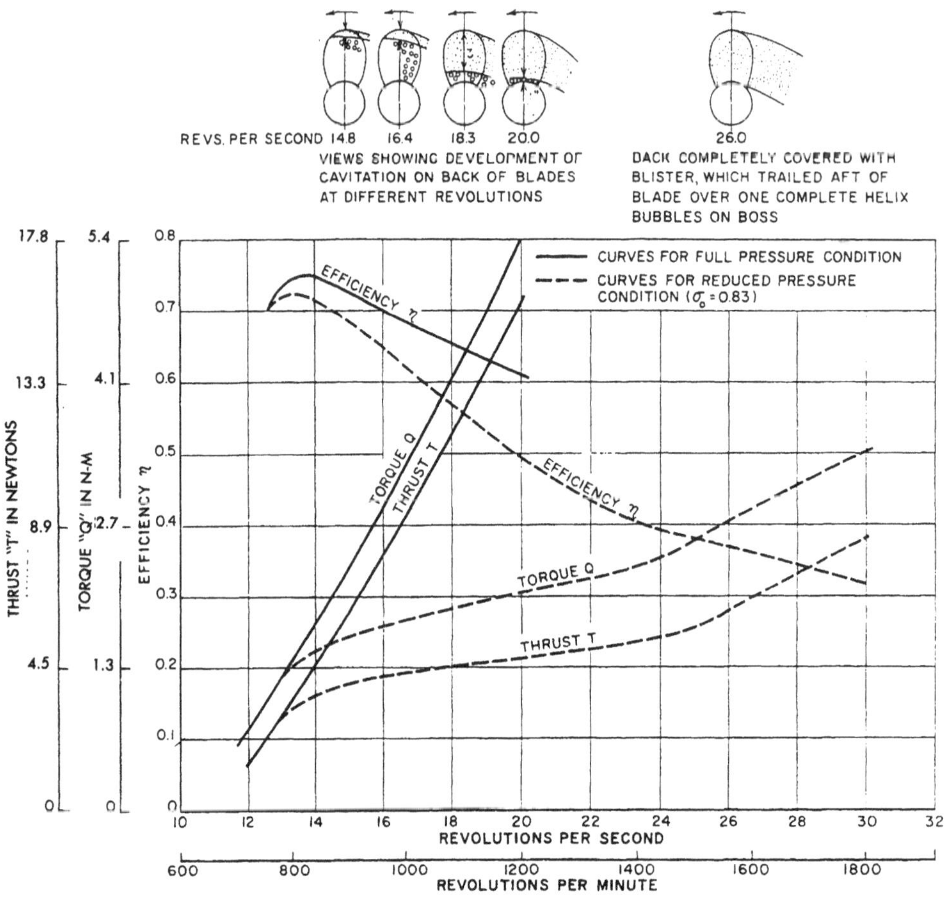

Fig. 92 Thrust, torque, and efficiency curves for a propeller working under supercavitating conditions

Propeller diameter = 8 in (20 cm) Blade area ratio = 0.444
Pitch ratio = 1.08 Number of blades = 3

hull efficiency. Such propellers have been used for many years on torpedoes to balance the reaction torques and prevent the torpedo body from rotating.

Early design procedures consisted of treating each screw as a separate propeller and relying on subsequent model experiments to check the design and find the pitches necessary for power balance. The circulation theory has been applied to the design of such propellers, originally for aircraft, and later to marine propellers. Lerbs (1955) developed a design method using his induction factors. Later Morgan described the practical application of Lerbs' theory to the successful design of such propellers. He gave a numerical example and showed that there was good agreement between theoretical predictions and the results of model tests. Morgan concluded that Lerbs' theory resulted in propellers with good torque balance and higher efficiencies than single propellers, and by giving more accurate values of the inflow velocities to each section of the propeller blades gave better control over the cavitation performance of contrarotating propellers.

The results of a model investigation into the relative merits of twin, single, tandem and contrarotating propellers for a very large merchant ship of large block coefficient were reported by Hadler, et al (1964). The size and speed of such ships calls for powers of 20,000 kW or more per shaft (26,800 hp). There is no great difficulty in absorbing such powers on one screw of a multiscrew ship where the propellers are working in a rather uniform wake. The trouble arises in single-screw ships because of the dangers of cavitation and propeller-excited vibration where the propeller has to operate in a highly uneven inflow pattern.

The model chosen for the experiments was that of the 106,000-dwt tanker *Manhattan,* with an installed power of 32,000 kW (42,880 hp) which, because of the resultant heavy propeller loading, was built as a twin-screw ship. The twin-screw model results were available for comparison, and the large power and relatively low speed would give a heavily loaded single screw for comparison with tandem and contrarotating arrangements.

The forebody and parallel part of the hull were unchanged, but the afterbody was altered to give a clear-water stern with an aperture suitable for a 7.6 m (24.9

ft) diameter propeller and sufficient fore-and-aft length to accommodate tandem and contrarotating propellers. As a result of these changes the length was increased by 6 m (19.68 ft).

The propeller designs were based on a P_D of 32,000 kW (42,880 hp) with a maximum diameter of 7.6 m (24.9 ft). Those for the 5 and 9-bladed, single screws were done by Lerbs' induction factor method (Lerbs, 1952) that for the contrarotating ones by the method of Morgan while a new method had to be developed for the tandem screws. In all cases extensive use was made of computer programs. The blade-area ratio for the 5-bladed design was based on Burrill's 5 percent back-cavitation criterion (Section 7.7 and Fig. 45). The blade area for the other screws was chosen to give the same degree of cavitation based on a comparative local σ at 0.7 radius. Tests in the cavitation tunnel in a simulated wake showed that no face cavitation or loss in efficiency was present in any screw. There was slight back cavitation near the tips at about 19 knots, there being little difference between the propellers.

The results of the test show that there was little difference between the P_D for the original twin-screw design and the single screw with 5 blades—the latter required some 0.4 percent less power. The 9-bladed screw required about 4 percent more power while the tandem arrangement was almost exactly halfway between them in efficiency. The contrarotating arrangement, on the other hand, where most of the rotational energy in the wake was regained, required 7 percent less power than the twin-screw design and 6½ percent less than the 5-bladed single screw. This gain in propulsive efficiency, together with the improved vibration characteristics of the smaller diameter screws and higher blade frequency, make the contrarotating propellers very attractive from the hydrodynamic point of view. But against this must be set the added weight and complication of the gearing, coaxial shafting, and sealing problems of the contrarotating arrangement.

10.8 Super-Cavitating Propellers. When the cavity on the back of a propeller blade has spread until it covers the whole of the back, which is then no longer wetted, the propeller is said to be operating in the fully- or super-cavitating regime. After the back of the section has become completely denuded of water, further increase in revolutions per minute cannot reduce pressure there any more, and so no additional lift can be generated by the back. On the face, however, pressure continues to increase with higher revolutions and so does the total thrust, although at a slower rate than before cavitation began.

One advantage of such propellers is the absence of back erosion, because the cavitation bubbles no longer collapse on the back of the blades. Also, the unsteady forces resulting from intermittent cavitation will be much reduced so that less vibration may be expected (Todd, 1946).

Parsons observed propellers working in this regime in his small tunnel in connection with the *Turbinia* design, and they have long been used on high-speed racing motor boats, the characteristics being determined by extensive trial-and-error experiments. One of the pioneers in this research was V. L. Posdunine, of the Moscow Academy, who directed attention to the use of super-cavitating propellers in a paper presented in London (Posdunine, 1944). He gave a theoretical model of the flow and deduced expressions for the thrust and ideal efficiency. It is of interest to note that he spoke at this time of "wedge-shaped blades," and in other papers showed sections quite similar to those used in such propellers today. He also stated that the action of such propellers was so different from that of conventional propellers that no single theory would cover both kinds.

In the discussion on Posdunine's paper some results of experiments carried out in the NPL water tunnel at Teddington in 1939 were given by Todd. These were for a model of an ordinary marine-type screw run up into the super-cavitating range and are perhaps the first such published results, Fig. 92. The propeller was illuminated by stroboscopic lighting and sketches of the cavitation are reproduced in the figure. At full atmospheric pressure in the tunnel there was no cavitation. With the pressure reduced to give a value of $\sigma = 0.83$, the thrust T and torque Q departed from the full pressure curves at about 800 rpm as back cavitation began to spread from the blade tips. At 1200 rpm, with cavitation covering nearly all the back of the blades, the efficiency was 0.49 as compared with 0.61 at full pressure without any cavitation. From 1500 rpm upwards, in the fully-cavitating zone, there was some recovery of the propeller as a thrusting mechanism, the rate of increase of thrust with rpm increasing, T/n^2 being practically constant, while the fall in efficiency became less pronounced. The cavity extended downstream for about one complete turn of a helix.

Since the back of the section when fully cavitating is no longer in contact with the water, the emphasis in design must be on a section shape which will ensure clean separation of the flow at the leading and trailing edges, at the same time providing good values of L/D ratio to give high efficiency. The theoretical determination of the lift and drag characteristics of super-cavitating sections in two-dimensional flow at zero cavitation number and a method of choosing the optimum shape for the face were presented by Tulin

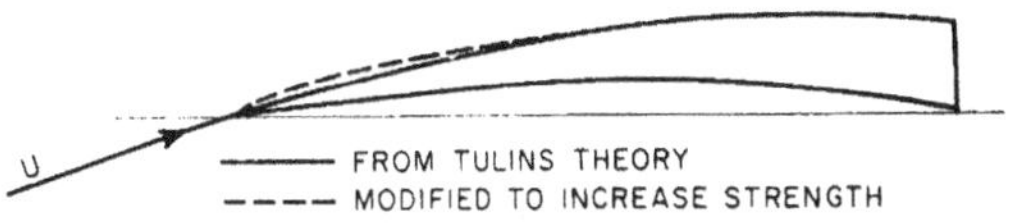

Fig. 93 Wedge sections for super-cavitating propeller

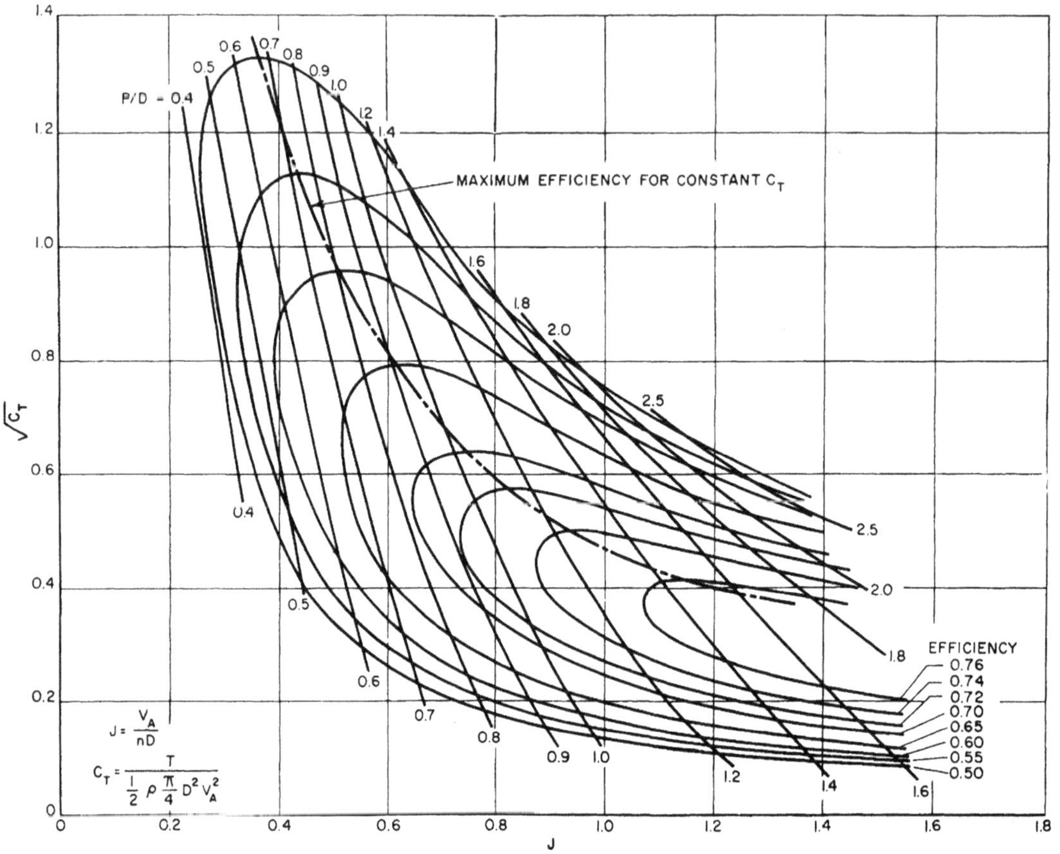

Fig. 94 Propeller design chart for super-cavitating propellers

(1955). The shape of the cavity can also be calculated, and that of the back of the section can then be chosen at will to give adequate strength so long as it remains within the cavity. This naturally led to a wedge-shaped section with an extremely fine leading edge to ensure separation, Fig. 93.

Lerbs next showed how the use of the Tulin sections for fully-cavitating marine propellers could lead to higher efficiencies than hitherto thought possible under such conditions of operation.

Tachmindji and Morgan (1958) applied circulation theory to the design of such propellers, using Tulin's theoretical results for the shape of the pressure face and the calculation of lift and drag, and wedge-shaped sections. Early experience with models showed that these sections with very thin leading edges could give rise to vibration and failure, and that the edge must be made thicker or the angle of attack greater. The experiments also indicated that in order to achieve satisfactory fully-cavitating operation the value of σ for the section at 0.7 radius should be equal to or less than 0.045. Using this fact, the authors calculated the performance of a systematic series of 3-bladed propellers and prepared design charts, an example of which is shown in Fig. 94.

Tachmindji, et al (1958) also gave a chart showing the regions in which super-cavitating propellers become practical, Fig. 95. Using the design charts and

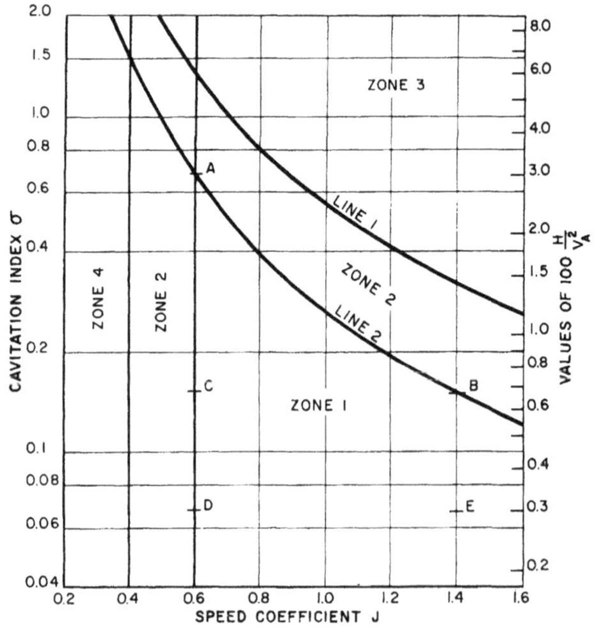

Fig. 95 Chart showing areas for practical use of super-cavitating propellers

$$\sigma = 2g\frac{H}{V_A^2} \text{ (metric units)}$$

Zone 1 Best region for fully-cavitating propellers
Zone 2 Marginal region, with some cavitation on all propellers
Zone 3 Best region for conventional propellers
Zone 4 Region of low efficiency for all propellers ($\eta_o < 0.64$)

Table 28—Comparison of Normal and Super-Cavitating Propellers (Performance of Normal Propellers Assumes no Thrust-loss Due to Cavitation).

Type of Ship	Passenger Liner		Destroyer		Motor Boat	
V, knots	40		39		36.5	
P_D, kW	186250		74500		5960	
No. of screws	4		2		4	
Type	NC	SC	NC	SC	NC	SC
Diameter, m	5.67	3.05	3.84	3.05	0.73	0.73
RPM	200	688	350	655	1800	2570
Pitch ratio	1.23	0.96	1.18	0.94	1.16	0.90
Efficiency	0.68	0.61	0.66	0.62	0.67	0.64

NC = noncavitating design. SC = super-cavitating design.

basing the estimates on conditions at point A, where the use of super-cavitating propellers is marginal, Table 28 indicates in a general way the possible fields of usefulness in ships. The figures for conventional screws assume that there is no cavitation, or at least insufficient to affect the thrust. This qualification is, of course, critical, since in the cases used in illustration cavitation is already a problem with conventional propellers, many of which would suffer from erosion. The estimates in Table 28 show that in order to achieve the low σ-values necessary for full cavitation the revolutions per minute have to be much higher and the pitch ratio lower than for the conventional propeller. For the liner the super-cavitating propeller is some 10 percent less efficient than the conventional, while in the destroyer and motor boat the difference is appreciably less, and in these cases the fully-cavitating propeller has the added advantage of not suffering from erosion. As still higher speeds are considered, the difficulties with the conventional propeller increase and the fully-cavitating type may be the only alternative for speeds in the range from 40 to 80 knots. It would not be possible to design conventional propellers to suit such conditions even if the necessary power plants were available. With super-cavitating screws there may, however, be serious strength problems (Allison, 1978). Nevertheless, the figures bring out the possibilities of adopting lighter, faster-running engines with such propellers.

Strength problems may be caused by the combination of high thrust and thin leading edge sections. In addition, as super-cavitating propellers achieve their highest efficiency at partial submergence, large varying forces and pressures may be expected as each blade enters, traverses through and exits from the water. Fig. 96 shows a pressure-time history based on full-scale tests for a SES at two locations on the propeller blade. Very high pressures show up at the moment of entry of the blade. Fig. 97 shows the measured strain near the blade root of the same propeller.

In the early fully-cavitating propellers the sections were of airfoil or crescent (hollow-faced) type and had relatively low efficiencies. Later work has shown that with proper choice of sections this loss of efficiency can to a considerable extent be avoided and the propellers made to give reasonable efficiencies under very advanced design conditions, thus extending the range over which marine propellers may continue to be used for high-speed ship propulsion.

Recent research at DTRC has confirmed the possibility of reaching efficiencies at least approaching those of the best subcavitating propellers if suitable rake is applied (Moore, et al, 1973). Fig. 98 shows that increased efficiency may be obtained by using inclined shafts (Crown et al 1971). This is confirmed by the results of Rutgersson (1978) for lower pitch ratios. For

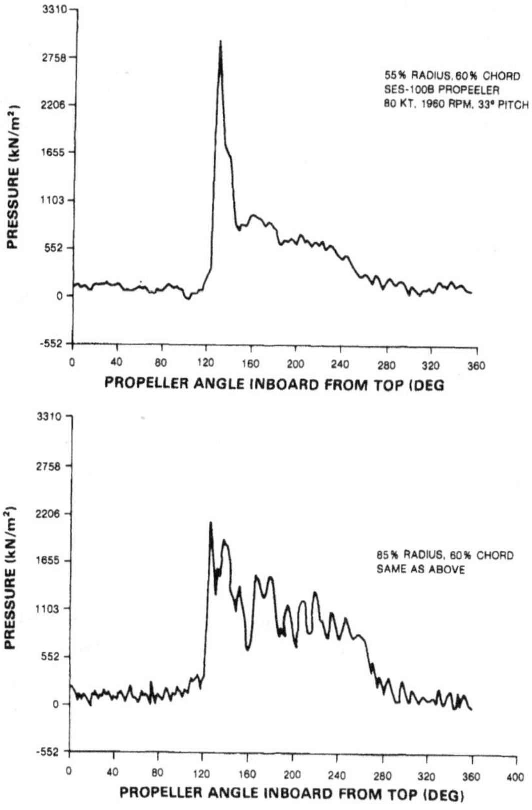

Fig. 96 Measured pressure-time histories

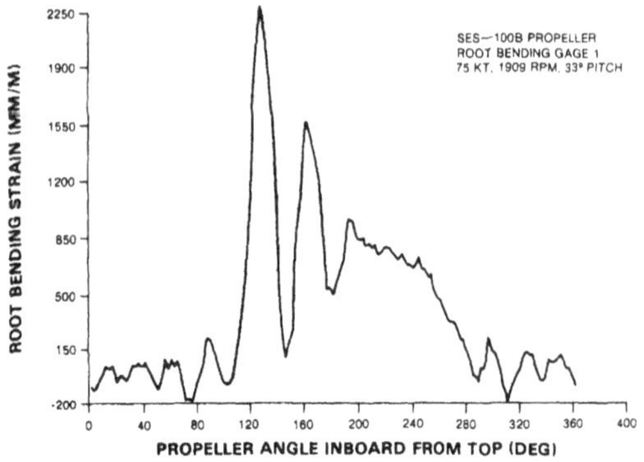

Fig. 97 Measured strain versus angle of rotation

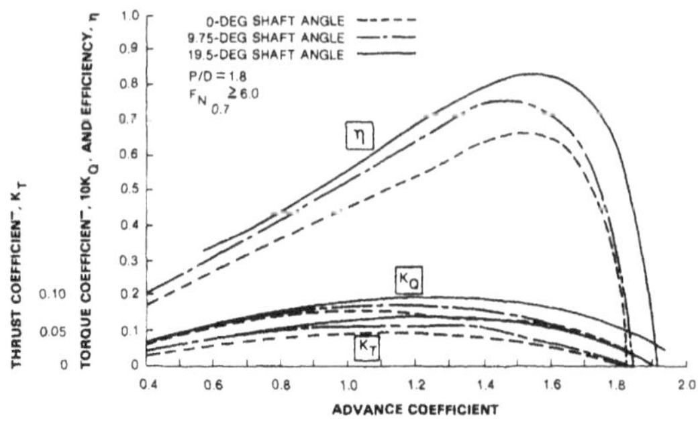

Fig. 98 Propeller characteristics showing effect of shaft angle

higher pitch ratios inclination may have adverse effects in some cases. Rutgersson also investigated the effect of bottom clearance on the propeller characteristics. Both torque and thrust were found to drop with decreasing clearance and attained lowest values of about 92 percent of the corresponding open-water values. Though the scatter was large, shaft inclination apparently had little influence. The presence of a rudder may be beneficial for thrust (some 10 percent) and efficiency (about 2 percent), especially for small clearances between rudder and propeller. Adverse effects were observed for a twin-rudder arrangement.

To check the adequacy of the design method and of the charts described by Tachmindji, a propeller was made with a diameter of 45.7 cm (18 in.), pitch ratio 1.533, and an expanded blade-area ratio of 0.444, which when run in the cavitation tunnel under fully-cavitating conditions gave an efficiency of 0.685 at the design conditions. The calculated series charts gave a pitch ratio of 1.57 and an efficiency of 0.70 for an area ratio of 0.50. Other similar tests have indicated agreement within some 3 percent (Hecker, et al, 1964). The ordinary conventional propeller charts based on model results call for a great deal of experimenting, and are costly both in time and money. For cavitating propellers this work is multiplied several-fold because a range of σ-values has to be covered with each model. The charts for cavitating propellers were made using theory and a computer only, and the foregoing comparisons were sufficiently close to make their use adequate at least for feasibility studies and preliminary design purposes. Further calculations were therefore made and design charts prepared for 2 and 4-bladed propellers (Caster, 1963). Together with the original charts, these now cover a wide range of propeller parameters:

Number of blades........	2, 3 and 4
Blade-area ratio	0.3 to 0.5 (2 blades)
	0.4 to 0.6 (3 blades)
	0.5 to 0.7 (4 blades)
J-values	0. to 1.6
C_T-values.................	0.3 to 2.25

The charts such as Fig. 94 give curves of pitch ratio and efficiency on a grid of J and $\sqrt{C_T}$ They apply to the case of σ = o, and correction factors are given for other values.

A series of model propellers operating partly in the fully-cavitating regime has been run in the Vosper cavitation tunnel (Newton, et al 1961). The series comprised twelve 3-bladed propellers suitable for high-speed craft, covering pitch ratios from 1.05 to 2.04 and blade-area ratios from 0.48 to 0.95. Under partial cavitation conditions, a propeller with hollow-faced sections has better performance characteristics than one with flat faces. The sections were therefore made with faces in accordance with the NACA (a = 1) mean line, with a quasi-elliptical thickness form superposed. The results therefore give information on fully-cavitating propellers with other than wedge-shaped sections. Newton and Rader concluded that the circulation theory as developed for conventional propellers can also be used in the design of cavitating propellers.

The performance of super-cavitating propellers was discussed at length by Venning, et al (1962). They considered that such propellers can meet the demand for high thrust and at the same time lessen blade erosion. The same thrust could not be supplied by conventional propellers under the conditions of high-speed, high RPM and small submergence occurring in certain high-speed craft. They offer no advantage over conventional propellers in noncavitating regions and should not be used unless the proper relation between V and n can be established. Also, a low value of σ on the blade sections is essential for success. Line 1 in Fig. 94 is based upon the incipient cavitation diagrams given by Eckhardt, et al (1955) and shows the beginning of performance breakdown due to cavitation on conventional propellers. Line 2 is based on an attempt to ensure that the local σ-value on a section at 0.7 radius is not greater than 0.045 as originally set down as a criterion for full cavitation by Tachmindji.

There are still many problems to be investigated in the application of fully-cavitating propellers. These have been reviewed by Venning, et al (1962) and Tulin (1964). Among these problems, the following may be mentioned briefly.

In a hydrofoil craft the thrust at take-off is almost the same as that when flying at full speed, although the speed is only about one half. This, combined with the accompanying change in immersion and the need to meet certain engine relationships between rpm and power, poses a difficult design problem. Also, such propellers are not suitable for running for extended periods at off-design conditions, such as a lower cruising speed, since the efficiency will be low and if they experience face cavitation they will probably suffer from erosion.

A discussion of this matching problem, along with two simplified design examples applied to a Surface Effect Ship (SES), may be found in Allison's work.

With present design knowledge super-cavitating conditions are not always achieved, and the thrust and efficiency are then usually low. The large cavities on the blades increase the interference between them and have a blockage effect, and the hull efficiency elements are changed, the thrust-deduction fraction being much reduced. One way of ensuring super-cavitating performance is by introducing air to the backs of the blades, either naturally or by pressure, so giving a ventilated propeller. Apart from ensuring full cavitation, this also enables such propellers to be used at lower speeds and perhaps even to simulate cavitation on propellers in ordinary self-propulsion tests in the towing tank (Bavin, et al, 1963). There is little difference in efficiency between fully-cavitating and ventilated propellers once the cavity is formed, providing the performance is compared on a basis of cavitation number and the latter is based on cavity pressure.

10.9 Overlapping propellers. Pien and Strøm Tejsen introduced a stern arrangement with a twin-screw propulsion system having overlapping propeller fields in 1967.

From tests performed with ship models with overlapping propellers both at the DTRC and at MARIN, it can be concluded that the reductions in power due to the overlapping twin screw arrangement are 5 to 8 percent compared to the conventional single screw solution and 20 to 25 percent compared to the conventional twin screw arrangement.

At MARIN Kerlen, et al (1972) found that the longitudinal distance between both propeller planes as well as the direction of rotation of the aft propeller and the pitch of the aft propeller are of relatively minor importance on the power requirements. The influence of the center distance (degree of overlapping) proved to be more distinct. In this particular case the optimum center distance was equal to approximately 0.7D. The maximum degree of overlapping does not lead to minimum power for these tests as the wake fraction of the aft propeller is so strongly decreased that the efficiency drop cannot be balanced by the increased efficiency of the forward propeller, which experiences a higher wake fraction.

Kerlen also investigated the propeller induced vibratory forces. They found thrust and torque variations of about 10 percent of the average thrust and torque. Compared to a conventional four-bladed propeller these variations are about 50 percent higher. However, if proper vibration analyses are performed this will probably not lead to excessive problems. The variations were not dependent on the relative phase position of the propellers.

The propeller shaft bending moments are determined by the eccentricity of the thrust (i.e. the point

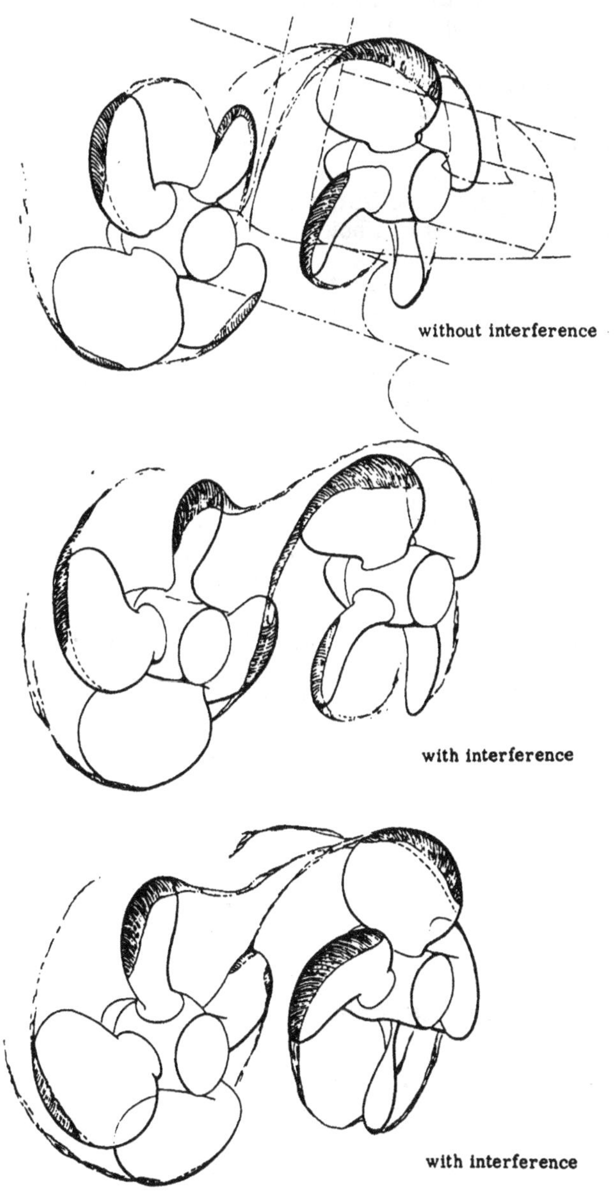

Fig. 99 Sketch of the observed interference of the tip vortices of overlapping propellers.

of application of the thrust), by the magnitude of the thrust and by the magnitude of the lateral propeller forces (in the vertical propeller plane). The latter force can be considered to work through the center of the propeller as any shift of this force vector is included in the torque of the propeller. For overlapping propellers it was found that both lateral forces are approximately equal and opposite in direction. Also they are not excessive in magnitude. The mean thrust eccentricity and the lateral bending moments have a value comparable to conventional single or twin screw arrangements. The phase position of the propellers was again almost irrelevant. The resulting bending moments have acceptable values. They were also found to be nearly independent of the relative angular propeller position. Hence no beating phenomena will occur. The difference in speed of the propellers when driven by independent prime movers will, however, lead to alternating vertical and torsional excitations of the hull, caused by pressure fluctuations. The cavitation properties of both propellers appear to be comparable to those of the respective conventional single-screw configurations. Unfavorable interaction of cavitating tip vortices can however occur, Fig. 99. As no unfavorable phenomena such as bubble cavitation were observed, the risk of erosion is considered negligible. More recently Muntjewerf, et al (1981) reported on results obtained at MARIN with an overlapping propeller arrangement applied to a 425,000-dwt tanker. For this ship the power saving amounted to 16 percent when compared to the original conventional single screw version.

10.10 Partially Submerged Propeller. Partially submerged or surface piercing propellers may be attractive due to the following aspects:

- They can be located directly behind the ship. Therefore, the additional resistance of shafts and shaft brackets is small, whereas the propeller diameter can be chosen larger than possible below the ship's bottom where often restrictions are posed.
- Most propeller designs will produce large cavities in the high-speed range (above 40 knots). When these cavities collapse in the neighborhood of the propeller blade, erosion damage can be unavoidable. Cavities filled with air do not collapse so vehemently.
- As the danger for erosion decreases, smaller blade area ratios than for a conventional propeller become possible. This reduces friction, and affects the efficiency positively.
- The skin friction of the propeller blade itself is reduced slightly when compared with a pure water environment.

With respect to efficiency in the 40-knot speed range the publications of Brandt (1970), Fleischer (1974), Hadler et al (1968), Hecker (1973), Rains (1981) and Scherer (1977) illustrate that an open water efficiency above 60 percent must be possible. Opposite to the aforementioned advantages stand two disadvantages:

- Propeller strength.
- Effect of cavitation/ventilation on propeller torque.

Partially submerged propellers suffer from larger stress variations than submerged propellers. Blade loading will vary roughly between zero in top position and maximum in bottom position, which is more than for conventional propellers. This implies that some attention must be paid to fatigue.

Because fast vessels generally operate in a large range of speed, and ship resistance is characterized by a considerable hump, conditions occur where speed is relatively low but propeller torque exceeds the design torque. Reasons for this are the difference in submergence of the propellers and the related strong influence of cavitation and ventilation on propeller thrust and torque. It is therefore recommended to choose the pitch ratio such that at lower speeds the torque values remain within reasonable limits which must be verified by experiments in a free surface cavitation tunnel or a depressurized towing tank. Initially, experiments with partially submerged propellers were carried out under atmospheric pressure (Hadler, et al, 1968) but later it was realised that the pressure above the water surface has to be scaled to meet the cavitation characteristics (Brandt, 1973).

10.11 Other Devices. Muntjewerf (1983) has presented a review of the efforts directed towards obtaining energy savings. Some of the mentioned devices have been covered in the previous sections. The use of low-RPM high-diameter arrangements has been mentioned in Section 8.2. A device suggested by Grim in 1966 in the *vane-wheel*, (Grim, 1983) and (Blaurock, 1983). The idea is simply to place behind the driven propeller of practically conventional design a second freely rotating vane wheel on the same shaft with a diameter larger than that of the propeller. The blades of this vane wheel are designed in such a way that the vane wheel is absorbing energy from the propeller race at its inner radii, which it transmits immediately in thrust at its outer radii outside the race of the propeller. In this way the required impulse to produce the total thrust of the propeller plus vane wheel is transmitted through a larger mass of water and hence at a lower required power than for the single conventional propeller of same thrust.

Full-scale tests on a research vessel with this type of propulsor were published by Grim in 1982. Tests were carried out both with a conventional propeller and a propeller combined with a vane wheel. The latter system was found to be 9 percent more efficient. Grim also states that if optimization of the propeller and vane wheel is carried out in advance (as opposed to the retrofitted vane wheel) gains of up to 12 percent could be possible. A comparable low RPM-large diameter propeller of the same diameter as the vane wheel turning at about a 20 percent lower RPM than the original propeller, would lead to some 8 percent

Table 29—Energy Saving Propulsion Arrangements

	Savings in required power in percent of power with conventional single propeller
Low rpm—large diameter	5–18
Skew or stern tunnel	1– 5
Propeller retrofit	2–10
Propeller with vane wheel	9–12
Reaction fins	4– 8
Additional thrusting fins	2– 5
Asymmetric gondola	5–10
Asymmetric aft body	5– 8
Ducted propeller	5–12
Duct ahead of propeller	5–10
Overlapping twin screw	5–16
Overlapping twin screw + low rpm	10–21→?

power saving (Muntjewerf, 1983). Probably the reduction of the rotational energy in the propeller slipstream resulting from the action of the vane wheel is responsible for the high performance of the propeller-vane wheel combination.

The advantages of the propeller and vane wheel combination over a low RPM propeller are evident, the combination can operate at normal shaft rpm so that for larger ships reduction gears are not necessary. Since the tip speeds of the vane wheel will be somewhere in between 12 to 18 m/sec (49.2 ft/sec) as compared to around 30 m/sec (100 ft/sec) for the low RPM propeller, cavitation and hull pressure fluctuations will be no problem for the propeller and vane wheel combination, while smaller tip-hull clearances can be accepted than with the low RPM propeller.

Recently another device called *reaction fin* was introduced (Takekuma, et al, 1981). The object of the fin is to introduce a swirling flow forward of the propeller to counteract the vortex set up by the screw itself, and this is achieved by placing fins in an essentially radial pattern in the stern aperture. The principle of this device is stated to be similar to that of a contra-propeller. To gain maximum effect the incidence angles of the fins must be optimized by tank testing. For strength reasons the tips of the fins are linked by a ring. The hull form in the after body influences the effectiveness of the reaction fin to a marked degree and it is accepted that they are best suited to high block-coefficient ships.

Sea trials were carried out on three large bulk carriers, both with and without a reaction fin on one ship, with a fin on a second ship and without a fin on a third vessel, sister ship of the second. The results show a reasonable agreement between tank predictions and measured power savings, namely around 7 to 8 percent in the ballast condition. The tests and trials would appear to show that for suitable hull forms there are savings to be made by adding reaction fins to the propeller arrangement. The sea trials as mentioned were carried out in the ballast condition, and the agreement with predictions has led to estimated potential savings of 4 to 6 percent in loaded operations, based on the model tests in that condition.

Improvements in noise and vibration are also claimed from acceleration measurements on one of the vessels. Maneuverability is said to be little affected by the addition of the fins. At model size, no significant effect on the extent of cavitation is recorded, nor on propeller induced vibration.

Since from the 10 percent maximum potential saving in energy from the total elimination of rotational losses, the contra-propeller action of the ship rudder alone is already contributing 3 to 4 percent, it will be clear from the above that part of the 4 to 8 percent power savings obtained with the reaction fin has to be attributed to an improvement in the flow on the after body and in the propeller-hull interaction.

Finally, some attention will be paid to a variation of ducted propeller design wherein the duct is located forward of the propeller and is integrated into the hull, the *integrated duct propeller* (Narita, et al, 1981). A drawback of conventional ducted propellers may be the erosion on the innerside of the duct caused by the heavy implosion pressures of cavitation. A duct forward of the propeller can be of advantage in this respect. The hydrodynamic investigations into this arrangement led further to the following conclusions:

- Application of the integrated duct propeller results in decreased hull resistance. Also the flow into the propeller is homogenized and stabilized.
- The duct thrust significantly increases in the behind condition due to inward radial flow around the stern.
- The effective wake fraction and the thrust deduction factor are dependent on the longitudinal location of the duct.
- The asymmetrical integrated duct propeller with larger chord length at the top performed better compared to the symmetrical duct.

Full-scale trials on a 250,000-dwt tanker indicated power savings of five percent. Cavitation observations with and without the device showed some reduction in sheet cavitation and an elimination of cloud cavitation. The hull pressure fluctuations were found to be reduced also by some 50 percent for both the sixth (blade frequency) and twelfth (twice the blade frequency) harmonics.

Finally, comparison of maneuvering tests at full scale showed some improved maneuverability with the Integrated Duct Propeller. These test results, however, were not corrected for wind and current.

To obtain some idea of the expected power savings with the use of the devices mentioned in this and previous sections, the table of Muntjewerf has been reproduced in Table 29. Some of the mentioned arrangements need more detailed investigations to gain further insight into the fields of application.

Section 11
Ship Standardization Trials

11.1 Purpose of Trials. Standardization trials are carried out to establish the relationship between the speed, shaft horsepower and propeller revolutions per minute under specified conditions of displacement, draft and trim.

They are run for one or more of the following purposes:

(*a*) To fulfill contractual obligations between builders and owners relating to speed, horsepower, and fuel consumption;

(*b*) To obtain performance data on full-sized ships for use in future designs;

(*c*) To obtain model-ship correlation allowances (C_A) for different types of ships and different hull surface conditions in order to improve the prediction of ship powers from the results of model tests;

(*d*) To determine the relationship between ship's speed and RPM of the propeller for use by the ship as an aid to navigation after making any necessary corrections that may be established in service.

11.2 General Plan of Trials. The most commonly used method of conducting standardization trials has been to make several consecutive runs, alternating in direction, over a measured distance at substantially constant RPM, observing the speed, power and rpm over each such run. Such observations of fixed landmarks or buoys give the speed over the ground, whereas we are interested in the speed of the ship through the water. These two speeds differ because of the effects of currents in the water due to tides or other causes. It is to eliminate such effects that the consecutive runs are necessary. Generally, three runs should be made at each of a number of engine settings, although two may be acceptable where the current conditions are favorable. Each series of runs at the same RPM should be made in uninterrupted sequence, so that the subsequent analysis will give a close approximation to the true speed through the water.

The number of such groups of runs needed in any particular case will depend upon the speed range to be covered, but should not be less than four, made at engine settings from half to full speed ahead. The intervals in rpm between the groups should be approximately equal. In the case of high-speed ships, such as destroyers, where there are definite humps and hollows in the resistance curve, additional points should be taken to delineate these features.

To eliminate uncertainties in the interpretation of the results, and in particular to obtain reliable model-ship correlation allowances, the ship on trial should be newly cleaned and painted, and the trials should be run under near-ideal conditions with little wind and sea.

The basic data measured on trial should include the following:

(*a*) The time taken to cover the measured course.

(*b*) The total propeller revolutions over the full length of the course.

(*c*) The thrust on the propeller (in ships fitted with a recording thrustmeter).

(*d*) The power transmitted to the propeller, preferably determined by measuring the torque in the propeller shaft by means of a torsionmeter. This is the only satisfactory way of obtaining the P_D. If no torsionmeter is fitted, then the power can be determined by measuring the electrical power input to the motors in electrically propelled ships or the mean pressures in the cylinders of diesel or steam engines. The instruments used to measure the pressures are called indicators and the corresponding power is called the indicated power. This is not likely to be very accurate, especially in motorships, and if reliable results are required a torsionmeter should always be used.

(*e*) Torsionmeter zero; i.e., the meter reading when there is no torque in the shaft. Before the installation of the meter, the shaft will have come to rest under the influence of the friction in the sterntube, shaft bearings and engine, so that there is usually an unknown residual torque in the shaft. Hence, the zero setting on the meter when fitted to the shaft does not correspond to zero torque.

On turbine ships the recommended procedure for determining the torsionmeter reading for zero shaft torque is by means of *drag shaft* tests. These are conducted at the beginning and end of each day's trials. However, it is desirable that there should be no prolonged endurance runs or crash stops between the finish of the speed runs and the drag tests to avoid any chance of the torsionmeter moving on the shaft. Such movement should not occur if the meter is fitted properly in the first place, but any doubts on this point will cast suspicion on the power measured on the trials.

The drag shaft tests are made in both the ahead and astern directions. With the ship operating steadily at about one-third ahead speed, the throttle is closed and the shaft allowed to coast to a stop without the use of astern power. Torque readings are taken at 10-sec intervals until the shaft stops. The engine is then backed at about two-thirds speed until steady astern conditions are reached, when the engine is again stopped. Torque readings are taken while the shaft coasts to a stop, no ahead power being used. The shaft usually stops more quickly under astern conditions, so that the torque in this case is usually read at 5-sec intervals.

The torque readings will be high at first, falling to a reasonably steady minimum value and than increasing slightly just before the shaft stops. The mean value of the average minimum ahead and astern torque readings is taken as the zero reading for the meter. The torque readings taken during the speed runs are corrected by adding or subtracting this zero reading, depending on its sign.

In conducting a drag shaft test on ships having more than one propeller, all engine operations and all torque readings are made simultaneously.

In direct-drive diesel and steam reciprocating-engined ships, the zero torque reading has to be obtained by means of the turning gear with the ship at rest. The torque measured is then assumed to be that due to friction in the sterntube and in the shaft bearings astern of the torsionmeter. If this is assumed to be equal for both directions of turning, the zero value will be given by the mean value of the ahead and astern readings.

Drag shaft tests or turning-gear tests should not be conducted until the shafting has warmed up to operating temperature.

(*f*) Clock time at the commencement of each run, direction of each run, and time of slack water.

(*g*) Sufficient data to determine the displacement and trim of the ship on each run. Drafts should be taken with the ship stationary in sheltered waters before the trial. The density of the water should be obtained at the same time by taking samples of water from about half-draft at each end of the ship. The displacement and trim during the trials should be calculated from the initial values by allowing for consumption of fuel and stores together with any changes made in ballasting arrangements. The final displacement should be checked after the trials by reading the drafts, if this is at all possible.

(*h*) Sea-water temperature and density on the course.

(*i*) Apparent wind speed and direction for each run, measured by anemometer.

(*j*) Depth of water on course, by echo sounder, especially if there are nearby shallow areas which might affect the ship's speed.

(*k*) Record of any unusually large rudder angles used during the measured runs.

(*l*) The size and direction of sea waves.

The date of the last dry-docking previous to the trial should be noted, together with the condition of the shell and paint, and the location of the ship since undocking should be recorded. When possible, a definition of the paint system together with records of the hull surface roughness should be made in order to assist in the derivation of meaningful model-ship correlation allowances, since such roughness is one of the prime causes of additional resistance, and can be the reason for otherwise inexplicable differences, even between sister ships.

The condition of the propeller-blade surfaces should also be noted at the time of docking.

Much information on the details of how to conduct standardization trials and the precautions to be observed will be found in a publication issued by The Society of Naval Architects and Marine Engineers (SNAME, 1973).

11.3 Measurement of Speed. For many years the common method of conducting standardization trials has been to use a measured course. With the continuing increase in size and speed of the larger ships, the available measured courses around the coasts of the United States, except for those at Guantanamo in Cuba and Point Vincente in California, have become inadequate for trials with such ships. These ships and many others are now often standardized in open water using electronic tracking systems, which are more convenient, reduce the time and cost of the trials and give results accurate enough for commercial purposes.

A measured-mile course is usually defined by two pairs of beacons ashore, the lines passing through the two pairs being parallel, the perpendicular distance between them usually being 1 nautical mile.

The minimum depth of water required in any given ship trial can be calculated by the methods set out in Section 5.5, Chapter V. In cases where an inadequate depth has to be accepted, these same methods will also allow an estimate to be made of the shallow-water effect and hence of the speed the ship would attain in deep water. Some figures of the loss of speed for dif-

Table 30—Speed Loss in Shallow Water

Ship	Percentage speed loss in water of depth 110 m	60 m	45 m	40 m
65,000-ton tanker at 16½ knots	0	0.9	2.0	2.7
16,000-ton tanker at 15½ knots	0	0.1	0.5	0.9
137-m cargo liner at 17 knots	0	0	0.2	0.5
174-m passenger ship at 22 knots	0	0.1	0.6	1.1
108-m cross-channel ship at 22½ knots	0	0	0.1	0.3
56-m trawler at 13½ knots	0	0	0	0

Table 31—Distance Required for Acceleration of Ships

Ship	Displacement ton	Full speed, knots	Distance required for acceleration from rest to 99.8 percent of full speed, nautical miles: With constant torque	With constant power
Low-speed cargo ship	11,850	10.5	2.68	2.27
Cargo liner	15,850	16.0	3.00	2.35
Passenger ship	48,158	24.0	3.01	2.60
Fast cross-channel ship	2,093	24.5	0.43	0.39

ferent types of ships in different depths of water have been given by Lackenby, Table 30, which are useful for general guidance, (Lackenby, 1963).

The ship should approach the measured mile on the prescribed course at steady propeller RPM for a sufficient distance to ensure that acceleration has ceased before entering the mile. The length of the approach run depends on the ship's displacement, the resistance and power characteristics, the change of speed required and the manner in which the machinery is operated. A run of 2 or 3 miles is ordinarily sufficient for large seagoing vessels of moderate speed, although this may prove inadequate in very large ships such as tankers. Estimates can be made of the length of approach run required by a method given by Lackenby (1952).

As examples, he gave the length of run required to accelerate a ship from rest to 99.8 percent of full speed, using either the full-speed constant torque or full-speed constant power throughout (Table 31).

In these examples the ships have been assumed to start from rest. Similar calculations can be made for other initial speeds, but initial speed has a quite small effect upon the distance necessary to reach full speed. Starting with a speed as high as 0.75 of the full speed, the distance required is still some 0.85 of that needed from rest. In view of the substantial loss of speed during the turn between successive mile runs, the distance needed to reach steady speed from rest may be taken as a guide to the length of a suitable approach run (Lackenby, 1952).

The use of pitot and propeller logs as an alternative to the measured mile is not particularly satisfactory, because of the difficulty of getting them sufficiently far out from the hull to be clear of the boundary layer. Even then, they are affected by the potential and wave flow at the point in question. They are useful for monitoring service performance if calibrated first on measured-mile trials, although the calibration may alter with time as the ship becomes rougher and the boundary layer thicker.

Modern electronic tracking and positioning systems offer a number of different ways of conducting speed trials that are economical and convenient.

There are two broad classifications of such systems, one based on the timing of a radio pulse (as is radar itself), and one based on a phase comparison of electrical waves, the latter being intrinsically capable of far greater accuracy than the former.

Because of the adoption of electronic tracking for many merchant ship trials, extensive studies have been made of the optimum procedure for such trials, and it has been determined that accuracy does not increase with length of run indefinitely; instead, there is an optimum length of run for optimum accuracy.

By far the greatest inaccuracies in such trials are due to the vagaries of local ocean currents. It has been found that the water in two areas only a few miles apart is often moving in quite different directions, and that these areas of motion are themselves in motion, in a sort of very large-scale, very mild turbulence. A ship making two long runs in opposite directions through such water will not be returning through the same water that it traversed on the first run, and the discussion of "means" in Section 11.4 is not applicable. The optimum length of run, for optimum accuracy, depends on the accuracy of the tracking system. An analysis of probable errors, including probable variations in current, leads to the conclusion that while 30 to 40 miles may be necessary for the pulse type of systems, for the more accurate phase-comparison system a much shorter run, with higher probability of traversing the same water in both directions and therefore more accurate "averaging," will result in higher accuracy.

In those areas where there is an electronic positioning network available, the time taken by the ship to cover a known distance can be observed, and an average speed over the ground obtained. This will not be the true water speed.

Another method is to run the ship directly away from and then towards a station on shore. Alternating runs carried out in this way enable the tidal currents to be dealt with as on a measured course, if the course is not too long, while the ship can remain well offshore, in deep water, since visual observation is no longer required.

Still a third possibility is to use either a radar reflector buoy or a special buoy containing the necessary phase-comparison equipment, floating freely in the sea,

Table 32—Standardization Trials of Single-Screw Ship *Old Colony Mariner*

(This is a Mariner-Class ship, and the trials were carried out on the measured-mile course at Rockland, Maine; and by electronic methods off Cape Cod.)

Ship Conditions:			
Length BP	161.08 m	Draft, mean	7.32 m
Length, LWL at trial draft	159.49 m	Trim by stern	1.00 m
Beam, molded	23.16 m	Displacement	16716 ton
Propeller particulars (manganese bronze):			
Diameter	6.71 m	MWR	0.277
Pitch at 0.7 radius	6.96 m	BTF	Variable
Number of blades	4	Proj. area ratio	0.418

Trial Courses:		
Course	Rockland, Maine	Cape Cod
Date of trial	October 15, 1952	October 17, 1952
Depth of water	100–120 m	160–180 m
Temperature of water	54.2F (12.3C)	55.5F (13.1C)
Density of water, ρ	1026.8 kg/m^3	1026.6 kg/m^3
Sea state	Slight	Moderate

(1) Run No. and direction	(2) Time of day at start	(3) Time on mile (average of 4)	(4) Observed speed in knots over ground V_o	(5) Relative wind direction
	pm	m s		
14 N	2.49	2 40	22.50	33° S
15 S	3.14	2 40.4	22.44	24° P
16 N	3.41	2 38.9	22.66	39° S
Mean of means			22.51	

(6) Relative wind velocity, V_R	(7) Wind direction coefficient, k	(8) Power correction for wind, δP (kW)	(9) Slope of P_E curve, $\delta P/\delta V$	(10) Speed correction for wind, δV, knots
25.6	1.17	542		0.21
31.7	1.19	844	2563	0.33
26.0	1.06	506		0.20

(11) Corrected speed over ground, V_G	(12) rpm n	(13) P_s	(14) $\frac{n}{V}$ average	(15) Corrected V through water, V_W
22.71	109.11	17621		22.76
22.77	109.23	17641	4.792	22.79
22.86	109.09	17702		22.76
22.78	100.16	17651		22.78

(16) Current $V_W - V_G$	(17) Time of day at middle of run
	pm
0.05 S	2.50
0.02 N	3.15
0.10 N	3.42

Table 33—Summary of Trial Results for *Old Colony Mariner*

(Displacement = 16,716 tons. Details in Table 32)

(1) Runs	(2) Mean speed in knots V corrected for wind and tide	(3) Mean rpm n	(4) Mean P_s	(5)	(6) $\frac{P_s}{n^3} \times 10^2$
		Rockland—October 15, 1952			
1–3	14.60	62.18	2,792	0.897	1.162
4–6	18.13	78.93	5,892	0.989	1.198
7–9	20.21	89.75	8,896	1.028	1.231
10–12	21.52	99.80	13,027	1.307	1.311
14–14 [a]	22.78	109.16	176.50	1.493	1.353
		Cape Cod—October 17, 1952			
1–3	14.70	62.73	2,941	0.926	1.191
4–6	20.71	92.68	9,907	1.115	1.244
7–9	22.53	107.29	16,667	1.457	1.349

[a] Run No. 13 was discarded, as ship had not reached steady conditions.

carrying out successive runs towards and away from it. In this case, if the trials are carried out in the ocean, well away from land, and, again, if the course is not too long, the buoy and ship may be considered as subject to the same current and rate of drift, so that the speed measured will be the speed through the water. Repeat runs at the same speed in the opposite direction are no longer necessary, so greatly reducing the total time required for the trials. The use of a radar reflector buoy in conjunction with the ship's own radar will generally require too long a run to be practicable for standardization runs. Much information on the conduct of speed trials by electronic tracking methods and the analysis of the data will be found in papers by Comstock, et al (1952) and Verstelle (1953).

11.4 Analysis of Speed Trials. There are several ways of analyzing speed trials, that described here being the one used at DTRC.

During the trials, running plots are kept of the following:

(*a*) Speed V against RPM n.

(*b*) Shaft power, P_S, against speed V.

(*c*) Ratio of V/n against clock time as a check on the tidal current.

(*d*) Ratio of P_S/n^3 against n.

The spots for runs in the same direction should fall on fair curves—any departure will suggest some inaccuracy in the data, which can be investigated at the time and corrected.

The analysis is based on Eggert's power method as described by Pitre (1932). It attempts to evaluate the effects of wind and current in order to reduce the data to standard model-basin conditions of zero current and zero air resistance. The procedure is best described by applying it to an actual example. Table 32 shows part of the trial results for the high-speed cargo ship *Old Colony Mariner*. Trials were conducted over the measured mile at Rockland and also by electronic means off Cape Cod so that the results given by the two methods can be compared.

The analysis of the highest speed group of 3 runs at Rockland is shown in detail in Table 32 and the final results over the whole speed range and for both trials are summarized in Table 33.

The procedure is described by reference to the numbered columns in Table 32:

Col (1) shows the run numbers and direction.

Col (2) lists the time of day for the start of each measured-mile run.

Col (3) gives the elapsed time on each run, the value being an average from four independent observers.

Col (4) shows the resulting observed speed V_0 over the ground.

Cols (5,6) give the direction and velocity of the relative wind, as observed by anemometer.

Cols (7,8) give the estimated wind resistance, based upon experiments carried out in a wind tunnel upon a superstructure model of the ship.

The air resistance for a head wind was expressed in the form (see also Section 5.2, Chapter V).

$$R_{AA} = k \times \text{coefficient} \times A_T \times V_R^2 \qquad (92)$$

where

A_T is above-water cross-sectional area of ship, m²

V_R is relative wind velocity, m/sec.

k is a wind-direction coefficient representing ratio of increase in axial resistance for any angle of relative wind to that for an ahead wind. In this case, k was obtained from specific model experiments.

The increase in P_E necessary to overcome the wind resistance is then given by

$$\delta P_E = R_{AA} \times V_0 \qquad (93)$$

For ships where no model tests have been made to determine the wind resistance, close estimates can be made by the methods described in Section 5.2, Chapter V. For this particular ship $A_T = 457$ m^2.

Col (9) shows the rate of increase in P_E with increase in speed, $\delta P / \delta V$. Values of $\delta P / \delta V$ are derived from the model P_E curve and plotted to a base of speed. For this ship the value at the mean speed of these three runs is 2563 kW per knot.

Col (10) gives the speed correction δV for the wind, being equal to

$$\delta V = \frac{\delta P}{\delta P / \delta V} = \frac{542}{2563} = 0.21 \text{ knot}$$

Col (11) gives the speed over the ground, V_G, after correction for wind and air resistance. It is seen that this effect is quite large. It varies with strength and direction of wind, type of vessel and speed, and is greatest in slow ships with large superstructures (see Section 5.2, Chapter V). It is highest when the relative wind is about 30 deg off the bow. The correction used in the analysis is only an approximation, and trials should not be run in high winds entailing large corrections.

Wind also means carrying rudder, and so gives rise to additional resistance.

The Sea Trial Code (SNAME, 1973) suggests the following maximum true wind speeds for the successful conduct of speed trials:

High-power, heavy vessels	25 knots
Large passenger vessels	20 knots
Smaller ships	15 knots

The speeds V_G in column (11) are still influenced by the current existing over the trial course during the runs. This current effect is eliminated as far as possible by taking the "mean of means" of the three runs on the measured mile at the same RPM but in different directions. If these speeds are V_1, V_2, and V_3, the mean speed V_m is found as follows:

	Mean speeds	Mean of means speed
V_1		
V_2	$\dfrac{V_1 + V_2}{2}$	
V_3	$\dfrac{V_2 + V_3}{2}$	$\dfrac{V_1 + 2V_2 + V_3}{4} = V_m$

This method of weighing gives the correct ship speed only if the speed of the tide is increasing or decreasing uniformly with time.

Similarly, the mean of two runs will give the correct speed only if the tide is constant, while that of 4 runs, $V_m = (V_1 + 3V_2 + 3V_3 + V_4)/8$ will be correct only if the tide speed is varying parabolically with time.

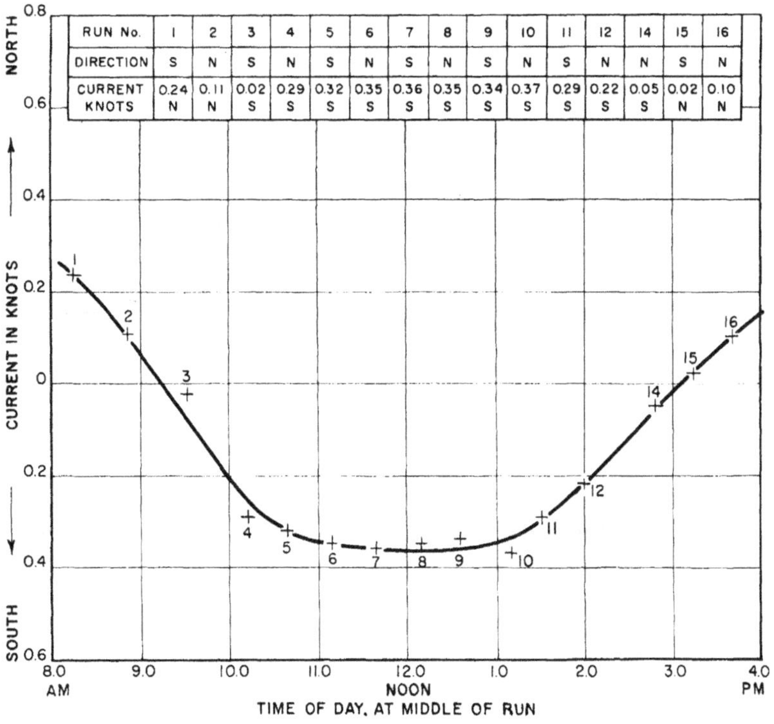

Fig. 100 Tidal current variation

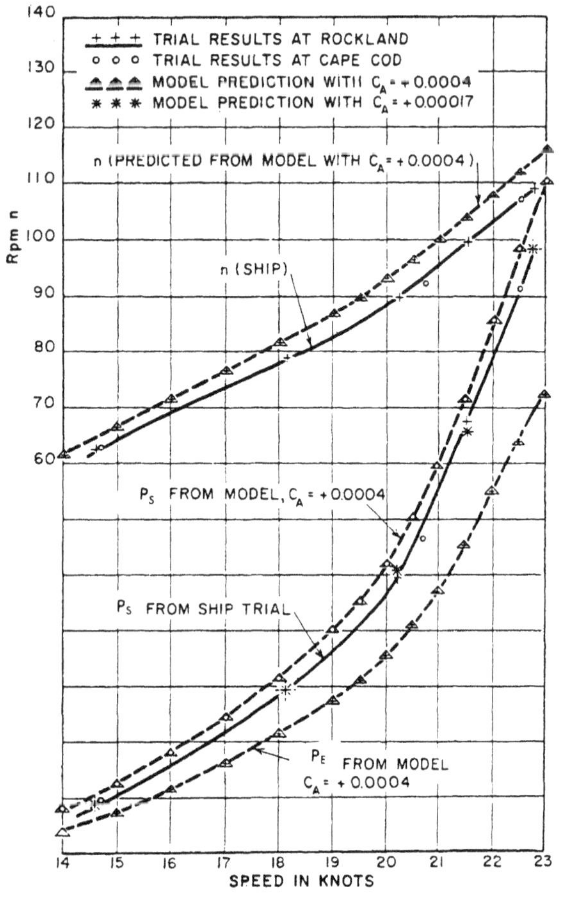

Fig. 102 Trial analysis of *Old Colony Mariner*

The mean of means speed for runs 14, 15 and 16 is shown at the bottom of column (11).

Cols (12,13) The RPM, n, and the P_s for the three runs are given in these columns, together with the mean of means values.

Cols (14,15) The RPM are not the same for all the three runs. For small changes in RPM, the ratio of n/V is constant, and this may be used to find the true speed through the water, V_W.

The average ratio of n/V is $109.16/22.78 = 4.792$, as given in column (14).

Then the true speed through the water for any individual run will be $V_{14} = n_{14}/(n/V) = 109.11/4.792 = 22.76$. These figures are given in column (15).

Cols (16,17) The current speed on any particular run is given by $(V_W - V_G)$; i.e., by [Col (15) − Col (11)]. These values of the current are shown in column (16) and the time of day at the middle of the corresponding run in column (17). By plotting the current speed against this time of day, a tidal-current curve can be drawn as a check upon the consistency of the results, Fig. 100.

The collected results for the complete trials at Rockland and Cape Cod are given in Table 33, and plotted in Figs. 101 and 102. The curves of P_D/V^3 and P_D/n^3 are chiefly of use for fairing purposes. The former follows in a general way the shape of the © curve, while the curve of P_D/n^3 is almost level so long as the slip does not change materially. At higher speeds, when slip begins to increase, the curve of P_D/n^3 also rises.

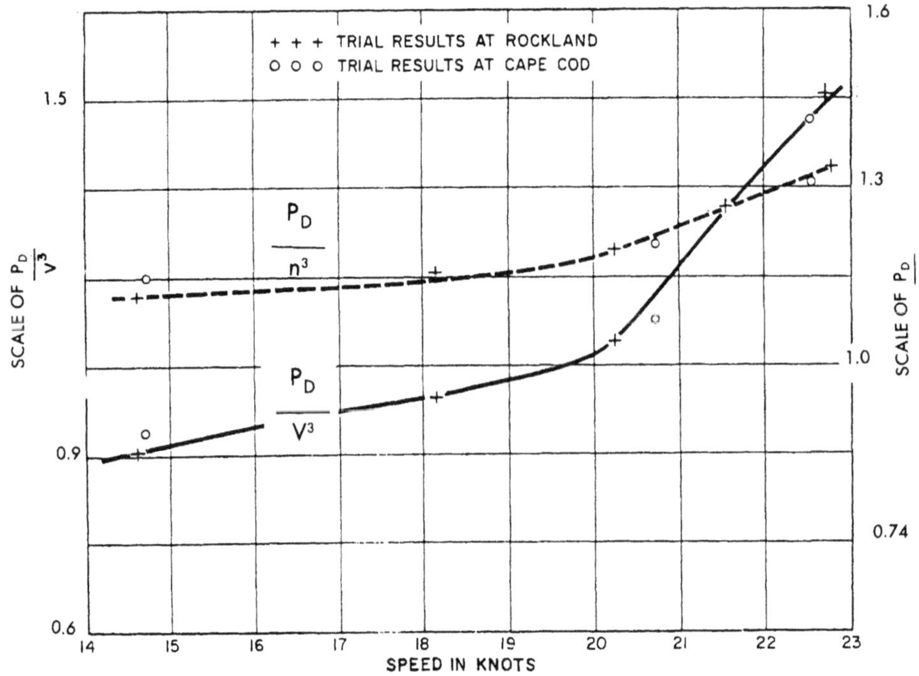

Fig. 101 Trial analysis for *Old Colony Mariner*; Curves of P_D/V^3 and P_D/n^3

Figs. 101 and 102 show the curves of ship P_D and n to a base of speed, together with curves of P_E, P_s and n as predicted from the model, and curves of P_D/V^3 and P_D/n^3. For the latter purpose, special experiments were run, after the trials, with the model in a condition corresponding exactly with that of the ship on trial. The predictions were made by the usual ATTC method, including an allowance of $C_A = +0.0004$.

Two points are of interest in this plot. The results from the Cape Cod trials, where the ship was a considerable distance offshore and in deep water, and made alternate runs towards and away from a land-based electronic station, are almost identical with those obtained on the measured mile. Such a method appears, therefore, to be an acceptable alternative to running trials over a measured course.

The second point is that both the predicted P_s and RPM, using the allowance of +0.0004, are higher than the values obtained on the trial, indicating that in this particular case the allowance is too high to give good correlation between model and ship. For more details concerning the allowance the reader is referred to Section 6.4 of Chapter V.

The above method of analysis is generally directed towards making a comparison between the powers measured on the ship and those predicted from the model. For this reason the effect of air resistance, whether due to the ship's own motion or to wind, or both, is as far as possible taken out of the comparison. The resultant ship speeds and powers are therefore equivalent to those we would expect if the ship were continually operating in a following wind of her own speed. To obtain powers for the ship in still air, it is necessary to add the resistance due to a relative head wind equal in velocity to the ship's own speed. Alternatively, the still air resistance may be considered as being included in the correlation allowance coefficient C_A.

The analysis of full-scale trials can be carried a little further to determine the analysis wake, for comparison with that measured on the model. This analysis wake is determined from the full-scale power and thrust measurements and the model open-water propeller curves.

The P_D at the propeller can be obtained from the measured P_S by allowing for the friction losses in bearings and sterntube. As stated in Section 1.4, the shaft transmission efficiency is normally taken as 0.98 for ships with engines aft and 0.97 for those with engines amidships.

The P_D can be written in the form

$$P_D = \frac{2\pi n Q}{60}$$

where Q is the torque in kNm and n is the RPM.

The torque coefficient K_Q is then

$$K_Q = \frac{3.6 \times 10^6 Q}{\rho n^2 D^5} \text{ with } n \text{ in RPM}$$

or

$$K_Q = \frac{60 \times 3.6 \times 10^6 \times P_D}{\rho 2\pi n^3 D^5}$$

Similarly, if the thrust T has been measured on the ship,

$$K_T = \frac{3.6 \times 10^6 T}{\rho n^2 D^4}$$

Entering the open-water performance curves for the propeller with the values of K_T and K_Q, the true advance coefficients J_{tT} and J_{tQ} can be found, where

$$J_t = \frac{30.87\, V_A}{nD}$$

with V_A in knots, D in m.

The apparent advance coefficient J_a is given by

$$J_a = \frac{30.87\, V}{nD}$$

Hence the analysis wake fractions w_T and w_Q can be determined, since

$$1 - w_T = V_A/V = J_{tT}/J_a \text{ and } 1 - w_Q = V_A/V = J_{tQ}/J_a$$

In the present example, thrust was not measured on the ship, so that only the torque wake w_Q can be found.

For the average speed runs 14, 15 and 16, the analysis using the open-water diagram of Fig. 103 leads to $w_Q = 0.261$.

The analysis of full-scale trials can finally be carried a little further to allow for current effects under ideal

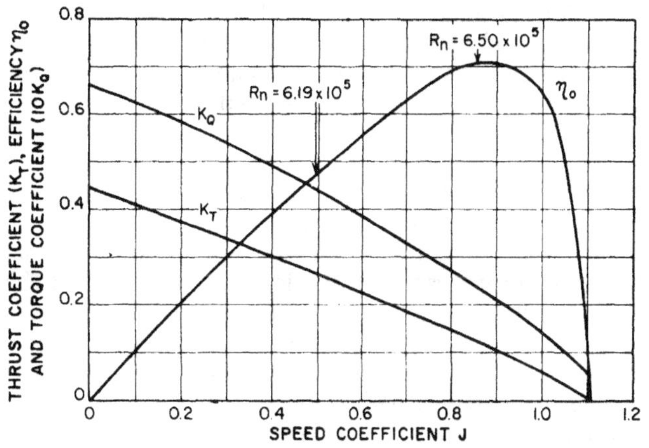

Fig. 103 Trial analysis of *Old Colony Mariner*
Propeller characteristic curves in open water

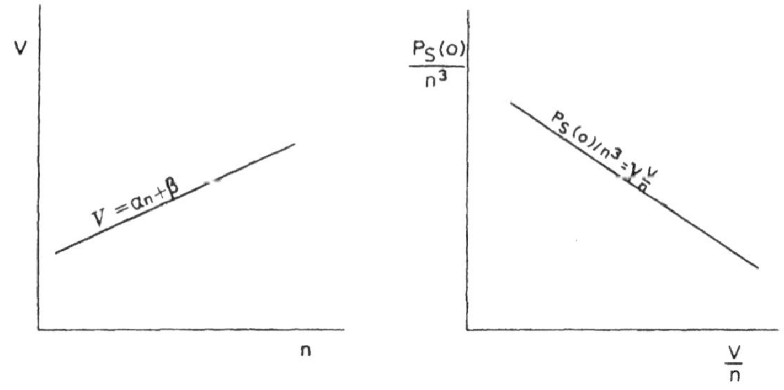

Fig. 104 Plot of Equations (94) and (95)
V=αn+β

weather conditions, by making use of the following relations between V, n and $P_{S(O)}$ given by Jourdain (1964)

$$V = \alpha n + \beta \qquad (94)$$

$$P_{S(O)}/n^3 = \gamma \frac{V}{n} + \delta \qquad (95)$$

As will be shown below, these two functions are practically linear over a wide range of operating conditions. They express the relation between ship speed through the water V and rate of rotation of the propeller n, and between the distance travelled per revolution V/n and the power coefficient $P_{S(O)}/n^3$.

If Q is the shaft torque, then instead of $P_{S(O)}/n^3$ the coefficient Q/n^2 or the torque coefficient K_Q could also be plotted as a function of V/n or the more logical dimensionless coefficient $J = V/nD$. The range of J-values corresponding to the range of operation from about half to full power being rather limited, the K_Q values can well be approximated by a linear function of J and hence the above mentioned linearity of Equation (95) will normally be valid.

A plot of Equations (94) and (95) shows the tendencies given in Fig. 104. If two speed points have been determined from mean of means analysis of sufficient runs per speed point, the Equations (94) and (95) can be solved and plotted. These plots will be found to be the same as the ones obtained from two arbitrary runs at different speeds if the latter could have been made in the absence of current.

The lines in the above figure drawn through the mean of means results can be used for various verifications. In the first place it is possible now to check for each run that the speed through the water V obtained from intersecting the left figure by n and the right figure $P_{S(O)}/n^3$ will be the same for both cases. If this is not the case it is certain that there is something wrong with the measurements or that the number of runs per speed point was not sufficient for eliminating the effect of current.

This last supposition can be verified easily now by determining the V for each run from Equation (94) and by deducing the current speed by subtracting V from the speed over the ground as measured on the mile for each run. These current speeds plotted as a function of clock time at the middle of each run will permit the following observations:

- The dispersion of the measuring points gives an indication of the accuracy of the measurements, and anomalous results can be detected and eliminated legitimately by estimating their real value by interpolation from the other points.
- A verification of the validity of the current function allowed in the mean of means approach can be made and if it is shown that appreciable differences exist between the actual and the allowed current function an iterative analysis of the current effect can be carried out by reintroducing the speed through the water found through use of the first current curve into Equation (94) and by repeating the process until convergence occurs.

Finally it should be noted, that the above analysis method can be used to advantage to complete series of runs per speed point that are incomplete due to mistakes made during the trials, while even single runs can be used now as long as they are properly surrounded by speed points determined from normal series of runs.

References

Abbott, I.H., Von Doenhoff, A.E. and Stivers, L.S. (1945), "Summary of Airfoil Data," NACA Report 824.

Aertssen, G. and Van Sluijs, M.F. (1972), "Service Performance and Seakeeping Trials on a Large Containership," *Trans.*, RINA, Vol. 114.

Allison, J.L. (1978), "Propellers for High-Performance Craft," *Marine Technology,* Vol. 15.

American Bureau of Shipping (1987), *Rules for Building and Classing Steel Vessels.*

Anonymous (1979), "Design and Evaluation of a Highly Skewed Propeller," MarAd Report No. MA-RD-940-79015, Washington DC.

Barnaby, S.W. (1897), "On the Formation of Cavities in Water by Screw Propellers at High Speed," *Trans.*, INA, Vol. 39.

Barnaby, S. W. (1900), "Marine Propellers," page 175.

Bavin, V.F. and Miniovich, I.J. (1963), "Experimental Investigation of Interaction Between Hull and Cavitating Propeller," Contribution of Leningrad Ship Model Basin, USSR, to 10th ITTC, London.

Betz, A. (1919), "Screw Propellers with Minimum Loss of Kinetic Energy," Reprinted in L. Prandtl and A. Betz, "Vier Abhandlungen zur Hydro- und Aerodynamik," Göttingen, 1927.

Betz, A. (1927), "Tragflügel und Hydraulische Maschinen," *Handbuch der Physik.*

Bienen, Th. and Von Karman, Th. (1924), "Zur Theorie der Luftschrauben," *Zeitschrift des Vereins Deutscher Ingenieure,* Nos. 48, 51 (Vol. 68) and No. 25 (Vol. 69, 1925).

Blaurock, J. (1983), "Propeller Plus Vane Wheel, An Unconventional Propulsion System," International Symposium on Ship Hydrodynamics and Energy Saving, El Pardo, Spain.

Boswell, R.J. (1969), "Static Stress Measurements on a Highly Skewed Propeller Blade," DTNSRDC Report No. 3247.

Boswell, R.J. (1971), "Design, Cavitation Performance, and Open-Water Performance of a Series of Research Skewed Propellers," DTNSRDC Report No. 3339.

Boswell, R.J. and Cox, G.G. (1973), "Design and Model Evaluation of a Highly-Skewed Propeller for a Cargo Ship," paper presented at the Chesapeake Section of SNAME. Also DTNSRDC Report No. 3672 (1974). Also *Marine Technology,* Vol. 11.

Boswell, R.J., Nelka, J.J. and Kader, R.D. (1975), "Experimental Spindle Torque and Open-Water Performance of Two Skewed Controllable-Pitch Propellers," DTNSRDC Report No. 4753.

Boswell, R.J., Miller, M.L. and Kadder, R.D. (1976), "Static Stress Measurements on a Series of Skewed Propeller Blades With and Without Forward Rake," DTNSRDC Report SPD-712-01.

Boswell, R.J., Jessup, S.D. and Nelka, J.J. (1979), "Experimental Unsteady and Time-average Loads on the Blades of a CP Propeller Behind a Model of the DD-963 Class Destroyer," Propellers '78 Symposium, SNAME S-6.

Bragg, E.M. (1922), "A Study of the Wake of Certain Models by Means of a Current Meter," SNAME *Transactions,* Vol. 30.

Brandt, H. (1973), "Modellversuche mit Schiffspropellern an der Wasseroberfläche," *Schiff und Hafen,* Vol. 25 (p. 415 and p. 493).

Burrill, L.C. (1943), "Developments in Propeller Design and Manufacture for Merchant Ships," *Trans.,* IME, London, Vol. 55.

Burrill, L.C. (1944), "Calculation of Marine Propeller Performance Characteristics," *Trans.,* NECI, Vol. 60.

Burrill, L.C. (1951), "Sir Charles Parsons and Cavitation," *Trans.,* IME, London, Vol. 63.

Burrill, L.C. and Emerson, A. (1953), "Propeller Cavitation—Some Observations from 16 Inch Propeller Tests in the New King's College Cavitation Tunnel," *Trans.,* NECI, Vol. 70.

Burrill, L.C. (1955), "The Optimum Diameter of Marine Propellers; A New Design Method," *Trans.,* NECI, Vol. 72.

Burrill, L.C. and Emerson, A. (1962–1963), "Propeller Cavitation—Further Tests on 16 Inch Propeller Models in the King's College Cavitation Tunnel," *Trans.,* NECI, Vol. 79.

Caster, E.B. (1963), "TMB 2 ,3 and 4-Bladed Supercavitating Propeller Series," DTMB Report 1637.

Comstock, J.P. and Hastings, C.E. (1952), "Raydist Speed-Measuring Equipment on the S.S. *United States* Sea Trials," SNAME *Transactions,* Vol. 60.

Cox, G.G. (1961), "Corrections to the Camber of Constant Pitch Propellers," *Trans.* RINA, Vol. 103.

Crown, D.E. and Hecker, R. (1971), "Inclined Shaft Performance of Two Partially Submerged Propellers Behind a Test Boat," DTNSRDC Report 249-H-14.

Cumming, R.A., Morgan, W.B. and Boswell, R. J. (1972), "Highly Skewed Propellers," SNAME *Transactions,* Vol. 80.

Cumming, R.A. and Morgan, W. B. (1974), "Propeller Design Aspects of Large High-Speed Ships," Symposium on High Powered Propulsion of Large Ships, NSMB Publication No. 490, Wageningen, Holland.

De Groot, D. and Hoffmann, F. J. (1949), "Designing of Screws of Large Blade Area by Means of the Gawn Series in B_{p-w} Form," NSMB, Pub. 85.

Denny, S.B. (1967), "Comparisons of Experimentally Determined and Theoretically Predicted Pressures in the Vicinity of a Marine Propeller," DTNSRDC Report No. 2349.

Dickmann, H.E. and Weissinger, J. (1955), "Beitrag zur Theorie Optimaler Düsenschrauben (Kortdüsen)," *Jahrbuch* der Schiffbautechnischen Gesellschaft, Vol. 49.

Dyne, G. (1973), "On the Scale Effect on Thrust Deduction," *The Naval Architect,* RINA, London.

Eisenberg, P., Preiser, H.S. and Thiruvengadam, A. (1965), "On the Mechanism of Cavitation Damage and Methods of Protection," SNAME *Transactions,* Vol. 73.

Ellis, A.T. (1965), "Parameters Affecting Cavitation and Some New Methods for Their Study," California Institute of Technology Hydrodynamics Laboratory, Report E.115.1.

Ellis, A.T. (1966), "On Jets and Shockwaves from Cavitation," 6th Naval Hydrodynamics Symposium, Washington DC., ONR

English, J.W. (1962), "The Application of a Simplified Lifting Surface Technique to the Design of Marine Propellers," National Physical Laboratory Report SHR 30/62.

Ficken, N.L. (1966), "Conditions for the Maximum Efficiency Operation of Cycloidal Propellers," Presented at the Chesapeake Section, SNAME.

Ficken, N.L. and Dickerson, M.C. (1969), "Experimental Performance and Steering Characteristics of Cycloidal Propellers," DTNSRDC Report 2983.

Fleischer, K.P. (1974), "Untersuchungen über das Zusammenwirken von Schiff and Propeller bei Teilgetauchten Propellern," *Jahrbuch* Schiffbautechnischen Gesellschaft, Vol. 68.

Garner, H.C. (1948), "Methods of Approaching an Accurate Three-Dimensional Potential Solution for a Wing," Aeronautical Research Council, R & M 2721.

Garner, H.C. (1949), "Theoretical Calculations of the Distribution of Aerodynamic Loading of a Delta Wing," Aeronautical Research Council, R & M 2819.

Gawn, R.W.L. (1937), "Results of Experiments on Model Screw Propellers with Wide Blades," *Trans.*, INA, Vol. 79.

Gawn, R.W.L. (1953), "Effect of Pitch and Blade Width on Propeller Performance," *Trans.*, INA, Vol. 95.

Gawn, R.W.L. and Burrill, L.C. (1957), "Effect of Cavitation on the Performance of a Series of 16 Inch Model Propellers," *Trans.*, INA, Vol. 99.

Gent, W. van and Oosterveld, M.W.C. (1983), "Ducted Propeller Systems and Energy Saving," *Proceedings* International Symposium on Ship Hydrodynamics and Energy Saving, Madrid. Also NSMB Publication No. 756 a.

Gent, W. van and Oossanen, P. van (1973), "Influence of Wake on Propeller Loading and Cavitation," Second Lips Propeller Symposium, Drunen, Holland; also NSMB Publication No. 442; also *International Shipbuilding Progress*, Vol. 20.

Giblon, R.P. (1975), "Service Margins and Power Plant Selection," *Proceedings,* First Ship Technology and Research Symposium, SNAME.

Godfrey, D.J. (1959), "Cavitation Damage—A Review of Present Knowledge," *Chemistry and Industry.*

Goldstein, S. (1929), "On the Vortex Theory of Screw Propellers," *Proceedings* of the Royal Society, Vol. 63, London.

Goldsworthy, E.C. (1939), "The Voith-Schneider System of Propulsion," *Engineering,* London.

Greenhill, A.G. (1888), "A Theory of the Screw Propeller," *Trans.*, INA, Vol. 29.

Grim, O. (1982), "Propeller und Leitrad auf dem Forschungsschiff *Gauss,* Ergebnisse und Erfahrungen," *Jahrbuch* der Schiffbautechnischen Gesellschaft, Vol. 76.

Grim, O. (1983), "Propeller and Vane Wheel, Contrarotating Propeller," 8th West European Graduate Education in Marine Technology, Gothenburg.

Groot, D. de and Hoffmann, F.J. (1949), "Designing of Screws of Large Blade Area by Means of the Gawn Series in $B_{p\,w}$Form," NSMB, Pub. 85.

Guilloton, R. (1949), "The Calculation of Ship Screws," *Trans.,* INA, Vol. 91.

Guilloton, R. (1955), "Application de la Courbure Induite au Calcul des Hélices Marines," Bulletin Association Technique Maritime et Aeronautique, N. 54.

Guilloton, R. (1955), "La Courbure Induite des Ailes Minces," Bulletin Association Technique Maritime et Aeronautique, No. 54.

Gutsche, F. (1933), "Versuche über die Profileigenschaften der Blattschnitten von Schiffschrauben," Mitteilungen der Preuszischen Versuchsanstalt, Berlin.

Gutsche, F. (1938), "Einfluss der Gitterstellung auf die Eigenschaften der im Schiffschrauben Entwurf benutzten Blattschnitt," *Jahrbuch* Schiffbautechnischen Gesellschaft.

Haberman, W.L. and Harley, E.E. (1961), "Performance of Vertical-Axis (Cycloidal) Propellers Calculated by Taniguchi's Method," DTMB Report 1564.

Haberman, W.L. and Caster, E.B. (1962), "Performance of Vertical-Axis (Cycloidal) Propellers According to Isay's Theory," *International Shipbuilding Progress,* Vol. 9.

Hadler, J.B., Morgan, W.B. and Meyers, K.A. (1964), "Advanced-Propeller Propulsion for High-Powered Single-Screw Ships," SNAME *Transactions,* Vol. 72.

Hadler, J.B. (1965), "Experimental Determination of Propeller Unsteady Forces at DTMB," *Proceedings,* First Conference on Ship Vibration, DTNSRDC Report 2002.

Hadler, J.B. and Cheng, H.M. (1965), "Analysis of Experimental Wake Data in Way of Propeller Plane of Single and Twin-Screw Ship Models," SNAME *Transactions,* Vol. 73.

Hadler, J.B. and Hecker, R. (1968), "Performance of Partially Submerged Propellers," Seventh Symposium on Naval Hydrodynamics ONR.

Hadler, J.B., Neilson, R.P., Rowen, A.L., Sedat, R.D., Seibold, F. and Zubaly, R.B. (1982), "Large-Diameter Propellers of Reduced Weight," SNAME *Transactions,* Vol. 90.

Hagen, G.R., Comstock, E.N. and Slager, J. (1986), "Investigation of Design Power Margin and Correlation Allowance for Surface Ships," *Marine Technology,* Vol. 23.

Hammer, N.O. and McGinn, R.F. (1978), "Highly Skewed Propellers—Full Scale Vibration Test Results and Economic Considerations," *Proceedings,* Ship Vibration Symposium, SSC and SNAME.

Hanaoka, T. (1962), "Hydrodynamics of an Oscillating Screw Propeller," 4th Symposium on Naval Hydrodynamics, Washington DC., ONR.

Harvald, S.A. (1950), "Wake of Merchant Ships," *Danish Technical Press,* Copenhagen.

Harvald, S.A. (1967), "Wake and Thrust Deduction

at Extreme Propeller Loadings," SSPA Publication No. 61.

Harvald, S.A. (1980), "Wake Distribution and Wake Measurements," *Trans.*, RINA, Vol. 123.

Hecker, R., Peck, J.G. and McDonald, N.A. (1964), "Experimental Performance of TMB Supercavitating Propellers," DTMB Report 1432.

Hecker, R. (1973), "Experimental Performance of a Partially Submerged Propeller in Inclined Flow," SNAME Spring Meeting.

Henderson, R.E., McMahon, J.P. and Wislicenus, G.F. (1963), "The Design of Pumpjets," Ordnance Research Laboratory, Technical Memorandum TM 506.3810-02.

Henderson, R.E., McMahon, J.P. and Wislicenus, G.F. (1964), "A Method for the Design of Pumpjets," Ordnance Research Laboratory Serial No. NOw 63-0209-c-7.

Hoekstra, M. (1977), "An Investigation into the Effect of Propeller Hull Interaction on the Structure of the Wake Field," Det norske Veritas Symposium, Oslo.

Holden, K.O., Fagerjord, O. and Frostad, R. (1980), "Early Design-Stage Approach to Reducing Hull Surface Forces Due to Propeller Cavitation," SNAME *Transactions*, Vol. 88.

Holtrop, J. and Mennen, G.G.J. (1982), "An Approximate Power Prediction Method," ISP, Vol. 29.

Holtrop, J. (1984), "A Statistical Re-analysis of Resistance and Propulsion Data," ISP, Vol. 31.

Horn, F. and Amtsberg, H. (1950), "Entwurf von Schiffdüsen Systemen (Kortdüsen)," *Jahrbuch* der Schiffbautechnischen Gesellschaft, Vol. 44.

Huse, E. and Wang, G. (1982), "Cavitation-Induced Excitation Forces on the Hull," SNAME *Transactions*, Vol. 90.

Hylarides, S. (1978), "Some Hydrodynamic Considerations of Propeller-Induced Ship Vibrations," Ship Vibration Symposium, SSC and SNAME; also NSMB Publication No. 609.

Isay, W.H. (1957), "The Voith-Schneider Propeller in the Wake of a Ship," *Ingenieur-Archiv*, Vol. 25.

Isay, W.H. (1958), "Ergänzungen zur Theorie des Voith-Schneider Propellers," *Ingenieur-Archiv*, Vol. 26.

Jobst, A.L. and Pandler, P.A. (1972), "A Flow Visualization Study of a High Pitch Cycloidal Propeller," Engineering Research Associates, Report No. 60-2.

Jonk, A. and Rem, A. (1980), "Investigations into the Propulsive Performance of Simplified Hull Forms," Netherlands Maritime Institute Monograph, No. M 39.

Jourdain, M. (1964), "L'essai de Vitesse," Bulletin d'Association Technique Maritime et Aeronautique No. 64.

Kawada, S. (1928), "On the Fundamentals of the Vortex Theory of Airscrews," International Congress, Rome.

Kawada, S. (1936), "Induced Velocity by Helical Vortices," *Journal of the Aeronautical Sciences*, Vol. 3: also report of the Aeronautical Research Institute, Tokyo, 1939.

Keller, W.H. Auf'm (1966), "Enige Aspecten bij het Ontwerpen van Scheepsschroeven," *Schip en Werf.*

Kerlen, H., Esveldt, J. and Wereldsma, R. (1972), "Propulsion-, Kavitations- und Vibrationsverhalten von Überlappenden Propellern für ein Containerschiff," *Jahrbuch* der Schiffbautechnischen Gesellschaft, Vol. 64; also *International Shipbuilding Progress*, Vol. 19.

Kerwin, J.E. (1961), "The Solution of Propeller Lifting Surface Problems by Vortex Lattice Methods," Massachusetts Institute of Technology, Naval Architecture Department Report.

Kerwin, J.E. (1963), "Linearized Theory for Propellers in Steady Flow," Massachusetts Institute of Technology, Naval Architecture Department Report.

Kerwin, J.E. and Leopold, R. (1963), "Propeller Incidence Correction Due to Blade Thickness," *Journal of Ship Research*, Vol. 7.

Kerwin, J.E. and Leopold, R. (1964), "A Design Theory for Subcavitating Propellers," SNAME *Transactions*, Vol. 72.

Kim, H.C. (1964), "Hydrodynamic Aspects of Internal Pump Jet Propulsion," University of Michigan, Ann Arbor.

Kirsten, F.A. (1928), "A New Type of Propeller," *Journal of the Society of Automotive Engineers.*

Knapp, R.T., Daily, J.W. and Hammitt, F.G. (1970), *Cavitation*, McGraw-Hill Book Company.

Koelbel, J.G., Stolz, J. and Beinert, J.D., "How to Design Planing Hulls," Motor Boating Ideal Series, Vol. 49.

Kooij, van der J. (1979), "Experimental Determination of Propeller-Induced Hydrodynamic Hull Forces in the NSMB Depressurized Towing Tank," *Proceedings*, Symposium on Propeller Induced Ship Vibration, London.

Kort, L. (1934), "Der neue Düsenschrauben-Antrieb," *Werft-Reederei-Hafen*, Jahrgang 15.

Kramer, K.N. (1939), "The Induced Efficiency of Optimum Propellers Having a Finite Number of Blades," NACA TM 884.

Kresic, M. and Haskell, B. (1983), "Effects of Propeller Design-Point Definition on the Performance of a Propeller/Diesel Engine System with Regard to In-Service Roughness and Weather Conditions," SNAME *Transactions*, Vol. 91.

Küchemann, D. and Weber, J. (1953), *Aerodynamics of Propulsion*, McGraw-Hill Book Co., New York.

Kuiper, G. (1981), "Cavitation Inception on Ship Propeller Models," NSMB No. 655, Wageningen.

Lackenby, J. (1952), "On the Acceleration of Ships," *Trans.*, IESS.

Lammeren, W.P.A., van Manen, J.D. van and Oosterveld, M.W.C. (1969), "The Wageningen B-Screw Series," SNAME *Transactions*, Vol. 77.

Lerbs, H.W. (1945), "Applied Theory of Free Running Ship Propellers," AEW Report No. 42/48.

Lerbs, H.W. (1951), "On the Effects of Scale and Roughness on Free Running Propellers," *Naval Engineers Journal,* Vol. 63.

Lerbs, H.W. (1952), "Moderately Loaded Propellers with a Finite Number of Blades and an Arbitrary Distribution of Circulation," SNAME *Transactions,* Vol. 60.

Lerbs, H. (1955), "Contra-Rotating Optimum Propellers Operating in a Radially Non-Uniform Wake," DTMB Report 941.

Lerbs, H. (1958), "On Supercavitating Propellers," Second Symposium on Naval Hydrodynamics, ONR.

Lerbs, H. and Rader, H. (1962), "Über den Auftriebsgradienten von Profilen im Propellerverband," *Schiffstechnik,* Vol. 9.

Levine, G.H. and Hawkins, S. (1969), "Comments on Service Margins for Ships," SNAME Technical and Research Report R-8.

Lewis, E.V. (1958), "The Problem of Service Power Margins," *Shipbuilding and Shipping Record.*

Lewis, F.M. (1935), "Propeller Vibration," SNAME *Transactions,* Vol. 43.

Ludwieg, H. and Ginzel, J. (1944), "Zur Theorie der Breitblattschraube," Aerodynamische Versuchsanstalt, Göttingen, Report 44/A/08.

Luttmer, B.R.L., Hylarides, S. and Keller, J. Auf'm (1984), "Effect of Skew on Stresses in Backing Fixed-Pitch Propellers," Propellers 1984 Symposium, SNAME.

Magnus, H.G. (1853), "On the Deflection of Projectiles," *Poggendorff's Annual.*

Manen, J.D. van (1954), "Open Water Test Series with Propellers in Nozzles," *International Shipbuilding Progress,* Vol. 1.

Manen, J.D. van (1957), "Fundamentals of Ship Resistance and Propulsion," Part B, Publication No. 132a, NSMB; also published in *International Shipbuilding Progress,* Vol. 4.

Manen, J.D. van (1957), "Recent Research on Propellers in Nozzles," *International Shipbuilding Progress,* Vol. 4.

Manen, J.D. van (1958), "Wake-Adapted Screw Series Designed by the Induction Factor Method," *International Shipbuilding Progress,* Vol. 5.

Manen, J.D. van and Superina, A. (1959), "The Design of Screw Propellers in Nozzles," *International Shipbuilding Progress,* Vol. 6.

Manen, J.D. van (1962), "Effect of Radial Load Distribution on the Performance of Shrouded Propellers," *Trans.,* RINA, Vol. 104.

Manen, J.D. van (1963), "Bent Trailing Edges of Propeller Blades of High Powered Single Screw Ships," *International Shipbuilding Progress,* Vol. 10; also NSMB Publication No. 215.

Manen, J.D. van (1965), "The Choice of the Propeller," *Marine Technology,* Vol. 3.

Manen, J.D. van (1966), "Results of Systematic Tests with Vertical Axis Propellers," *International Shipbuilding Progress,* Vol. 13; also *Jahrbuch der* Schiffbautechnischen Gesellschaft, Vol. 57, 1963: also NSMB Publication No. 235.

Manen, J.D. van and Oosterveld, M.W.C. (1966), "Analysis of Ducted Propeller Design," SNAME *Transactions,* Vol. 74.

McCormick, B.W. and Eisenhuth, J.J. (1962), "Design and Performance of Propellers and Pumpjets for Underwater Propulsion," 17th Annual Meeting, American Rocket Society.

Miller, M.L. (1969), "Experimental Determination of Unsteady Propeller Forces," 7th Symposium on Naval Hydrodynamics, ONR.

Miller, R.T. (1963), "A Note on Sinkage and Squat," *Bureau of Ships Journal.*

Minsaas, K. and Slattelid, O.H. (1971), "Lifting Surface Corrections for 3-bladed Optimum Propellers," *International Shipbuilding Progress,* Vol. 18.

Moor, D.I., Parker, M.N. and Pattulo, R.N.M. (1961), "The BSRA Methodical Series—an Overall Presentation," *Trans.,* INA, Vol. 103.

Moore, D.H. and Peck, J.G. (1973), "Inclined Shaft Propeller Performance Characteristics," SNAME Spring Meeting.

Morgan, W.B., Silovic, G. and Denny, S.B. (1968), "Propeller Lifting Surface Corrections," SNAME *Transactions,* Vol. 76.

Morgan, W.B. and Caster, E.B. (1968), "Comparison of Theory and Experiment on Ducted Propellers," 7th Symposium on Naval Hydrodynamics, Rome, Italy.

Mueller, H.F. (1955), "Recent Developments in the Design and Application of the Vertical Axis Propeller," SNAME *Transactions,* Vol. 63.

Müller-Graf, B. (1978), "Systematische Untersuchung von Verstellpropellern," Mitteilungen der Versuchsanstalt fur Wasserbau und Schiffbau, Heft 1954, Berlin.

Multhopp, J. (1955), "Methods for Calculating the Lift Distribution of Wings (Subsonic Lifting Surface Theory)," Aeronautical Research Council, R & M 2884.

Muntjewerf, J.J. and Oosterveld, M.W.C. (1981), "Energy Saving Propulsion Arrangements," Lloyd's List Workshop, "Why Burn Money" at the SHIP-ASIA Conference, Hong Kong.

Narita, H., Yagi, H., Johnson, H.D. and Breves, L.R. (1981), "Development and Full-Scale Experimences of a Novel Integrated Duct Propeller," SNAME *Transactions,* Vol. 89.

NAVSEA (1984), "Prediction of Smooth-Water Powering Performance for Surface Displacement Vessels," Design Data Sheet 051-1.

Newton, R.N. and Rader, H.P. (1961), "Performance Data of Propellers for High-Speed Craft," *Trans.,* INA, Vol. 103

Newton, R. N. (1960), "*Proceedings* of Ninth ITTC," Paris.

Nishiyama, T. and Nakajima, Y. (1961), "Lifting Surface Theory of Widely Bladed Propellers," *Trans.*, JSNA, Vol. 109.

O'Brien, T.P. (1957), "Some Effects of Blade Thickness Variation on Model Screw Performance," *Trans.*, NECI, Vol. 73.

Oossanen, P. van and Kooij, J. van der (1972), "Vibratory Hull Forces Induced by Cavitating Propellers," *Trans.*, RINA, Vol. 115.

Oossanen, P. van (1974), "Calculation of Performance and Cavitation Characteristics of Propellers Including Effects of Non-Uniform Flow and Viscosity," NSMB No. 457, Wageningen.

Oosterveld, M.W.C. (1968), "Model Tests with Decelerating Nozzles," ASME Symposium on Pumping Machinery for Marine Propulsion.

Oosterveld, M.W.C. (1970), "Wake Adapted Ducted Propellers," NSMB No. 345.

Oosterveld, M.W.C. and Oossanen, P. van (1975), "Further Computer-Analyzed Data of the Wageningen B-Screw Series," *International Shipbuilding Progress*, Vol. 22.

Parsons, C.A. and Cook, S.S. (1919), "Investigations into the Causes of Corrosion or Erosion of Propellers," *Trans.*, INA, Vol. 61.

Pien, P.C. (1961), "The Calculation of Marine Propellers Based on Lifting Surface Theory," *Journal of Ship Research*, Vol. 5.

Pitre, A.S. (1932), "Trial Analysis Methods," SNAME *Transactions*, Vol. 40.

Posdunine, V.L. (1944), "On the Working of Supercavitating Propellers," *Trans.*, INA, Vol. 86.

Prandtl, L. and Betz, A. (1927), "Vier Abhandlungen zur Hydrodynamik und Aerodynamik, Göttingen.

Prandtl, L. (1979), "Application of Modern Hydrodynamics to Aeronautics," NACA Report No. 116; also in R.T. Jones (1979), "Classical Aerodynamic Theory," NASA Reference Publication RP-1050, Washington DC.

Rains, D.A. (1981), "Semi-Submerged Propellers for Monohull Displacement Ships," SNAME Propellers '81 Symposium.

Rankine, W.J.M. (1865), "On the Mechanical Principles of the Action of Propellers," *Trans.*, INA, Vol. 6.

Reech, F. (1852), "Cours de Méchanique." First announced in 1832, see E.G. Barrillon, North East Coast Institution of Engineers and Shipbuilders, Vol. 45, 1928–1929.

Rutgersson, O. (1978), "Supercavitating Propeller Performance Influence of Propeller Geometry and Interaction between Propeller, Rudder and Hull," ASCE-IAHR/AIHR-ASME Symposium on Design and Operation of Fluid Machinery.

Rutgersson, O. (1982), "Propeller-Hull Interaction Problems on High-Speed Ships; on the Influence of Cavitation," Symposium on Small Fast Warships and Security Vessels, RINA.

Sacks, A.H. and Burnell, J.A. (1962), "Ducted Propellers—A Critical Review of the State of the Art," *Progress in Aeronautical Sciences*, Vol. 3.

Scherer, J.O. (1977), "Partially Submerged and Supercavitating Propeller Systems," 18th ATTC.

Schoenherr, K.E. (1934), "Recent Developments in Propeller Design," SNAME *Transactions*, Vol. 42.

Schoenherr, K.E. (1963), "Formulation of Propeller Blade Strength," SNAME *Transactions*, Vol. 71.

Sladky, J. (1976), *Marine Propulsion*, The American Society of Mechanical Engineers, New York.

SNAME (1973), "Code for Sea Trials," Publication No. C-2.

Sparenberg, J.A. (1959), "Application of Lifting Surface Theory to Ship Screws," *Proceedings*, Kon. Nederlandse Academie van Wetenschappen, Series B, Vol. 62, Amsterdam.

Sparenberg, J.A. (1960), "On the Efficiency of a Vertical-Axis Propeller," Third Symposium on Naval Hydrodynamics, ONR.

Sparenberg, J.A. and Graaf, R. de (1967), "On the Optimum One-Blade Cycloidal Ship Propeller," Groningen University Report TW-50.

Stephens, H.G. (1974), "Open Water Performance of a Controllable-Pitch (C-P) Propeller Series," DTNSRDC Report SPD-011-13.

Strecheletzky, M. (1950), "Hydrodynamische Grundlagen zur Berechnung der Schiffsschrauben," Verlag G. Braun, Karlsruhe.

Strecheletzky, M. (1955), "Berechnungskurven für Dreiflügelige Schiffsschrauben," Verlag G. Braun, Karlsruhe.

Strom-Tejsen, J. and Porter, R.R. (1972), "Prediction of Controllable-Pitch Propeller Performance in Off-Design Conditions," Third Ship Control Systems Symposium, Bath, England.

Tachmindji, A.J. and Milan, A.B. (1956), "The Calculation of Goldstein Factors for Three, Four, Five and Six Blades," DTMB Report No. 1934.

Tachmindji, A.J. and Morgan, W.B. (1958), "The Design and Estimated Performance of a Series of Supercavitating Propellers," 2nd Symposium on Naval Hydrodynamics, ONR.

Takekuma, K., Tsuda, S., Kawamura, A. and Kawaguchi, N. (1981), "Development of Reaction Fin as a Device for Improvement of Propulsive Performance of High Block Coefficient Ships," *Journal*, JSNA, Vol. 150.

Tani, M., Okamoto, H., Hirabayashi, K. and Koyama, M. (1979), "Propulsion Systems for RORO Containerships with Particular Reference to the Controllable-Pitch Propeller," *Marine Technology*, Vol. 16.

Taniguchi, K. (1960), "Studies on a Trochoidal Propeller," Doctor of Engineering Thesis, Tokyo University.

Taniguchi, K. (1962), "Sea Trial Analysis of the Vertical-Axis Propellers," Fourth Symposium on Naval Hydrodynamics, ONR.

Timman, R. and Vossers, G. (1955), "On Michell's Expression for the Velocity Potential of the Flow Around a Ship," *International Shipbuilding Progress,* Vol. 2.

Todd, F.H. (1946), "The Fundamentals of Ship Propulsion," *Trans.,* Institute of Marine Engineers Vol. 58, London.

Triantafyllou, M.S. (1979), "Computer-Aided Propeller Design Using the B-Series," *Marine Technology,* Vol. 16.

Troost, L. (1938), "Open-Water Test Series with Modern Propeller Forms," *Trans.,* NECI, Vol. 54; also NSMB No. 33, Wageningen.

Troost, L. (1940), "Open-Water Test Series with Modern Propeller Forms," *Trans.,* NECI, Vol. 56.

Troost, L. (1951), "Open-water Test Series with Modern Propeller Forms," *Trans.,* NECI, Vol. 67.

Troost, L. (1956), "The Pitch Distribution of Wake Adapted Marine Propellers," SNAME *Transactions,* Vol. 64.

Troost, L. (1957), "A Simplified Method for Preliminary Powering of Single-Screw Merchant Ships," SNAME *Transactions,* Vol. 65.

Tsakonas, S., Breslin, J. and Miller, M. (1967), "Correlation and Application of an Unsteady Flow Theory for Propeller Forces," SNAME *Transactions,* Vol. 75.

Tsakonas, S. and Jacobs, W.R. (1969), "Propeller Loading Distributions," *Journal of Ship Research,* Vol. 13; also Davidson Laboratory Report No. 1319, 1968.

Tulin, M.P. (1955), "Supercavitating Flow past Foils and Struts," National Physical Laboratory Symposium on Cavitation in Hydrodynamics.

Tulin, M.P. (1964), "Supercavitating Propellers—History, Operating Characteristics and Mechanisms of Operation," Hydronautics Technical Report 127-6.

Van Manen; see Manen, J.D. van.

Venning, E. and Haberman, W.L. (1962), "Supercavitating Propeller Performance," SNAME *Transactions,* Vol. 70.

Verstelle, J.Th. (1953), "Methods of Conducting Ships' Speed Trials," *Journal of the Institute of Navigation,* Vol. 6, London.

Volpich, J. and Bridge, I.C. (1954–57), "Paddle-Wheels—Part I, II and III," IESS, 1954/55, 1955/56 and 1956/57.

Wald, Q.R. (1970), "Analysis of the Integral Pumpjet," *Journal of Ship Research,* Vol. 14.

Watts, P. (1909), "Speed Trials of HM Torpedo-Boat Destroyer *Cossack* at Skelmorlie and the Maplin Sands," *Trans.,* INA, Vol. 51.

Weinig (1932), "Strömung durch Profilgitter und einige Anwendungen auf die Strömung in Propeller," To be found in G. Kempf and E. Förster, *Hydromechanische Probleme des Schiffsantriebs,* page 171.

Weissinger, J. and Maass, D. (1968), "Theory of the Ducted Propeller; A Review," 7th Symposium on Naval Hydrodynamics, ONR.

White, W.H. (1882), *A Manual of Naval Architecture,* Second Edition, p. 652.

Wrench, J.W. (1957), "The Calculation of Propeller Induction Factors," DTMB Report 1116.

Yamazaki, R. (1962), "On the Theory of Screw Propellers," Fourth Symposium on Naval Hydrodynamics, ONR.

Zhu, D.M. (1981), "A Computational Method for Cycloidal Propellers," *International Shipbuilding Progress,* Vol. 28.

CHAPTER VII

William S. Vorus

Vibration

Section 1
Introduction

1.1 General. One of the problems in the design of all modern ships is the avoidance of objectionable elastic vibration of the hull structure in response to external or internal forces. Such vibration can cause discomfort to passengers and interfere with performance of crew duties, and damage or adversely affect the operation of mechanical and electrical equipment on board.

Since mechanical vibration can be defined generally as the oscillatory motion of rigid, as well as elastic, bodies, the subject of ship vibration is actually very broad in scope. In fact, all of the ship dynamics problems of primary interest to the naval architect, excluding maneuvering, involve some form of vibration.

For convenience, the overall response of a ship can be separated into two parts: One is the ship's motion as a rigid body in response to a seaway; the other is the elastic or flexural response of the hull or other structure to external or internal forces. Rigid body motions are considered under the general subject of seakeeping and are therefore not usually referred to as vibration. See Chapter VIII, Vol. III on motions in waves. Flexural vibration can be excited in the form of vertical and horizontal bending, torsion, and axial modes of the elastic structure of the hull girder, as well as in the form of local vibration of sub-structures and components. Such vibration that is excited by the ship's propellers is a particularly troublesome problem, and it will be the principal subject of this Chapter. Flexural vibration can also be excited directly by forces internal to rotating machinery, and by the external forces of sea waves encountered by the ship. Vibration excited by sea waves (referred to as springing and whipping) is considered under both Motions in Waves (Chapter VIII Vol. III), and Strength (Chapter IV Vol. I), although many of the basic principles of hull vibration covered in this chapter are directly applicable.

Concern about propeller-induced ship vibration has existed since the marine screw propeller was first developed in the mid 19th century; the French text book *Theorie du Navire* (Pollard and Dudebout, 1894)[1] included a chapter on propeller-induced ship vibration. In the early days the relatively few blades per propeller and the low propeller RPM excited ships at low frequency in a characteristically beam-like hull flexure. The early analytical work, such as that by Schlick (1884–1911) and by Krylov (1936) therefore concentrated on the application of beam theory in developing methods to help in avoiding propeller-induced ship hull vibration problems.

As ships have evolved the character of propeller-induced vibration has become more complex and vibration trouble has become more frequent. The greatest problems have occurred in the modern generation of oceangoing merchant ships. This has been due, in large part, to two aspects of design evolution that, aside from a consideration of vibration, qualify as technological advancements. These two aspects are the location of engine rooms and accommodations aft into the immediate vicinity of the propeller(s), and the increase in ship power. The increased use of diesel engines has also contributed to the increased frequency of vibration problems, but not to as great a degree. Ship vibration has also become a greater problem in recent years because of tightening of standards of acceptable vibration. Most commercial ship specifications now establish criteria on acceptable vibration; compliance must be demonstrated by the measurement of vibration on the vessel builder's trials. Today, exhaustive studies, employing both experimental and analytical methods, are conducted during the design of almost all large ships in attempting to avoid vibration troubles.

The object of this chapter is to discuss the basic theory and the practical problems of flexural vibration of ships' hulls, and of their sub-structures and com-

[1] Complete references are listed at end of chapter.

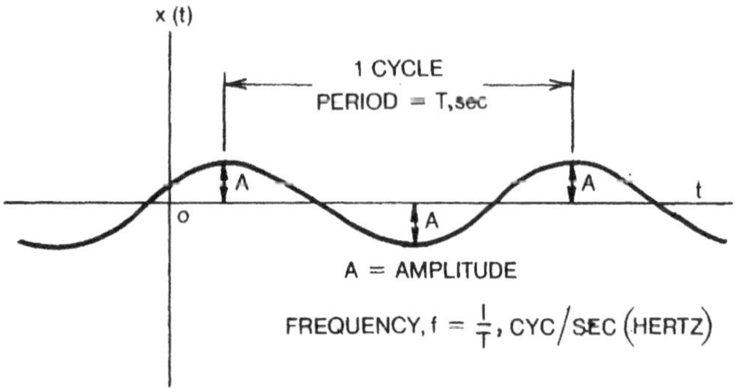

Fig. 1 Vibration displacement

ponents, with particular attention to propeller-excited vibration. Machinery-excited vibration is covered to a lesser extent.

A working knowledge of ship vibration requires the reader to be reasonably well versed in mathematics and engineering mechanics, as well as in a set of "tricks of the trade" with which naval architects, and engineers in general, usually feel more comfortable. However, a comprehensive knowledge of ship vibration theory is not necessarily required in order to work effectively with the subject as certain levels. Hence, this chapter has been organized so that readers with different interests and backgrounds can find material to meet their needs.

Section 2, Theory and Concepts, provides depth in understanding the fundamental concepts of ship vibration, as well as a foundation for further study of the techniques employed in vibration analysis. It is intended primarily for those whose theoretical tools are relatively close to the surface of their working knowledge.

The naval architect or shipyard engineer, interested more in design methods, can avoid some of the risk of becoming bogged-down in theory by proceeding to the third section, Analysis and Design. This section is self-contained and deals with practical solutions to potential vibration problems that should be addressed during the design stage.

The last section, Criteria, Measurements and Post-Trial Corrections, provides material for establishing whether or not vibration characteristics of a completed ship are satisfactory and how to make corrections, if necessary. The ship owner or operator, typically not particularly interested in design procedures, and not at all interested in vibration theory and concepts, may proceed directly to this section.

1.2 Basic Definitions. The following basic definitions are provided for the uninitiated. The definitions are loose, and aimed at the context most needed and most often used in the theory of vibration of ships.

Vibration—Vibration is a relatively small amplitude oscillation about a rest position. Fig. 1 depicts the variation in vibratory displacement with time.

Amplitude—For vibration of a fixed level of severity (steady-state periodic vibration), amplitude is the maximum repeating absolute value of the vibratory response, i.e., displacement, velocity, acceleration. Displacement amplitude for steady-state vibration is denoted as A on Fig. 1. For transient vibration, a time dependent amplitude sometimes may be defined.

Cycle—One cycle of vibration is the time between successive repeating points; refer to Fig. 1. The time required for completion of one cycle is its period.

Frequency—Frequency is the number of vibration cycles executed per unit time; it is the inverse of the vibration period.

Natural Frequency—A natural frequency is a frequency at which a system vibrates when stimulated impulsively from the rest position. The requirement for natural vibration is that the system possess both mass and stiffness. For continuous mass and stiffness distributions, the system possesses an infinite number of natural frequencies, even though only a relatively small number are usually of practical significance. On impulsive stimulation from rest, the continuous system will vibrate at all of its natural frequencies, in superposition; the degree of vibration at any particular natural frequency will depend on the characteristics of the impulsive stimulus.

Mode—Each different natural frequency of a system defines a mode of system vibration. The modes are ordered numerically upward from the natural frequency with the lowest value.

Mode Shape—A mode shape is a distribution of relative amplitude, or displacement shape, associated with each mode. Fig. 2 depicts mode shapes typical of a ship hull girder. The three vertical plane mode shapes shown correspond to the first three vertical plane flexural bending modes; two lower modes, with mode shapes corresponding to rigid body heave and pitch, occur at lower natural frequencies.

Node—A node is a null point in a distribution of vibratory displacement, or in a mode shape. In general, the number of nodes in a mode shape increases with modal order (natural frequency). This is the case of

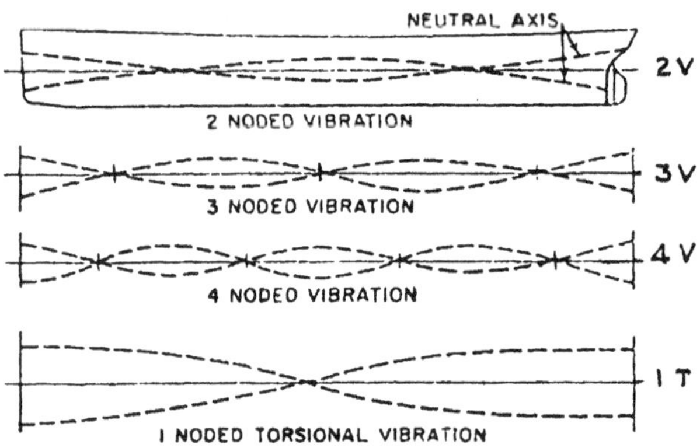

Fig. 2 Modes of hull girder vibration

the ship hull girder vibration depicted in Fig. 2; modes 2V, 3V and 4V have successively higher natural frequencies.

Excitation—Vibratory excitation is an applied time-dependent stimulus (force or displacement) that produces vibration. Excitation can be transient (e.g. impulsive), random, or periodic. A steady-state periodic excitation, such as approximately produced by a steadily operating ship propeller, produces a steady-state periodic forced vibration of the character of that depicted in Fig. 1.

Exciting Frequency—For a steady-state periodic excitation the exciting frequency is the number of cycles of the excitation completed per unit time, which is the inverse of the excitation period. Under steady-state conditions the frequency of the vibration is always equal to the exciting frequency. However, the distribution of system vibration response at the steady-state exciting frequency can be viewed as a weighted superposition of the mode shapes of all the natural modes. The degree of participation of any mode is sensitive to the proximity of the natural frequency associated with that mode to the imposed exciting frequency.

Resonance—Resonance is the condition that occurs in steady state forced vibration when the exciting frequency coincides with any one of the system natural frequencies. The common frequency is then also called a *resonant frequency.* At resonance the vibration amplitude is limited only by system damping, ignoring non-linearities. The damping in engineering structures, including ships, is generally very light, so that resonant vibratory amplitudes are often disproportionately large relative to non-resonant levels. With the disproportionate amplification of one system mode at resonance, the distribution of system resonant vibration will often correspond closely to the mode shape of the resonant mode.

Beat—Beating is a characteristic of systems whose exciting frequency varies over a small range. The resulting response contains a low *beat frequency.* The value of the beat frequency also varies, but its maximum value is equal to the bandwidth of the exciting frequency variation.

Section 2
Theory and Concepts

2.1 Continuous Analysis. All systems that are capable of vibrating, including ships, have at least *piece-wise* continuous properties. That is, the mass, elasticity, damping and excitation properties are continuous within *pieces,* but may have jumps in value where these pieces connect. Unfortunately, piece-wise continuous mathematical models of piece-wise continuous systems that are at all general are of little use in vibration analysis because of the intractability of their solutions; discrete models are necessary for most practical purposes, as is shown in Section 2.2. However, simple continuous models, representing idealizations of real systems, are extremely valuable in understanding basic vibration concepts. Their simple solutions can often provide surprising insight into the behavior of the complex systems whose basic character they approximate.

The simple continuous model that has been used repeatedly over the years to demonstrate certain fundamental aspects of ship vibration (Kennard, 1955) and (Todd, 1961), is the uniform continuous beam model of the ship hull. This model is depicted in Fig. 3 for the case of vertical vibration.

Here the ship hull girder is represented by a uniform one-dimensional beam. The beam is supported by a uniform elastic foundation, of stiffness k per unit length, representing the buoyancy spring of the water (water specific weight times section beam). The foundation has a uniformly distributed damping coefficient, c, representing hydrodynamic damping. The uniform beam mass per unit length is μ (including hydrodynamic added mass) and its uniform stiffness is EI, where E is modulus of elasticity and I sectional moment of inertia. The beam is acted upon by the distributed forcing function, $f(x,t)$, which for purposes of example, represents the vibratory excitation due to the unsteady pressure field of a propeller.

The Fig. 3 model is, in a strict sense, a valid demonstration tool for propeller induced ship vibration occurring typically at relatively low propeller RPM. At higher exciting frequencies associated with modern ship propellers operating near design RPM, the dynamics of mass systems *sprung*[2] from the hull girder, deckhouses for example, become important. However, as the vibration of the basic hull girder retains at least a beam-like character at high frequency, the Fig. 3 model is still instructive, although incomplete.

The differential equation of motion governing vibration of the Fig. 3 model is available from almost any general reference on mechanical vibration. Denoting $w(x,t)$ as the vertical vibratory displacement of the beam, the governing equation is:

[2] i.e., connected by structure that acts as a spring.

$$EI \frac{\partial^4 w}{\partial x^4} + \nu I \frac{\partial^5 w}{\partial x^4 \partial t} + \mu \frac{\partial^2 w}{\partial t^2} + c \frac{\partial w}{\partial t} + kw = f(x,t) \quad (1)$$

Aside from the second term on the left-hand side, Equation (1) represents the standard *Euler beam* on an elastic foundation. The second term in (1) derives from the inclusion of a visco-elastic term in the stress-strain law for the beam material (Kennard, 1955), ν is the visco-elastic constant. The second term in (1), as well as the fourth, involve the first time derivative of the displacement, and therefore represent damping; c is the hydrodynamic damping coefficient of the elastic foundation, by previous definition; νI in the second term in (1) represents a material damping coefficient of the hull beam.

The Euler beam representation, Equation (1), can be easily extended to the *Timoshenko beam* by including beam rotational inertia and shear flexibility in the derivation. However, the additional terms introduced add substantial complexity to the equation as well as to the complexity of its possible analytic solutions. Since the purpose of this sub-section is only to establish concepts and the formulas to be derived are not intended for actual application, inclusion of shear flexibility and rotational inertia in the equation of motion would not serve the purpose well.

End conditions on the equation of motion are required for uniqueness of its solution. The end conditions on Equation (1), corresponding to zero end moment and shear, are:

$$\frac{\partial^2 w}{\partial x^2} = \frac{\partial^3 w}{\partial x^3} = 0 \quad \text{at } x = 0 \text{ and } x = L \quad (2)$$

(*a*) *Steady-State Response to Periodic Excitation.* In propeller-induced ship vibration the steady propeller excitation is, in reality, a random excitation that remains stationary while conditions are unchanged. However, it is approximately periodic with fundamental frequency equal to the propeller RPM times the number of blades. The excitation is therefore approximately expressible as a Fourier series in the time variable. With steady state vibratory response to the periodic excitation being the interest, $w(x,t)$ is likewise expressible in a Fourier series.

The procedure for solving the equation of motion, Equation (1), for the steady state vibration is to substitute the two Fourier series representations for $w(x,t)$ and $f(x,t)$ into the equation. The time dependency is then cancelled out, and the resultant series of ordinary differential equations in x are solved term by term for the unknown coefficients of the displacement series.

For demonstration purposes assume a one-term Fourier series (i.e., simple harmonic) representation for

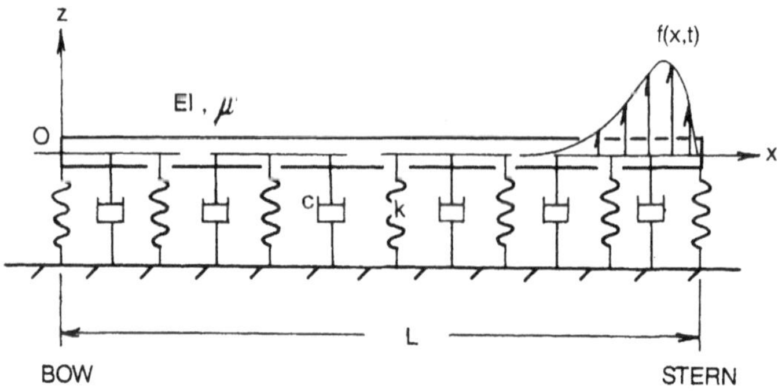

Fig. 3 Ship hull beam model

the excitation force distribution in time. Then optional forms are:

$$f(x,t) = F(x) \cos(\omega t - \alpha) \quad (3)$$

$$= \text{Re } F(x) e^{i\omega t} \quad (4)$$

where, by identity,

$$e^{i\omega t} \equiv \cos \omega t + i \sin \omega t$$

and Re denotes "real part of." $F(x)$ is the amplitude distribution of the excitation force along the length of the ship, and ω is its frequency. Defining ω as *blade-rate* frequency, $N\Omega$, where Ω is the propeller angular velocity and N the number of blades, Equation (3) would be a valid approximation of $f(x,t)$ provided that the *fundamental harmonic* of the excitation is dominant, i.e., provided that the excitation at multiples of blade-rate frequency is relatively insignificant. This is often true, particularly in cases where propeller cavitation does not occur.

For steady-state vibration in response to $f(x,t)$, $w(x,t)$ will have the similar form,

$$w(x,t) = W_c(x) \cos \omega t + W_s(x) \sin \omega t$$

$$= \text{Re } W(x) e^{i\omega t} \quad (5)$$

where, in view of Equation (5),

$$W(x) = W_c(x) - iW_s(x) \quad (6)$$

$W(x)$ is the unknown *complex* amplitude, which includes phase as well as amplitude information. $W(x)$ is to be determined by solution of the equation of motion.

Substitution of Equations (4) and (6) into Equation (1) and end conditions, Equation (2), with cancellation of the time dependency, produces:

$$EI\left(1 + i\frac{\omega\nu}{E}\right)\frac{d^4 W}{dx^4} - (\omega^2\mu - i\omega c - k)W = F(x) \quad (7)$$

with,

$$\frac{d^2 W}{dx^2} = \frac{d^3 W}{dx^3} = 0 \quad \text{at } x = 0 \text{ and } L \quad (8)$$

It is convenient to non-dimensionalize the variables in Equations (7) and (8) before considering solutions for $W(x)$. Redefine the variables in non-dimensional form as,

$$W = \frac{W}{L}, \quad x = \frac{x}{L}, \quad F = \frac{F}{EI/L^3}$$

Also define,

$$\kappa^4 = \left(\frac{\omega}{\Omega_f}\right)^2 - 2i\zeta_c\left(\frac{\omega}{\Omega_f}\right) - \left(\frac{\Omega_r}{\Omega_f}\right)^2$$

where in κ^4,

$$\Omega_r = \sqrt{\frac{kL}{\mu L}}$$

A Characteristic Rigid-Body Frequency

$$\Omega_f = \sqrt{\frac{EI/L^3}{\mu L}}$$

A Characteristic Flexural Frequency

$$\zeta_c = \frac{c}{2\mu\Omega_f}$$

Hydrodynamic Damping Factor

and denote,

$$\zeta_\nu = \frac{\nu\Omega_f}{2E}$$

Structural Damping Factor

Equation (7) then becomes:

$$\left(1 + 2i\,\zeta_\nu\,\frac{\omega}{\Omega_f}\right)\frac{d^4 W}{dx^4} - \kappa^4 W = F(x) \qquad (9)$$

This is the non-dimensional equation for steady-state vibration amplitude in response to harmonic excitation. Its end conditions are:

$$\frac{d^2 W}{dx^2} = \frac{d^3 W}{dx^3} = 0 \qquad \text{at } x = 0 \text{ and } 1$$

(*b*) *Undamped End-Forced Solution—Demonstrations.* The simplest meaningful solution of Equation (9) is obtained by specializing $F(x)$ to be a concentrated end force and discarding the damping terms. This solution, obtained by direct inversion of the reduced equation, is,

$$W(x) = \frac{F}{2\kappa^3}\cdot\frac{1}{1 - \cosh\kappa\cos\kappa}$$
$$\cdot\,[(\sinh\kappa - \sin\kappa)(\cos\kappa x + \cosh\kappa x)$$
$$-\,(\cosh\kappa - \cos\kappa)(\sin\kappa x + \sinh\kappa x)] \qquad (10)$$

Here, the force is concentrated at the stern, $x = 1$ (Fig. 3). With zero damping $W(x)$ is pure real and κ is given by,

$$\kappa^4 = (\omega/\Omega_f)^2 - (\Omega_r/\Omega_f)^2 \qquad (11)$$

The solution, Equation (10), permits several relevant observations. These are developed as follows:

1. *Resonant Frequencies—Added Mass and Buoyancy Effects.* The undamped solution, Equation (10), implies infinite vibration amplitude at the values of ω which make $\cosh\kappa\cos\kappa$ equal to unity. These values of ω are therefore the system *resonant* frequencies, which are denoted as ω_n. Denoting $\kappa = \kappa_n$ at values of ω equal to ω_n, the resonant frequencies correspond to the infinity of roots of,

$$\cosh\kappa_n\cos\kappa_n = 1 \qquad (12)$$

where, from Equation (11),

$$\kappa_n^4 = (\omega_n/\Omega_f)^2 - (\Omega_r/\Omega_f)^2 \qquad (13)$$

The first root of Equation (12) is obviously $\kappa_0 = 0$. This implies, from (13), that,

$$\omega_0 = \Omega_r = \sqrt{\frac{kL}{\mu L}}$$

This is just the rigid body heave, or pitch, resonant frequency; the two are the same for a ship with uniform, or longitudinally symmetric, mass and buoyancy distributions. At the low frequency of the rigid body resonance corresponding to κ_0, the mass distribution μ is frequency dependent due to the surface wave effects in the hydrodynamic component of μ. The frequency dependence of μ diminishes as the vibratory frequency increases. In reality, ship hydrodynamic added mass in essentially invariant with frequency at frequencies corresponding to the full flexural modes.

The second root of Equation (12) is $\kappa_1 = 4.73$, which corresponds to the first hull flexural mode. All subsequent κ_n are greater than κ_1. Therefore, assuming μ to be independent of frequency for $n \geq 1$, Ω_f and Ω_r are constants in Equation (13), and the first flexural mode resonant frequency, and all of those above it, are directly available from Equation (13) as,

$$\omega_n = \Omega_f\sqrt{\kappa_n^4 + (\Omega_r/\Omega_f)^2};\ n = 1, \ldots, \infty \qquad (14)$$

with κ_n determined from Equation (12).

Now, for ships, the ratio Ω_r/Ω_f is typically on the order of 1, and therefore much smaller than κ_n^4 in (14). This demonstrates the fact that the effect of buoyancy in stiffening a ship hull in vertical flexural vibration exists, but is insignificant in normal circumstances. Discarding Ω_r/Ω_f in Equation (14), the beam resonant frequencies are approximated by,

$$\omega_n = \kappa_n^2\Omega_f = \kappa_n^2\sqrt{\frac{EI/L^3}{\mu L}};\ n \geq 1 \qquad (15)$$

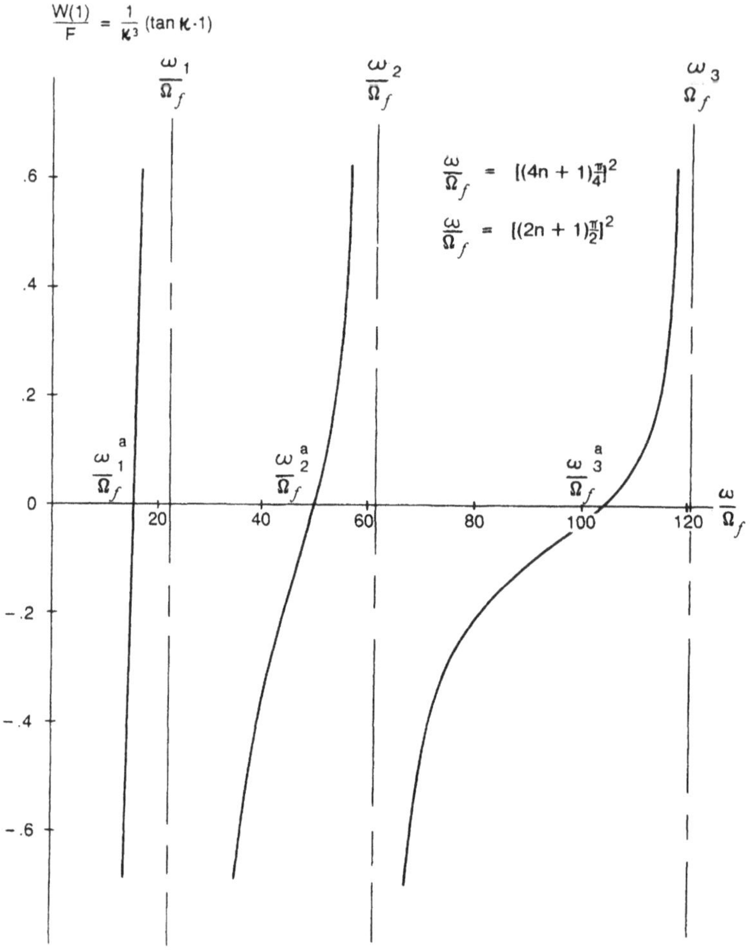

Fig. 4 Hull beam response characteristics

Being typically negligible, the effects of buoyancy will be discarded in all subsequent considerations of flexural vibration; Ω_r will be deleted in the definition of κ so that the existence of non-zero rigid-body modes ($n = 0$) is ignored. Furthermore, Ω_f appearing in κ will be taken as frequency independent, since the hydrodynamic added mass in Ω_f is a constant at high frequency.

Note that although in the case of wave-excited vibration both rigid-body and flexural vibration occur, the two responses are essentially independent superpositions.

2. *Stern Vibration Level.* Consider the vibration at the position of the concentrated excitation force by setting $x = 1$ in (10):

$$\frac{W(1)}{F} = \frac{\sinh\kappa\cos\kappa - \cosh\kappa\sin\kappa}{\kappa^3\,(1 - \cosh\kappa\cos\kappa)} \qquad (16)$$

For exciting frequencies in the range of the beam flexural resonant frequencies, the corresponding values of κ, as arguments of the hyperbolic functions, can be considered as large. That is, for large κ,

$$\sinh\kappa \approx \cosh\kappa \approx \frac{1}{2}\,e^{\kappa}$$

Therefore, at high frequency,

$$\frac{W(1)}{F} \approx \frac{\cos\kappa - \sin\kappa}{\kappa^3\,(2e^{-\kappa} - \cos\kappa)} \approx \frac{1}{\kappa^3}\,(\tan\kappa - 1) \qquad (17)$$

Equation (17) implies that, for a forcing function of fixed amplitude, the end vibration generally decreases with frequency as κ^{-3}, or $\omega^{-3/2}$. Zero vibration at the stern occurs at the *anti-resonant* frequencies, $\omega_n{}^a$, corresponding approximately to,

$$\tan\kappa_n{}^a = 1$$

or, from Equation (15),

$$\omega_n{}^a/\Omega_f = [(4n + 1)\,\pi/4]^2; \quad n = 1, \ldots, \infty \qquad (18)$$

Large vibration occurs only in the immediate vicinity of the resonant frequencies, the flexural values of which, from Equation (17), correspond approximately to,

$$\tan \kappa_n = \pm \infty$$

or,

$$\omega_n / \Omega_f = [(2n + 1)\, \pi / 2]^2; \quad n = 1, \ldots, \infty \tag{19}$$

As ω increases, Equation (17) implies a limiting state where the vibration is zero except at the resonances. But the resonant frequencies, (19), at which the vibration is infinite, occur in the limit (large n), infinitely far apart. The trend toward this limiting case is exhibited in Fig. 4, which is a plot of Equation (17) in the frequency range of the first few flexural modes.

With regard to the relationship of (17) to actual ship vibration, it is not true, in general, that the spacing of the hull girder resonances increases as frequency increases. The disagreement is due to the exclusion of shear and rotational inertia in the beam model, as well as to the exclusion of the effects of local vibratory subsystems sprung from the hull beam. These effects become influential in ship hull girder vibration at high frequency.

The vibration is also, in reality, certainly not infinite at the resonant frequencies; this prediction is, of course, due to the deletion of damping in Equation (9).

It is likewise not true that propeller-induced vibration has a generally decreasing trend with frequency, as (17) implies. In reality, however, the amplitude of the propeller excitation, in this case F, increases with frequency, roughly as frequency squared. With an ω^2 variation of F in (17), $W(1)$ then increases generally as $\omega^{1/2}$, which is more realistic than decreasing as $\omega^{3/2}$.

3. *Relative Vibration of Bow and Stern.* Setting $x = 0$ in (10), the vibration amplitude at the beam end opposite that to which the excitation is applied is,

$$\frac{W(0)}{F} = \frac{\sinh \kappa - \sin \kappa}{\kappa^3 (1 - \cosh \kappa \cos \kappa)} \tag{20}$$

Using (16) and (20), the ratio of the end displacements is,

$$\frac{W(0)}{W(1)} = \frac{\sinh \kappa - \sin \kappa}{\sinh \kappa \cos \kappa - \cosh \kappa \sin \kappa} \tag{21}$$

Again, replacing the hyperbolic functions by the exponential for large κ,

$$\frac{W(0)}{W(1)} = \frac{1 - 2 \sin \kappa\, e^{-\kappa}}{\cos \kappa - \sin \kappa} \simeq \frac{1}{\cos \kappa - \sin \kappa}$$

$$= \frac{1}{\sqrt{2} \cos (\kappa + \pi / 4)} \tag{22}$$

At the anti-resonant frequencies, (18), $W(0)/W(1)$ becomes infinite since $W(1) = 0$ by definition of the anti-resonance. At the resonant frequencies, (19), $W(0)/W(1) = \pm 1$, by (12). The minimum absolute value of the displacement ratio occurs at $\cos (\kappa + \pi / 4) = \pm 1$; its value is $\min | W(0)/W(1) | = 1/\sqrt{2}$. The frequencies at which this minimum value occurs are,

$$\frac{\omega}{\Omega_f} = [(4n + 3)\, \pi / 4]^2; \quad n = 1, \ldots, \infty \tag{23}$$

This prediction is definitely contradictory to observations of ship vibration at high frequency. The simple undamped end-forced solution predicts that the vibration level at the ship bow should never be less than roughly 70 percent of the vibration at the stern. In reality, propeller-induced ship hull girder vibration is known to concentrate at the stern at high propeller RPM, with the vibration diminishing rapidly forward and often being hardly detectable in the vessel forebody.

A reconciliation of theory and observation as to this particular point requires a more general solution to Equation (9), which includes damping as well as a less restricted characterization of the propeller excitation. However, the direct analytic solution procedure used to produce Equation (10) is no longer suitable for providing the desired insight in the more general case.

(*c*) *A More General Solution: Modal Expansion.* The modal, or *eigenfunction* expansion technique allows damping as well as an arbitrary excitation character to be handled with relative ease. Basically, modal expansion is an expression of the fact that the vibration can be viewed as a superposition of the independent natural modes. The solution to the equation of motion is expressed as an infinite series, versus the alternative closed-form possibility represented by (10). The series is expanded in terms of the infinite set of normal modes of the unforced, undamped system.

1. *Natural Frequencies and Mode Shapes.* Returning to the equation of motion for the Fig. 3 uniform beam, Equation (9), the unforced, undamped system in this case corresponds to (9) with zero damping and excitation:

$$\frac{d^4 W}{dx^4} - \kappa^4 W = 0 \tag{24}$$

$$\frac{d^2 W}{dx^2} = \frac{d^3 W}{dx^3} = 0 \qquad \text{at } x = 0 \text{ and } 1$$

where κ is defined by Equation (11). The solution to the homogeneous differential, (24), is, for $\kappa \neq 0$,

$$W(x) = C_1 \sin \kappa x + C_2 \cos \kappa x + C_3 \sinh \kappa x + C_4 \cosh \kappa x \tag{25}$$

Applying the two end conditions at $x = 0$ eliminates two of the four constants in (25) as,

$$C_2 = C_4, \qquad C_1 = C_3 \tag{26}$$

Application of the remaining end conditions at $x = 1$ gives the following simultaneous equations for determining C_3 and C_4,

$$\begin{bmatrix} \sin\kappa - \sinh\kappa & \cos\kappa - \cosh\kappa \\ \cos\kappa - \cosh\kappa & -\sin\kappa - \sinh\kappa \end{bmatrix} \begin{vmatrix} C_3 \\ C_4 \end{vmatrix} = \begin{vmatrix} 0 \\ 0 \end{vmatrix} \quad (27)$$

or,

$$[B]\,|C| = |0|$$

Then by inversion,

$$|C| = [B]^{-1}\,|0|$$

Therefore, unless $[B]$ is singular, the only solution to Equations (27) is $|C| = |0|$. But this implies that $W(x) = 0$, which is not of interest. Non-zero $|C|$, and non-zero $W(x)$, therefore require that $[B]$ be singular. $[B]$ is singular only if its determinant is zero. From Equation (27),

$$\det [B] = -2\,(1 - \cos\kappa \cosh\kappa) \quad (28)$$

Denote the values of κ which make $\det [B] = 0$ as κ_n; these values are the system *eigenvalues*. The infinite set of eigenvalues are determined so that Equation (28) is zero, i.e., so that,

$$\cos\kappa_n \cosh\kappa_n = 1, \qquad n = 1, \ldots \infty. \quad (29)$$

But this is just Equation (12), which established the system resonant frequencies. From (11), ignoring the Ω_r term,

$$\kappa_n^4 = (\omega_n / \Omega_f)^2 \quad (30)$$

where the ω_n were identified as the *resonant* frequencies. But under present considerations the ω_n are the frequencies corresponding to unforced and undamped, or *natural,* system vibration; the system resonant frequencies are therefore synonomous with the system natural frequencies.

Non-zero values of C_3 and C_4 from (27) therefore exist only at values of ω satisfying (29). However, the values of the constants, while not zero, are indeterminant, since the coefficient determinant is zero at these frequencies. The fact that the determinant of the coefficients is zero at the natural frequencies implies that the two Equations (27) are linearly dependent at the natural frequencies. That is, two independent equations from which to determine the two constants are not available. The only information available from (27) is the relationship between C_3 and C_4 at the natural frequencies. Either one of the two equations can be used for this purpose; the same result will be obtained because of the linear dependency. From the second equation of (27),

$$\frac{C_3}{C_4} = \frac{\sin\kappa_n + \sinh\kappa_n}{\cos\kappa_n - \cosh\kappa_n} \quad (31)$$

Substitution of (31) and (26) back into the homogeneous solution (25) gives the beam vibration amplitude at the natural frequencies, as a function of x, except for a constant factor. This *relative* amplitude distribution at the natural frequencies is called the *eigenfunction,* or mode shape, and is denoted by ψ_n. From (25), in terms of arbitrary C_4, the mode shape for the Fig. 3 beam is,

$$\psi_n(x) = C_4 \Big[\cos\kappa_n x + \cosh\kappa_n x + \frac{\sin\kappa_n + \sinh\kappa_n}{\cos\kappa_n - \cosh\kappa_n}(\sin\kappa_n x + \sinh\kappa_n x)\Big] \quad (32)$$

Equation (32) is the beam mode shape for $\kappa \neq 0$. This function has the character of the vertical mode shapes depicted on Fig. 2 of Section 1.2; increasing n corresponds to increasing node number.

For $\kappa = 0$ the solution to the homogeneous system, Equation (24) is,

$$W(x) = C_1 + C_2 x + C_3 x^2 + C_4 x^3 \quad (33)$$

The end conditions at $x = 0$ reduce (33) to,

$$W(x) = C_1 + C_2 x \quad (34)$$

which satisfies the end conditions at $x = 1$ identically. The mode shape identified with (34) is therefore the zeroth order rigid body heave/pitch mode, whose corresponding natural frequency was previously identified as Ω_r by Equations (12) and (13); Ω_r has been assumed to be zero in consideration of the flexural modes.

2. *Vibratory Displacement.* Modal expansion expresses the solution of the equations of motion, Equation (9), as a weighted summation of the infinite set of mode shapes,

$$W(x) = \sum_{n=1}^{\infty} A_n\,\psi_n(x) \quad (35)$$

Back substituting Equation (35) into (9) and utilizing the *orthogonality* property of the mode shapes, the A_n in Equation (35) are determined as,

$$A_n = \frac{F_n / K_n}{1 - (\omega/\omega_n)^2 + 2i\,\zeta_n\,\omega/\omega_n} \quad (36)$$

where F_n, K_n, and ζ_n in Equation (36) represent the following:

$$F_n = \int_{x=0}^{1} F(x)\,\psi_n(x)\,dx \qquad \text{Modal exciting force} \quad (37)$$

$$K_n = \left(\frac{\omega_n}{\Omega_f}\right)^2 \int_{x=0}^{1} \psi_n^2(x)\,dx \qquad \text{Modal stiffness} \quad (38)$$

$$\zeta_n = \zeta_v\,(\omega_n/\Omega_f) + \zeta_c\,(\Omega_f/\omega_n)$$

Modal damping factor (39)

Substitution of (36) into (35) gives the complex vibration amplitude,

$$W(x) = \sum_{n=1}^{\infty} \left[\frac{F_n / K_n}{1 - (\omega/\omega_n)^2 + 2i\zeta_n \omega/\omega_n} \right] \psi_n(x) \tag{40}$$

Substitution of this result into (5), and using a trigonometric identity, gives the vibration displacement at any point x along the beam at any time,

$$w(x, t) = \sum_{n=1}^{\infty} \left[\frac{F_n / K_n}{\sqrt{[1 - (\omega/\omega_n)^2]^2 + (2\zeta_n \omega/\omega_n)^2}} \right] \psi_n(x) \cos(\omega t - \alpha_n) \tag{41}$$

The modal phase angle, α_n, relative to F_n, is,

$$\alpha_n = \tan^{-1} \left[\frac{2\zeta_n \omega/\omega_n}{1 - (\omega/\omega_n)^2} \right]$$

The form of Equation (41) demonstrates that modal expansion can be viewed as just a superposition of the independent responses of an infinite number of equivalent one-degree-of-freedom systems. The stiffness, damping, and excitation of each equivalent system are the modal values corresponding to Equations (37), (38), and (39). The equivalent mass would be the modal mass, $M_n = K_n/\omega_n^2$. The responses of each of the single degree-of-freedom systems is distributed according to the mode shapes of the respective modes.

3. *Relative Vibration of Bow and Stern.* The reasons for the rapid attenuation of hull girder vibration on moving forward from the stern, which were left unexplained by the simple theory of the last subsection, can now be reconsidered with the aid of the modal expansion, Equation (40).

It is first convenient, although not at all necessary, to *normalize* the eigenfunction set, Equation (32), by assigning specific values to the constant C_4. Choosing a value of unity of the $\psi_n(x)$ at the forcing end, i.e.,

$$\psi_n(1) = 1; \qquad n = 1, \ldots, \infty \tag{42}$$

C_4 in Equation (32) is evaluated as,

$$C_4 = \frac{\cos \kappa_n - \cosh \kappa_n}{2 \sin \kappa_n \sinh \kappa_n} \tag{43}$$

Then from (43) and (32), the eigenfunction at $x = 0$ has the values,

$$\psi_n(0) = (-1)^{n+1}; \qquad n = 1, \ldots, \infty \tag{44}$$

It will also be notationally convenient to define, $W_n(x) \equiv A_n \psi_n(x)$, where A_n is given by Equation (36). Equation (40) is then, alternatively,

$$W(x) = \sum_{n=1}^{\infty} W_n(x) \tag{45}$$

By Equations (42), (44), and (45), the displacements at the two ends of the beam are,

$$W(1) = \sum_{n=1}^{\infty} W_n(1)$$

and (46)

$$W(0) = \sum_{n=1}^{\infty} (-1)^{n+1} W_n(1)$$

Equation (46) shows that the absolute values of the displacement components from each mode are the same at the beam ends. Differences in the sums must therefore be due only to the alternating form of the series for $W(0)$, associated with phase changes occurring within the displacement components at the forcing end. In fact, this character of the displacement series, Equation (46), is the basis for understanding the reasons for the rapid decay of hull girder vibration forward from the stern. Fig. 5 is intended as an aid in this purpose. Fig. 5 is composed of sketches of the W_n components for six modes, arbitrarily, and their summations, for three different cases.

The first column in Fig. 5 depicts the displacement for the undamped beam with the concentrated force applied at the extreme end. This was the case studied in the last sub-section and for which the minimum ratio of end displacements was predicted to be never less than $1/\sqrt{2}$. The second column in Fig. 5 represents the case where damping remains zero, but the concentrated force is applied at a position $x = x_0$ slightly less than 1, corresponding to a typical propeller position. In the third column on Fig. 5, the force has been replaced at the beam end, but damping has been assumed to be non-zero and significant.

The exciting frequency is assumed to lie arbitrarily between modes 3 and 4 in Fig. 5. The value of ω can be considered as that given by Formula (23) with $n = 4$; Equation (23) predicts the frequencies at which $\min |W(0)/W(1)| = 1/\sqrt{2}$ occurs for the undamped, end-forced case.

Consider the three cases of Fig. 5 individually.

Case 1—Undamped, End forced. From Equation (40), with $x_0 = 1$ and for $\zeta_n = 0$,

$$W_n(x) = \frac{F/K_n}{1 - (\omega/\omega_n)^2} \psi_n(x) \tag{47}$$

In general, the modal forces for the three cases of Fig. 5 are $F_n = F\psi_n(x_0)$, by Equation (37). For $x_0 = 1$ in the first case $F_n = F$ for all n since $\psi_n(1) = 1$ by Equation (42). For ω between two resonant frequencies, ω_{N-1} and ω_N, the beam end displacements can be written from (46) and (47) as,

$$W(1) = -\sum_{n=1}^{N-1} |W_n(1)| + \sum_{n=N}^{\infty} |W_n(1)|$$

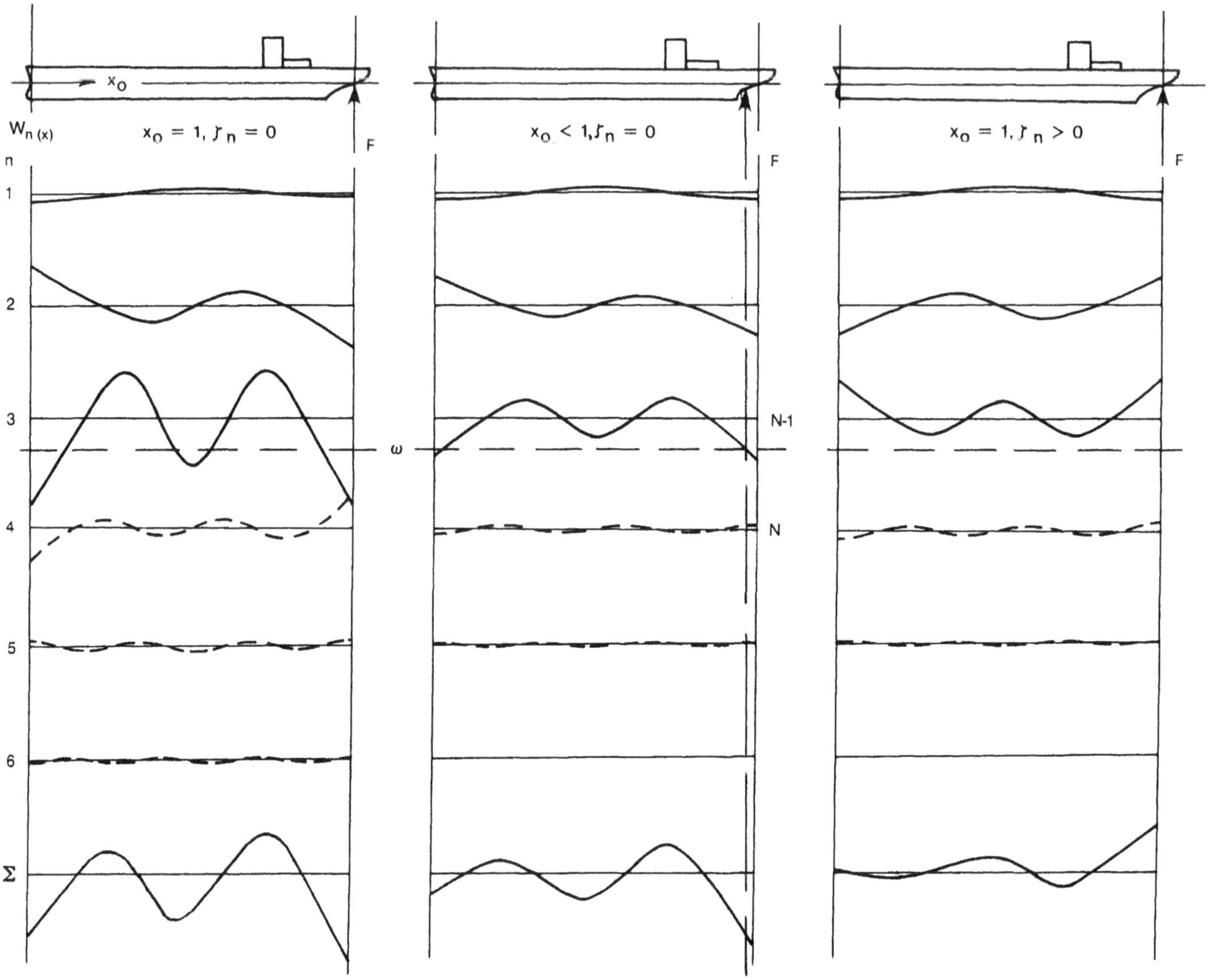

Fig. 5 Hull mode superpositions

$$W(0) = -\sum_{n=1}^{N-1} (-1)^{n+1} |W_n(1)| + \sum_{n=N}^{\infty} (-1)^{n+1} |W_n(1)| \quad (48)$$

Here, the sign change occurring in the denominator of Equation (47) at $n = N$ has been explicitly assigned. At the end $x = 1$, all of the modes below ω are of the same sign, but of opposite sign to the modes above ω. Imperfect cancellation occurs, with the lower modes dominating the upper. At $x = 0$ on the other hand, interferences occur among the groups of modes both below and above ω due to the alternating signs shown in Equation (48). The dominant terms immediately above and below ω, i.e., $W_{N-1}(0)$ and $W_N(0)$, have the same signs, however, and support rather than cancel. As a result, $W(0)$ is relatively large. In fact, the ratio $|W(0)/W(1)|$ occurring at ω for $x_0 = 1$ and $\zeta_n = 0$, which has the value $1/\sqrt{2}$, is a maximum value of the minimum ratio, as both repositioning the excitation force forward and allowing non-zero damping result in a more rapid attenuation of displacement away from the forcing point.

Case 2—Undamped, $x_0 < 1$. Considering the case where $x_0 < 1$, which corresponds to the second column on Fig. 5, the modal force is,

$$F_n = F\,\psi_n(x_0)$$

in Equation (40). The modal forces now converge with increasing n, since, as the aftermost beam nodal point moves aft toward the forcing point with increasing n, the $\psi_n(x_0)$ decrease. Thus the higher modes become less *excitable* by the concentrated force. The result is a decrease in the cancellation in $W(1)$, by Equation (48), as the net displacement produced by the modes above ω decreases relative to the net contribution from below. Also, a weakening of the modes above ω reduces the support of the large N^{th} mode in $W(0)$, relative to the $N - 1^{\text{th}}$. This results in a relative decrease in $W(0)$, with respect to $W(1)$, and a larger difference in the end displacements. This decreasing propeller excitability of the higher hull girder modes by virtue of convergence of the modal force series was the explanation given by Baier and Ormondroyd (1952) for the rapid attenuation of propeller induced hull girder vibration forward from the stern region.

Case 3—Damped, End-forced. Turning to the case of non-zero damping, but with $F_n = F$, the terms in the displacement series are:

$$W_n(x) = \frac{F / K_n}{1 - (\omega / \omega_n)^2 + 2i\zeta_n \, \omega / \omega_n} \psi_n(x) \quad (49)$$

If the modal damping factor, ζ_n in Equation (49) increases with n, then the convergence of the displacement series is accelerated, with the same effects as produced by convergence of the modal forces just considered. Damping also modifies the relative phases of the modes. This occurs most strongly for modes in the immediate vicinity of the exciting frequency, since the damping in the denominator of (49) is relatively strongest for ω / ω_n in the vicinity of 1. For zero damping, the modes below the exciting frequency are 180 deg out of phase with the modes above due to the sign change in the denominator of (47). Damping spreads the phase shift. If the damping is strong enough the most dominant modes to either side of the exciting frequency can be approximately in phase and 90 deg out of phase with the exciting force. This is the situation depicted on Fig. 5, where damping has delayed the phase shift in the two modes below ω. The result is increasing modal interference with distance away from the forcing point.

The effect depicted in Column 3 of Fig. 5 is contingent upon a modal damping factor which increases with modal order and/or is relatively large in the modes in the vicinity of the exciting frequency. In this regard reconsider the modal damping factor which arose in the derivation of the uniform beam modal expansion, (39),

$$\zeta_n = \zeta_v \, (\omega_n / \Omega_f) + \zeta_c \, (\Omega_f / \omega_n)$$

The structural damping factor, ζ_v, is a constant, by Equation (9). The hydrodynamic damping factor, ζ_c, has not been specifically defined, but it actually has a decreasing magnitude with frequency. Furthermore ω_n / Ω_f is large for all n. Therefore, for n large,

$$\zeta_n \simeq \zeta_v \, (\omega_n / \Omega_f) = (\nu / 2E) \, \omega_n$$

ζ_n therefore increases with n, and becomes large at large n corresponding to ω_n at high frequency excitation. The ζ_n developed with the idealized beam model therefore appears to meet the requirements for the effects of damping exhibited on Fig. 5. High hull damping in the frequency range of propeller excitation was the explanation suggested by Kennard (1955) for the concentration of vibration in the stern of vessels when operating at high propeller RPM.

This discussion with regard to Fig. 5 should help to avoid the common misconception that the concentration of propeller-induced vibration in the stern of a vessel is evidence that the vessel is exhibiting something other than beam-like vibration. To the contrary, sternward concentration of vibration at high frequency is due to interference in the beam modes at the bow and support at the stern. As shown, this occurrence is due both to increasing modal damping and decreasing modal excitability as modal order, and exciting frequency, increase.

2.2 Discrete Analysis. (*a*) *Mathematical Models.* Modern day ship vibration analysis employs mathematical models that are non-uniform and discrete, rather than uniform and continuous, almost exclusively. Such models represent the continuous mass, stiffness, damping and excitation characteristics of the physical structure at a discrete number of points, which are called *nodal points.* The equivalent nodal point properties are translated in terms of an assemblage of discrete, or *finite,* elements; the finite elements inter-connect the nodal points of the structural model. (Note that these nodal points are not the same as the nodes defined in Section 1.2).

In analyzing the discrete model all forces and displacements are referred to the model nodal points. In general, six components of displacement, consisting of three translations and three rotations, and six corresponding components of force, can exist at each nodal point of the model. The model is usually constrained, however, so that less than the possible six displacements are allowed at any nodal point. The number of such displacements allowed at any point are referred to as its *degrees of freedom.* If mass, or mass moment of inertia, is associated with a particular nodal point displacement, then that displacement defines a *dynamic* degree of freedom. Otherwise, the degree of freedom is *static.* While the total number of degrees of freedom of continuous systems is always infinite, the total number of degrees of freedom of a discrete model is finite, being the sum of the numbers assigned to each of the model nodal points.

Discrete analysis of ship vibration can be performed to any arbitrary level of detail, with model complexity limited primarily by available computing facilities. Often, the ship hull girder, as considered in the last subsection, is modeled along with its sprung substructures, i.e., deckhouses, decks, double-bottoms, etc., in a single discrete model (Sellars and Kline, 1967) (Reed, 1971) and (Kagawa, 1978). In many cases meaningful estimates of substructure vibration characteristics can also be obtained using only a discrete model of the substructure, with approximate boundary conditions applied at its attachment to the hull girder (Sandstrom and Smith, 1979).

Discrete analysis is conveniently demonstrated by an idealized example of the latter approach noted above. Consider the simple finite element model for a ship deckhouse shown in Fig. 6. Here, the house is

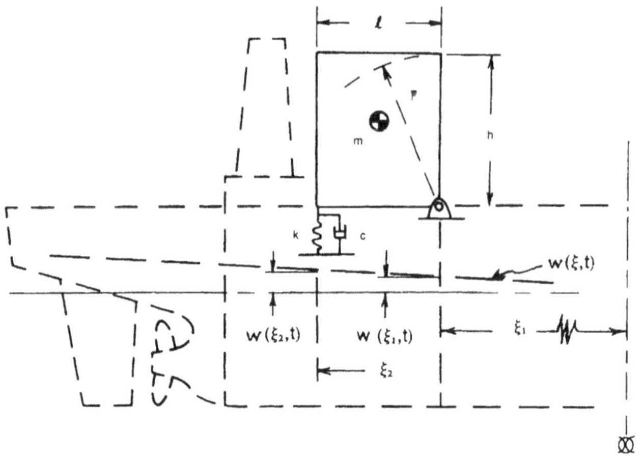

Fig. 6 Ship deckhouse vibration model

modeled two-dimensionally as a rigid box of mass m and radius of gyration $\bar{r}$. The house front is taken, typically, as a continuation of the forward engine room bulkhead; the connection at main deck is assumed to act as a simple pin allowing completely free rotation. The parallel connection of finite elements with axial stiffness and axial damping represent the supporting structure along the house after bulkhead. This structure would be composed, typically, of pillars erected within the engine room cavity. The house is base excited by the vertical vibratory displacement of the hull girder, $w(\xi, t)$, ξ now being the axial coordinate along the hull girder. The applied base displacements, $w(\xi_1, t)$ and $w(\xi_2, t)$, are the hull girder displacements at the forward engine room bulkhead and at the base of the after foundation; $w(\xi_1, t)$ and $w(\xi_2, t)$ are assumed at this point to be specified in advance.

Use of the Fig. 6 model for serious vibration analysis is not entirely valid in two respects. Primarily, the typical deckhouse does not truly act as if rigid at propeller excitation frequencies. While the underdeck supporting structure is quite often the predominant flexibility in propeller induced deckhouse vibration, the bending and shear flexibilities of the house itself can usually not be considered as unimportant. Some degree of interaction of the house with the hull girder also occurs. Because of this the base displacements are not easily prescribed with accuracy in advance. In spite of these shortcomings, the simple Fig. 6 deckhouse model is instructive; it captures the basic characteristics of fore-and-aft deckhouse vibration in the spirit of the simple uniform beam model for hull girder vertical vibration studied in the last sub-section.

Proceeding as described, the degree of freedom assignments of the Fig. 6 finite element model are shown on Fig. 7. Here x_j is used to denote generalized displacement, i.e., rotation or translation. In view of the assumed house rigidity, all displacements in the vertical/fore-and-aft plane can be specified in terms of the three assigned on Fig. 7. All other possible displacements at the two nodal points of the Fig. 7 model are assigned zero values by virtue of their omission. Of the three total degrees of freedom assigned on Fig. 7, two are dynamic degrees of freedom. These are x_1 and x_2, as they are associated with the house mass moment of inertia and house mass, respectively. x_3 is a static degree of freedom. Also, two of the three degrees of freedom are specified:

$$x_2 \equiv w(\xi_1, t) \qquad \text{and} \qquad x_3 \equiv w(\xi_2, t)$$

from Fig. 6.

Once x_1 is determined the vertical and fore-and-aft displacements at any point (ξ, η) on the house are available, respectively, as,

$$w(\xi, \eta, t) = w(\xi_1, t) - x_1(t)(\xi - \xi_1)$$
$$u(\xi, \eta, t) = x_1(t)\eta \tag{50}$$

(*b*) *Equations of Motion.* The equations of motion governing the general finite element model are derived as follows:

It is first required that the model be in dynamic equilibrium in all of its degrees of freedom simultaneously. Application of Newton's Law in each degree of freedom in turn produces,

$$[m]\{\ddot{x}\} = -\{f_s\} - \{f_d\} + \{f\} \tag{51}$$

where, for M total degrees of freedom,

$[m]$ is the $M \times M$ model mass matrix,
$\{\ddot{x}\}$ is the $M \times 1$ nodal point acceleration vector and,
$\{f_s\}$, $\{f_d\}$, and $\{f\}$ are the $M \times 1$ nodal point stiffness, damping, and excitation force vectors, respectively.

The characteristics of the model finite elements are established in advance to satisfy compatibility and material constitutive requirements on the local level. Satisfaction of these requirements for linear behavior leads to the following relations between the nodal point internal forces and the nodal point displacements,

$$\{f_s\} = [k]\{x\} \qquad \{f_d\} = [c]\{\dot{x}\} \tag{52}$$

Here $[k]$ is the model stiffness matrix and $[c]$ is the model damping matrix, both of which are square matrices of order M.

Substitution of (52) into (51) produces the linear equations of motion governing the general discrete model,

$$[m]\{\ddot{x}\} + [c]\{\dot{x}\} + [k]\{x\} = \{f\} \tag{53}$$

This $M \times M$ system of equations can be readily solved for the unknown nodal point displacements once $[m]$, $[c]$, $[k]$, and $\{f\}$ are specified.

Actually, the equations of motion can be interpreted as a general statement and conveniently used to determine their own coefficients. For example, if the accelerations and velocities are set to zero, Equation (53) reduces to,

$$[k]|x| = |f|$$

In expanded notation,

$$\begin{vmatrix} f_1 \\ f_2 \\ \cdot \\ \cdot \\ \cdot \\ f_M \end{vmatrix} = \begin{bmatrix} k_{11} & k_{12} & \cdot & \cdot & \cdot & k_{1M} \\ k_{21} & k_{22} & & & & \\ \cdot & & \cdot & & & \\ \cdot & & & \cdot & & \\ \cdot & & & & \cdot & \\ k_{M1} & \cdot & \cdot & & \cdot & k_{MM} \end{bmatrix} \begin{vmatrix} x_1 \\ x_2 \\ \cdot \\ \cdot \\ \cdot \\ x_M \end{vmatrix}$$

The subscripts refer to the numbers assigned to the nodal point degrees of freedom. Now for purposes of defining the k_{ij}, require in addition to zero velocities and accelerations, that all displacements, x_i, be zero except for $i = j$, and set $x_j = 1$. Then for any degree of freedom i, multiplication gives,

$$f_i = k_{ij}$$

The k_{ij} is therefore defined as the *force in degree of freedom i* due to a unit *displacement in degree of freedom j*, with all other degrees of freedom completely restrained. Complete restraint means restraint from acceleration, velocity, and displacement. Also, as to the matter of signs, the designation force in degree of freedom i is interpreted as the force required at i in order to *accomplish* the degree of freedom assignment at i.

The corresponding definitions of m_{ij} and c_{ij} are similarly derived from the general Equations (53) by making the appropriate degree of freedom assignments. Definitions for m_{ij} and c_{ij} identical to that above for k_{ij} result, but with unit accelerations and velocities, respectively, replacing the unit displacements.

In calculating the components of the excitation force vector, f_i, the model is completely restrained in all degrees of freedom. f_i is then the resultant of the applied forces tending to *overcome* the restraint in degree of freedom i.

In this connection consider again the simple model of Fig. 7. The displacements in the 3 degrees of freedom are x_1, x_2, and x_3 with x_1 to be determined and the other two specified. By applying zero and unit accelerations, velocities, and displacements in the 3 degrees of freedom, in turn, the mass, damping, stiffness and excitation force matrices are assembled by the rules stated above as,

$$[m] = \begin{bmatrix} m\bar{r}^2 & -m\bar{\xi} & 0 \\ -m\bar{\xi} & m & 0 \\ 0 & 0 & 0 \end{bmatrix}$$

$$[c] = \begin{bmatrix} c\ell^2 & -c\ell & c\ell \\ -c\ell & c & -c \\ c\ell & -c & c \end{bmatrix}$$

$$[k] = \begin{bmatrix} k\ell^2 & -k\ell & k\ell \\ -k\ell & k & -k \\ k\ell & -k & k \end{bmatrix}$$

$$|f| = \begin{vmatrix} 0 \\ f_2 \\ f_3 \end{vmatrix} \qquad (54)$$

The force components f_2 and f_3 in the excitation force vector above are the unknown forces associated with the known displacements x_2 and x_3; f_1 is zero as no external moment is applied at the pin connection.

This example demonstrates the general case. Either the applied external force, f_i, or displacement, x_i, must be specified for each degree of freedom; both will never be known prior to solution of the system equations. The equations corresponding to the known forces are first solved for the unknown displacements. The unknown forces are then calculated using the then completely known displacements with the equations corresponding to the unknown forces.

For the Fig. 7 system the first part of the operation described above produces a single equation of motion for determining the single unknown displacement x_1. It is,

$$m\bar{r}^2 \ddot{x}_1 + c\ell^2 \dot{x}_1 + k\ell^2 x_1 = m\bar{\xi}\, \ddot{x}_2 + c\ell(\dot{x}_2 - \dot{x}_3) + k\ell(x_2 - x_3) \qquad (55)$$

On solving this equation for x_1, the unknown force components, f_2 and f_3, are then determined from the two remaining equations by multiplication as,

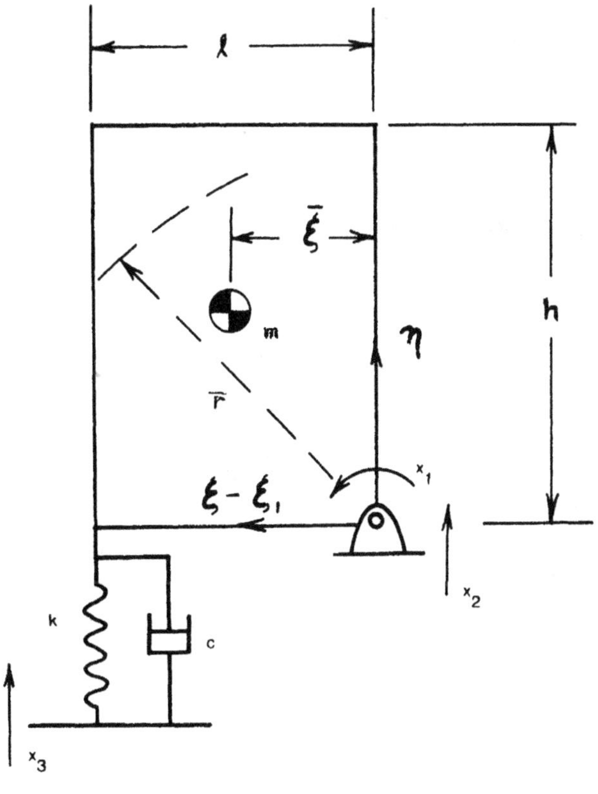

Fig. 7 Deckhouse model, degrees of freedom

$$\begin{vmatrix} f_2 \\ f_3 \end{vmatrix} = \begin{bmatrix} -m\xi & m & 0 \\ 0 & 0 & 0 \end{bmatrix} \begin{vmatrix} \ddot{x}_1 \\ \ddot{x}_2 \\ \ddot{x}_3 \end{vmatrix} + \begin{bmatrix} -c\ell & c & -c \\ c\ell & -c & c \end{bmatrix} \begin{vmatrix} \dot{x}_1 \\ \dot{x}_2 \\ \dot{x}_3 \end{vmatrix} + \begin{bmatrix} -k\ell & k & -k \\ k\ell & -k & k \end{bmatrix} \begin{vmatrix} x_1 \\ x_2 \\ x_3 \end{vmatrix} \quad (56)$$

(*c*) *Solutions.* For L of the total M model nodal point displacements unknown, L governing differential equations, in the general form of Equation (53), must be solved. The $L \times 1$ force vector in Equation (53) will be completely known in terms of the L applied force components and the $M - L$ applied displacements.

The same basic solution procedure applied in the continuous analysis of the last section is also followed here. The approximate periodicity of the propeller excitation allows the time variable to be separated from the differential equations by the use of Fourier Series. For propeller angular velocity Ω and blade number N, define $\omega \equiv mN\Omega$ as the mth harmonic propeller exciting frequency. Then for $|F|$ and $|X|$ representing the mth harmonic complex force and displacement amplitude vectors, the equations of motion, (53), can be satisfied harmonic by harmonic by solving

$$\{-\omega^2[m] + i\omega[c] + [k]\}\,|X| = |F| \quad (57)$$

Define the system dynamic matrix as $[D]$,

$$[D] = -\omega^2[m] + i\omega[c] + [k] \quad (58)$$

Equation (57) is then,

$$[D]|X| = |F| \quad (59)$$

with solution, $|X| = [D]^{-1}|F|$ (60)

Returning to the Figs. 6 and 7 deckhouse model, with

$$|x| = \text{Re}|X|\,e^{i\omega t}$$

the system dynamic matrix, from (55), is,

$$[D] = -\omega^2 m\bar{r}^2 + i\omega c\ell^2 + k\ell^2 \quad (61)$$

which is a 1×1 matrix on the single unknown complex amplitude, X_1. Likewise, the complex exciting force vector in (55) is,

$$|F| \equiv F_1 = -\omega^2 m\xi X_2 + (i\omega c\ell + k\ell)(X_2 - X_3)$$

The inversion required by (60), using (61), is then simply,

$$X_1 = \frac{-\omega^2 m\xi X_2 + (i\omega c\ell + k\ell)(X_2 - X_3)}{-\omega^2 m\bar{r}^2 + i\omega c\ell^2 + k\ell^2} \quad (62)$$

Equation (62) can be written in the standard form for vibration of systems with one dynamic degree of freedom by writing its numerator as,

$$F_1 = F_1^R + iF_1^I = \text{mod}\,F_1 e^{-i\beta}$$

and the denominator as,

$$\frac{(1/K)e^{-i\alpha}}{\sqrt{[1 - (\omega/\omega_n)^2]^2 + (2\zeta\,\omega/\omega_n)^2}}$$

so that,

$$x_1 = \text{Re}X_1 e^{i\omega t} \quad \text{is,}$$

$$x_1(t) = \frac{(\text{mod}\,F_1/K)\cos(\omega t - \alpha - \beta)}{\sqrt{[1 - (\omega/\omega_n)^2]^2 + (2\zeta\omega/\omega_n)^2}} \quad (63)$$

where, in the above,

$$\text{mod}\,F_1 = \sqrt{(F_1^R)^2 + (F_1^I)^2}$$

$$\beta = \tan^{-1}[-F_1^I/F_1^R]$$

$$K = k\ell^2$$

$$\omega_n = \sqrt{K/m\bar{r}^2}$$

$$\zeta = \frac{c\ell^2}{2m\bar{r}^2\,\omega_n}$$

$$\alpha = \tan^{-1}\left[\frac{2\zeta\,\omega/\omega_n}{1 - (\omega/\omega_n)^2}\right] \quad (64)$$

In the general case, an analytic closed form inversion of the system equations like that performed above for the simple one dynamic degree-of-freedom system is not possible. Two alternatives exist. The most obvious is just a direct numerical inversion of Equation (60). Powerful numerical algorithms are readily available for inverting systems of linear simultaneous algebraic equations. However, direct numerical inversion can be disadvantageous in several respects, which are described further on. The alternative solution procedure is, again, *eigenvector*, or *modal, expansion.* Modal expansion is the series solution of the equations of motion, (57), in terms of the natural frequencies and mode shapes of the discrete model.

1. *Natural Frequencies and Mode Shapes.* By definition, natural frequencies are frequencies of vibration of the free, or unforced, and undamped system. From Equation (57), the equations of motion for the free, undamped discrete model are,

$$\{-\omega^2[m] + [k]\}\,|X| = |0|$$

Denote,

$$[D^*(\omega)] = [D]|_{[c]=0} = -\omega^2[m] + [k]$$

Then,

$$[D^*]|X| = |0| \tag{65}$$

This equation implies that $|X| = 0$ unless $[D^*]$ is singular. But by definition of natural vibration, $|X|$ is not zero. Therefore, the frequencies ω which make $[D^*(\omega)]$ singular are the system natural frequencies; $[D^*]$ is singular if its determinant is zero. Define,

$$P(\omega) \equiv \det [D^*(\omega)] \tag{66}$$

$P(\omega)$ is called the *characteristic polynomial.* For N system dynamic degrees of freedom, $P(\omega)$ is a polynomial of order N in ω^2; it has N positive roots in ω. The N positive values of ω which make $P(\omega) = 0$ are the natural frequencies, ω_n,

$$P(\omega_n) = 0 \qquad n = 1, \ldots, N \tag{67}$$

While the number of natural frequencies possessed by continuous systems is always infinite, the *number of natural frequencies of the discrete model is equal to its number of dynamic degrees of freedom.* In this regard, it is worth repeating that all real physical systems are at least piece-wise continuous. Therefore, discrete *systems* can be viewed only as discrete *models* of continuous systems; this distinction is not unimportant.

Proceeding, with the N *model* natural frequencies in hand, a return to Equation (65) gives,

$$[D^*(\omega_n)]|X| = |0| \tag{68}$$

Now, $|X|$ is not necessarily zero at $\omega = \omega_n$ since $[D^*(\omega)]$ is singular at these frequencies, but it is undefined. Just as with the continuous analysis, the singularity of the coefficient matrix of Equation (68) implies a linear dependency within the L equations. That is, only $L - 1$ linearly independent equations exist at $\omega = \omega_n$, $n = 1, \ldots, N$, for determining the L unknown components of $|X|$. All that is available from Equation (68) are the relative amplitudes, called *mode shapes,* or *eigenvectors,* at each of the N natural frequencies.

The $L \times 1$ mode shape vector is denoted $|\psi_n|$, $n = 1, \ldots N$. It is determined by assuming any one of its L components as known. Then the $L - 1$ equations on the remaining $L - 1$ mode shape components at each n are solved in terms of the one presumed known. That is, assuming arbitrarily that the Lth mode shape component is known, Equation (68) is written,

$$\begin{bmatrix} D^*_{11} & D^*_{12} & \cdot & \cdot & \cdot & D^*_{1\,L-1} \\ D^*_{21} & D^*_{22} & & & & \\ \cdot & & \cdot & & & \\ \cdot & & & \cdot & & \\ \cdot & & & & \cdot & \\ D^*_{L-11} & D^*_{L-12} & \cdot & \cdot & \cdot & D^*_{L-1\,L-1} \end{bmatrix} \begin{vmatrix} \psi_{1n} \\ \psi_{2n} \\ \cdot \\ \cdot \\ \cdot \\ \psi_{L-1n} \end{vmatrix} = -\psi_{Ln} \begin{vmatrix} D^*_{1L} \\ D^*_{2L} \\ \cdot \\ \cdot \\ \cdot \\ D^*_{LL} \end{vmatrix} \tag{69}$$

The $(L - 1) \times (L - 1)$ system of linear algebraic equations, (69), is then solved by standard numerical methods for the $(L - 1)$ component $|\psi_n|$ for some or all of the N modes of interest.

For the Fig. 6 deckhouse example the above is simple since both L and N are one. The $[D^*]$ matrix from Equation (61) is,

$$[D^*] = -\omega^2 m\bar{r}^2 + k\ell^2$$

which is also the characteristic polynomial $P(\omega)$. $P(\omega_n) = 0$ gives the natural frequency,

$$\omega_n = \sqrt{k\ell^2/m\bar{r}^2} \qquad \text{with } n = 1.$$

The mode shape $|\psi_n|$ is ψ_{11}, which has an arbitrary value.

2. *Modal Expansion.* At this point in the development of the solution for the uniform beam of the last sub-section, a brief description of the modal expansion solution procedure was followed simply by its statement, for that simple case. Here, it is considered worthwhile to develop the solution in order to illustrate a special difficulty which occurs in the more general case.

As before, the complex displacement amplitude vector is first written as a series of the mode shapes weighted by unknown coefficients, A_n:

$$|X| = \sum_{n=1}^{N} A_n |\psi_n| \tag{70}$$

Substitute Equation (70) back into the governing equations (57),

$$\sum_{n=1}^{N} \{-\omega^2[m]|\psi_n| + i\omega[c]|\psi_n| + [k]|\psi_n|\} A_n = |F| \tag{71}$$

now multiply Equation (71) by some $|\psi_m|^T$, T denoting *transpose,* with m not necessarily equal to n.

$$\sum_{n=1}^{N} \{-\omega^2|\psi_m|^T[m]|\psi_n| + i\omega|\psi_m|^T[c]|\psi_n| + |\psi_m|^T[k]|\psi_n|\} A_n = |\psi_m|^T|F| \tag{72}$$

But due to *orthogonality,*

$$|\psi_m|^T[m]|\psi_n| = 0 \qquad \text{for } m \neq n$$

Define, for $m = n$,

$$|\psi_m|^T[m]|\psi_m| = M_m \quad (73)$$

as the mth mode *modal* mass.

By Equation (73), the summation of the matrix products involving $[m]$ in (72) is reduced to a single constant, M_m. Similar reduction of the products involving $[k]$ in (72) is accomplished as follows:

By Equation (68),

$$[D^*(\omega_n)]|\psi_n| \equiv \{-\omega_n^2[m] + [k]\}|\psi_n| = |0|$$

Multiply by $|\psi_m|^T$,

$$-\omega_n^2|\psi_m|^T[m]|\psi_n| + |\psi_m|^T[k]|\psi_n| = |0|$$

Therefore, in view of Equation (73),

$$|\psi_m|^T[k]|\psi_n| = \begin{cases} 0 & n \neq m \\ \omega_m^2 M_m & n = m \end{cases}$$

Define,

$$K_m = \omega_m^2 M_m \quad (74)$$

as the mth mode *modal stiffness*, such that,

$$\omega_m = \sqrt{K_m / M_m}$$

Also define,

$$|\psi_m|^T|F| = F_m \quad (75)$$

as the mth mode *modal exciting force*.

Substitute Equations (73), (74), and (75) back into (72),

$$(-\omega^2 M_m + K_m) A_m + i\omega \sum_{n=1}^{N} |\psi_m|^T[c]|\psi_n| A_n = F_m \quad (76)$$

Now, if orthogonality can be employed to reduce the damping term in Equation (76) similarly as with the mass and stiffness, then the A_m required in the solution (70) are determined. However, the orthogonality on the damping matrix does not, in general, exist for $N > 1$. It exists only in special cases. For example, if $[c]$ is proportional to $[m]$ and/or $[k]$, then orthogonality exists (c was proportional to both k and m in the simple distributed model of subsection 2.1; that provided the mode shape orthogonality required at (36)). That is, for,

$$[c] = \gamma_n[k] + \delta_n[m] \quad (77)$$

where γ_n and δ_n are constants which are allowed to vary only from mode to mode, then, in Equation (76),

$$|\psi_m|^T[c]|\psi_m| = \gamma_m K_m + \delta_m M_m \equiv C_m \quad (78)$$

C_m is called the *modal damping coefficient*. Presuming C_m to exist, the A_m are then, from (76),

$$A_m = \frac{F_m}{-\omega^2 M_m + i\omega C_m + K_m}$$

or

$$A_m = \frac{F_m / K_m}{1 - (\omega/\omega_m)^2 + 2i\zeta_m \omega/\omega_m} \quad (79)$$

where,

$$\zeta_m = \frac{C_m}{2M_m\omega_m} = \frac{\gamma_m \omega_m}{2} + \frac{\delta_m}{2\omega_m} \quad (80)$$

ζ_m is the mth mode *modal damping factor*. Substituting Equation (79) into (70) completes the derivation,

$$|X| = \sum_{n=1}^{N} |X_n| e^{-i(\alpha_n + \beta_n)}$$

with

$$|X_n| = \frac{\operatorname{mod} F_n / K_n}{\sqrt{[1 - (\omega/\omega_n)^2]^2 + (2\zeta_n \omega/\omega_n)^2}} |\psi_n| \quad (81)$$

$|x(t)|$ follows as,

$$|x(t)| = \operatorname{Re}|X|e^{i\omega t}$$

$$= \sum_{n=1}^{N} |X_n| \cos(\omega t - \alpha_n - \beta_n) \quad (82)$$

Here, α_n and β_n are the modal phase angles,

$$\beta_n = \tan^{-1}[-F_n^I / F_n^R]$$

$$\alpha_n = \tan^{-1}\left[\frac{2\zeta_n(\omega/\omega_n)}{1 - (\omega/\omega_n)^2}\right]$$

Equations (81) and (82) are general equations that have wide application to ship vibration problems, as discussed in Section 3. These equations again confirm that modal expansion can be viewed as just a superposition of the responses of N equivalent one degree of freedom systems representing each of the N modes of the discrete model. The only difference between the above solution for continuous versus discrete models is the length of the series. The continuous case, having infinite degrees of freedom, generates an infinite series.

The restriction imposed upon damping in Equation (77) for $N > 1$ must also be observed in continuous analysis; this difficulty did not appear explicitly in the last sub-section because the beam with uniform properties, in fact, possesses proportional damping automatically.

The restriction on damping is severe. For the internal material damping of structural systems a damping matrix proportional to stiffness can be justified; the simple theory used in the last sub-section for allowing for material damping of the continuous beam leads to this conclusion. However, where other sources of damping are also present, proportionality is usually destroyed and, in such cases, the modal expansion, Equation (81), does not exist, theoretically.

Nevertheless, temptations exist for applying the modal formula to models where proportional damping cannot, in reality, be justified. Ship vibration is a typical example. Three rather significant advantages of modal expansion over the direct numerical inversion approach, Equation (60), exist:

(1) The solution, Equation (81), is in terms of arbitrary exciting frequency, ω. A summation must merely be performed to evaluate the model response at any frequency of interest; the direct inversion requires complete numerical reanalysis of each variation of ω.

(2) In general, a discrete model of a continuous system is accurate for only the system modes within a limited frequency range. That is, while typically the lowest modes of an N degree of freedom model should represent the same modes of the continuous system with accuracy, the Nth mode of the discrete model would be expected to bear no resemblance whatever to the Nth mode of the continuous system. A direct inversion theoretically includes the responses of all N model modes. While including the erroneous model modes may not actually contaminate the results of the analysis, it is certainly inefficient to carry them. In modal analysis, the series can be truncated at levels where modeling inaccuracy becomes pronounced without sacrificing the accuracy of the analysis within the frequency range for which the model was constructed. This means that only a relative few of the N natural frequencies and mode shapes of the discrete model need be evaluated in order to predict the system vibration characteristics of concern.

(3) The semi-analytical form of the modal expansion provides insight into the relative contributions of the elements of mass, stiffness, damping, and excitation influencing a particular vibration problem. This visibility is not available with a purely numerical inversion of the model equations.

Returning again to the Figs. 6 and 7 deckhouse model, the modal expansion of the one dynamic degree of freedom model is just the analytic solution, Equation (63), as comparison with Equations (81) and (82) confirms. The deckhouse response predicted by the simple one degree of freedom model is interesting, however. For simplicity, assume that the hull girder vibration in way of the house, Fig. 6, is rather "flat." That is, assume that the aftermost hull girder nodal point at frequency ω is far enough forward of the house that $w(\xi_1, t) \approx w(\xi_2, t)$ on Fig. 6; the house base experiences a pure vertical translation. Then, in Equation (62), with $X_2 = X_3 \equiv X$,

$$F_1 = -\omega^2 m \xi X$$

Taking X real (which implies a reference phase of zero),

$$\text{mod } F_1 = \omega^2 m \xi X \qquad \text{and} \qquad \beta = \pi.$$

Also assume that the house is in resonance at ω. The house rocking vibration, by Equation (63), is then,

$$x_1(t) = \frac{\omega_n^2 m \xi X / K}{2\zeta} \cos(\omega t - 3\pi/2) \qquad (83)$$

with $\alpha = \pi/2$ at resonance. By Equation (50) the fore-and-aft vibration displacement at the house top is,

$$u(h, t) = x_1(t) h$$

Substituting Equation (83),

$$\frac{u(h,t)}{X} = \frac{\omega_n^2 m \xi h}{2K\zeta} \cos(\omega t + \pi/2)$$

But $\omega_n^2 = K / m\bar{r}^2$, which gives,

$$\frac{u(h,t)}{X} = \frac{\xi h}{2\zeta \bar{r}^2} \cos(\omega t + \pi/2) \qquad (84)$$

Taking as typical values of the data in Equation (84), $\xi/\bar{r} = 3/4$, $h/\bar{r} = 4/3$, and $\zeta = 0.05$, the house-top fore-and-aft displacement is,

$$\frac{u(h,t)}{X} = 10 \cos(\omega t + \pi/2) \qquad (85)$$

This simple analysis implies that the fore-and-aft vibration at the house top can be 10 times the vertical vibration on main deck at resonant conditions. This is not at all out of line with observations. Unacceptable fore-and-aft vibration levels in deckhouses, accompanied by relatively low level vibration of the hull girder, and elsewhere in a ship, is a common occurrence.

2.3 Propeller Exciting Forces. The propeller excitation in the foregoing has been characterized as a simple force concentrated at some point near the aft end of the hull girder. This is acceptable only for elementary demonstration purposes. Propeller excitation is a complicated combination of concentrated forces and moments acting at the propeller hub, plus a distribution of fluctuating pressure acting over the after hull surfaces. The concentrated propeller bearing forces and moments are largely responsible for the vibration of main propulsion machinery and shafting systems, but are not unimportant, in general, as a source of hull vibratory excitation. The usually dominant hull excitation of modern ships is, however, the propeller-induced hull surface pressure field. This is particularly true when any degree of fluctuating sheet cavitation occurs on the propeller blades, which is more

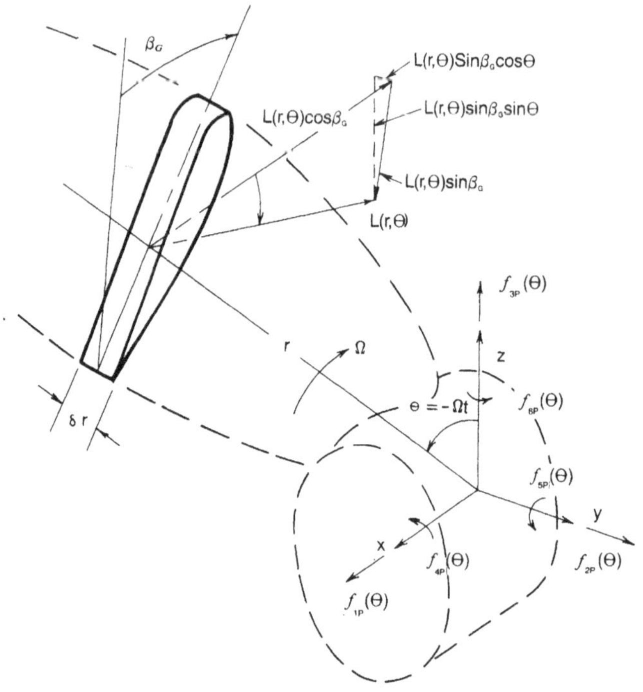

Fig. 8 Propeller blade-element forces

often the rule than the exception. The fundamental concepts and theory of propeller bearing forces and propeller induced hull surface forces are treated in the following.

(*a*) *Propeller Bearing Forces.* Consider Fig. 8, which depicts a propeller blade rotating with angular velocity Ω in the clockwise direction, looking forward. By virtue of the rotation through the circumferentially non-uniform wake the spanwise blade lift distribution, $L(r, \theta)$, fluctuates with time, or with blade position angle $\theta = -\Omega t$. It is of interest to determine the three force and three moment components in the propeller hub produced by the time varying lift distributions of all N blades acting simultaneously. Toward this purpose, define the complex function,

$$g(r, \theta; p) = -L(r, \theta) e^{ip\theta} e^{j\beta_G} \tag{86}$$

Here i and j both denote $\sqrt{-1}$, but they are to be treated as independent in all algebraic manipulations; the reason for this artifice is only for compactness of notation. β_G in Equation (86) is the geometric pitch angle of the blade section at r and p is an integer to be assigned later.

The function $g(r, \theta; p)$ by Equation (86) represents a *pseudo lift* distribution on one blade of the N-bladed propeller. The effects of all N blades acting simultaneously is effected by replacing θ by $\theta + 2\pi(k - 1)/N$ in Equation (86) and summing over k. This operation defines a new complex lift function representing the collective effects of the N blades:

$$G(r, 0; p) \equiv \sum_{k=1}^{N} g(r, \theta + 2\pi(k - 1)/N; p)$$
$$= -\sum_{k=1}^{N} L(r, \theta$$
$$+ 2\pi(k - 1)/N) e^{ip\theta} e^{\frac{2\pi i(k_m - 1)}{N}} e^{j\beta_G} \tag{87}$$

Now, the circumferential wake non-uniformity appears from the blade to be very nearly periodic in time, with fundamental period $T = 2\pi/\Omega$. With the assumption of linearity, the lift distribution, $L(r, \theta) \equiv L(r, t)$, is also periodic with the same period. $L(r, \theta)$ can therefore be written in the Fourier Series,

$$L(r, \theta) = L_o(r) + \text{Re} \sum_{q=1}^{\infty} L_q(r) e^{iq\theta} \tag{88}$$

Here $L_q(r)$ is the qth harmonic complex lift amplitude of the blade section at radius r; $L_o(r)$ is the steady lift distribution associated with steady thrust and torque. A choice in procedures for determining the $L_q(r)$ harmonics is available on specification of the corresponding harmonics of the wake inflow (see Section 3). It is presumed at this point that a sufficient number of the $L_q(r)$ harmonics are available from some source.

An alternative representation of Equation (88), which is useful for insertion into (87), is,

$$L(r,\theta) = L_0(r) + \frac{1}{2} \sum_{q=1}^{\infty} [L_q(r) e^{iq\theta} + \overline{L}_q(r) e^{-iq\theta}] \tag{89}$$

where the bar denotes complex conjugate. Discarding the steady lift and substituting Equation (89) into (87) produces,

$$G(r,\theta;p) =$$
$$-\frac{1}{2} e^{j\beta_G} \sum_{q=1}^{\infty} \left[L_q(r) e^{i(q+p)\theta} \sum_{k=1}^{N} e^{\frac{i(q+p)2\pi(k-1)}{N}} + \overline{L}_q(r) e^{-i(q-p)\theta} \sum_{k=1}^{N} e^{\frac{-i(q-p)2\pi(k-1)}{N}} \right] \tag{90}$$

But it can be easily verified that the k summations appearing in (90) are equal to zero if $q \pm p$ is not some integer multiple of N, say mN, and the summations are equal to N for $q \pm p = mN$. Utilizing these facts, Equation (90) reduces to,

$$G(r,\theta;p) = -\frac{N}{2} e^{j\beta_G} \sum_{m=1}^{\infty} [L_{mn-p}(r) e^{imN\theta} + \overline{L}_{mn+p}(r) e^{-imN\theta}] \tag{91}$$

The bearing forces $f_{ip}(\theta)$, i = 1, . . . , 6, Fig. 8, are now given in terms of $G(r,\theta;p)$ from (91) as:

$$f_{1p}(\theta) = \mathrm{Re}_j \int_{r=r_h}^{R} G(r,\theta;\, 0)\, dr$$

$$f_{2p}(\theta) = \mathrm{Re}_i\, Im_j \int_{r=r_h}^{R} G(r,\, \theta;\, 1)\, dr$$

$$f_{3p}(\theta) = Im_i\, Im_j \int_{r=r_h}^{R} G(r,\theta;\, 1)\, dr$$

$$f_{4p}(\theta) = Im_j \int_{r=r_h}^{R} r\, G(r,\theta;\, 0)\, dr$$

$$f_{5p}(\theta) = \mathrm{Re}_i\, \mathrm{Re}_j \int_{r=r_h}^{R} r\, G\, (r,\theta;\, 1)\, dr$$

$$f_{6p}(\theta) = Im_i\, \mathrm{Re}_j \int_{r=r_h}^{R} r\, G(r,\theta,\, 1)\, dr \qquad (92)$$

The subscripted prefixes Re and Im refer to the real and imaginary parts of the complex quantities involving i and j; the complex lift harmonic is $Lq = Lq^R + iLq^I$ in this regard.

As an example, consider the vertical bearing force, f_{3p}. Equations (91) and (92) give,

$$f_{3p}(\theta) = Im \left\{ -\frac{N}{2} \sum_{m=1}^{\infty} \int_{r=r_h}^{R} \sin \beta_G\, [L_{mN-1}(r)\, e^{imN\theta} + \overline{L}_{mN+1}(r)\, e^{-imN\theta}]\, dr \right\}$$

Using the facts that,

$$Im\, Z = -\, ReiZ$$

and

$$Im\, \overline{Z} = \mathrm{Re} iZ,$$

$$f_{3p}(\theta) = -\, \mathrm{Re} \left\{ \frac{N}{2i} \sum_{m=1}^{\infty} e^{imN\theta} \int_{R=r_h}^{R} [L_{mN-1}(r) - L_{mN+1}(r)] \sin \beta_G\, dr \right\}$$

This formula differs from that given by Tsakonas, Breslin, et al (1967), for example, only in sign. The sign difference is due to the fact that positive lift is here taken as that with forward axial component, in the usual sense. This is opposite to the above reference convention.

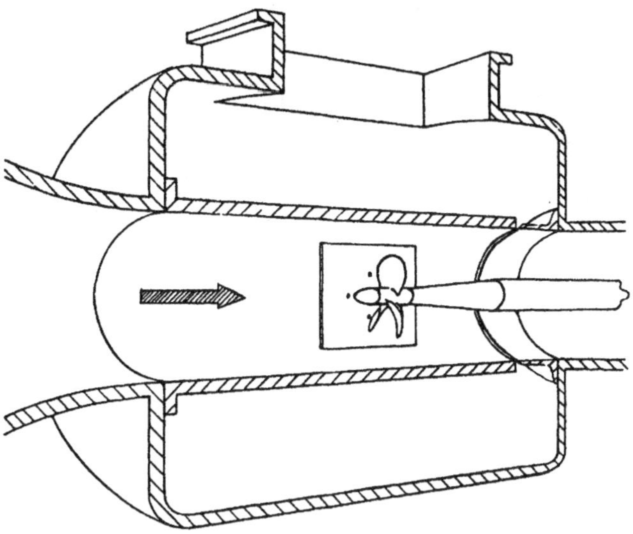

Fig. 9 Flat-plate pressure measurements

The following important facts should be observed from Equations (91) and (92):

1. Propeller bearing forces are periodic with fundamental frequency equal to the propeller angular velocity times the number of blades. The fundamental frequency, $N\Omega$, is called *blade-rate* frequency. The bearing forces, as written in Equations (91) and (92), are composed of terms at blade-rate frequency, plus all of its integer multiples, or harmonics, $mN\Omega$.

2. Only certain terms, or harmonics, of the unsteady blade lift, and therefore of the hull wake, contribute to the bearing forces. While the forces on a single blade consist of components corresponding to all wake harmonics, a filtering occurs when the blade forces superimpose at the propeller hub. Equations (91) and (92) show that the unsteady thrust and torque, f_{1p} and f_{4p}, depend only on the lift, or wake, harmonics that are integer multiples of blade number. The lateral forces and moments, on the other hand, are produced entirely by the wake harmonics corresponding to integer multiples of blade number, plus and minus one.

(*b*) *Propeller-Induced Hull Surface Pressures and Forces.* A thorough understanding of the hull surface pressure distributions produced by a propeller, and the integration to resultant hull surface forces, is only attained with a considerable expenditure of effort. The subject is very complex. Nevertheless, much has been accomplished since the pioneering experimental work of F. M. Lewis (1973), in both understanding hull surface excitation and developing methods for predicting it.

1. *Uniform inflow conditions.* It is useful to begin with the simplest possible case: the pressure induced on a flat plate by a propeller operating in a

uniform inflow. This is depicted in Fig. 9, which is a sketch of the water tunnel arrangement from which the data shown in Fig. 10 were measured (Denny, 1967). Two different 3-bladed propellers were used in the experiments. The propellers were identical in all respects, including performance, except one had blades of double the thickness of the other. With the assumption of linearity, this allowed the independent effects of blade thickness and blade lift to be distinguished from the experimental data recorded with the two propellers. The left-side plots in Fig. 10 show the amplitude and phase of the plate pressure induced by blade thickness; the right-side plots correspond to blade lift. The predictions of theories available in the late 1960's are also shown in Fig. 10.

The pressure data shown on Figure 10 correspond

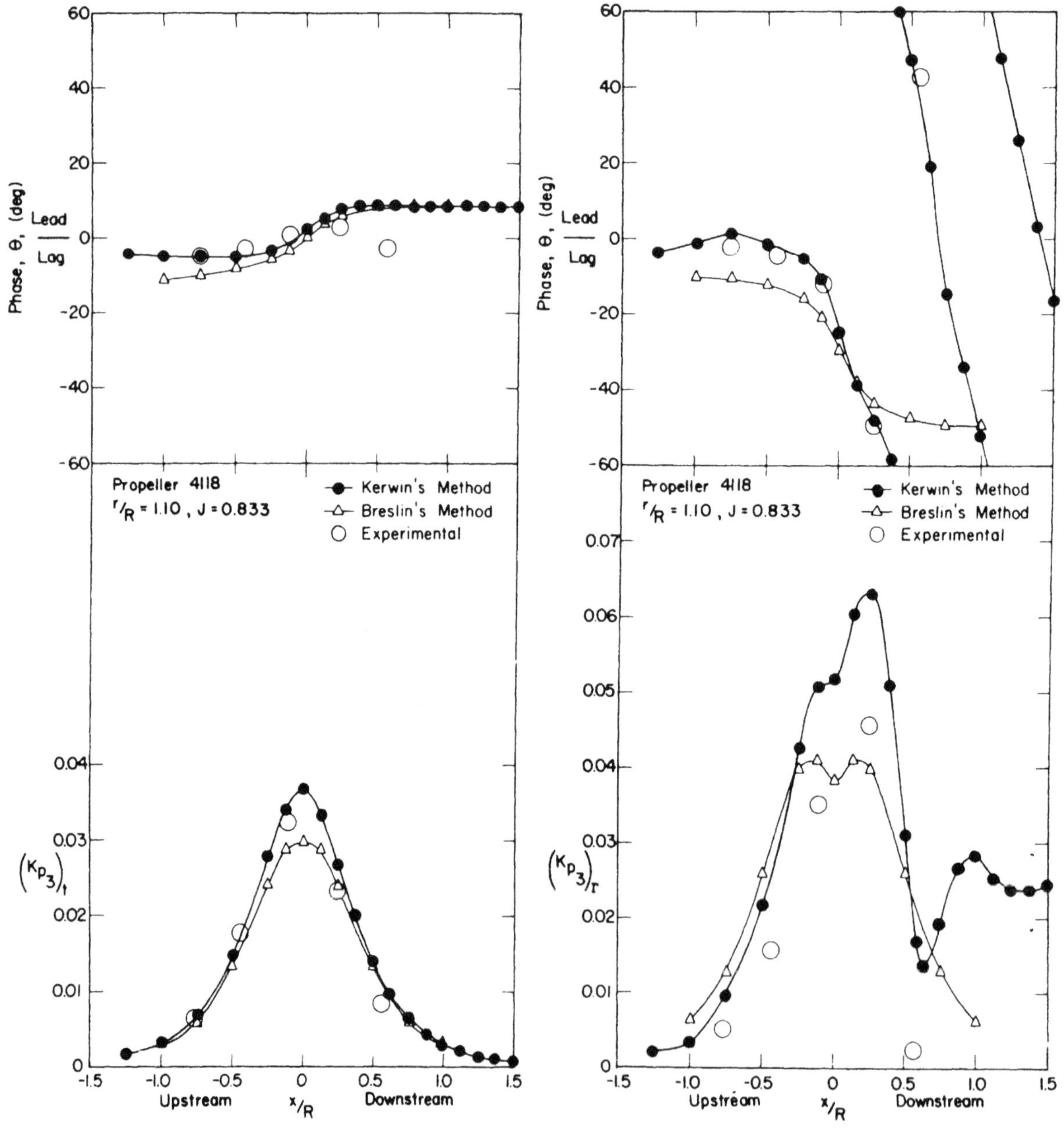

(a) — Comparisons of Theoretical and Experimental Values, Thickness Contribution, $r/R = 1.10$, $J = 0.833$

(b) — Comparisons of Theoretical and Experimental Values, Loading Contribution, $r/R = 1.10$, $J = 0.833$

Fig. 10 Flat-plate pressure amplitude and phase distributions

to blade rate frequency. Just as in the case of bearing forces, all multiples of blade-rate frequency also occur, but the higher harmonics become negligible quickly for the uniform wake case. The phase indicated on Fig. 10 is defined as the position angle of the propeller blade nearest the plate (Fig. 9) when the pressure is positive (compressive) maximum; positive angle is defined as counterclockwise, looking forward. With this definition, the phase relative to a single cycle of the three-cycle per revolution blade rate signal is obtained by multiplying the phase angles on Fig. 10 by 3. This quickly confirms that the blade thickness pressure is approximately in-phase up and down stream of the propeller; it is an even function in x, approximately. On the other hand, a large phase shift occurs in the pressure due to blade lift up and downstream; it behaves as an odd function in x, approximately. This behavior suggest some substantial cancellation in the lift associated pressure, at least, on integration to the net resultant vertical force on the plate. Actually, if the plate is infinite in extent, the thickness pressure, as well as the lift pressure, both independently integrate to produce identically zero net vertical force on the plate. This fact is a demonstration of the *Breslin Condition* (Breslin, 1959). This was established by integrating theoretical pressures induced by a non-cavitating propeller operating in uniform inflow over the infinite flat plate, and showing the identically zero result.

Fig. 11 is a contour plot of the blade rate pressure amplitude from a similar, but different, uniform wake, flat plate experiment (Breslin and Kowalski, 1964). Here, only amplitude is shown; the phase shift distribution responsible for the cancellation on integration is not apparent from Fig. 11. Figs. 10 and 11 clearly imply that propeller-induced hull surface pressure is highly localized in the immediate vicinity of the propeller; the pressure is reduced to a small percentage of its maximum value within one propeller radius of the maximum. There is a tendency, on the basis of this observation, to draw the false conclusion that resultant forces occurring in the general ship case should be similarly concentrated on the hull in the near region of the propeller. This common misconception is explained by the considerations of the following subsection.

2. *Circumferentially non-uniform wake effects.* It was shown in the propeller bearing force theory that only certain *shaft rate* harmonics of the non-uniform wake contribute to the *blade rate* bearing force harmonics. In the case of the propeller induced hull-surface excitation, the entire infinity of shaft rate wake harmonics contribute to each blade rate excitation harmonic. But particular wake harmonics are nevertheless dominant, with the degree of dominance depending primarily on hull form. This will be considered in more detail further on.

The pressure distribution corresponding to the wake operating propeller (without cavitation) has a very similar appearance to the uniform wake case. Fig. 12, from Vorus (1974), shows calculated and measured blade-rate pressure amplitude at points on a section in the propeller plane of a model of the DE1040. It was assumed in both of the pressure calculations shown that the hull surface appeared to the propeller as a flat plate of infinite extent.

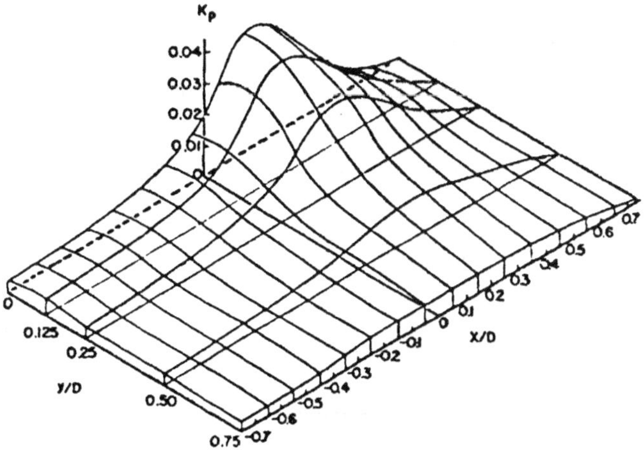

Flat plate at $J = 0.6$

Fig. 11 Flat-plate pressure contours

The upper part of Fig. 12 shows the measured pressures produced by the wake operating propeller, along with the corresponding calculated results. Both blade rate pressure calculations include the uniform wake effects of steady blade lift and blade thickness (Fig. 10 and 11), plus the contributions from the circumferentially non-uniform part of the wake. The nonuniform wake contribution is represented by wake harmonics 1 through 8 (the "zeroth" wake harmonic component referred to in Fig. 12 is equivalent to the steady blade-lift and blade-thickness components).

The lower part of Fig. 12 shows a breakdown of the calculated blade rate pressure distribution from above, as indicated, into contributions from the uniform wake components (steady blade lift and blade thickness) and non-uniform wake components (sum of unsteady lift harmonics 1 to 8). The important point is that the pressure is dominated by the uniform wake effects; the pressure associated with the uniform wake from the lower part of Fig. 12 is essentially identical to the total from above. The non-uniform wake contribution to the blade rate pressure is buried at a low level within the large uniform wake component.

Interestingly, the integral of the pressure to a vertical force on the relatively flat stern has an entirely different character with regard to the relative contributions of the uniform and non-uniform wake components. This is shown in Fig. 13, also from Vorus (1974). The second column on Fig. 13 shows the total blade rate vertical hull surface force calculated on the DE1040. The succeeding 10 columns show the contri-

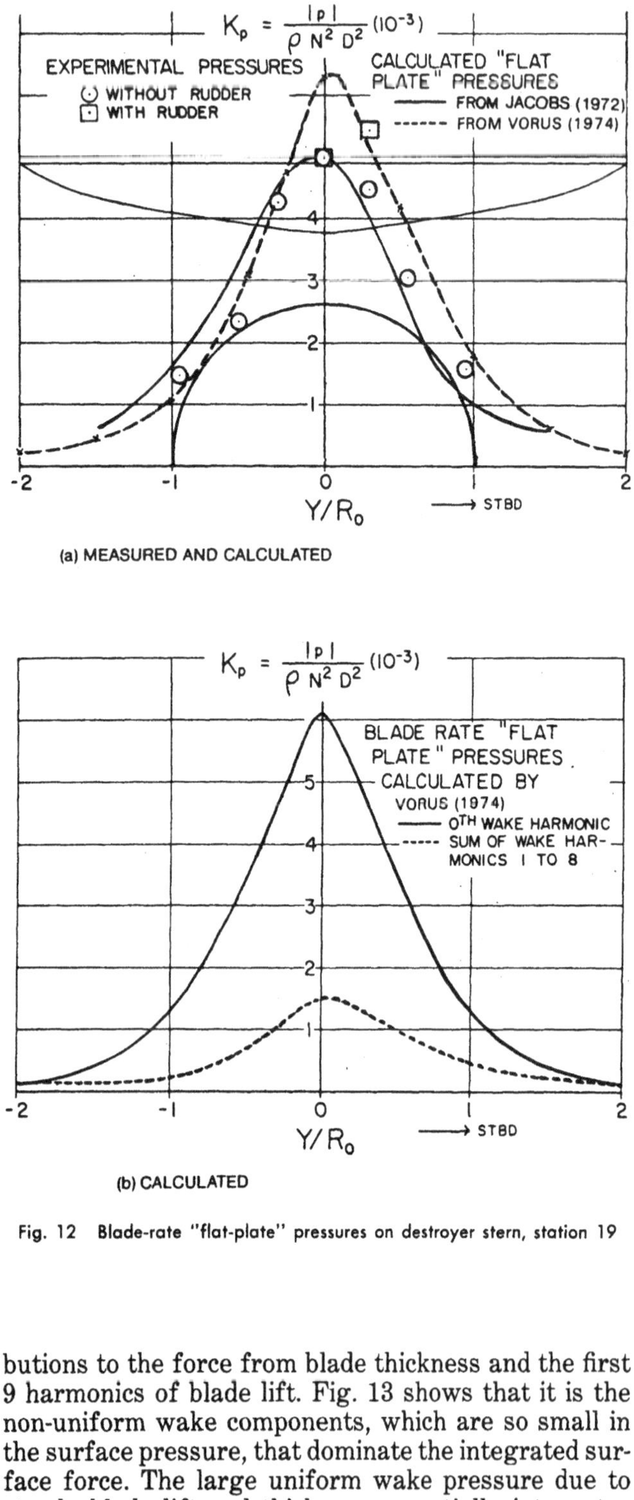

Fig. 12 Blade-rate "flat-plate" pressures on destroyer stern, station 19

butions to the force from blade thickness and the first 9 harmonics of blade lift. Fig. 13 shows that it is the non-uniform wake components, which are so small in the surface pressure, that dominate the integrated surface force. The large uniform wake pressure due to steady blade lift and thickness essentially integrates to zero over the flat stern surface (the Breslin Condition), leaving a blade-rate exciting force due almost entirely to the wake harmonics of orders in the vicinity of blade number (the DE1040 propeller has 5 blades).

Actually, the Breslin condition, as established by Breslin (1959) for the uniform inflow case, can be generalized to cover the non-uniform inflow case as well. It can be stated that, for the case of the general non-cavitating propeller, *the unsteady vertical force induced on an infinite plate above the propeller is equal and opposite to the unsteady vertical force acting on the propeller; the net vertical force on the plate-propeller combination is identically zero.* This, of course, covers the uniform inflow case since the vertical forces on the plate and propeller are both individually zero. The DE 1040 example of Fig. 13 is a good approximate demonstration of the non-uniform inflow case. It was shown by Equations (91) and (92) that the vertical bearing force is produced exclusively by the blade order multiple harmonics of the wake, plus and minus one. For the propeller operating in a wake under an infinite flat plate, the vertical force on the plate, being equal but opposite to the vertical bearing force, must also have to be composed exclusively of the blade order wake harmonics, plus and minus one. These harmonics are obvious in the DE 1040 vertical surface force spectrum of Fig. 13; the DE 1040 stern would be characterized as flat plate-like. With 5 blades, the 4th and 6th harmonics dominate the vertical blade rate surface force, along with the 5th. Amplification of the 5th harmonic is due to the presence of the water surface off the water-plane ending aft.

With regard to the degree of cancellation in the net vertical force on the DE 1040, the bearing force amplitude was calculated to be 0.00205. Its vector addition with the surface force of 0.0015 amplitude produced a net force of amplitude equal to 0.00055, which reflects substantial cancellation. It is noteworthy that F. M. Lewis (1963) measured a net vertical force of amplitude 0.0004 on a model of the same vessel at M.I.T. In the case of the DE 1040, the surface force is smaller in amplitude than the bearing force, but this is not a generality.

At any rate, the characteristics demonstrated in Figs. 11, 12 and 13 clearly indicate that measured surface pressure is a very poor measure of merit of propeller vibratory excitation; hull vibration is produced largely by the surface pressure integral, whose severity is not necessarily well represented by the magnitude of the surface pressure distribution. This fact also implies the level of difficulty that one should expect in attempting to evaluate hull surface forces by numerically integrating measured hull surface pressure. The measurements would have to be extremely precise so as to accurately capture the details of the small non-uniform wake pressure components imbedded in the large, but essentially inconsequential, uniform wake pressure component.

One other relevant aspect with regard to this last point deserves consideration. Returning to Fig. 12, it was noted that the hull was assumed to be an infinite flat plate for purposes of the pressure calculation. This assumption might be expected to result in reasonable satisfaction of the hull surface boundary condition in the very near field of the propeller. So long as the pressure decays rapidly within the propeller near field, reasonably accurate estimates of the pressure maxima might therefore be expected with the flat plate assumption. This is confirmed by Fig. 12; all of the pressure measurement points, where good agreement with calculation is shown, are relatively close to the propeller, and well inside the waterplane boundaries. Outside the waterplane boundaries the relief effects of the water free-surface impose a very different boundary condition than that of a rigid flat plate. Hull surface pressure in the vicinity of the waterplane extremities would therefore be poorly approximated by the infinite flat plate assumption (Vorus, 1976). The overall validity of the flat plate assumption should therefore depend on the relative importance of surface pressure near the waterplane extremities, outside the immediate propeller near field.

From the point of view of the pressure maxima, the very rapid decay of the dominant uniform wake part justifies the flat plate assumption. On the other hand, accuracy of the integrated hull surface forces depend on accurate prediction of the small non-uniform wake pressure components. It is a fact that while these components are relatively small, they also decay much more slowly with distance away from the propeller. It is obvious from Fig. 12 that the pressure persisting laterally to the water surface (which is assumed to be a continuation of the flat plate in the calculations) is due entirely to the non-uniform wake components.

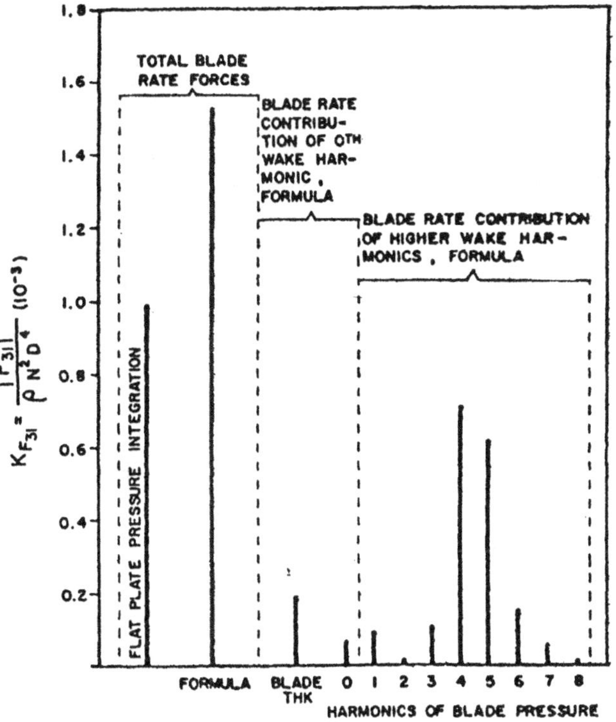

Fig. 13 Calculated blade rate vertical hull surface forces on destroyer stern (DE 1040)

These small pressures persist over large distances and integrate largely in-phase to produce the hull surface forces.

The flat-plate assumption should therefore be less reliable for the prediction of hull surface forces, than for hull surface pressure maxima. This is supported by Fig. 13. The first column on Fig. 13 represents the vertical force amplitude calculated by integrating the calculated "flat plate" pressures over the DE1040 afterbody. The second column in Fig. 13 is the vertical force calculated using a reciprocity principle (Vorus, 1974) which satisfies the hull and water surface boundary conditions much more closely than does the flat plate approximation. While some slight differences in the wake used in the two calculations were discovered, the main difference in the two total force levels shown is due primarily to misrepresentation of the water surface in the flat plate calculation.

The fact that the most important non-uniform wake part of the surface pressure acts over a large surface area actually suggest that total integrated hull surface forces are not the best measure of hull vibratory excitation either. It is the scalar product of pressure distribution and vibratory mode shape represented in the generalized forces of Equation (41), or (82), that would properly allow for "propeller excitability," in the context of the discussion of Fig. 5 (Vorus, 1971.)

3. *Cavitation effects.* The propeller cavitation of concern from the standpoint of vibratory excitation is fluctuating sheet cavitation which expands and collapses on the back of each blade in a repeating fashion, revolution after revolution (Figure 14). The sheet expansion typically commences as the blade enters the region of high wake in the top part of the propeller disc. Collapse occurs on leaving the high-wake region in a violent and unstable fashion, with the final remnants of the sheet typically trailed out behind in the blade tip vortex. The sheet may envelope almost the entire back of the outboard blade sections at its maximum extent. For large ship propellers, sheet average thicknesses are on the order of 10 cm, with maxima on the order of 25 cm occurring near the blade tip just before collapse.

The Fig. 14 type of cavitation, while of catastrophic appearance, is usually not deleterious from the standpoint of ship propulsive performance. The blade continues to lift effectively; the blade suction-side surface pressure is maintained at the cavity vapor pressure where cavitation occurs. The propeller bearing forces may be largely unaffected relative to non-cavitating values for the same reason. The cavitation may or may not be errosive, depending largely on the degree of *cloud cavitation* (a mist of small bubbles) accompanying the sheet dynamics. The devastating appearance of fluctuating sheet cavitation is manifested consistently only in the field pressure that it radiates. The level of hull surface excitation induced by a cavitating propeller can be easily an order of magnitude larger

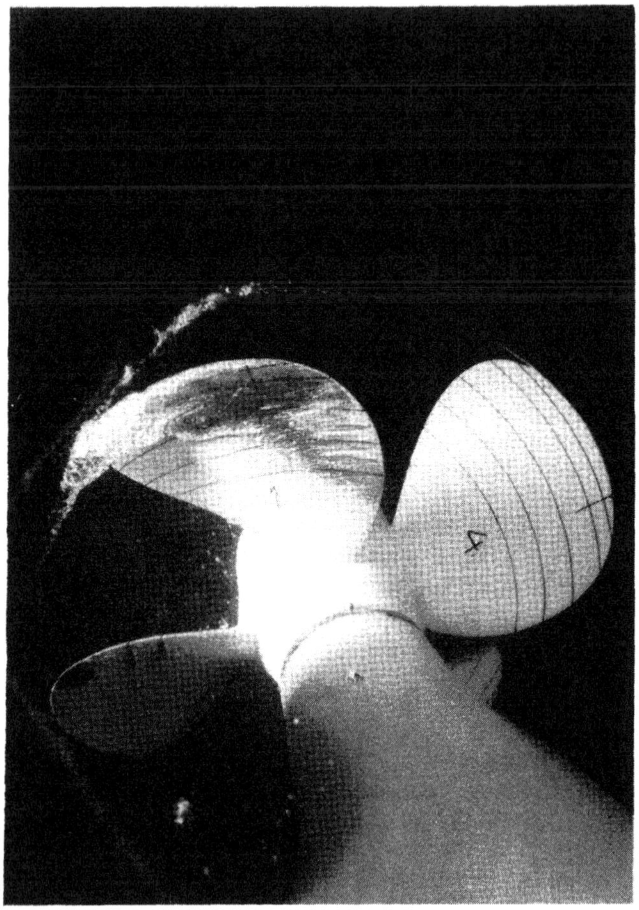

Fig. 14 Fluctuating sheet cavitation

than typical non-cavitating levels. The Breslin Condition does not apply in the cavitating case, and vertical hull surface forces due to unsteady cavitation typically exceed vertical propeller bearing forces by large amounts.

Fluctuating sheet cavitation can be characterized as an unsteady blade thickness effect from the standpoint of field pressure radiation. Any unsteady blade thickness effects associated with the non-cavitating propeller are higher order. Furthermore, the steady average cavity thickness (zeroth harmonic) produces field pressure on the order of that produced by the bare blade. It is the source-like volume expansion associated with the cavity unsteadiness that produces the large blade rate radiated pressures, and its multiples.

Just as with the unsteadiness of blade lift in the non-cavitating case, the cavitating hull forces are produced primarily by the pressure components associated with the higher cavitation harmonics of order near blade number, and its multiples. For the same maximum cavity volume, the shorter the duration of the cavitation, the higher its high harmonic content. Strength in the high harmonics of the cavitation spectrum results in significant excitation at the blade rate multiples; slow convergence of the blade rate excitation series is a characteristic of cavitating propellers.

In view of the importance of the various sets of harmonics involved in propeller excitation, one important distinction between the cavitating and non-cavitating cases should be recognized at this point. In the non-caviting case a one-to-one relationship exists between the harmonics of the circumferentially non-uniform wake and the harmonics of blade lift; the assumption of linearity which makes each blade lift harmonic a function of only the corresponding wake harmonic has been established as valid because of the typically small flow perturbation in the non-cavitating case. Such a linear relationship does not exist between the wake harmonics and the cavitation volume harmonics. Certainly it is the non-uniform wake that almost solely produces the fluctuating sheet cavitation. But sheet cavitation growth has been found theoretically to be most responsive only to the first few harmonics of the wake. The sheet cavitation, which is produced mainly by the low harmonic content of the wake, typically completes its cycle within a relatively small fraction of one propeller revolution. The volume associated with this rapid expansion and collapse has much more strength in its high harmonics than does the part of the wake that produces it.

As an aside, it may some day prove to be a fortunate circumstance that cavitation effects, which are most important in the propeller vibratory excitation problem, depend most strongly on only the gross features (low harmonics) of the non-uniform wake. Unlike the fine detail of the wake to which non-cavitating forces are most sensitive, some hope may be held for rational prediction of gross wake characteristics.

The character of the cavitation-induced hull pressure field differs from the non-cavitating case in one important respect. The multiple blade-rate pressure components produced by the higher cavity harmonics, which are dominant in the integrated forces, are no longer mere "squiggles" imbedded in a vastly larger zeroth harmonic field. The now large pressure components from the cavitation unsteadiness should be more accurately captured in measurements of total pressure signals. For this reason, measurements of cavitation-induced point pressures would be expected to be a more meaningful measure of vibratory excitation than are non-cavitating pressures. However, the filtering action of the hull surface on integration still appears to be capable of producing inconsistencies between point pressure and integrated force levels. Higher order cavitation harmonics with strength in the pressure distribution will be modified in strength by the surface integration, to different degrees. Different weightings of the various pressure harmonic components could logically result in a superposition of drastically different character in the force resultants. Measured pressures of levels inconsistent, from case to case, with the levels of the forces that they integrate to produce should not be unexpected.

Greater accuracy should also be achievable in numerically integrating measured cavitation-induced pressures to attain hull surface force estimates. This is, again, because the size of the important pressure components is relatively greater than in the non-cavitating case. However, coverage of a large area of the model surface with pressure transducers should be required in view of the very slow attenuation of the cavitation induced pressure signal. In this regard, whether forces or pressures are the interest, it is no doubt most important that boundary conditions be modeled accurately, either in analysis or experiments. Theory indicates, for example, that due to the slow spatial pressure attenuation associated with the cavitation volume source strength, surface pressures, even in the immediate propeller vicinity, can be overestimated by a factor on the order of four in typical cases if the rigid wall boundary condition is employed at the water surface.

Approximate formulas for evaluating propeller-induced forces are proposed in Section 3.

Section 3
Analysis and Design

3.1 Introduction. More and more the ship designer is being faced with the requirement to deal effectively with propeller and machinery induced vibration in his design work. One may feel uncomfortable, if not bewildered, by the seemingly endless complexity of the problem and the myriad of physical interrelationships influencing one's decisions.

Indeed, a mere description, without accompanying quantitative analysis, presents an imposing problem. Excluding effects of the seaway, the ship hull is excited mechanically by rotating machinery systems and hydrodynamically by its propeller(s). These excitation sources are essentially periodic, but they are not, in general, simple harmonic, i.e., purely sinusoidal. Because of this, excitations also occur at all multiples of a fundamental exciting frequency associated with each excitation source. The strengths of the various excitations, and their *harmonics*, are often highly sensitive to the details of design and fabrication. Moderate propeller cavitation, for example, which may be acceptable in all other respects, can produce hull vibratory excitation forces on the order of tens of tons, persistent at frequencies out to several multiples of the blade-rate fundamental (RPM times number of blades).

The infinity of excitations stimulate the ship to vibrate in generally all directions. The degree to which the ship responds to the excitations is sensitive to its *natural* vibration characteristics, or natural vibratory *modes*. The coincidence of the *natural frequency* identified with some natural mode and the *exciting frequency* of some excitation component corresponds to a condition of *resonance*. At resonance, rigidity is counterbalanced by inertia and limitless vibratory amplification by the excitation is opposed only by damping, to first order. Since in ships, as in most engineering structures, damping is small, resonance is in general a condition that would be desirable to avoid.

Unfortunately, resonances cannot be avoided. The infinity of excitation frequencies overlies an even more dense infinity of natural frequencies. The natural modes vary in character from the overall lateral bending, axial, and torsional modes of the hull girder to highly localized vibration of plating panels, piping, handrails, and a plethora of others. Transmission paths of the vibration through the ship structure are highly influenced by distributions of local resonances, or near resonances; impedance to vibration transmission is reduced in regions where local resonances occur, and vice-versa. The propagation of low-level, generally non-resonant vibration through a ship provides the base excitation capable of resonating local elements; this can often be observed in regions far removed from the source of the responsible excitation. The seeming complexity of it all is amplified upon recognition of the existence of dynamic as well as static coupling; excitations occurring in one direction can produce resonant vibration in other directions through the directional coupling of intervening structure. Substructures, or subregions, of a vessel which are treated as independent of one another in more conventional design considerations, can be dynamically coupled to a significant degree. For example, longitudinal resonance in the main propulsion system can produce foundation dynamic forces and moments large enough to excite objectional fore-and-aft rocking/bending of the vessel deckhouse, depending upon the compliance of the intervening structure.

It is fortunate, in view of the above limitless latticework of unavoidable resonances, that, as frequency level increases, the various series of excitation harmonics do converge, the relevant natural vibratory modes become more difficult to excite, and the predominant damping mechanisms increase in strength.

Ship vibration is, in practice, not as difficult to deal with as the preceding description might suggest. With patience, the complexities can be systematically sorted out, more or less understood, and dealt with in a rea-

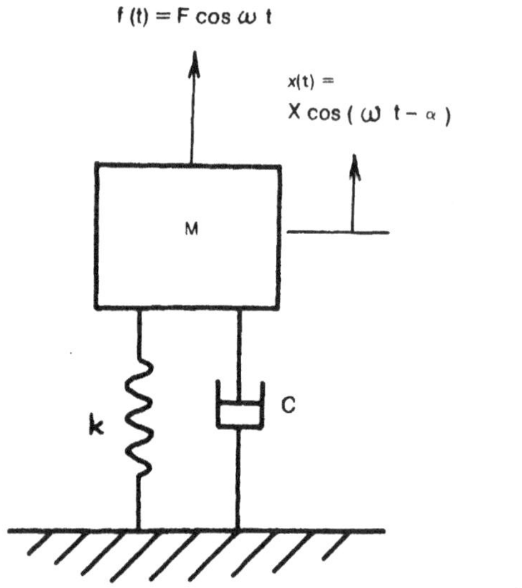

Fig. 15 Steady-state harmonic vibration of one-mass system

sonably effective way through the basic vehicle of rational mechanics. Indeed, the general response formulas developed in the preceding section, Equations (41) or (82), contain the near totality of possibilities for influencing any vibration. These formulae predict the vibratory displacement of continuous (41), or discrete, (82), mathematical models of vibratory systems. In either case, the system displacement is written as a superposition of displacements of a set of equivalent one-mass systems. The mass, stiffness, damping, and excitation force elements of each of the equivalent one-mass systems are constructed as explained in Section 2. The vibratory behavior of any complex system can therefore be dealt within terms of the collection of equivalent one-mass systems vibrating simultaneously. For this reason, much insight into the various sensitivities of the vibration of any particular system, whether simple or complex, can be gained by applying a few simple observations from the theory for one-mass systems.

(a) Basic Considerations. The general one-mass system is depicted in Fig. 15. The M, K, and C denote the mass, stiffness, and damping of the system, respectively, and $f(t)$ is the simple harmonic exciting force of amplitude F and frequency ω. The values of M, K, C, and F can be considered as independent of time, but vary, in general, with the exciting frequency, ω.

Either of the general response formulas of Section 2, Equation (41) or (82), reduce to the following simple formula on application to the Fig. 15 one-mass system:

$$x(t) = \frac{F/K}{\sqrt{[1 - (\omega/\omega_n)^2]^2 + (2\zeta\,\omega/\omega_n)^2}} \cos(\omega t - \alpha)$$
$$= X \cos(\omega t - \alpha) \qquad (93)$$

Here X is the amplitude of the steady-state simple harmonic vibration displacement at frequency ω, and α is the displacement phase angle relative to $f(t)$,

$$\alpha = \tan^{-1} \frac{2\zeta\omega/\omega_n}{1 - (\omega/\omega_n)^2} \qquad (94)$$

By Equations (74) and (80), the ω_n and ζ in (93) and (94) are:

$$\omega_n = \sqrt{K/M} \quad \text{natural frequency} \qquad (95)$$
$$\zeta = C/2M\omega_n \quad \text{damping factor}$$

When considered in light of the general response formula, (82), the one-mass system displacement $x(t)$, can, by (93), be alternatively viewed as the contribution of one of the set of system modes to the complete system vibration. In this view, the M, K, C, and F are the *modal* values whose magnitudes vary from mode to mode depending on the distributions of system mass, stiffness, damping, and excitation relative to the *mode shape* of the particular mode; this is according to Equations (73) through (78) of Section 2.

Fig. 16 is the familiar plot of $X/(F/K)$ and α from Equations (93) and (94) versus frequency ratio, ω/ω_n. Note that the function F/K is the vibratory displacement amplitude that would be predicted by quasi-static analysis. $X/(F/K)$ can therefore be viewed as a correction factor on the quasi-static displacement for dynamic effects. This ratio is called the dynamic *Magnification Factor*. It is apparent from Fig. 16 that the Magnification Factor can act to reduce the quasi-static displacement amplitude as well as to magnify it.

While Fig. 16 displays the basic character of the one-mass system vibration of interest, some care must be exercised in its interpretation. As noted, the M, C, K, and F are frequency dependent, in general. The curves of Fig. 16 can therefore be misleading with regard to the variation of vibratory amplitude and phase angle with frequency. For example, for an exciting force amplitude increasing as ω^2, such as in the case of a rotating machinery unbalance, multiplication of the Fig. 16 response characteristic with ω^2 is required in order to represent the correct frequency dependence of the actual displacement.

The Fig. 16 curves are instructive. However, the possibilities for influencing vibration are most directly apparent from Formula (93) for one-mass system response. All possibilities lie in only four variables:

1. Excitation, F
2. Stiffness, K
3. Frequency Ratio, ω/ω_n
4. Damping, ζ

It is obvious from (93) that any of the following contribute to vibration reduction:

1. Reduce exciting force amplitude, F. In propeller induced ship vibration the excitation is reduced by

changing the propeller unsteady hydrodynamics. This may involve lines or clearance changes to reduce the non-uniformity of the wake inflow or it may involve geometric changes to the propeller itself. Specifics in this regard are identified in the subsection on propeller excitation.

2. Increase stiffness, K. Stiffness, which is defined as spring force per unit deflection, cannot be considered independently of frequency ratio, ω/ω_n, since $K = \omega_n^2 M$. However, Equation (93) clearly shows that stiffness should be increased rather than decreased when variations in natural frequency are to be accomplished by variations in stiffness. It is bad practice, in general, to reduce system stiffness in attempts to reduce vibration.

3. Avoid values of frequency ratio near unity; $\omega/\omega_n = 1$ is the resonant condition. From (93) at resonance,

$$X = (F/K)/2\zeta \tag{96}$$

Here the excitation is opposed only by damping; note the peak in the frequency response curve of Figure 16 at resonance. Obviously, ω/ω_n can be varied by varying either ω or ω_n. The spectrum of ω can be changed by changing the RPM of a relevant rotating machinery source, or, in the case of propeller-induced vibration, by changing the propeller RPM or its number of blades. ω_n is changed by changes in system mass and/or stiffness, by Equation (95); increasing stiffness is the usual and preferred approach. Specific measures for resonance avoidance in ships are considered in the next subsection.

4. Increase damping, ζ. Damping of structural systems in general, and of ships in particular, is small; $\zeta \ll 1$. Therefore, except very near resonance, the vibratory amplitude is approximately,

$$X = \frac{F/K}{|1 - (\omega/\omega_n)^2|}$$

which is damping independent. Damping is therefore relatively unimportant except in resonant vibration, by Equation (96). Furthermore, damping is difficult to increase significantly in systems such as ships; ζ is the least effective of the four parameters available to the designer for implementing changes in ship vibratory characteristics.

(b) Recommended Design Approach. While the basic vibratory behavior of ships is described qualitatively by the simple one-mass system formula, implementation of more general formulas, such as (82), is necessary if attempts at detailed quantification of ship vibratory response are to be made, as recommended by Noonan (1981) for high-powered naval ships. However, it is widely accepted that the present state of the art does not provide accurate enough definition of all of the system parameters, principally damping, to make detailed calculations of ship vibration response, per se, worth the effort as a routine design exercise. Of course, the designer should keep in touch with continuing research developments that may affect this picture.

Accurate beforehand prediction of ship vibration response for, say, comparison against established habitability criteria, would be desirable. However, experience has proved that such is not only impossible, but also unnecessary in designing ships with consistently acceptable vibration characteristics. Four elements were identified in the preceding as being influential in determining ship vibratory response, and their relationship to vibration reduction was identified. While quantification of all four elements is required in predicting vibratory response level, acceptable results can usually be achieved with reasonable effort by focusing attention in design on only two of the four elements, and deemphasizing the importance of vibration response calculations, except in special cases. The two of the four elements of crucial importance are excitation and frequency ratio. The achievement in design of two objectives with regard to these elements has proved to result in ships with consistently acceptable vibration characteristics:

1. Minimization of the dominant vibratory excitations within the normal constraints imposed by other design variables, and

2. avoidance of resonances involving active participation of major subsystems in frequency ranges where the dominant excitations are strongest.

Fortunately, unlike vibration response, the excitation and frequency ratio elements involved in these objectives are predictable with reasonable reliability. Detailed hydrodynamic calculation procedures in conjunction with model testing have been established in excitation analysis, at least to the level of reliable relative predictions. Natural frequencies involving the ship hull and its major subsystems are predictable using modern structural analysis methods. Hence, the accuracy levels achievable in predictions of propeller

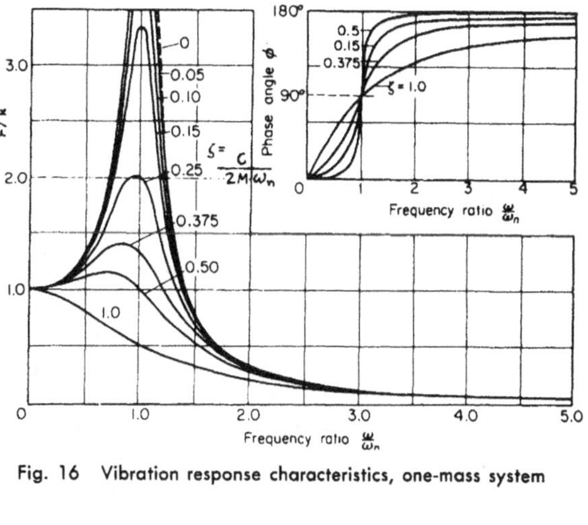

Fig. 16 Vibration response characteristics, one-mass system

and engine excitation and of ship natural frequencies have been found to be high enough to consistently achieve the two objectives cited above. Approximate methods are discussed subsequently.

The detailed calculations and experiments required in assuring excitation minimization and resonance avoidance are usually performed by specialist groups or model basins, and are usually not the immediate responsibility of the ship designer. A main function of the ship designer in this regard is, however, to establish a preliminary design to serve as the subject of the detailed investigations. The quality of the preliminary design will be reflected in the number of detailed iterations required for achieving an acceptable final design. For the purpose of establishing high quality preliminary designs, which require a minimum of expensive and time consuming calculations and model testing, the designer is desirious of both guidance as to the areas of his design likely to be in most need of attention, and some simple methodology for identifying the critical areas. As suggested by Johannessen and Skaar (1980), attention to vibration in preliminary design of large ships can usually be limited to the following main items:

1. Hull girder vertical vibration excited by a diesel main engine.
2. Main machinery longitudinal vibration excited by the propeller.
3. Superstructure longitudinal vibration excited by hull girder vertical vibration and/or main machinery longitudinal vibration.

Many good sources of material are available for seeking help in resonance avoidance and excitation minimization with regard to these three critical items, e.g. Bourceau and Volcy (1970), Breslin (1970), Norske Veritas (1979), Johannessen and Skaar (1980), and Ward (1982). The remainder of this section is directed specifically to the same need; the focus is on providing additional insight into, and facility in using, methodology of established effectiveness for approximate estimates of natural frequencies and exciting force levels for the three critical items cited above.

3.2 Approximate Evaluation of Hull Girder Natural Frequencies. The vertical beam-like modes of vibration of the hull girders of modern ships are dangerous in two respects:

(a) They can be excited to objectionable levels by resonances with the dominant low frequency excitations of slow-running diesel main engines.

(b) Vertical vibration of the hull girder in response to propeller excitation is a direct exciter of objectionable fore-and-aft superstructure vibration.

The propeller is generally incapable of exciting the hull girder modes themselves to dangerous levels. This is primarily because the higher hull girder modes whose natural frequencies fall in the range where propeller excitation is significant have low excitability (refer to the discussion of Fig. 5 in Section 2). However, the low-level vertical hull girder vibration that does occur serves as the base excitation for excessive vibration of superstructures and other attached subsystems which are in resonance with the propeller exciting frequencies. The mechanics of this excitation is demonstrated by the base-excited deckhouse example of Section 2.

The natural frequencies corresponding to the two-noded vertical bending modes of conventional ship hulls can be estimated with reasonable accuracy using either the Burrill (1934-35) or Todd (1961) formulas, of which the latter can take account of the effect of long superstructures. A later formulation was given by Kumai (1968).

Kumai's Formula for two-noded vertical bending is:

$$N_{2v} = 3.07 \cdot 10^6 \sqrt{\frac{I_v}{\Delta_i L^3}} \quad \text{cpm} \tag{97}$$

where

I_v = moment of inertia (m^4)

$\Delta_i = \left(1.2 + \frac{1}{3} \cdot \frac{B}{T_M}\right) \Delta$ = displacement = including virtual added mass of water (tons)

L = length between perpendiculars (m)

B = breadth amidships (m)

T_M = mean draft (m)

Table 1—Comparison, 2-noded hull vertical vibration, Hz

Ship No.	Type	Size (t)	Kumai	Finite-Element Method	Deviation, %
1	reefer	15 000	1.54	1.51	+ 2
2	RO/RO	32 000	1.46	1.16	+26
3	RO/RO	49 000	1.49	1.6	− 7
4	RO/RO	42 000	1.04	0.94	+10
5	chemical tanker	33 000	1	0.93	+ 8
6	bulk carrier	73 000	0.63	0.64	
7	multi-purpose	15 500	2	1.62	+23

Table 1, from Johannessen and Skaar (1980), gives an indication of the accuracy that can be expected from Equation (97). The table compares the prediction of the 2-noded vertical hull bending natural frequency by Kumai's formula with the predictions of detailed finite element calculations performed on seven different ships.

The 2-noded hull vertical bending natural frequencies actually lie well below the dangerous exciting frequencies of either typical diesel main engines or propellers, and are therefore of little consequence in these considerations. As will be demonstrated further

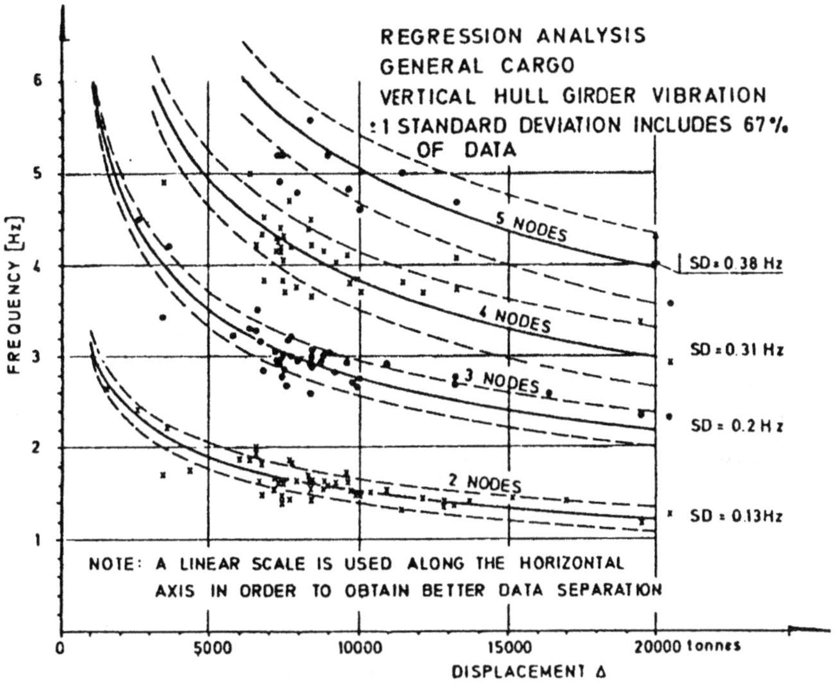

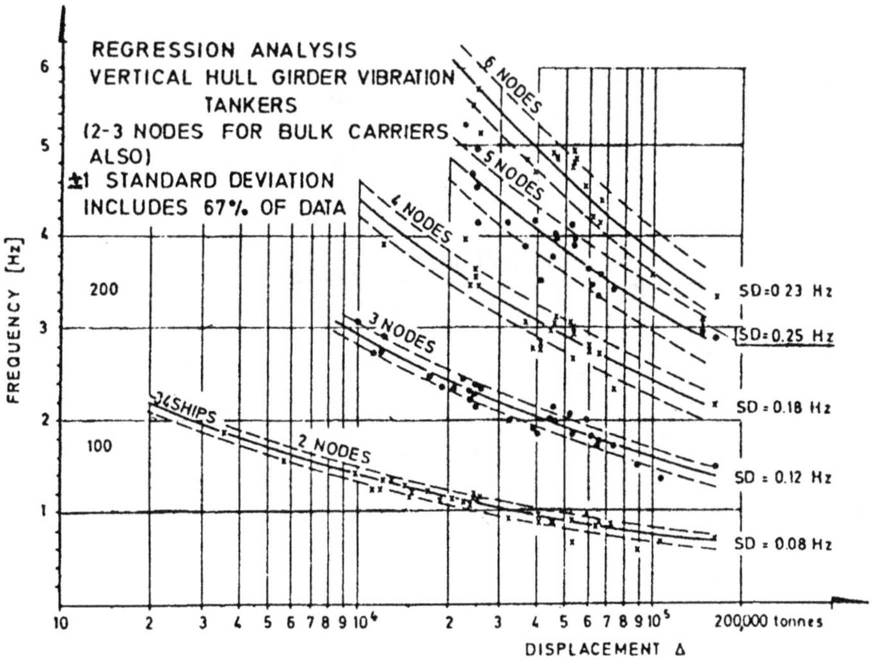

Fig. 17 Natural frequencies of vertical hull vibration

on, it is hull girder modes with typically a minimum of 4 or 5 nodes that can be excited excessively by the diesel main engine. In the case of the propeller, the hull girder vertical bending modes that fall near full-power propeller blade-rate excitation are typically more than 7-noded. Full-power blade rate excitation of large ships is usually in the range of 8 to 12 Hz; as indicated in Table 1, the 2-noded vertical hull bending mode, on the order of 1 to 2 Hz, is well below the blade-rate excitation frequency level during normal operation.

It is observed that hull girder natural frequencies increase more or less linearly with node number from the 2-noded value for the first few modes. The data

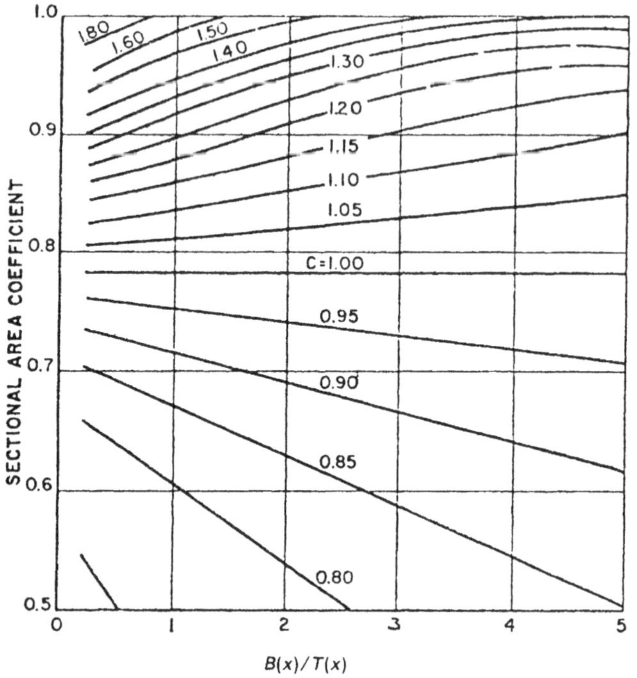

Fig. 18 Virtual-mass coefficients

shown on Fig. 17, from Johannessen and Skaar (1980), provide estimates of the natural frequencies of the first four vertical bending modes of general cargo ships and of the first five vertical bending modes of tankers. Note the good agreement between the Table 1 data and Figure 17 for 2-noded cases. Also note that the 6 Hz maximums represented by Figure 17 still lie well below typical full-power propeller excitation frequencies, and the accuracy of the data fits indicated on the figure is deteriorating rapidly as modal order increases. The primary reason for the increasing data scatter with node number is the increasing influence of local effects, i.e., approaching resonances of deckhouses, decks, etc., on the basic beam modes still identifiable.

The Kumai (1968) formula, in conjunction with Figure 17, is, however, useful in preliminary steps to avoid resonances with a main diesel engine. The following formula, from Johannessen and Skaar (1980), representing the Figure 17 data, expresses the first few vertical bending natural frequencies in terms of the 2-noded value:

$$N_{nv} \approx N_{2v}(n - 1)^{\alpha} \tag{98}$$

$\alpha = 0.845$ General Cargo Ships

1.0 Bulk Carriers

1.02 Tankers

Here N_{2v} is the 2-noded vertical natural frequency and n is the number of nodes; n should not exceed 5 or 6 in order to remain within the range of reasonable validity of Equation (98). Note the approximate proportionality of N_{nv} to node number in Equation (98); this is also evident in Fig. 17.

More accurate estimates of the lower hull girder modes can be obtained by modeling the hull girder as a non-uniform beam. The basic model required is essentially that used in static calculations of longitudinal strength. The natural frequency analysis should therefore be within the capability of the conventional design office, which typically engages in computerized longitudinal strength calculations. The non-uniform beam dynamic analysis differs from the static analysis in one major respect, however. A hydrodynamic added mass distribution must be estimated and superimposed on the vessel mass distribution in order to obtain natural frequency estimates with any degree of realism. Estimation of the required added mass distribution for use in calculating the hull girder vertical modes by way of non-uniform beam analysis is the subject of the next subsection.

3.3 Hydrodynamic Added Mass. Ships are unlike most other vehicles in respect to the substantial inertial effects to which they are subjected by the high density medium in which they operate. The water inertia forces, being proportional to ship surface accelerations, imply an equivalent or effective fluid mass imagined to accelerate along with the ship mass. This effective mass is termed hydrodynamic added mass.

Hydrodynamic added mass is usually large. For example, in the case of a deeply submerged circular cylinder in heave motion, ideal fluid theory predicts an added mass per unit length of the cylinder equal to the mass per unit length of displaced fluid. The corresponding value for a sphere is one-half the mass of the fluid displaced. Added mass effects cannot be ignored in ship vibration analysis.

The calculation method of F. M. Lewis (1929) remains today the most popular method for estimating the added mass distribution of a vertically vibrating ship. By Lewis, the hydrodynamic added mass per unit length at longitudinal position x along the vertically vibrating ship is:

$$m(x) = (\pi/8)\rho B^2(x) C(x) J_n \quad \text{t/m} \tag{99}$$

where,

ρ = density of water, t/m^3

$B(x)$ = section beam, m

$C(x)$ = section 2-dimensional added mass coefficient.

J_n = *Lewis J-Factor,* representing a reduction factor on the 2-dimensional added mass for 3-dimensionality of the vibration-induced flow.

The 2-dimensional section added mass coefficient, $C(x)$, is determined using the so called *Lewis-form* conformal mapping of the ship sections.

This transformation, of the form,

$$Z = z + \frac{a(x)}{z} + \frac{b(x)}{z^3} \qquad (100)$$

transforms a unit circle from plane, z, into the ship section plane Z. The shape of the particular ship section is represented in Equation (100) by the mapping parameters $a(x)$ and $b(x)$. $a(x)$ and $b(x)$ are determined so that the section area coefficient and beam/draft ratio are preserved in the transformation. On specification of a and b, the ideal fluid solution for the unit circle manipulates to give the 2-dimensional added mass for the ship section. All of this is concisely represented on Fig. 18, which was constructed by Todd (1935). $C(x)$ can be extracted from Fig. 18 on specification of the section area coefficient, $A(x)/[B(x)\,T(x)]$, and the section beam-draft ratio. Typical Lewis-form section shapes are shown in Chapter VIII, Vol. III.

The J-Factor in Equation (99) is most easily defined by rewriting the equation as,

$$m(x) = m_{2-D}(x)\,J_n \qquad (101)$$

where

$$m_{2-D} = (\pi/8)\rho B^2 C \qquad (102)$$

is just the 2-dimensional added mass per unit length at section x. Then, integrating Equation (102) over the ship length,

$$M = J_n \int m_{2-D}(x)dx \equiv M_{2-D}\,J_n$$

giving,

$$J_n = M/M_{2-D} \qquad (103)$$

J_n is therefore the ratio of the total added mass in n-noded vibration to the total value assuming 2-dimen-

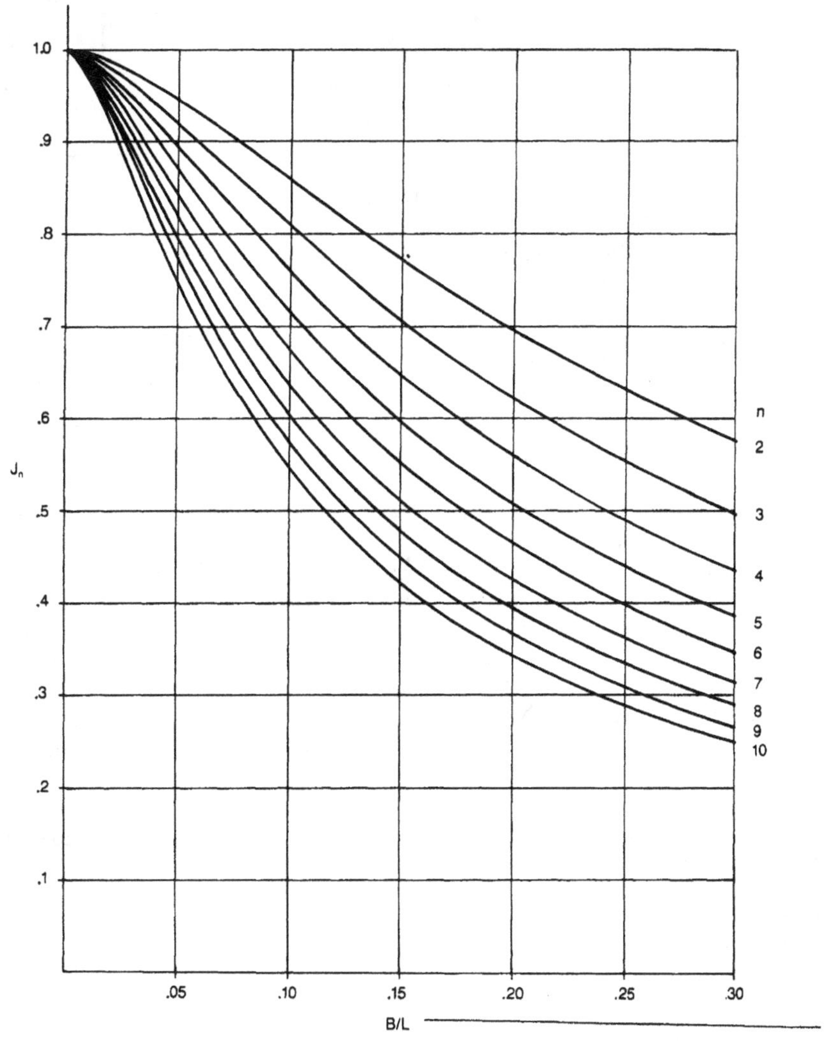

Fig. 19 Lewis J-factor (n is mode number)

sional flow section by section. Lewis assumed that this ratio for a ship was approximately equal to that for a spheroid (ellipse of revolution) of the same beam/length ratio; the exact value of M, as well as that of M_{2-D}, are available for the spheroid, assuming ideal fluid flow.

J_n so determined from the spheroid calculations (Vorus and Hylarides, 1981), can be extracted from Fig. 19; B/L on Fig. 19 involves the midship beam, B, rather than the section beam, $B(x)$, used on Fig. 18. The J_n are functions of the number of nodes, n, in the hull girder vibration; the J_n, and therefore the added mass, vary mode by mode. Note from Fig. 19 that the J_n decrease with increasing n; this is due to the increasing 3-dimensionality of the flow in the higher modes.

For example, consider a ship section with $A(x)/[B(x)T(x)] = 0.9$, $B(x)/T(x) = 2$, and $B/L = 0.15$. From Fig. 18, $C(x) = 1.17$. Assume that the 7-noded hull vertical natural frequency is of interest. For 7 nodes, from Fig. 19,

$$J_7 = 0.515$$

With $\rho = 1$ t/m^3 for SW., the added mass per unit length at the section is, from Equation (100):

$$m(x) = 0.237\ B^2 \text{ t/m}$$

3.4 Approximate Evaluation of Superstructure Natural Frequencies. Propeller induced vibration of stern-mounted superstructures has become one of the naval architects greatest concerns. As proposed in the preceding subsection, a non-uniform beam model which ignores the dynamics of sprung substructures produces useful estimates of the hull girder lower natural frequencies for purposes of resonance avoidance with a main diesel engine. It is indeed fortunate that the lower rocking/bending natural frequencies associated with stern superstructures, which usually fall in the range of propeller blade-rate exciting frequencies, can, conversely, be estimated with useful accuracy by ignoring the dynamics of the hull girder. This is the case when the mass of the superstructure is small relative to the effective mass of the hull girder near the coupled natural frequencies of interest. Any consideration of vibratory response, versus natural frequencies alone, must, on the other hand, allow for the dynamic coupling. This is obvious on consideration of the fact that, in the preponderance of cases, superstructure vibration is excited by the hull girder vibration at its base.

The superstructure vibration mode of primary concern is a fore-and-aft rocking/bending mode excited through vertical vibration of the hull girder; an idealization of this mode was developed for conceptual purposes in the second section of the chapter.

For obtaining preliminary estimates of superstructure fore-and-aft rocking/bending natural frequencies the semi-empirical method of Hirowatari and Matsumoto (1969) has great utility (Sandström and Smith, 1979). The Hirowatari Method was developed from correlations of simple analysis and measured fore-and-aft superstructure natural frequencies on approximately thirty ships. In this method the fore-and-aft "fixed base" natural frequency of the superstructure, i.e., superstructure cantilevered from the main deck, is determined according to deckhouse type and height. The fixed base natural frequency is then reduced by a correction factor to account for the rotational flexibility of the underdeck supporting structure. Specifically, the procedure of Hirowatari is as follows:

(*a*) Select superstructure type from Fig. 20.
(*b*) Determine superstructure height, h.
(*c*) Read f_∞ (fixed-base natural frequency) as a function of h from Figure 21.
(*d*) Read f_e/f_∞ (the correction factor) from Table 2.

Table 2—Flexible Base Correction Factors

Type	f_e/f_∞
A, C	.625
B	.602
D	.751

(*e*) Compute f_e (the expected deckhouse natural frequency in the first fore-and-aft mode) from the following formula,

$$f_e = f_\infty\,(f_e/f_\infty) \tag{104}$$

It is reported in Sandström and Smith (1979) that this procedure generally produces results that are within 15 percent of measurements from shaker tests. However, the method becomes inapplicable when the superstructure type varies significantly from those given in Fig. 20. Furthermore, there is some uncertainty regarding the use of the correction factors for superstructure support flexibility given in Table 2, since the supporting structure may vary from deep beams to column supports to structural bulkheads. Despite these difficulties, the method seems to work quite well in most cases considering the limited input that is required. This feature makes the Hirowatari Method particularly attractive in the early design stages when the design data are sparse or unknown.

In either design studies or in post design corrective investigations the best approach is often to develop, or to calibrate, a mathematical model from which to evaluate the effects of design changes. Proceeding with the idea of approximations using simple analysis, the two basic effects influencing the fundamental fore-and-aft superstructure natural frequency are exemplified in Hirowatari's approach:

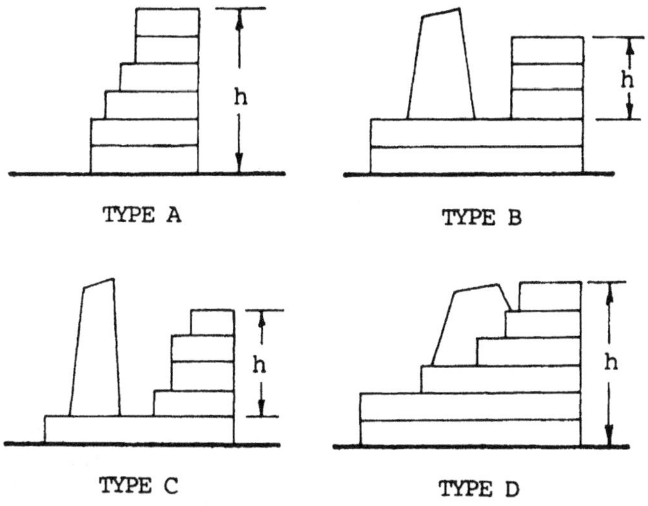

Fig. 20 Deckhouse types

(a) Cantilever (fixed base) bending and shear of the superstructure as a beam over its height *h*, Fig. 21.

(b) Rocking of the superstructure as a rigid box on the effective torsional stiffness of its supporting structure.

Ordinarily, one of the superstructure main transverse bulkheads will be a continuation of one of the two engine room transverse bulkheads. The intersection of the continuous bulkhead and the deck identified with the superstructure base, Fig. 20, can usually be taken as the axis about which the rocking of the house occurs.

The fore-and-aft natural frequency of the superstructure due to the combined effects of rocking and bending/shear can be estimated using Dunkerley's Equation (Thomson, 1973) as,

$$f_e = \sqrt{\frac{1}{1/f_\infty{}^2 + 1/f_R{}^2}} \qquad (105)$$

Here, f_∞ has been identified as the fixed base cantilever natural frequency, from Fig. 21, or by analysis. f_R in Equation (105) is the rocking natural frequency of the rigid superstructure, of height *h*, on its supporting stiffness,

$$f_R = 60/2\pi \sqrt{K_f/J} \quad \text{cpm} \qquad (106)$$

J is the mass moment of inertia of the superstructure about the rocking axis and K_f is the effective torsional stiffness of the superstructure foundation, also about the axis of rotation.

The Hirowatari procedure, in conjunction with Equations (105) and (106) has utility in design or post design corrective studies where estimates must be made as to the relative effects of structural changes. This is demonstrated by the following numerical example:

Assume that a conventional Type A superstructure, Fig. 20, has been designed, preliminarily. The house height, *h*, is 15 meters. Referring to Figs. 20 and 21 and Table 1,

$$f_\infty \simeq 800 \text{ cpm}$$

and

$$f_e/f_\infty = 0.625$$

which gives the estimated fore-and-aft house natural frequency, f_e, as

$$f_e = 0.625(800) = 500 \text{ cpm}$$

The rocking frequency is estimated from (105) as,

$$f_R = \sqrt{\frac{1}{1/f_e^2 - 1/f_\infty^2}} = 640 \text{ cpm}$$

From Equation (106) then,

$$K_f/J = (2\pi/60)^2 f_R{}^2 = 4490 \text{ rad}^2/\text{sec}^2 \qquad (107)$$

Now assume that the mass of the house, *m*, has been estimated as 300 tons. Also assume that the house front is a continuation of the engine room forward transverse bulkhead, so that the house effectively rotates about its front lower edge, Fig. 22. Assume a radius of gyration, $\bar{r}$, of the house about this axis of 10 meters. The house mass moment of inertia, *J*, is then,

$$J = m\bar{r}^2 = 3 \times 10^7 \text{ kg-m}^2$$

The effective rotational stiffness of the foundation is then estimated from Equation (107), as,

$$K_f = 1.35 \times 10^{11} \text{ N-m/rad} \qquad (108)$$

Proceeding with the senario, assume that stiffening is proposed in the form of two parallel pillars made-up

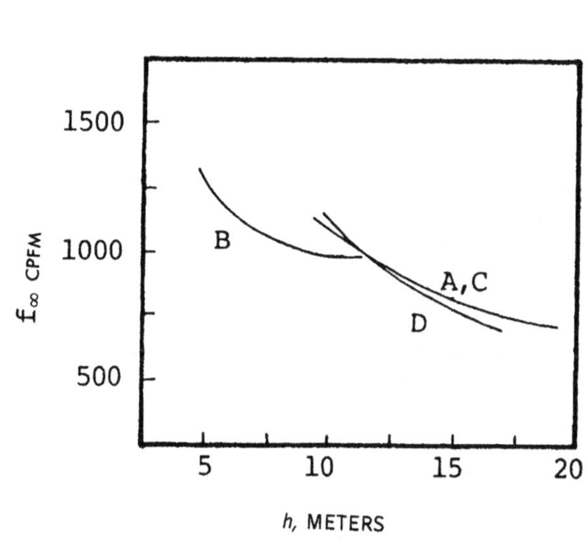

Fig. 21 Fixed base superstructure natural frequencies

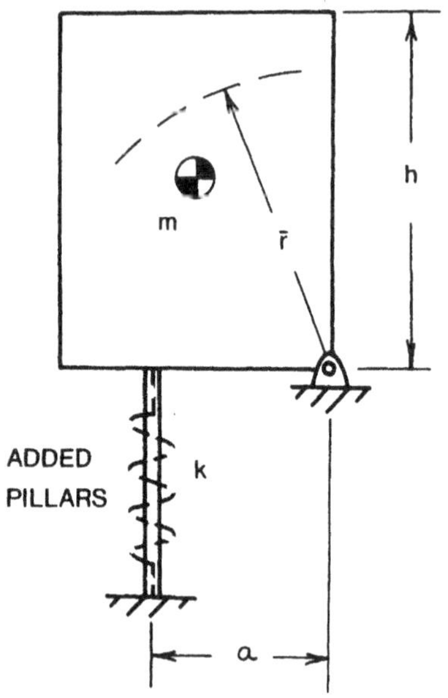

Fig. 22 Deckhouse stiffening

of 20 cm extra-heavy steel pipe, each 6 meters long, and located under the house side bulkheads as indicated on Fig. 22.

The effective axial stiffness of the parallel pillars is readily calculated to be,

$$k = 5 \times 10^8 \text{ N/m}$$

The pillars are located at a distance $a = 5$m aft of the forward bulkhead, so that the incremental rotational stiffness added by the pillars is,

$$\delta K_f = ka^2 = 1.25 \times 10^{10} \text{ N-m/rad}$$

The stiffness of the stiffened foundation is, in view of Equation (108),

$$K_f' = K_f + \delta K_f = 1.475 \times 10^{11} \text{ N-m/rad}$$

which represents a 9.3 percent increase. The new rocking frequency, from Equation (106), becomes,

$$f_R' = f_R \sqrt{1.093} = 669 \text{ cpm}$$

Then, from Equation (105), the house fore-and-aft natural frequency is raised to,

$$f_e' = \sqrt{\frac{1}{1/f_\infty^2 + 1/f_R'^2}} = 513 \text{ cpm}$$

which represents a 2.6 percent increase over the value of 500 cpm without the pillars.

In conclusion, the simple analysis in this example should have been at least useful for judging that the proposed pillars would not be very effective in raising the superstructure natural frequency.

3.5 Main Thrust Bearing Foundation Stiffness. In addition to stern superstructures, experience has shown the criticality of avoiding longitudinal resonance of the main propulsion machinery with propeller blade rate excitation. Not only may such a resonance result in excessive vibration of the machinery components, but transmission of the resonant alternating thrust into the hull structure through the machinery foundation can also excite hull structural elements and subsystems to excessive levels.

Just as with stern superstructures, significant dynamic coupling exists in general, between the vibratory responses of the main propulsion machinery, the hull girder and, in turn, between the hull girder and its sprung substructures and subsystems. But for basically the same reason as in the case of superstructures, the system natural frequencies identified with main propulsion machinery vibration can be estimated with the dynamics of the hull girder ignored, provided proper allowance is made for the stiffness of the local connecting hull structure, which in this case is the engine room bottom structure.

The conventional model for main propulsion longitudinal vibration, which is based on the above approximation, is clearly formulated in the companion volume, *Marine Engineering* (Harrington, 1971). The most difficult element of the model to estimate is the equivalent spring of the engine foundation and participating bottom structure. Furthermore, while the machinery elements are customarily the responsibility of the marine engineer, the naval architect is responsible for the design of the vessel bottom structure. Estimation of the stiffness of the structure constraining longitudinal vibration of the main propulsion machinery is therefore an appropriate subject for inclusion in this chapter.

A structural model of varying degree of refinement can be used for the calculation of the subject stiffness; the accuracy of the calculation should be expected to vary in some direct manner with the degree of refinement used. A relatively simple model, which is reported to produce results of satisfactory reliability, is that proposed by SNAME T&R Report R-15 (1972). This report models the main machinery foundation and an attached strip of engine room bottom as a non-uniform beam pinned at the engine room bulkheads. Fig. 23, extracted from the above reference, shows a typical foundation and bottom structure for a cargo ship with the thrust bearing mounted aft of the reduction gear. The dashed lines in the overhead view indicate the width of bottom assumed as effective in flexing with the foundation proper.

A fore-and-aft force applied statically along the shaft centerline at the thrust bearing produces, in addition to deflection of the thrust bearing elements, bending and shear of the members of the machinery foundation and bottom. The fore-and-aft deflection contributed by the foundation proper is primarily a shear deflection of the vertical plates aligned in the fore-and-aft direction. The deflection contributed by the bottom

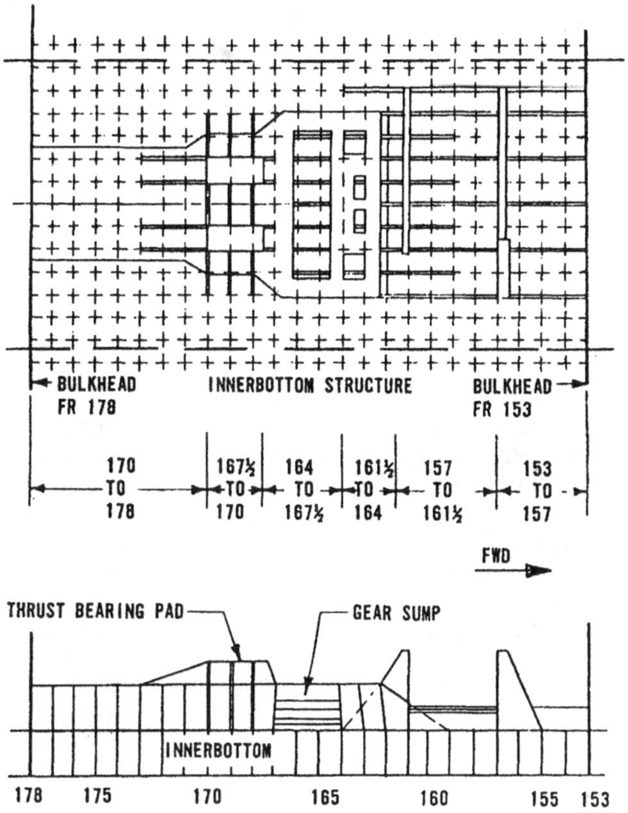

Fig. 23 Typical foundation for a cargo ship with an aft mounted thrust bearing

is primarily due to bending rotation; the deflected slope of the bottom projected through the vertical distance between the shaft centerline and the bottom neutral axis produces a fore-and-aft deflection at the bearing. This is depicted on Fig. 24, along with the beam moment of inertia distribution corresponding to the Fig. 23 equivalent beam cross-sections.

The calculation results for the Figure 23 foundation and bottom, from SNAME T&R Report R-15, are summarized in Table 3.

Stiffness is, in general, defined as force per unit deflection. Therefore, the axial force applied at the thrust bearing divided by the deflection that it produces at the bearing is the stiffness at the bearing. In the above summary, a unit fore-and-aft force was applied, so that the stiffness is the reciprocal of the calculated fore-and-aft deflection. The stiffness due to bottom rotation therefore adds serially to the stiffness of the fixed base foundation to produce the combined stiffness shown in Table 3. This combined stiffness would, in turn, be added serially to the stiffness of the thrust bearing elements to obtain the stiffness of the equivalent spring to be used in the mass-elastic model for estimating longitudinal natural frequencies of the main machinery system (Harrington, 1971).

Note from Table 3 that the deflection of the bottom is of equal importance to the deflection of the foundation proper in the combined stiffness.

3.6 Diesel Engine Excitation. Diesel engine vibratory excitation can be generally considered to be composed of three periodic force and three periodic moment components acting at the engine foundation. Actually, the periodic force component along the axis of the engine is inherently zero, and some other components usually balance to zero depending on particular engine characteristics.

Two distinctly different types of forces can be associated with the internal combustion reciprocating engine. These are: *(a)* gas pressure forces due to the combustion processes, and *(b)* inertia forces produced by the accelerations of the reciprocating and rotating engine parts. As shown in Den Hartog (1956), the gas

Table 3—Summary of Stiffness Calculations for the Foundation Shown in Fig. 23

Bottom Rotation, rad.	2.13×10^{-10}
Deflection at Thrust Bearing Due to Bottom Rotation, m/N	1.61×10^{-10}
Bottom Stiffness, N/m	6.21×10^{9}
Shear Deflection of Foundation, m/N	1.27×10^{-10}
Bending Deflection of Foundation, m/N	2.74×10^{-11}
Total Foundation Deflection (Bending + Shear), m/N	1.54×10^{-10}
Foundation Stiffness, N/m	6.48×10^{9}
Combined Stiffness, N/m	3.17×10^{9}

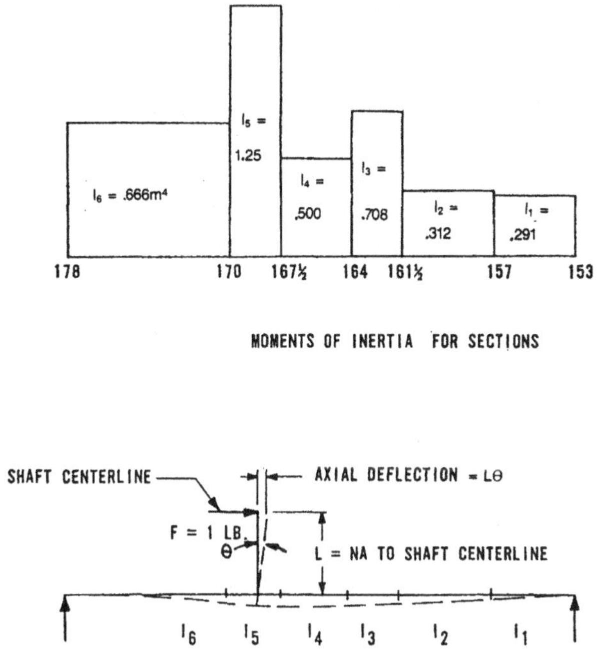

Fig. 24 Beam model for calculating axial deflection due to bottom rotation

pressures can produce only torsional moments about the engine fore-and-aft axis; the vertical and transverse gas pressure forces balance within the engine and, assuming engine rigidity, do not appear at the foundation. The vertical force and moment, which are of primary concern with regard to hull vibratory excitation, and the transverse force and moment as well, are due entirely to unbalanced inertial effects. But following the Den Hartog analysis it is readily seen that for engines of more than two cylinders, which is the case of interest, the vertical and transverse inertia force components also balance identically to zero at the engine foundation. This leaves only the vertical and transverse moments about which to be concerned. These moments can be written as:

$$m_y(t) = \text{Re}\, M_{y1}\, e^{i\Omega t} + \text{Re}\, M_{y2}\, e^{2i\Omega t}$$

and (109)

$$m_z(\text{t}) = \text{Im}\, M_{z1}\, e^{i\Omega t}$$

Here m_y is the vertical moment, about the transverse y axis, Figure 25, and m_z is the transverse moment, about the vertical z axis. Ω in (109) is the engine angular velocity, in rad per sec.

The complex notation in Equation (109) is for convenience in defining the moment amplitudes. By definition,

$$e^{ix} \equiv \cos x + i \sin x$$

where

$$i = \sqrt{-1}$$

The Re and Im in Equation (109) imply the use of only the real or imaginary part, respectively, of the complex numbers formed from the products of the complex moment amplitudes and the complex exponentials.

Equations (109) show that the vertical moment, m_y, has both once per revolution and twice per revolution components; the transverse moment occurs exclusively at the once per revolution engine RPM frequency.

The complex moment amplitudes in Equation (109) are given by the following formulas:

$$M_{y1} = (M_{\text{rec}} + M_{\text{rot}})\, r\Omega^2\, \ell_c \sum_{m=1}^{M} m e^{\frac{2\pi i(k_m - 1)}{M}}$$

$$M_{y2} = M_{\text{rec}} \frac{r^2}{\ell}\, \Omega^2\, \ell_c \sum_{m=1}^{M} m e^{\frac{4\pi i(k_m - 1)}{M}} \qquad (110)$$

$$M_{z1} = M_{\text{rot}}\, r\, \Omega^2\, \ell_c \sum_{m=1}^{M} m e^{\frac{2\pi i(k_m - 1)}{M}}$$

The variables in Equation (110) are, with the aid of Fig. 25:

M_{rec} and M_{rot} represent the equivalent masses experiencing the accelerations of the piston and crank pin, respectively, of one cylinder, due to the constant crank shaft angular velocity Ω. The mass M_{rec} is composed of the mass of the piston assembly and piston rod, plus a fraction of the mass of the connecting rod. The mass M_{rot} is composed of the balance of the connecting rod mass, plus an equivalent mass at the crank pin representing the weight eccentricity of the crank throw.

r is the crank radius. ℓ is the connecting rod length. ℓ_c is the longitudinal distance between the cylinder axes. M is the number of cylinders. k_m is the firing order of the mth cylinder; for $m=1, \ldots M$, k_m has generally non-consecutive integer values $1, \ldots, M$.

The real amplitudes of the moment components, which correspond to the maximum values of interest, are just the respective modulii of Equation (110). The values of the moment amplitudes are usually tabulated in the manual for a particular engine. They can also be calculated by Equation (110). This is demonstrated by the following example:

The majority of low speed marine diesels currently in service have 6 cylinders. A typical firing order for such engines is: 1–5–3–4–2–6. With M=6 and k_m=1,5,3,4,2,6 for $m=1, \ldots, 6$, the summations in Equation (110) are:

$$\sum_{m=1}^{M} m e^{\frac{2\pi i(k_m - 1)}{M}} = 0$$

and

$$\sum_{m=1}^{M} m e^{\frac{4\pi i(k_m - 1)}{M}} = 3 - 1.732i$$

The foundation moment amplitudes are therefore:

$$M_{y1} = 0$$

$$M_{y2} = -M_{\text{rec}}\,(r^2/\ell)\,\Omega^2 \ell_c (3 + 1.732\text{i})$$

$$M_{z1} = 0$$

This shows that in the case of the 6-cylinder engine of the above firing order, only the 2nd order vertical moment, aside from a torsion, exist to excite the hull. The magnitude of this moment, usually denoted as M_{2v}, is

$$M_{2v} = |M_{y2}| = 3.464\, M_{\text{rec}}\,(r^2/\ell)\,\Omega^2\,\ell_c \qquad (111)$$

Data for the Sulzer RND 76M engine, from Sulzer Bros. (1977) are,

Stroke: $2r$, m	1.55
Connecting rod length, l, m	3.775
and mass, t	3.22
Mass of piston, piston rod and cross-head, t	4.76
Crank-shaft length, m	9.6
RPM	122

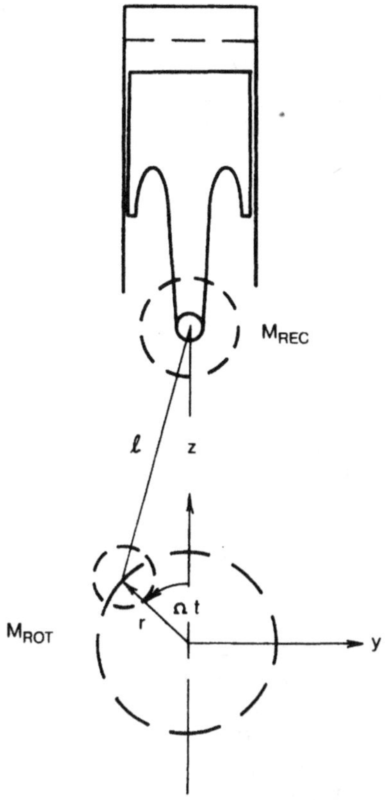

Fig. 25 Diesel engine excitation

Taking one-half of the connecting rod mass as effective in reciprocating with the piston, the data for use Equation (110) are:

$$M_{rec} = 6.37 \text{ t}$$
$$r = 0.775 \text{ m}$$
$$\ell = 3.775 \text{ m}$$
$$\Omega = 122\ (2\pi / 60) = 12.78 \text{ rad per sec}$$
$$\ell_c = 9.6 / 5 = 1.92 \text{ m}$$

Substitution gives,

$$M_{2v} = 112.3 \text{ ton-m}$$

The value from the data book for this engine is 119.4 t-m.

The 2nd order vertical moment of this example is the diesel engine excitation of most concern; it is larger than the first order moments, in general. One guideline (Johannessen and Skaar, 1980) recommends attention in cases where M_{2v} exceeds 50 t-m. The potential danger is in resonating one of the lower hull girder vertical modes with a large second order vertical moment. This possibility is demonstrated by the following example:

Table 1 of Sub-section 3.2 gives estimated values for the fundamental 2-noded vertical hull critical for a number of different ships. Sub-section 3.2 also suggests a simple extrapolation Equation (98) for estimating the first few higher-mode natural frequencies, given the fundamental value:

$$N_{nv} \approx N_{2v}(n - 1)^{\alpha} \ (n \text{ not to exceed} \sim 5)$$

Assuming the "reefer," ship number 1 of Table 1, for example, $N_{2v} = 92.4$ cpm, and $\alpha = 0.845$. The first four vertical hull-girder modes of this vessel would then be predicted to have corresponding natural frequencies of:

n	N (cpm)
2	92.4
3	166
4	234
5	298

The frequency of the 2nd order engine excitation from the previous example is $2 \times \text{RPM} = 244$ cpm. It is relatively close to the predicted natural frequency of the 4-noded third hull girder mode.

While the natural frequency estimate of 234 cpm is indeed rough, it should at least have been reliable enough to dictate further analysis to refine the hull girder natural frequency estimates in this particular example.

In the case of projected high excitability in resonant vibration with the diesel engine moments, which does develop in the course of design on occasion, the excitation moment components can usually be reduced effectively by the incorporation of compensators (Sulzer Bros., 1977). These devices consist of rotating counterweights usually geared directly to the engine. They are rotated at the proper rate and with the proper phase to produce cancellation with the undesirable 1st or 2nd order engine generated moment.

3.7 Propeller Excitation. Propeller excitation is far more difficult to quantify than the excitation from internal machinery sources. This is because of the complexity of the unsteady hydrodynamics of the propeller operating in the non-uniform hull wake. In fact, the non-uniform hull wake is the most complicated part; it is unfortunate that it is also the most important part. Propeller-induced vibration would never be a consideration in ship design if the propeller disk inflow was circumferentially uniform. Any treatment of propeller excitation must begin with a consideration of the hull wake.

(a) Hull Wake Characteristics. In reality, hull wakes are both time varying and circumferentially

nonuniform. Under steady ahead operation, which is the condition of primary interest when ship vibration is the concern, time variations of the wake are entirely random variations associated with the random character of boundary layer turbulence and the seaway in which the ship operates. Random vibration analyses of ships is therefore implied. However, the present state of the art allows for the circumferential nonuniformity of hull wakes, but assumes, for steady operation, that wake is time invariant in a ship fixed coördinate system.

Although several empirical formulations have been proposed for estimating time average ship wake distributions (Holden, et al, 1980), it is generally accepted, at this time, that model tests are required. The model nominal wake field, which is conventionally measured in the propeller disk by pitot tube survey with the propeller removed (Pien, 1958) is, however, not a completely accurate representation of the time average inflow. Both distortions of the rotational flow by the propeller and the diffraction flow component from the hull boundary are absent, and scaling is a problem. Efforts are usually made to apply corrections to model nominal wakes for these effects (Sasajima and Tanaka, 1966), (Dyne, 1974), (Huang and Groves, 1981). However, the uncorrected model nominal wakes are probably adequate for the relative types of evaluation often of primary interest in design studies.

Model nominal wake data are presented either as a contour plot or as curves of velocity versus angular position at different radii in the propeller disc. The latter representation for the axial and tangential velocity components for a conventional stern single screw merchant ship (Hadler and Cheng, 1965) is shown in Fig. 26; a radial component also exist, of course, but it has little influence on propeller vibratory forces.

The position angle, θ, in Fig. 26 is taken as positive counterclockwise, looking forward, and x is positive aft. The axial and tangential wake velocities in Fig. 26 are nondimensional on the ship forward speed, U.

Note from Figure 26 that the axial velocity is symmetric in θ about top-dead-center (even function) and the tangential velocity is asymmetric (odd function). This is a character of single screw ships due to the transverse symmetry of the hull relative to the propeller disk; such symmetries in the wake do not, of course, exist with twin-screw ships. Usually, velocities are measured only over the half-circles of the propeller disk in single screw cases, with opposite-hand projections on the basis of the required symmetry. Occasionally, the full circles are measured; the measured deviations from symmetry reflect the inaccuracies of model construction and instrumentation, as well as possible inconsistencies in obtaining true time averages.

The wake illustrated in Fig. 26 represents one of the two characteristically different types of single-screw ship wakes. The flow character of the conventional *cruiser* or *clearwater* stern in Fig. 26 is basically waterline flow; the streamlines are more-or-less horizontal along the skeg and into the propeller disk. The flow components along the steep buttock lines forward of the propeller disk are small. The dominant axial velocity field of the resultant wake has a substantial defect running vertically through the disk along its vertical centerline, at all radii. This defect is the shadow of the skeg immediately forward. The tangential flow in the propeller disk, being the component of the upward flow toward the free surface, is relatively small. The idealization of this wake is the two dimensional flow behind an infinitely long vertical strut placed ahead of the propeller. In this idealization, the axial velocity distribution is invariant vertically, and the tangential (and radial) velocities are symmetric about the transverse disk axis. The basic character can be detected in the Fig. 26 data.

A characteristically different wake flow is that associated with the *strut* or *barge-type* stern, which has a broad counter above the propeller disk and little irregularity forward. The flow character over this type of stern is basically along the buttock lines, versus the waterlines. Some wake non-uniformity may be produced by appendages forward, such as struts and bearings or by shaft inclination, but the main wake defect, depending on the relative disk position, will be that of the counter boundary layer overhead. In this case a substantial axial wake again exist, but only in the top of the disk. As generally only the blade tips penetrate the overhead boundary layer, the axial wake defect occurs only at the extreme radii near top-dead-center, rather than at all radii along the vertical centerline as in the case of the conventional single screw stern. Just as in the case of the conventional stern, the tangential disk velocity with the strut stern will be generally small; the vertically upward velocity ratio through the propeller disk will have average values on the order of the tangent of the sum of the buttock and shaft inclination angles. The idealization in the case of the barge stern, as a sequel to the vertical strut idealization of the wake of the conventional stern, is an infinite horizontal flat plate above the propeller. Here, the degree of axial wake non-uniformity depends on the overlap between the propeller disk and the plate boundary layer. The tangential and radial wake components are due entirely to the shaft inclination angle in this idealization, as the flat plate boundary layer produces only an axial defect.

1. *Hull-Propeller Clearance.* The distinction between the two different wake types is very important to understanding the importance of clearances between the propeller blades and local hull surfaces. First of all, it is helpful to consider the hull surface excitation as composed of two effects:

Wake effect—The effect of changing the wake inflow to the propeller according to a specified propeller

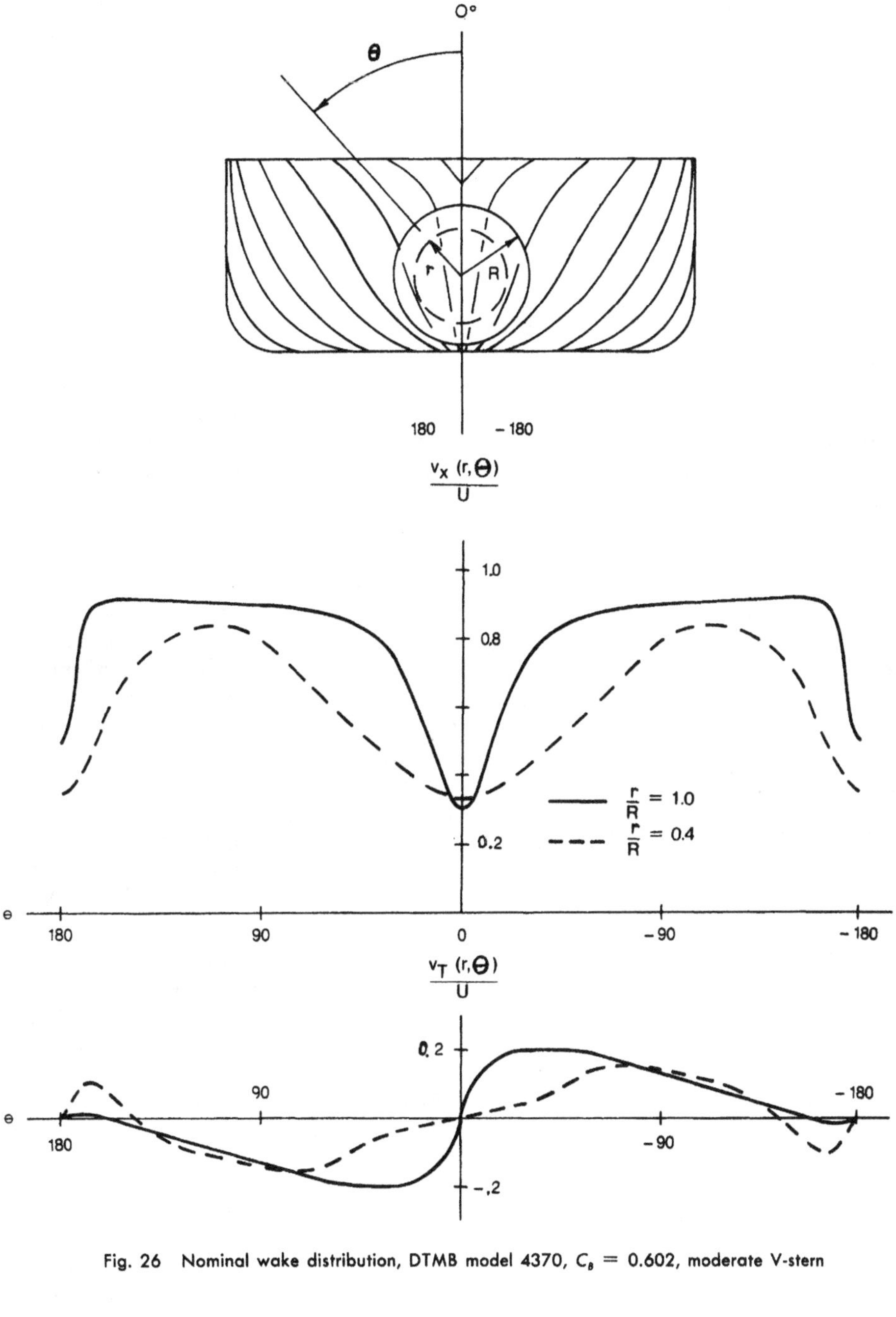

Fig. 26 Nominal wake distribution, DTMB model 4370, C_B = 0.602, moderate V-stern

relocation, but with the propeller actually fixed in position relative to the hull.

Diffraction effect—The effect of changing the propeller location relative to the hull but with the wake inflow to the propeller held fixed.

It is a common misconception that the cruciality of propeller-hull clearances has to do primarily with the diffraction effect. To the contrary, studies show that for wake inflow held invariant, propeller-induced excitation level is relatively insensitive to near field variations in propeller location. It is the high sensitivity of propeller blade pressures and cavitation inception to the variations in wake non-uniformity accompanying clearance changes that dictates the need for clearance minima. In general, the wake gradients become more extreme as propeller-hull clearances are decreased.

Reconsider the conventional and strut-stern wake types in light of the above fact. It is probable that too much emphasis is often placed on aperture clearances in conventional stern single-screw ships. From the point of view of vertical clearance, there is no significant boundary layer on the narrow counter above the propeller with this stern type. Furthermore, from the point of view of the vertical strut idealization of the skeg, the axial velocity distribution would be invariant with vertical disk position. The critical item with vertical tip clearance in the conventional stern case seems to be the waterline slope in the upper skeg region. Blunt waterline endings can result in local separation and substantially more severe wake gradients in the upper disk than suggested by the strut idealization. Fore-and-aft clearances in the conventional stern case should be even less ciritcal than the vertical clearances. Wakes attenuate very slowly with distance downstream. While increasing the fore-and-aft clearances between the blade tips and the skeg edge forward certainly acts to reduce the wake severity, the reduction will be marginally detectable within the usual limits of such clearance variation. An exception may exist in the case of separation in the upper disk due to local waterline bluntness. The closure region of the separation bubble exhibits relatively large gradients in axial velocity.

For the broad flat countered strut stern vessel the vertical tip clearance is a much more critical consideration. A relatively uniform wake will result if the propeller disk does not overlap the overhead boundary layer. This is, in general, the condition achieved on naval combatant vessels; the rule of thumb in U.S. Naval design practice is a minimum vertical tip clearance of one-quarter propeller diameter. Vibration problems are almost unheard of on naval combatants.

Some wake non-uniformity on strut stern ships does result from shaft struts and from the relatively high shaft inclinations often required to maintain the 25 percent overhead tip clearances. With proper alignment to the flow, shaft struts produce highly localized irregularities in the wake which are generally not effective in the production of vibratory excitation. The main effect of shaft inclination is a relative up-flow through the propeller disk. The cavitation that can result at the 3 and 9 o'clock blade positions has proved to be of concern with regard to noise and minor blade erosion, but the hull vibratory excitation produced has never been considered to be of significance.

The clearance minimum of 25 percent of a propeller diameter is more or less the standard in commercial practice as well as naval, although some classification rules show less. In conventional stern merchant ships the vertical clearances tend to average slightly less (~0.15D) and forward clearances from blade tips to skeg average slightly more (~0.3D). Some conservatism no doubt exists in these rules of thumb, otherwise they would not have withstood the test of time. However, the studies needed to provide a rationale for clearance selection on a case by case basis have never been conducted; guidance for deviating from the accepted standards is completely lacking at this time.

2. *Skew Considerations.* It was pointed out in the case of the conventional single screw ship wake, Fig. 26, that the shadow of the vessel skeg produces a heavy axial wake defect concentrated along the disk vertical centerline. The blades of conventional propellers ray-out from the hub, i.e., the blade midchord lines are more-or-less straight rays eminating from the hub centerline. Such "unskewed" blades abruptly encounter the axial velocity defect of the conventional stern wake at the top and bottom-dead-center blade positions. The radially *in-phase* character of the abrupt encounter results in high net blade loads and radiated pressures.

A more gradual progression of the blades through the vertical wake defect is accomplished by curving the blade midchord lines. Different radii enter and leave the wake spike at different times; cancellation results in the radial integrations to blade loads and radiated pressures, with the result of sometimes drastically reduced vibratory excitation. Two such successfully used *skewed* propellers are shown on Fig. 27, from Hammer and McGinn (1978).

Skew will obviously work less effectively with strut-stern wakes since the axial velocity defect tends to be concentrated more in the outer extreme radii. The more radially uniform distributions of the conventional stern case are not available with which to achieve as high a degree of *dephasing* and radial cancellation. Of course, the strut stern vessel is in less need of propeller design extremes, as vibration problems are already essentially eliminated by the stern form selection, provided proper clearances are incorporated.

Care must be taken in incorporating skew, particularly in replacement propellers, that adequate clearances between the blade tips and the rudder be maintained. As the blades are skewed in the pitch helix the tips move downstream, closing-up blade tip-to-rudder clearances. The consequences can be increased hull vibratory forces transmitted through the rudder, as well as rudder erosion caused by the collapsing sheet cavitation shed downstream off the blade tips as they sweep through the top of the propeller disk. The recourse is to incorporate *warp* into the blades along with the skew. Warping is a forward raking of the skewed blades back to (and sometimes beyond) the propeller disk. It is equivalent to skewing the blades in the plane of the disk rather than in the pitch helix.

It should be noted for the sake of completeness that skew has a beneficial effect in reducing the effects of vibration producing fluctuating sheet cavitation, even when such cavitation may be concentrated at the blade tips. The blade curvature is thought to result in a radially outward flow component in the vicinity of the blade tips which tends to sweep the cavity sheets into the tip vortex, where the critical collapse phase tends to occur more gradually.

3. *Harmonic Analysis.* It is necessary to progress beyond mere descriptive considerations of hull wake for assuring achievement of acceptably small propeller excitation. Harmonic, or Fourier, analysis of

(a) *Sea Bridge* Class (b) *San Clemente* Class

Fig. 27 Highly skewed propeller installations

the predicted wake is required in almost all procedures for assessing propeller excitation severity.

The axial and tangential wake velocity components, Fig. 26, can be written as the following Fourier series in position angle θ, for selected radii:

$$\frac{v_x(r,\theta)}{U} \approx \frac{V_a(\mathrm{r})}{U} + \mathrm{Re} \sum_{q=1}^{Q} C_{xq}(r)\, e^{iq\theta}$$

$$\frac{v_T(r,\theta)}{U} \approx \mathrm{Re} \sum_{q=1}^{Q} C_{T_q}(r)\, e^{iq\theta} \qquad (112)$$

In (112), $V_a(\mathrm{r})/\mathrm{U}$ is the steady circumferentially average axial velocity, which is just the radially varying speed of advance through the propeller disc; the steady average tangential velocity is taken as zero, since ship wakes have negligible steady *swirl.* The prefix Re in (112) denotes *real part* of the complex series, as previously noted, and $e^{iq\theta}$ is, by identity,

$$e^{iq\theta} \equiv \cos q\theta + i \sin q\theta$$

The complex coefficients $C_{xq}(r)$ and $C_{T_q}(r)$ are determined from the numerical wake data using the formula,

$$C_q(r) = \frac{1}{\pi} \int_{\theta=-\pi}^{\pi} \frac{v(r,\theta)}{U}\, e^{-iq\theta}\, d\theta \qquad (113)$$

The series in (112) are truncated at Q terms, and are therefore denoted as approximations. It has been found that a value of Q on the order of 10 reproduces wake contours, typically Fig. 26, with reasonable accuracy. For Q no higher than approximately 10 it has also been found that the coefficients in (112) can be calculated with acceptable accuracy by employing Simpson's Rule at Equation (113), with points spaced at even 5 degree increments. Using this procedure, the coefficients are calculated as,

$$C_q(r) = \frac{1}{108} \sum_{j=1}^{73} W_j \frac{v(r,\theta_j)}{U}\, e^{-iq\theta_j} \qquad (114)$$

where

$\theta_j = -\pi + (j - 1)\pi/36$ in radians, and the W_j are the Simpson's multipliers,

$W_j = 1,4,2,4, \ldots, 4,1.$

Given the wake data in the form of Fig. 26, Equation (114) is executed for $q = 1$ to 10 for typically 8 different radii in the propeller disc; the 8 radial stations are usually r/R = 0.25, 0.35, . . . , 0.95.

One executing Equation (113), or (114), it will be found that for single screw ships the $C_{xq}(r)$ are pure real, and the $C_{T_q}(r)$ are pure imaginary. This is due to the symmetry of single screw ship wakes. For multi-screw ship wakes both the real and imaginary parts of both sets of coefficients will be nonzero, in general.

(b) Approximate Selection of Propeller Blade Skew. As discussed in the last sub-section, if one is forced to accept the highly irregular wake imposed by the conventional cruiser type single screw stern, then propeller blade skew may be an alternative means for limiting the propeller induced vibratory excitation to acceptable levels.

Skew is normally specified as an angle in the projected plane of the blade. Skew angle is defined on Fig. 28 and denoted $\alpha_s(r)$; it is the angle at some radius r in the projected view between the ray bisecting the blade section at the hub, and the ray bisecting the blade section at r. Percentage skew is given by the formula,

$$\% \text{ skew} = \alpha_s(R)\, N \times 100/360 \qquad (115)$$

where $\alpha_s(R)$ is the skew angle at the blade tip and N is the number of blades. By Equation (115), 100 percent skew, for example, corresponds to the tip of one blade lying on the generator line through the root of the adjacent blade; both of the propellers shown on Fig. 27 have 100 percent skew. High skew is considered to be 50 percent or more. While skew is always made positive, or zero, at the blade tip, it is often negative at intermediate radii; the skew angle is conventionally positive counter-clockwise, looking forward.

The idea of skew, as previously explained, is to synchronize the lift fluctuations over the radii of the blades in such a way that cancellations occur in the radial integrations. By shifting the blade sections unequally along the helices, the sections can be made to enter the regions of wake concentration at different instants, with the result of reduced net forces.

Actually, rough judgments as to effective amounts of blade skew for a given application can be made from the wake data alone, without any explicit calculations of forces at all. This procedure was proposed by Cumming, Morgan, and Boswell (1972), and is as follows:

Referring to Fig. 28, the relative velocity normal to the blade section at (r,θ) is,

$$v_n(r,\theta) = -v_x \cos \beta_G + (\Omega r + v_t) \sin \beta_G \qquad (116)$$

where $v_x(r,\theta)$ and $v_t(r,\theta)$ are the axial and tangential nominal wake velocities and $\beta_G(r)$ is the blade geometric pitch angle. Normal velocity is important to the magnitude of the propeller blade pressure distribution; by linear theory the pressure distribution is, in fact, proportional to the normal velocity distribution, as is shown further on.

It is convenient to replace β_G by β, the hydrodynamic advance angle, in Equation (116) so that blade particulars are eliminated. From Fig. 28,

$$\tan \beta = \frac{V_a(r)}{\Omega r} = \frac{V_a(r)/U}{(\pi/J)(r/R)} \qquad (117)$$

where J is the advance ratio, U/nD. The difference in β and β_G is not within the accuracy of this exercise.

Substituting the wake Fourier series, Equation (112), into (116):

$$\frac{v_n(r,\theta)}{U} = -\frac{V_a(r)}{U} \cos \beta + \Omega r \sin \beta + \text{Re} \sum_{q=1}^{Q} V_{nq}(r)\, e^{iq\theta} \qquad (118)$$

The qth harmonic normal velocity complex amplitude in Equation (118) is,

$$V_{nq} = -C_{xq} \cos \beta + C_{T_q} \sin \beta \qquad (119)$$

with $C_{xq}(r)$ and $C_{T_q}(r)$ being the qth harmonic axial and tangential wake Fourier coefficients from Equation (114). Equation (118) can be written alternatively as,

$$\frac{v_n(r,\theta)}{U} \simeq \sum_{q=0}^{Q} \frac{v_{nq}(r,\theta)}{U}$$

where for harmonic q,

Table 4—Locus of Normal Velocity Maxima for 4th Harmonic of Series 60, $C_B = 0.6$, Wake $J = 0.834$

r/R	0.25	0.35	0.45	0.55	0.65	0.75	0.85	0.95
Va/U	0.522	0.640	0.736	0.791	0.813	0.818	0.828	0.845
β, deg, (117)	29.0	26.2	23.5	20.9	18.4	16.1	14.5	13.3
C_{x4} (114)	0.	−.0475	−.0747	−.0802	−.0705	−.0567	−.0460	−.0370
C_{T4}	0.0201i	0.0184i	0.0173i	0.0137i	0.0089i	0.0075i	0.0056i	0.0062i
$V_{n4} = -C_{x4} \cos \beta$	0.	0.0426	0.0685	0.0749	0.0669	0.0545	0.0445	0.0360
$+ C_{T4} \sin \beta$ (119)	0.0097i	0.0081i	0.0069i	0.0049i	0.0028i	0.0021i	0.0056i	0.0014i
γ_4, deg, (121)	−90.	−10.8	−5.8	−3.7	−2.4	−2.2	−1.8	−2.2
θ, deg, (122)								
$n = -1$	−112.5	−92.7	−91.5	−90.9	−90.6	−90.6	−90.5	−90.6
$n = 0$	−22.5	−2.7	−1.5	−.93	−.60	−.55	−.45	−.55
$n = 1$	67.5	87.3	88.5	89.1	89.4	89.5	89.6	89.5
$n = 2$	157.5	177.3	178.5	179.1	179.4	179.5	179.6	179.5

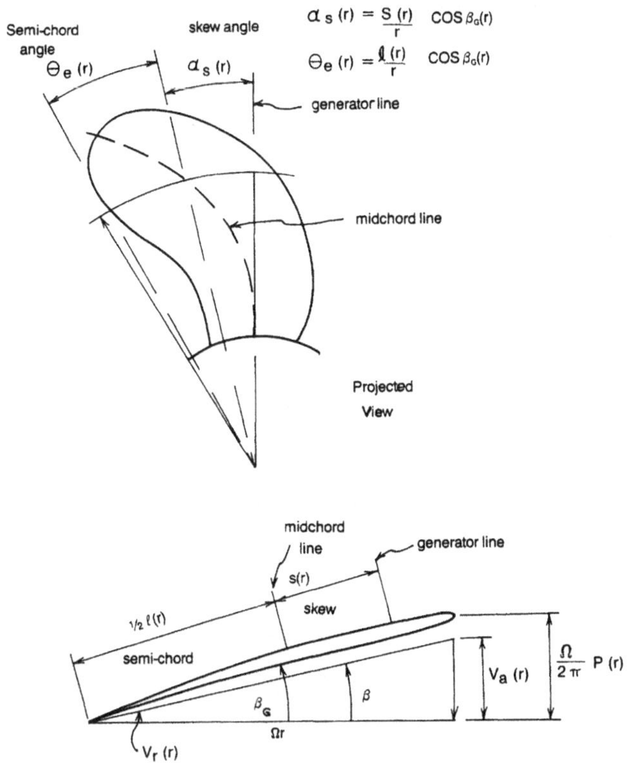

Fig. 28 Propeller geometry and nomenclature

$$\frac{v_{nq}(r)}{U} = |V_{nq}| \cos(q\theta - \gamma_q) \quad (120)$$

In Equation (120),

$$|V_{nq}| = \sqrt{(V_{nq}^R)^2 + (V_{nq}^I)^2}$$

and

$$\gamma_q = \tan^{-1}(-V_{nq}^I / V_{nq}^R) \quad (121)$$

with the superscripts R and I denoting real and imaginary parts.

Now the values of θ in Equation (120) corresponding to maximum values of the wake normal velocity occur for,

$$\cos(q\theta - \gamma_q) = 1$$

giving,

$$q\theta - \gamma_q(r) = 2n\pi;\ n = 0, 1, \ldots$$

The blade position angles for maximum normal velocity at the section r midchord line are therefore,

$$\theta(r) = (\gamma_q(r) + 2n\pi)/q;\ n = 0, 1, 2, \ldots \quad (122)$$

$\theta(r)$ can be plotted versus r to show contour lines of the normal velocity maxima in the propeller disk. A skew line of $\alpha_s(r)$ versus r can then be sketched-in that interferes appropriately with the $\theta(r)$ lines to imply the desired cancellation. This procedure is best illustrated with the following example:

Table 4 shows the computation of $\theta(r)$ from Equation (122) for the 4th harmonic of the Series 60, $C_B = 0.6$, wake (Stuntz, Pien, et al, 1960), which is essentially that depicted in Fig. 26. The computations are performed at 8 radii, as indicated.

A value of J of 0.834 was used. The same calculation for $q = 3$ was also performed. The $\theta(r)$ versus r curves by Equation (122) are plotted for $q = 3$ on Fig. 29 and for $q = 4$ on Figure 30. The solid lines on the figures are the normal velocity maxima corresponding to $n = -1, 0, 1$ for $q = 3$ and to $n = -1, 0, 1, 2$ for $q = 4$. The dashed lines are the normal velocity minima midway between the maxima.

Assume for purposes of example that alternating propeller thrust is of primary concern, and should therefore dictate the skew selection. Also assume that a 4-bladed propeller is to be used. As shown in Section 2, for N propeller blades, blade-rate alternating thrust, as well as torque, are produced exclusively by the Nth wake harmonic, i.e., $q = 4$ in this example. Therefore, based on the $q = 4$ curves of Fig. 30, a linearly varying skew of 40 deg, or 44 percent, is selected. This is the line denoted $\alpha_s(r)$ on Fig. 30. Rotating this line in θ shows that roughly equal parts of the blade fall in regions of positive and negative normal velocity at all times; this skew distribution should therefore produce the desired radial cancellation and result in significantly reduced unsteady thrust over that which would be developed by an unskewed blade.

The linear 44 percent skew distribution selected is also drawn on the 3rd harmonic wake contour plot on Fig. 29. Here, as the skew line is shifted in θ it is seen that some degree of radial interference also occurs, but not to the same degree as with $q = 4$. The critical mid-region of the blade, between approximately the 0.4 and 0.8 radii, still encounters roughly in-phase normal velocity maxima and minima with the 40 deg skew. Now, from the bearing force formulas of Section 2, Equations (92), the 3rd and 5th wake harmonics produce the lateral bearing forces and moments for 4 blades; the lower harmonic, in this case 3, usually dominates due to the convergence of the wake Fourier series. Therefore, in this example, the propeller designer might also expect some reduction in the lateral bearing forces and moments on incorporating 44 percent skew, but not as much as in the alternating thrust.

Actually, cases occur in the exercise demonstrated by this example where with any possible consecutive sequence of wake harmonics around blade number, some normal velocity contours are skewed (out-of-phase radially) and some are unskewed. Since the general rule is that the blade should be skewed if the wake is unskewed, and vice-versa, it may then be impossible in such cases to avoid actually increasing certain bearing force components on decreasing others with skew incorporation. This is often the case with the single-screw conventional stern merchant ship wake; it can be seen to a slight degree in the 3rd and 4th harmonics

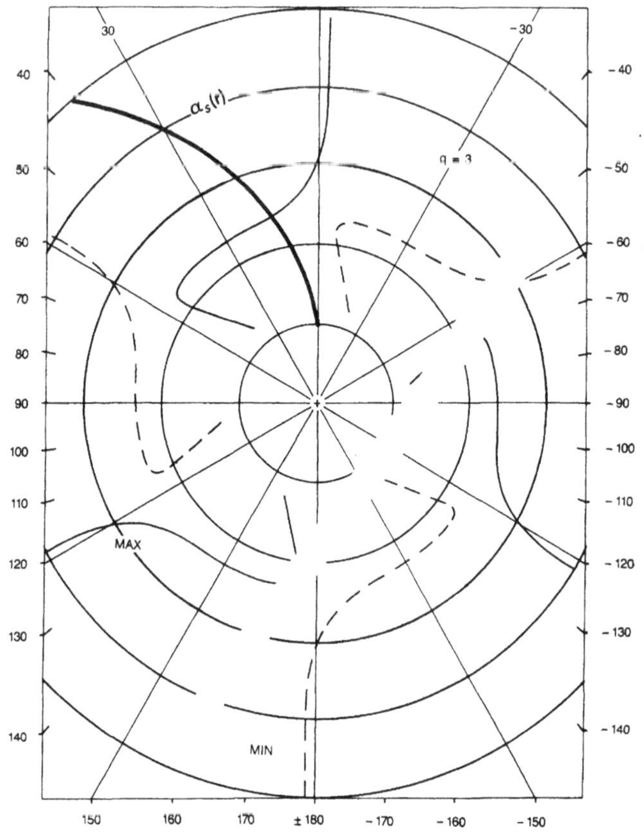

Fig. 29 Nominal wake maxima, Series 60, 3rd harmonic

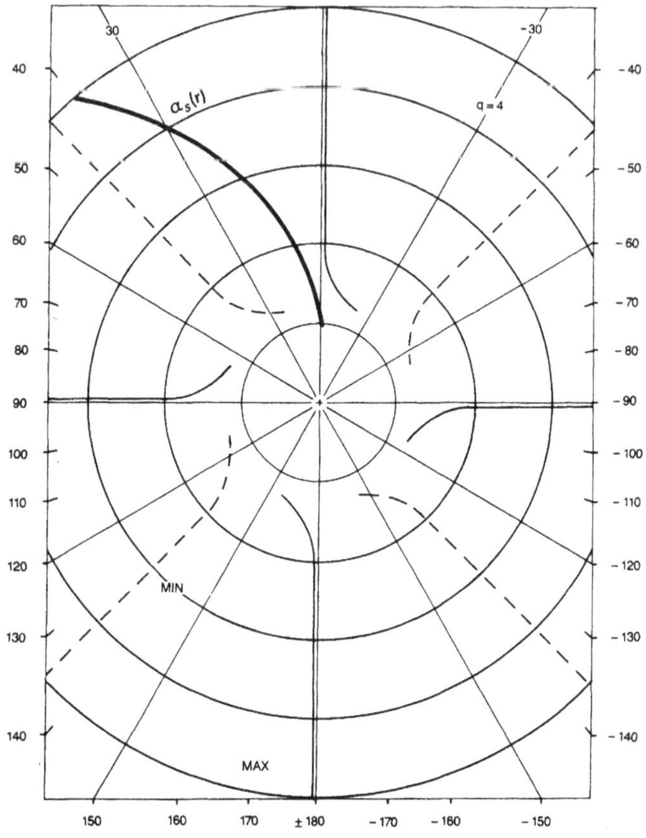

Fig. 30 Nominal wake maxima, Series 60, 4th harmonic

of the Series 60, $C_B = 0.6$ wake of Figs. 29 and 30. The even wake harmonics are characteristically unskewed and the odd harmonics are characteristically skewed in this wake type (Cumming, et al, 1972). An order of importance of the various force components must therefore be established beforehand in such cases to provide a rational basis for the skew selection.

(c) Estimation of Propeller Bearing Forces. As described to some depth in Section 2, the propeller excitation consists of a set of 3 force and 3 moment components acting in the propeller hub, plus a distribution of unsteady pressure over the after hull surfaces.

The propeller hub forces, or bearing forces, are the collective effects at the propeller hub of the unsteady blade pressures resulting from operation of the propeller in the circumferentially non-uniform wake. Formulas for calculating the bearing force components are developed in Section 2, as the Formulas (92). These formulas are written in terms of the radial distribution of unsteady blade lift, denoted *Lq(r)*. As discussed in Section 2, *Lq(r)* can be estimated by various approaches, with various levels of accuracy, and requiring various levels of effort. One procedure which is relatively simple to apply, and is at least accurate enough for meaningful relative evaluations for variations in design parameters, such as wake, and skew, is the 2-dimensional *gust theory* of von Kármán and Sears (1938), applied stripwise (Lewis, 1963). It has the following form,

$$Lq(r) = \rho U^2 R \, C_{L_q}(r)$$

with,

$$C_{L_q}(r) = \pi \frac{V_r(r)}{U} \frac{l(r)}{R} V_{nq}(r) \overline{C}_s(r,k^*) e^{-iq\alpha_s(r)} \qquad (123)$$

The variables in Equation (123) are the following:

- $Lq(r)$ — q^{th} harmonic complex lift amplitude distribution.
- ρ — fluid density.
- U — ship speed.
- R — propeller radius.
- $V_r(r)$ — relative velocity tangent to blade section pitch line. From Fig. 28, ignoring the propeller self-induced velocities,

$$V_r(r) = \sqrt{V_a^2 + \Omega^2 r^2}$$

or

$$\frac{V_r(r)}{U} = \sqrt{\left(\frac{V_a}{U}\right)^2 + \left(\frac{\pi}{J}\frac{r}{R}\right)^2} \tag{124}$$

with $J = U/nD$.

$V_{nq}(r)$ — q^{th} harmonic complex wake velocity normal to blade section pitch line at r. This is essentially Equation (119), but definition of the normal using the true geometric pitch rather than the hydrodynamic advance (Fig. 28) is recommended,

$$V_{nq}(r) = -C_{xq}\cos\beta_G + C_{T_q}\sin\beta_G \tag{125}$$

Here $C_{xq}(r)$ and $C_{T_q}(r)$ are q^{th} harmonic axial and tangential wake coefficients from Equation (114) (and Table 3), and β_G is the geometric pitch angle.
From Fig. 28,

$$\tan\beta_G(r) = \frac{P(r)}{2\pi r} = \frac{P(r)/D}{\pi(r/R)} \tag{126}$$

where $P(r)$ is the blade pitch distribution.

$l(r)$ — blade section chord length at r (in the expanded view), Fig. 28.

$\overline{C}_s(r,k^*)$ — complex conjugate of the Sears function, from Fig. 31. This is an Argand diagram which gives the real and imaginary parts of C_s as a function of the section reduced frequency, k^*.

$$C_s = C_s^R + i\,C_s^I$$

The reduced frequency is defined as,

$$k^*(r) = q\theta_e(r) \tag{127}$$

where q is the harmonic order and θ_e is the blade section projected semi-chord angle. In terms of the section chord length $l(r)$,

$$\theta_e(r) = \frac{1}{2}\left[\frac{(l/R)\cos\beta_G}{r/R}\right] \tag{128}$$

$\alpha_s(r)$ — blade section skew angle, in radians (Fig. 28). The f_{ip} of Equations (91) and (92) can be manipulated into the following form useful for computation,

$$f_{ip}(\theta) = \sum_{m=1}^{\infty} |F_{ipm}|\cos mN(\theta - \beta_{ipm}) \tag{129}$$

where the blade position angle $\theta = -\Omega t$ from Figure 8. N is blade number in Equation (129) so that the series represents the superposition of blade-rate harmonics. F_{ipm} is the mth blade rate harmonic complex amplitude of the ith force component. The real amplitude and phase angle in Equation (129) are given in terms of the real and imaginary parts of F_{ipm} as,

$$|F_{ipm}| = \sqrt{(F_{ipm}^{R})^2 + (F_{ipm}^{I})^2} \tag{130}$$

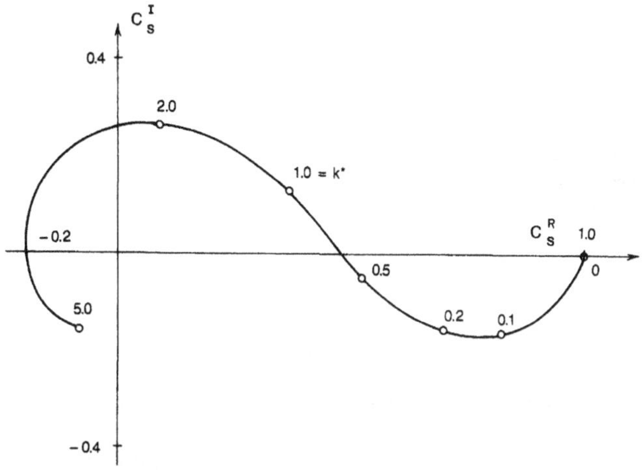

Fig. 31 Sears function $C_s(k^*)$

and,

$$\beta_{ipm} = \frac{1}{mN}\tan^{-1}\left(\frac{-F_{ipm}^{I}}{F_{ipm}^{R}}\right) \tag{131}$$

The β_{ipm} by Equation (131) corresponds to the position angle of the propeller blade nearest top-dead-center when the mth blade rate harmonic of the ith force component is positive maximum. The positive force directions are indicated on Fig. 8 and the propeller blade position angle, θ, is positive counterclockwise, looking forward.

The fundamental blade-rate harmonic of the bearing forces is usually predominant, so that attention can be restricted to $m = 1$ in Equations (129) through (131) in the majority of cases.

If one of the bearing force components must be singled out as most worthy of the designer's attention, it would be the alternating thrust, as the exciter of longitudinal vibration of main propulsion machinery. The hull surface excitation component, rather than the bearing forces, is the more critical director exciter of the hull, as will be considered further on.

Focusing attention on the alternating thrust ($i = 1$) specifically, the complex amplitude from Equations (91) and (92) can be written,

$$F_{1pm} = -N\int_{r=r_h}^{R} L_{mN}(r)\cos\beta_G(r)\,dr \tag{132}$$

Here, r_h denotes propeller hub radius and the lift harmonic order q in Equation (123) is mN. In terms of the non-dimensional lift coefficient of Equation (123), (132) can be rewritten as,

$$C_{1pm} \equiv \frac{F_{1pm}}{\rho U^2R^2}$$

$$= -N \int_{r = r_h/R}^{1.0} C_{LmN}(r) \cos \beta_G(r) \, dr \qquad (133)$$

For purposes of computation, a rectangle rule integration at eight equally spaced radial stations is of commensurate accuracy with that of the recommended formulas. For $rh/R = 0.2$, which is the usual case, Equation (133) becomes,

$$C_{1pm} = -N/10 \sum_{j=1}^{8} C_{LmN_j} \cos \beta_{G_j} \qquad (134)$$

The steps in the computation by Equation (139) are illustrated by the following example:

Consider a Troost Series B.4 propeller (Troost, 1937-1951) with 4 blades, operating in the Series 60, $C_B = 0.6$ wake of the previous example. A propeller which matches the Series 60 wake and the J of 0.834 has a $P(r)/D = 1.024$ at $r/R = 0.7$ and an expanded area ratio (EAR) of 0.471. Table 5 outlines the computation of the blade-rate alternating thrust coefficient, C_{1p1}, from Equation (134).

First of all, the phase angle, β_{1p1}, from Table 5 is sensible. $\beta_{1p1} = 42.4$ deg implies that the alternating thrust is maximum aft (positive) when a propeller blade is 42.4 deg before top-dead-center (with right-hand rotation clockwise looking forward). But the blade-rate alternating thrust executes four complete cycles, each of 90 deg duration, in one 360 deg revolution. The thrust therefore changes direction from maximum aft to maximum forward in 45 deg. The phase angle of 42.4 deg therefore corresponds to a blade at -2.6 deg, or effectively at top-dead-center, when the alternating thrust is maximum forward. This is intuitively satisfactory.

The force coefficient of $|C_{1p1}| = 0.152$ from Table 5 is a typical value for cases for which specific measures have not been taken to reduce vibratory excitation. Alternating thrust amplitude is often expressed as a percentage of steady thrust. Assume that the propeller and wake of Table 4 belong to a ship with 17,800 DHP and a speed of 20 knots. Assume that the propeller diameter is 6 meters. Then,

$$|F_{1p1}| = 0.152\rho \, U^2R^2 = 15.2 \text{ t}$$

Taking a QPC of 0.65 and a thrust deduction fraction of 0.1 as typical values, the steady thrust for this vessel would be,

$$T = \frac{550\text{DHP} \cdot QPC}{U(1-t)} = 134 \text{ t}$$

The alternating thrust in this example is therefore 11.3 percent of the steady thrust.

Now, the skew of the Troost series propeller blade is relatively low; it is 16.7 percent at the blade tip, as can be deduced from Table 5. It was judged in the example of the preceding subsection that a significant reduction in alternating over that of an unskewed propeller operating in the Series 60 wake could be achieved with a linearly varying skew out to 40 deg at the tip, or 44 percent with the 4-bladed propeller.

The Table 5 computation has been repeated in Table 6 with the above increased skew, and with all other propeller geometric data held fixed (a slight pitch adjustment would actually have to accompany the new skew distribution to maintain the same performance, but this is higher order to the unsteady force computations). Only the data which are different than that of Table 5 has been included in Table 6, as indicated thereon. A drop in $|C_{1p1}|$ from 0.152 to 0.112 on increasing skew from 16.7 percent to 44 percent represents a 26 percent reduction in alternating thrust. Greater increases in skew would result in yet greater reductions in alternating thrust. In fact, skew distributions can theoretically be found which result in zero alternating thrust. The 100 percent skew distributions which have been incorporated with the conventional single screw merchant ship wake on several occasions, Fig. 27, typically approach this limit. However, as was described relative to the example of the preceding subsection, skew distributions designed to accomplish reductions in a single bearing force component, such as alternating thrust, will generally not reduce the other force components by the same degree, and some increases may even occur.

To demonstrate this last point, the vertical bearing force corresponding to the Tables 5 and 6 propeller and wake was calculated. Formulas (92) for $i = 3$ was implemented in a similar tabular format as Tables 5 and 6. As indicated by Equations (92), the blade rate lateral forces and moments are due to the wake harmonics to either side of blade number, versus the blade number harmonic in the case of alternating thrust. It is convenient to write the respective complex amplitudes for $i = 2$, 3, 5, and 6, Fig. 8, in the following form:

$$\begin{aligned} F_{ipm} &= F_{ipm}^{\ +} + F_{ipm}^{\ -} \\ &= \rho U^2R^2 \left(C_{ipm}^{\ +} + C_{ipm}^{\ -}\right) \end{aligned} \qquad (135)$$

Here the + and − superscripts denote the contributions of the $mN + 1$ and $mN - 1$ wake harmonics, respectively. For blade-rate ($m = 1$) excitation with the 4-bladed example propeller, the lift harmonics corresponding to $q = 3$ and $q = 5$ were evaluated by Equation (123) and substituted into the respective $i = 3$ formula of (92). For the original Troost propeller of Table 4, this computation produced the following vertical force components:

$$\begin{aligned} C_{3p1}^{\ +} &= (0.5020 + 0.3740i)10^{-2} \\ C_{3p1}^{\ -} &= (-1.650 + 1.248i)10^{-2} \end{aligned} \qquad (136)$$

From (135),

Table 5—Blade-Rate Alternating Thrust Coefficient, Troost B.4 Prop, P/D = 1.02, Ear = 0.471 Series 60, C_B = 0.6 Wake

r/R Troost (1937-1951)	0.25	0.35	0.45	0.55	0.65	0.75	0.85	0.95	Σ
P/D	0.876	0.945	0.999	1.024	1.024	1.024	1.024	1.024	
l/R	0.418	0.465	0.500	0.518	0.521	0.497	0.428	0.305	
α_s, rad	0.023	0.060	0.105	0.143	0.160	0.190	0.232	0.275	
β_G, deg, (126)	48.12	40.68	35.24	30.65	26.64	23.49	20.98	18.94	
θ_e, rad, (128)	0.558	0.504	0.454	0.405	0.358	0.304	0.235	0.152	
k^*, (127)	2.23	2.02	1.82	1.62	1.43	1.22	0.94	0.61	
$\frac{V_a}{U}$, Table 3	0.522	0.640	0.736	0.791	0.813	0.818	0.828	0.845	
$\frac{V_r}{U}$, (124)	1.077	1.465	1.848	2.218	2.580	2.941	3.307	3.677	
$C_s = C_s^R + iC_s^I$ (Fig. 31)	0.018 0.226i	0.077 0.269i	0.133 0.262i	0.190 0.245i	0.245 0.220i	0.307 0.177i	0.385 0.109i	0.485 −.002i	
C_{x4} Table 3	0.	−.0475	−.0747	−.0802	−.0705	−.0567	−.0460	−.0370	
C_{T4}	0.0201i	0.0184i	0.0173i	0.0137i	0.0089i	0.0075i	0.0056i	0.0062i	
$V_{nq} = -C_{x4}\cos\beta_G + C_{T4}\sin\beta_G$ (125)	0. 0.0150i	0.0360 0.0120i	0.0610 0.0100i	0.0690 0.0070i	0.0630 0.0040i	0.0520 0.0030i	0.0430 0.0020i	0.0350 0.0020i	
$C_{L4} = (C_{L4}^R + iC_{L4}^I) \times 10^{-2}$ (123)	.7270 .1170i	2.006 −1.802i	5.151 −2.933i	8.193 −1.991i	9.304 −.2680i	8.539 2.533i	6.080 5.097i	2.460 5.658i	
$C_{1p1} = (C_{1p1}^R + iC_{1p1}^I) \times 10^{-2}$ (133)									−14.90 −2.720
$\|C_{1p1}\|$, (130)									0.152
β_{1p1}, deg, (131)									42.4

$$|C_{3p1}| = 0.0199$$

This coefficient corresponds to 1.49 percent of the 134 ton steady thrust of the example ship, and is a typical value for the conventional stern merchant ship wake. For the propeller with greater skew (Table 6), on the other hand,

$$C_{3p1}^{+} = (.1260 + 0.7920i)10^{-2}$$

$$C_{3p1}^{-} = (-1.886 - 1.590i)10^{-2}$$

Again, from (135),

$$|C_{3p1}| = 0.0193$$

The comparison predicts that the vertical bearing force decreases by 3 percent with the skew increase. The probability of a lateral force reduction less that achieved in alternating thrust (26 percent) was to be expected on the basis of Figs. 29 and 30, as described in the associated example.

While the relatively simple 2-dimensional approximation of lift by (123) is considered to be reliable for the types of relative evaluations represented by the preceding examples, a warning is in order with regard to the interpretation and use of the absolute magnitudes so predicted. Applying the 2-dimensional theory stripwise, as recommended, is equivalent to assuming that the propeller blade has infinite aspect ratio (span to chord ratio) in regard to the evaluation of the self-induced velocities, which is accomplished by the Sears function in Equation (123). This assumption results in a not insignificant overestimate of lift for aspect ratios typical of marine propeller blades. The approximate degree of overestimate can be judged with the aid of Fig. 32 (Breslin, 1970). This figure applies to rectangular wings, of aspect ratios 1 and 4, traversing sinusoidal gusts of reduced frequency k^*. The ordinate is the ratio of unsteady lift calculated by the 2-dimensional strip-wise approximation, Equation (123), to that calculated by a lifting surface theory which allows for

Table 6—Blade-Rate Alternating Thrust Coefficient, Repeat of Table 5 Computation With Increased Blade Skew*

r/R	0.25	0.35	0.45	0.55	0.65	0.75	0.85	0.95	Σ
α_s, rad	0.0436	0.1309	0.2182	0.3054	0.3927	0.4800	0.5672	0.6545	
$C_{L4} = (C_{L4}^R \times 10^{-2} + iC_{L4}^I)$	0.7153 0.1767i	2.431 −1.169i	5.916 −.3840i	7.729 3.371i	5.773 7.303i	1.087 8.840i	−3.577 7.081i	−5.521 2.756i	
$C_{1p1} = (C_{1p1}^R \times 10^{-2} + iC_{1p1}^I)$									−4.559 −10.27i
$\|C_{1p1}\|$									0.1124
β_{1p1}, deg									28.5

* Only data different from Table 5 included in Table 6.

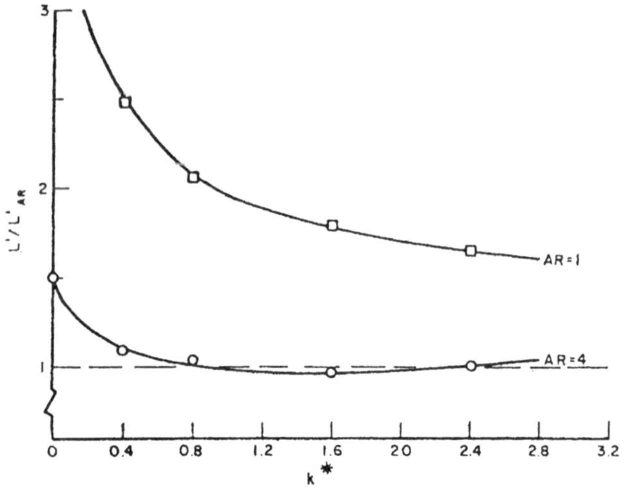

Fig. 32 Ratio of lift calculated using unsteady two-dimensional formula stripwise to exact unsteady result for aspect ratios 1 and 4

the finite aspect ratio effects. On consideration that the aspect ratios of marine propellers are typically on the order of 2 to 3 and reduced frequencies are on the order of 1 to 2, Fig. 32 suggest lift overestimates on the order of 30 to 50 percent by the 2-dimensional formulation. This is consistent with the conclusion of the comparative analysis of various propeller force calculation procedures reported in Boswell, Kim, et al (1983). However, the proposed 2-dimensional formulation incorporates all of the design variables other than aspect ratio in the correct physical way, and is therefore, as previously stated, useful in the design-type of trade trade-off investigations where the premium is on reliable relative evaluations. It is consistent with the proposed objective of minimizing the propeller excitation within the normal design constraints, which requires force evaluations with reasonably high relative, rather than absolute, accuracy.

An alternative simple method for calculating propeller vibratory bearing forces is that of Tanibayashi (1980). This method is essentially the quasi-steady method of McCarthy (1961), with semi-empirical modifications to allow for non-zero frequency effects. The comparisons of the latter reference suggest that the Tanibayashi method may have better absolute accuracy than the 2-dimensional unsteady strip method for some ranges of the variables. However, the Tanibayashi method, being less rational, does not appear to be as generally reliable in predicting the correct trends with changes in the variables. As discussed above, this characteristic is important to the relative accuracy required in many design considerations. For the types of design exercises illustrated by the preceding examples (as well as those to follow) the 2-D unsteady strip method is recommended over other methods of the simple type.

(d) Estimation of Propeller-Induced Hull Surface Excitation. Other than in the case of longitudinal vibration of main propulsion machinery and some main shafting vibration problems, the propeller bearing forces are of secondary importance to the hull surface excitation in propeller induced ship vibration. But this is only because of the common occurrence of some degree of moderate fluctuating sheet cavitation on the propeller blades. As discussed to some depth in Section 2, the bearing forces are relatively insensitive to fluctuating sheet cavitation, and it is usually ignored in their analysis. This is not the case, however, with the hull surface excitation; fluctuating sheet cavitation can amplify the propeller induced hull surface pressures and resultant forces by easily an order of magnitude over the non-cavitating levels. The occurrence of propeller cavitation cannot be ignored in attempts to quantify propeller induced hull surface excitation.

Unfortunately, while developing rapidly, the state-of-the-art has not yet produced a methodology for design stage estimation of hull surface excitation of relative accuracy and utility equal to that available for bearing force estimation. Wilson (1981) summarized the simple formulas and criteria currently available to the designer for dealing with hull surface excitation. Wilson compares the various proposed formulations against data from the few cases of recent U.S. Naval ship vibration problems, and concludes that none of the formulations appear capable of providing reliable indications about the likelihood of ship vibration trouble. Irrationality is no doubt responsible, in large part, for the inadequacy of the quick estimation techniques currently available; they are, for the most part, little more than "rules of thumb" based simply on intuition and empiricism.

Actually, experience has shown that if wake non-uniformity is at least not ignored in stern design, and if state-of-the-art propeller design is employed, which includes incorporating blade skew for bearing force control and maintenance of cavitation inception standards (Cox and Morgan, 1972), then serious vibration problems will seldom occur. Nevertheless, a simple, yet rational, though incomplete, formulation for hull surface force prediction is outlined in the remainder of this subsection. While the formulation should be of some limited utility to the designer in its present form, it is presented mainly as a rational framework to be filled-out as the state-of-the-art in this area matures.

As vertical hull vibration has been identified as the main girder vibration of concern, attention is focused here on the vertical component of the hull surface force. Hull surface force, rather than pressure, is considered to be the more appropriate measure of merit of hull surface excitation for minimization considerations, on the basis of the reasons cited in Section 2. The non-cavitating and cavitating cases are considered separately.

1. *Non-Cavitating Vertical Hull Surface Forces.* Vorus, Breslin and Tein (1978) derive the following formula for the complex amplitude of a non-cavitating hull vertical surface force coefficient:

$$C^{NC}_{3hm} = -C_{1pm}\, v^{*}_{30x} - i\,C^{-}_{3pm}\, v^{*}_{31\theta} + i\,C^{+}_{3pm}\, \bar{v}^{*}_{31\theta} \qquad (137)$$

The terms in Equation (137) are the following:

C^{NC}_{3hm}—m^{th} blade rate harmonic vertical (i = 3) non-cavitating hull surface force coefficient;

$$F^{NC}_{3hm} = \rho U^2 R^2 C^{NC}_{3hm}$$

C_{1pm}—m^{th} blade-rate harmonic alternating thrust coefficient, e.g., Tables 5 and 6.

C^{+}_{3Pm-}—m^{th} blade-rate harmonic vertical bearing force coefficients corresponding $mN + 1$ (+) and $mN - 1$ (−) wake harmonic contributions, e.g., Equation (135).

v^{*}_{30x} and $v^{*}_{31\theta}$—velocities induced in the propeller disk by vertically downward unit velocity motion of the bare hull, as described in Vorus, et al (1978). $\bar{v}^{*}_{31\theta}$ is the complex conjugate of $v^{*}_{31\theta}$.

Formula (137) applies to the stern type for which the breadth of the counter directly above the propeller can be characterized as large. It is unfortunately not applicable to the case of the cruiser type stern of conventional single screw merchant ships, whose counter is narrow. For the broad stern type, however, for which the propeller bearing forces have been estimated, Equation (137) can be used to estimate the non-cavitating vertical hull surface force provided that the bare hull induced velocity data is available.

Table 7—Hull Induced Velocity Data for Use Equation in (137) (Broad Countered Stern Forms)

	$v_{30x}{}^{*}$	$v_{31\theta}{}^{*}$
Single Screw	0.5	$-0.5i$
Twin Screw	0.3	$-0.4 - 0.45i$

Table 7 gives approximate average values of the required induced velocity data appropriate for use with single screw and twin screw ships.

The numbers in the table are approximate averages from detailed calculations of the induced velocities for many ship cases. The tangential velocity component, $v_{31\theta}{}^{*}$, is most sensitive to waterplane breadth over the propeller; $v_{31\theta}{}^{*}$ increases with waterplane breadth. The extreme value of $v_{31\theta}{}^{*}$ that has been encountered, approximately $-0.6i$, was for a single-screw, barge-stern laker where the ratio of waterplane breadth aft to propeller diameter approached 4.0. The more sensitive of the velocity components to stern geometry is, however, the axial component, v_{30x} in Equation (137). This component depends most strongly on the axial distance from the propeller to the waterplane ending. For propeller inset distance denoted x_0, the extreme values of $v_{30x}{}^{*}$ encountered have been 0.15 for a twin-screw naval cruiser with deep inset $D/x_0 \sim 0.5$, and 0.75 for the same barge stern single screw laker with a very shallow inset $D/x_0 \sim 2$. A few of the cases for which this data has been evaluated are described by Vorus, Breslin and Tein (1978).

In the following example, assume a broad countered single screw ship with a skeg, configured such that the Series 60 wake of Fig. 26 is reasonably representative (this is assumed for example only; a wake evaluated from model tests should be used in actual analysis). Also assume that the propeller is the Troost B.4 subject of the examples in the preceding subsection. The bearing force coefficients required in Equation (137) for blade-rate surface force evaluation are therefore the values from Table 5 and Equation (136). That is,

$$C_{1p1} = (-1.50 - 0.272i)10^{-2}$$
$$C_{3p1}{}^{+} = (0.502 + 0.374i)10^{-2}$$
$$C_{3p1}{}^{-} = (-1.65 + 1.25i)10^{-2}$$

Using the induced velocity values corresponding to the single screw ship case of Table 7, Equation (137) gives,

$$C^{NC}_{3h1} = 0.0132 - 0.00676 \qquad (138)$$

whose amplitude and phase are,

$$|C^{NC}_{3h1}| = 0.0149$$

$$\beta^{NC}_{3h1} = \frac{1}{4}\tan^{-1}\frac{0.00676}{0.0132} = 6.8 \text{ deg}$$

The non-cavitating vertical surface force calculated above is slightly smaller than the vertical bearing force, whose magnitude was calculated in Equation (136) as, $|C_{3p1}| = 0.0199$.

The most appropriate measure of hull girder vertical vibratory excitation should actually be the net vertical force, represented by the vector addition of the vertical bearing and vertical surface forces. Denoting the net vertical force coefficient as C^{NC}_{3Nm},

$$C^{NC}{}_{3N1} = C^{NC}{}_{3h1} + C_{3p1}$$

Substituting from Equations (136) and (138), for the subject example,

$$C^{NC}{}_{3N1} = (0.174 + 0.94i)10^{-2}$$

with amplitude and phase,

$$|C^{NC}_{3N1}| = 0.00956$$

$$\beta^{NC}_{3N1} = -19.9 \text{ deg}$$

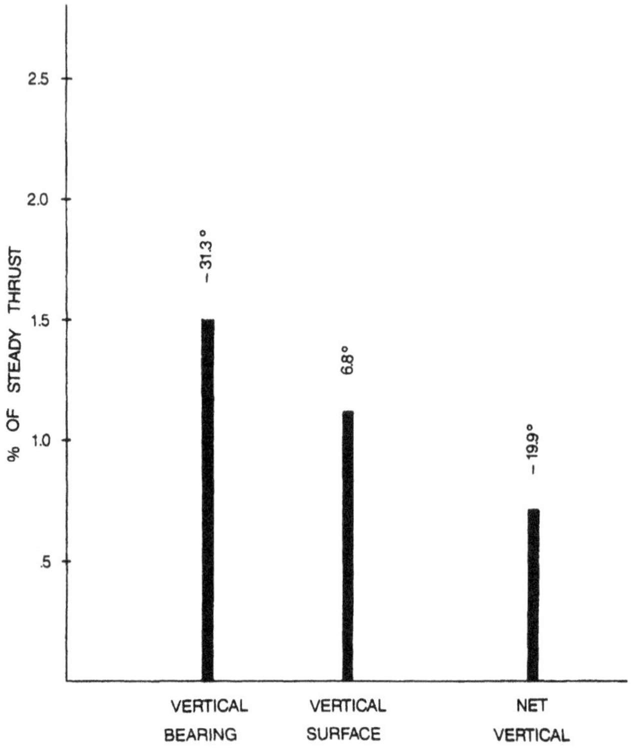

Fig. 33 Blade-rate vertical forces, Series 60, C_B = 0.60, Troost B-4 propeller

Thus, the net vertical force predicted in this example is smaller than both the individual vertical bearing force and vertical surface force components. This is to be expected in the case of the broad countered stern to which Equation (137) applies; refer to the discussion of the Breslin condition in Section 2.3. The comparison is shown on the bar graph, Fig. 33; the bar heights denotes the percentages of thrust of the example propeller and the numbers at the tops of the bars are the phase angles.

2. *Cavitating Vertical Hull Surface Forces.* Vorus, Breslin and Tein (1978) also derive a rational formula for the vertical hull surface force coefficient due to fluctuating sheet cavitation. It is:

$$C^c_{3hm} = imN^2 \left(\frac{\pi}{J}\right)\left(\frac{b_0}{R}\right)\left(\frac{\phi_{30}^*}{b_0}\right)\left(\frac{\dot{\forall}_{mN}}{UR^2}\right) \quad (139)$$

The terms in this formula are:

N—propeller blade number.
J—advance ratio, U/nD.
b_0—design waterline offset in the vertical plane of the propeller disk.
ϕ_{30}^*—velocity potential induced in the propeller disk by vertically upward unit velocity motion of the bare hull, as described in Vorus, Breslin and Tein (1978).
$\dot{\forall}_{mN}$—the mth harmonic of the cavitation volume velocity on one propeller blade.

Formula (139), like the non-cavitating Formula (137), is reduced from a general reciprocity formulation on the basis of broad waterplane aft. However, due to more rapid convergence characteristics of the hull induced potential, ϕ_{30}^*, in (139) versus the hull induced velocity components in (137), Formula (139) has been found to work quite well for vessels whose sterns are actually characterized as narrow. Furthermore, the function ϕ_{30}^* / b_0 has been found to vary only moderately from one stern to the next. In the many detailed calculations of ϕ_{30}^* that have been performed, the extreme values of ϕ_{30}^* / b_0 encounted have been approximately 0.4 and 0.7. However, most fall very close to the average of these extremes; a value of $\phi_{30}^* / b_0 =$ 0.5 for all cases should be consistent with the best accuracy achievable in estimating the cavity volume term in (139), and also consistent with the intended use of the formula.

The illusive term in (139) is the cavity volume velocity harmonic, $\dot{\forall}_{mN}$. It is for lack of data in this regard that the cavitating force formula (139) must be held in reserve at the present time. Work is in progress in this area, however, (Lee, 1979) and (Stern and Vorus, 1983) and hope can be held that the dynamics of fluctuating sheet cavitation will be quantified to the degree needed for, at least, reliable relative evaluations in the not too distant future. Some limited cavitation volume velocity data are, however, available at this time. For example, cavity volume dynamics were estimated by the numerical method of Stern and Vorus (1983) in the excitation force analysis of the Navy oiler documented in Vorus (1981). The data in this reference are of unsubstantiated accuracy, but it is, at any rate, useful here for demonstrating the character of the required term and the cavitating hull surface force computation by Equation (139). This is done in the context of an example.

Fig. 34 shows a cavity volume velocity curve calculated by the theory of Stern and Vorus (1983) for the 7-bladed highly skewed propeller of a naval oiler. The x's on the figure indicate the result of summing the Fourier series expansion of the curve using 21 terms; the series is of the form,

$$\dot{\forall}(\theta) \simeq \text{Re} \sum_{q=0}^{21} \dot{\forall}_q e^{iq\theta} \quad (140)$$

with the $\dot{\forall}_q$ calculated using the same general formula as used in the wake harmonic analysis, i.e., Equation (113). The 21 non-dimensional complex $\dot{\forall}_q$ coefficients of Equation (140) are tabulated in Table 8.

The oiler is single screw with a conventional merchant ship stern; $b_0/R = 0.587$ in (139). Other relevant data is:

$J = 1.032$
$N = 7$

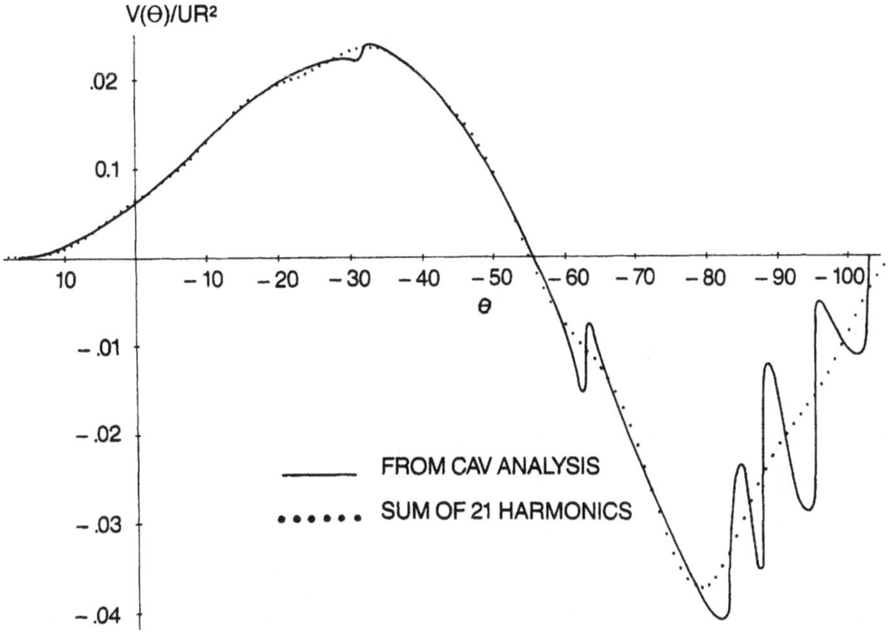

Fig. 34 Cavitation volume velocity, U.S. naval oiler

Table 8—Harmonic Coefficients of Cavitation Volume Velocity for a Naval Oiler

Harmonic Order q	$\dot{\forall}_q / UR^2 \times 10^{-2}$	
0	0.00713	+ i
1	0.355	$-0.288i$
2	0.733	$+0.196i$
3	0.285	$+0.774i$
4	-0.416	$+0.541i$
5	-0.433	$-0.0908i$
6	0.00608	$-0.238i$
7	0.149	$-0.0256i$
8	-0.0228	$-0.116i$
9	-0.0795	$-0.0139i$
10	0.0185	$-0.0573i$
11	0.0539	$+0.0125i$
12	0.0070	$+0.0455i$
13	-0.0287	$+0.0227i$
14	-0.0323	$-0.0127i$
15	-0.00333	$-0.0385i$
16	0.0376	$-0.0175i$
17	0.0239	$+0.0333i$
18	-0.0326	$+0.0271i$
19	-0.0333	$-0.0313i$
20	0.0241	$-0.0388i$
21	0.0384	$-0.0141i$

Table 9—Cavitation Induced Vertical Surface Forces

Blade-Rate Harmonic m	$\dot{\forall}_{m7} / UR^2$ (10^{-2}) (Table 8)	$\lvert C^c_{3hm} \rvert$ (141)	β_{3hm} (deg)	$\lvert F^c_{3hm} \rvert / T$
1	$0.149 + 0.0256i$	0.0832	11.5	0.0776
2	$-0.0323 - 0.0127i$	0.0382	-8.0	0.0356
3	$0.0384 - 0.0141i$	0.0675	5.3	0.0675

$D = 6.4$ m
$U = 21.4$ knots
thrust, $T = 139$ t

Taking $\phi_{30}^* / b_0 = 0.5$ (the actual calculated value was 0.63), and substituting in Equation (139) the vertical force coefficient for blade-rate harmonic m is,

$$C^c_{3hm} = -55im(\dot{\forall}_{m7} / UR^2) \quad (141)$$

This coefficient, along with the corresponding fractions of steady thrust, are listed in Table 9 for the first three blade-rate harmonics.

The 7.8 percent vertical blade rate force calculated above is not unusually large, as cavitation induced forces go. Values on the order of 30 percent of steady thrust are not unheard of. It is, however, seven times larger than the noncavitating surface force from the example of the preceding subsection, Fig. 33. The naval oiler of this example did, in fact, not have a particularly severe vibration at blade-rate frequency.

The perhaps more alarming aspect of the Table 9 data are the high multiple blade-rate force components. This substantial high harmonic content is a characteristic of the excitation induced by cavitating propellers. It is due, mathematically, to the slow convergence of the volume velocity Fourier series, as is obvious from Table 8. Physically, it is due to the rapid expansion and collapse of the cavitation (Fig. 34). The strong higher blade order excitation harmonics of cavitating propellers are quite capable of producing excessive vibration, and also, because of the higher frequencies, excessive noise. The subject naval oiler did, in fact, suffer more from an objectionable noise problem, which was attributed to propeller cavitation.

Section 4
Criteria, Measurements, and Post-Trial Corrections

4.1 Criteria of Acceptable Vibration. It has recently become more the rule than the exception that new ship specifications require measurement of vibration on builders' trials, and place contractual limits on acceptable vibration levels. The vibration of primary concern is that occurring within habitable spaces, principally within deckhouses and engine rooms, and criteria are consequently based primarily on habitability standards. Limits on levels of equipment vibration, from an operability standpoint, are sometimes involved in specifications, particularly for naval vessels. The applicable standard covering equipment vibration is: Military Standard, *Mechanical Vibration of Shipboard Equipment* (1969).

Most of the criteria established by the classification societies for merchant ships, which then reappear as limits in ship specifications, are at least consistent with Fig. 35, if not based directly upon it. Fig. 35, from SNAME (1980) is a plot of vibration response amplitude versus its frequency. The zones identified on Fig. 35 represent different levels of vibration severity; they are defined as:

Zone I—Vibration levels in this zone are low enough so that adverse comments from personnel would not be expected.

Zone II—Vibration levels in this zone indicate that while vibration is noticeable, few adverse comments would be expected.

Zone III—In this zone vibration levels and human response increase rapidly in severity and adverse comments would be expected.

The "response" of Fig. 35 can be chosen as displacement response, velocity response, or acceleration response, as indicated by the three different scales on the figure. For simple harmonic vibration, which Fig. 36 assumes, a simple relationship exists among the scales of displacement, velocity, and acceleration on Fig. 37. That is, for vibration displacement response at a point occurring as,

$$x(t) = X \cos \omega t \tag{142}$$

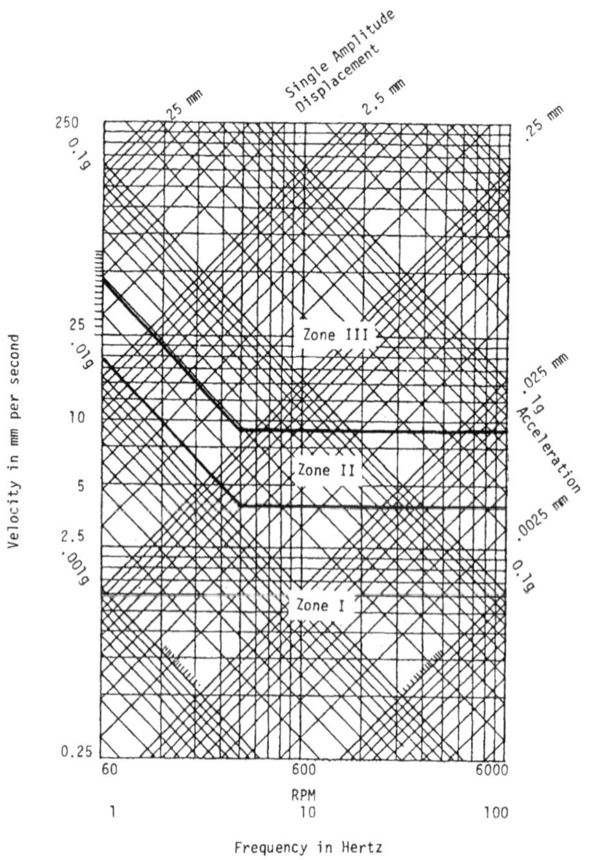

Fig. 35 Guidelines for Ship Vibration (Vertical and Horizontal, Single Amplitude)

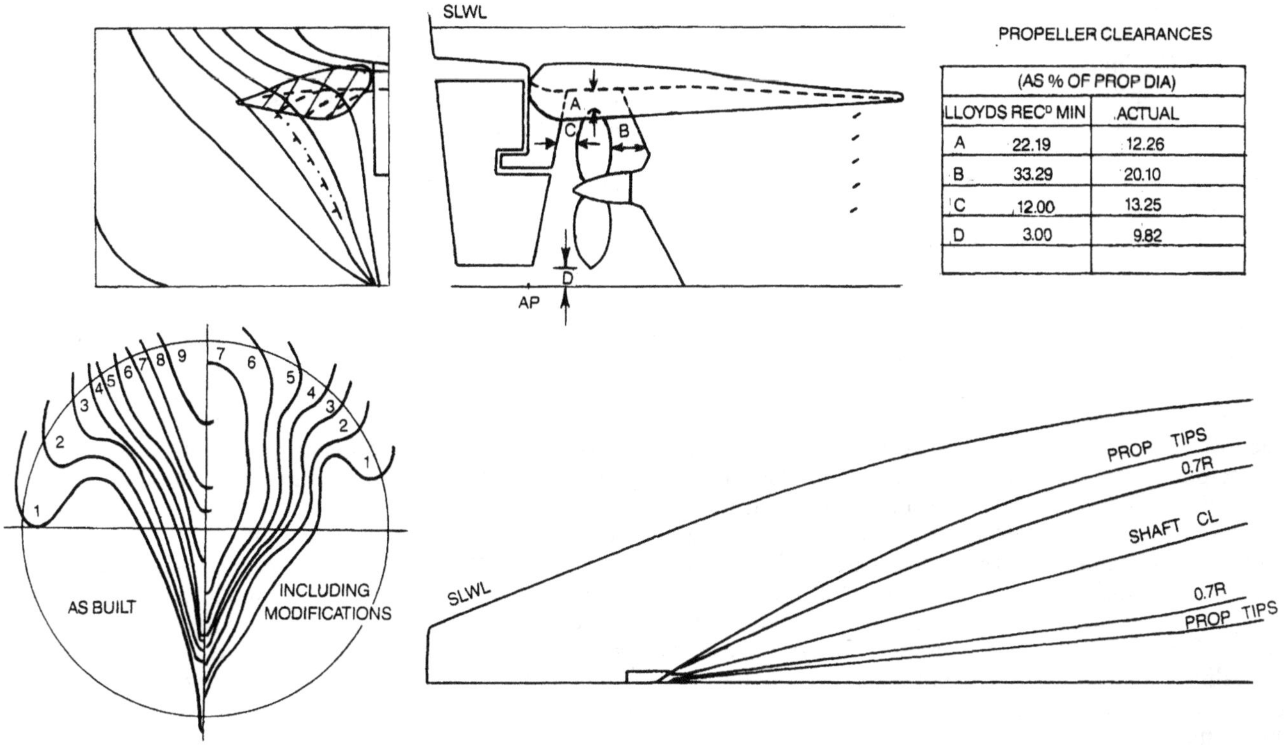

Fig. 36 Wake improvement by means of special stern appendages, 7000-t pallet cargo ship

the displacement response amplitude is X. The velocity response, on the other hand, is, from Equation (142),

$$v(t) = \dot{x}(t) = \omega X \cos(\omega t + \pi/2) \qquad (143)$$

The velocity response amplitude is therefore ωX, with ω being the vibratory frequency in radians per unit time. The acceleration response amplitude is similarly, $\omega^2 X$, by differentiation of Equation (143).

As an example of the interchangeability, consider the vibration response corresponding to a displacement amplitude of $X = 1$ mm at a frequency of 5Hz. The velocity amplitude is,

$$X\omega = 1(5)2\pi = 31.4 \text{ mm per sec}$$

and the acceleration amplitude is,

$$X\omega^2 = 1[(5)2\pi]^2 = 996 \text{ mm per sec}^2$$

Of course, these are all the same point on Fig. 35.

Velocity has replaced displacement in recent years as the popular unit for referring to ship vibration level. Full-power propeller blade-rate excitation frequency for the modern large ship is, for example, on the order of 10Hz. Zone 2 of Fig. 35, whose vibration levels would be noticed by exposed personnel, has extremes of 4 and 9 mm/sec in the 10Hz range. The vibration limits for habitable spaces imposed by most ship specifications seem generally to lie within this band.

For example, criteria that appear to have been adopted in a number of ship specifications, both naval and commercial, set an objective of 6.4 mm/sec and 3.8 mm/sec maximum vertical and horizontal, respectively, for vibration velocity of the hull girder. The design objective on major substructure maximum vibration, such as deckhouses, is 7.6 and 5.1 mm/sec vertical and horizontal, respectively. Note that all of these values fall within Zone II of Fig. 35, for typical propeller blade-rate frequencies. The criteria further recommend that maximum acceptable limits be set at 150 percent of the above values. The resulting vertical upper limits can then be seen to fall in lower Zone III on Fig. 35, with the corresponding horizontal limits falling in upper Zone II.

Fig. 35, and other criteria like it, are readily applied when the vibration can be at least approximately characterized as simple harmonic, i.e., periodic at a single frequency. However, as noted in the last section, in general, ship vibration is not simple harmonic; it is not even periodic. Ship vibration is random, i.e., it is composed of components at all frequencies, rather than at a single one. The random character of ship vibration is clearly evident in records from underway vibration surveys. But the data from such complex records must often be compared with simple criteria, such as Fig. 35, to quantify its severity.

In propeller-excited ship vibration cases where cavitation is not heavily involved, propeller input spectra are narrow-band around blade-rate frequency, as mentioned in the last section. Furthermore, for structural resonance, or near resonance, at blade-rate frequency, the band of the vibration response spectra around the resonant frequency is further narrowed. In such cases, which are not uncommon, analog records unmistakably display a dominant blade-rate frequency characteristic. The RMS vibratory response amplitude is then usually evaluated from the records, either by "eye-ball," or more precisely by spectral analysis, and matched with blade-rate frequency, as in Equation (142), for comparison with the established limit criteria.

In the other extreme, where cavitation is heavily involved in non-resonant vibration, measured vibration records still usually exhibit a basically periodic character, but components at more than one discrete frequency are clearly evident. The component frequencies are the strong blade-rate multiples of the slowly convergent hull surface excitation associated with the cavitation intermittency. In this case, with significant component vibration occurring simultaneously at several different frequencies, it is not always clear how guidelines such as Fig. 35 are to be used. Instructions, if provided at all, are usually to evaluate the several component responses completely independently of one another at their respective frequencies.

A realistic and yet concise standard for the specification of propeller-induced ship vibration limits or criteria appears, in some respects, to be hardly less elusive than some parts of the design methodology needed to provide assurance in meeting such standards. However, the reader should keep in touch with continuing work on the problem of criteria, including that of the International Standards Organization (ISO).

4.2 Vibration Measurement.

(a) Design Verification. The SNAME Code for Shipboard Vibration Measurement (SNAME, 1975) recommends a very comprehensive program, and instrumentation package, for shipboard vibration surveys. The Code is invoked in many ship vibration specifications.

The instrumentation package proposed in the above Code is more than adequate to establish compliance, or non-compliance, with the typical ship vibration specification. The instrumentation consists of a set of 12 inductance-type velocity pick-ups, with signal processing through an equal number of integrating amplifiers, and with a permanent record of the resultant vibration displacement signature recorded graphically on a multi-channel recorder. In providing the capability for simultaneous multi-point vibration measurement, this instrumentation can be used to establish vibration frequency, amplitude, and local relative displacement (mode shape). Evaluation of ship vibration for purposes of comparison against the typical specification will require the measurement of frequency and amplitude at the pre-established survey points, but not the phase relationships between points. Amplitude/frequency information can be obtained with acceptable accuracy using relatively simple portable instruments, with the measurements at the survey points performed in some sequence, rather than simultaneously.

Of course, if the simple vibration survey of a new ship should establish that the specification limits are badly exceeded, then the type of instrumentation package proposed in SNAME (1975) may become absolutely essential to expeditious rectification. In this respect, invoking the SNAME Code in design specifications may be considered an insurance worth the resulting extra investment.

(b) Post-Trial Corrective Investigations. The approach to resolving a ship vibration problem, as with most engineering problems, involves two steps: The first step is to clearly establish the cause of the problem, and the second step is to implement the changes required to eliminate it in an efficient manner.

In about 80 percent of cases the basic cause of a ship vibration problem is its propeller. This fact seems to be very elusive to the vibration analyst familiar only with land-based power plant-oriented vibration problems; ship vibration is indeed a case of "the tail wagging the dog" most of the time. Whether or not the vibration of a particular ship has its source in the propeller is easily established from underway vibration measurements. If at some shaft RPM the measured frequency of the vibration is predominantly RPM times propeller blade number, and varies directly with shaft RPM, then the propeller is definitely the exciting source. If blade-rate frequency, or its multiples, is not strongly detectable in the records, then it is almost certain that the propeller is not the primary excitation, unless the records exhibit a strong shaft-rate frequency, which could indicate propeller unbalance difficulties, but these are rather rare.

Once the excitation frequency has been established from the underway measurements, it is next in order to establish whether or not resonance with structural natural frequencies play a significant role in the magnitude of the vibration problem. For non-cavitating propellers, excessive hull vibration should be expected to be resonant vibration. Resonant vibration is established by varying shaft RPM in steps and recording vibration amplitude successively at each RPM in the region where the problem has been identified as being most intense. If a plot of displacement amplitude versus RPM shows a definite peak with increasing RPM, followed by decline, then resonant vibration is established and the position of the peak establishes the natural frequency of the resonant structural mode. If the amplitude/RPM characteristic does not peak, but has an increasing trend as roughly RPM squared in the upper power range, then structural resonance is not playing a major role. If, alternatively, the amplitude RPM characteristic increases very rapidly only in the immediate vicinity of full power, without estab-

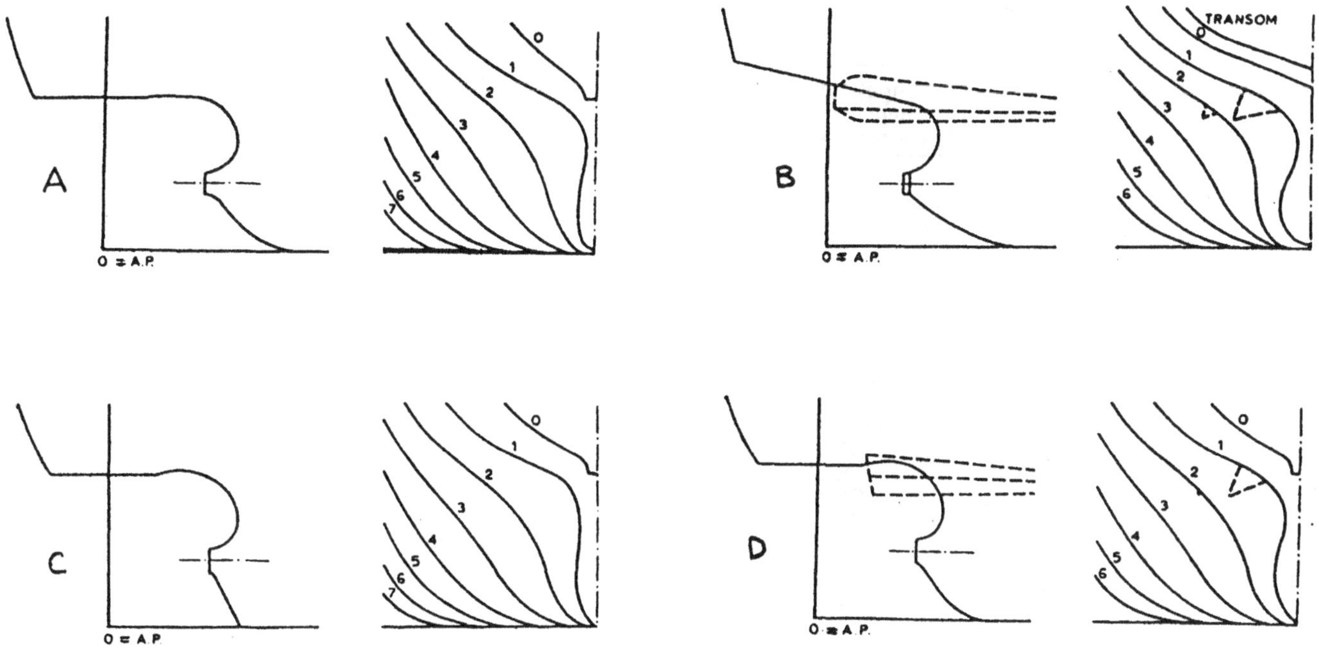

Fig. 37 Model sterns for force measurements, Table 9

lishing a definite peak up to the maximum obtainable RPM, a full power resonance may or may not be indicated. This exhibition can be entirely the manifestation of the onset of propeller cavitation, which tends to produce an almost discontinuous amplification of the hull surface excitation at the onset RPM. The sudden appearance of strong harmonics of blade-rate frequency in the vibration records, accompanied by violent pounding in spaces above the counter, are good indications of a full power non-resonant vibration problem caused by excessive propeller cavitation.

If non-resonant vibration due to propeller cavitation is established, then the underway survey could probably be discontinued, with attention then turned to hydrodynamic design changes in the stern/propeller configuration. This course of action is considered in the next subsection.

If the problem is established as highly localized resonant vibration of plating panels, piping, and the like, then the vibration survey likewise need go no further. It is usually quite obvious in such cases how natural frequency changes, through local stiffening, can be effectively and expediently accomplished to eliminate the locally resonant conditions.

If, on the other hand, the vibration problem is established as a resonant condition of a major substructure, such as a deckhouse, which is all too often the case, then the vibration survey had best proceed to obtain mode shape information in the interest of an expeditious correction program.

4.3 Post-Trial Corrections. Just as in developing a vibration-sufficient ship design, all possibilities for correcting a vibration-deficient one are explicitly reflected in the general response Formula (82) of Section 2. Practically speaking, there are three possibilities: (*a*) reduce vibratory excitation, (*b*) change natural frequencies to avoid resonance, or (*c*) change exciting frequencies to avoid resonance. Except in the rather rare case of excessive diesel engine-excited hull vibration, which can usually be corrected by moment compensators (Sulzer Bros., 1977) achievement of any of the three correction possibilities identified above will almost always involve modifications in either stern/propeller hydrodynamics or hull structure.

(a) Hydrodynamic Modifications. The most effective way to reduce propeller vibratory excitation is to reduce the circumferential non-uniformity of the wake in which the propeller operates. In the design stage acceptable wakes can be achieved by taking proper care with stern lines. (Refer to Sub-section 3.7). In a post-design corrective situation basic lines changes are, of course, not possible. However, with good luck in the case of poor stern lines, considerable improvements in wake can be accomplished by back fitting one of the several types of wake adapting stern appendages. The partial tunnel, Fig. 36, has been the most broadly applied of the wake adapting appendages, which also include vortex generators and, more recently, wake adapting nozzles. The partial tunnel was apparently first retrofitted for vibration reduction purposes by Baier and Ormondroyd (1952) on the laker *Carl D. Bradley* in 1951. The idea is to divert the upward flow along the buttock lines forward longitudinally into the upper propeller disc to reduce the wake

Table 10—Vertical Surface Force Measurements on Models of Fig. 37

Ship identification	amplitude of 1st, 2nd and 3rd harmonic of vertical hull force in percent of mean thrust: before application of partial tunnel	after application of partial tunnel	full-scale vibration level before application of partial stern tunnel	full-scale vibration level after application of partial stern tunnel
A	6.8/4.2 2.0/2.0 x) 1.9/0.5	—	unacceptable	acceptable
B	12.6 12.4 7.7	15.3 4.0 0.9	unacceptable	acceptable
C	32.7/12.1 13.6/14.6 x) 5.4/ 5.7	—	unacceptable	?
D	28.0 6.9 4.4	41.9 3.4 1.4	unacceptable	?

x) different propellers

spike near top-dead-center. This device will obviously work most effectively on the buttock-flow type of stern; the partial tunnel has been applied successfully over the years on the Great Lakes ore carriers, most of which have barge-type sterns with very steep buttock angles. On the other hand, for sterns which exhibit a basically waterline-flow character, the partial tunnel would be expected to be more or less ineffective due to the lack of upward flow to divert. However, the effectiveness of the partial tunnel cannot always be accurately judged by simply classifying a prospective application as one of the two limiting cases of buttock versus waterline flow. For example, the stern shown on Fig. 36, from Rutherford (1978-79) might be classified as more of a waterline flow, yet the modifications shown produced significant improvement in the nominal wake, as exhibited by the before and after axial velocity contours. The Fig. 36 modifications however include vortex generators as well as the partial tunnel, and the contributions of each to the wake improvements shown are not known. A more direct indication of the effectiveness of the partial stern tunnel in reducing vibratory excitation is given on Fig. 37 and in Table 10, from Hylarides (1978). Fig. 37 shows the stern lines of four ships on which partial stern tunnels were fitted as a result of post-trial corrective studies conducted at the Netherlands Ship Model Basin, now MARIN. Pressures were measured on model sterns in the Cavitation Tunnel, integrated, and then harmonically analyzed to produce the first three harmonics of blade-rate vertical hull surface force. The force amplitudes, as percentages of steady thrust, are listed in Table 10 for each case. In the two cases where the outcome of the tunnel retro-fit is indicated, the vibration was judged to be acceptable.

The force results of Table 10 for the two cases where measurements are listed both before and after the tunnel addition are surprising in one respect. In both cases significant reductions in the 2nd and 3rd blade-rate amplitudes are attributed to the tunnels, but an increase in the blade-rate forces are indicated. This is not impossible, yet it seems unlikely. In spite of the success of such model test programs in solving vibration problems associated with propeller hydrodynamics, it is difficult to have high confidence in the accuracy of force predictions of the type listed in Table 10. This is for the general reasons cited in Section 2. In view of the advances that are being made in the development of analytical/numerical hydrodynamic models, it seems certain that in the near future, hybrid schemes, exploiting the best features of numerical and experimental analysis, in combination, will be available to replace the purely experimental programs typical of that which produced the data of Table 10.

The decision to retro-fit a wake adapting stern appendage should not be made lightly without quantification of the advantages and disadvantages; a price is usually paid for appendages in increased hull resistance. As a minimum, model tuft-tests with and without the appendage should be performed to observe the change in stern surface flow. The absence of any noticeable smoothing may be misleading, however; a wake survey can show improvements in the propeller plane not discernable in the tuft behavior. Furthermore, aside from nominal wake considerations, it has been found that greatest wake improvements are sometimes achieved through propeller/appendage interaction (Hylarides, 1978). This implies that model tuft-tests should be conducted both with and without the operating propeller. In these cases, the best indicator of significant effective wake improvements from the standpoint of vibratory excitation may be an improvement, by several percentage points, in the propulsive efficiency from model SHP test conducted with and without the wake adapting appendage, as explained in Hylarides (1978).

Aside from wake improvements, the only recourse for reducing propeller excitation is modification or replacement of the propeller. Some instances of successful modifications of troublesome propellers have been reported. For example, trimming blade tips by several centimeters to reduce wake severity at the extreme propeller radii can produce improvements, but some degree of RPM increase must then be tolerated. Successful modifications of existing propellers are rare because of the usually unacceptable trade-offs of performance degradation against vibration improvement. The same disadvantages exists in propeller replacement considerations. Replacement propellers, with modified features such as changed blade number, reduced diameter (for increased hull clearance), increased blade area, reduced pitch in the blade tips, etc., may relieve the vibration problem, but often for a dear price in vessel performance. It is unfortunate that, with the exception of blade skew, essentially all of the measures available in propeller design for reducing vibratory excitation, once the stern lines are established, act also to reduce propeller efficiency (refer to Chapter V for propeller design considerations). It cannot be emphasized strongly enough that the greatest insurance against propeller-induced vibration problems and the persistent difficulties which then almost always ensue is to place high emphasis on wake uniformity in making trade-offs at the original establishment of vessel lines.

(b) Structural Modifications. The most cost effective approach for eliminating structural resonances is usually to shift natural frequencies through structural modifications; the alternative is to shift exciting frequency by changes in engine RPM or number of propeller blades.

Just as with hydrodynamics related problems, the most intelligent way to approach the correction of a vibration problem that promises to involve significant structural modifications is through the use of the tools of rational mechanics. A structural math model should first be calibrated to approximately simulate the existing vibration characteristics. Modification possibilities are then exercised with the model, and their probability of success is established on paper. In this way the probability of a "one-shot" success when shipboard modifications are subsequently implemented is maximized. The alternative and unenlightened "cut-and-try" approach to the solution of serious ship vibration problems is fraught with frustration, and with the real possibility of expending vast amounts of time and money and never achieving complete success.

Of course, the paper studies proposed as a tool for use in correcting a serious ship vibration problem must be concluded quickly; several months, or even several weeks, is not available when delivery of a vessel is stalled, awaiting the resolution of vibration deficiencies. This places a premium on formulation of the simplest possible structural models which still retain adequate realism to provide the basis for the required judgments as to the relative effects of vessel modifications. This is where the collection of thorough trial vibration data can pay for itself. Measurement of vibratory mode shape data is often a near necessity for securing guidance in formulating calibration models of the desired simplicity, but with sufficient accuracy. This is illustrated by the following simple example:

Assume that excessive vibration of a Type A deckhouse, Fig. 20, occurs on the builder's trials of a vessel. Vibratory displacement amplitude data are recorded with phase calibrated pick-ups mounted at points on the house and on the main deck. The records establish the following information.

1. The vibration occurs at predominantly blade-rate frequency, confirming the propeller as its exciting source.
2. The vibration amplitude peaks at 100 RPM, and the propeller has five blades. A resonance of the house at 500 cycles per min (cpm) is therefore established.
3. Vibration records recorded at 100 RPM show that the vibration of the house is predominantly fore-and-aft, with fore-and-aft amplitude increasing with a quasi-linear characteristic from low levels at main deck to a maximum of 0.75 mm at the house top. The house top is 15 meters above main deck.
4. The 100 RPM record also shows that the amplitude of the vertical vibration at main deck is approximately uniform at 0.1 mm over the house length. The vertical vibration amplitude is also approximately constant at this same level up the house front, which is a continuation of the forward engine room bulkhead.

The above characteristics are judged to allow the use of the simple rocking/bending house model in conjunction with the Hirowatari method, Figs. 20 and 21 and Table 2, Section 3.

(c) Determination of Model Constants. For a Type A house with h = 15 meters, the fixed-base fundamental house natural frequency is estimated from Fig. 21 as,

$$f_\infty = 800 \text{ cpm}$$

Using the Dunkerley formula, Equation (105), with the measured house natural frequency of $f_e = 500$ cpm, the effective rocking frequency is,

$$f_R = \sqrt{\frac{1}{1/f_e^2 - 1/f_\infty^2}} \tag{144}$$

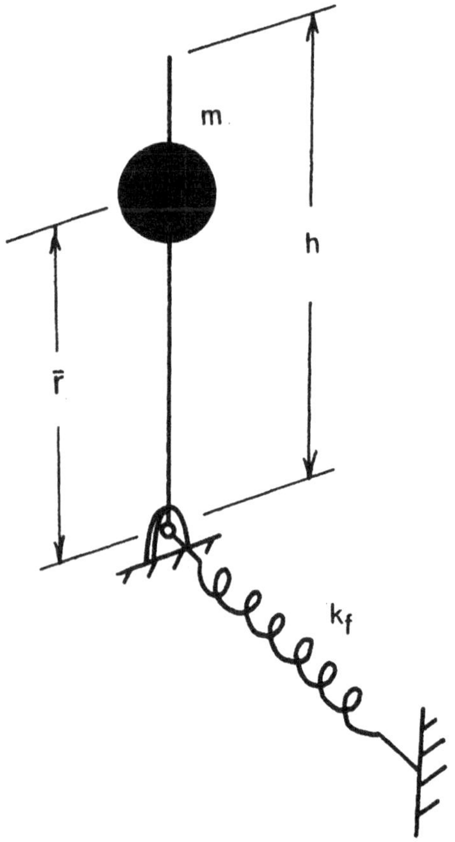

Fig. 38 Mass-elastic model of deckhouse and support structure

$$= 640 \text{ cpm}$$

The two frequencies f_∞ and f_R can be used to determine the effective stiffnesses of the house and its underdeck supporting structure for use in the simple model of Fig. 38.

For a house mass established as m=300t, with a radius of gyration, $\bar{r}$, about the house forward lower edge of 10 m, the effective torsional stiffness of the under-deck supporting structure is, from Formula (106):

$$K_f = (2\pi / 60)^2 f_R^{\,2}$$
$$= 1.35 \times 10^{11} \text{ N-m per rad} \quad (145)$$

where

$$J = m\bar{r}^2 = 3 \times 10^7 \text{ kg-m}^2$$

An approximate effective bending/shear stiffness of the house is obtained by first lumping the house mass at the radius of gyration above the assumed pin support on the main deck at the forward bulkhead; this preserves the mass moment of inertia in the Fig. 38 model. Then, for the house base fixed,

$$k_H = (2\pi / 60)^2 f_\infty^2\, m = 2.1 \times 10^9 \text{ N per m} \quad (146)$$

The effective combined torsional stiffness for use in the equivalent reduced one-mass system of Fig. 39 is then,

$$K = \frac{1}{1/K_f + 1/(k_H \bar{r}^2)}$$
$$= 0.822 \times 10^{11} \text{ N-m per rad} \quad (147)$$

This combined stiffness can also be deduced directly from the measured natural frequency and the house mass moment of inertia as,

$$K = (2\pi / 60)^2 f_e^2\, J$$
$$= 0.822 \times 10^{11} \text{ N-m per rad} \quad (148)$$

The effective exciting moment due to the vertical hull girder vibration can be estimated using the formula developed in the simple rigid box deckhouse analysis of the second section. Referring to the development of Formula (83), the amplitude of the exciting moment is,

$$M_e = \omega^2\, m\bar{\xi}X \quad (149)$$

Here, $\bar{\xi}$ is the longitudinal coordinate to the house *CG*, measured aft from the house front, say, 5 m, and X is the 0.1 mm amplitude of the main deck vertical vibration. In terms of arbitrary hull girder vibration frequency ω,

$$M_e = 150\, \omega^2 \text{ N-m} \quad (150)$$

The final remaining element of the Fig. 39 equivalent 1-mass model, the damping factor ζ, is estimated using the measured 0.75 mm house top vibration amplitude. With $|\Theta|$ being the amplitude of the equivalent vibratory rocking rotation of the house, the fore-and-aft amplitude of the house top is approximated as,

$$U = |\Theta|\, h$$

where h is the 15m house height above main deck. Substituting the response formula for the Fig. 39 model, Equation (93),

$$U = \frac{h\, M_e / K}{\sqrt{[1 - (\omega/\omega_n)^2]^2 + (2\,\zeta\, \omega/\omega_n)^2}} \quad (151)$$

But at resonance, $\omega = \omega_n$, so that,

$$U = \frac{h\, M_e / K}{2\,\zeta} \quad (152)$$

or

$$\zeta = \frac{h\, M_e}{2\, K\, U} \quad (153)$$

For $\omega = \omega_n = 2\pi fe/60 = 52.4$ rad per sec in (150),

$$M_e = 4.12 \times 10^5 \text{ N-m}$$

The damping factor is then, from Equation (153),

$$\zeta = \frac{15\,(4.12 \times 10^5)}{2\,(0.822 \times 10^{11})\,(0.75 \times 10^{-3})} \quad (154)$$
$$= 0.05$$

With the calibrated model so established as an equivalent one degree of freedom system, with constants, J, K, ζ, and M_e the above formula can be reused to evaluate changes in the house-top vibratory displacement, U, resulting from selected changes in the array of design variables included in the simple formulation.

(d) Structural Modifications. To demonstrate this procedure, assume that stiffening in the form of the added parallel pillars of the Sub-section 3.4 and Fig. 22 example is contemplated. Following that example, the torsional stiffness of the underdeck supporting structure is raised from the above value of 1.35×10^{11} N-m per rad to 1.475×10^{11} N-m per rad by the pillar addition. Re-substituting into Equations (147) and (148), the increased combined stiffness of $K = 0.866 \times 10^{11}$ N-m per rad results in a 2.6 percent increase in natural frequency from the measured value of 500 cpm to 513 cpm.

Continuing with the scenario, assume that the full power RPM of the vessel is 105, which corresponds to a full power blade-rate exciting frequency of 525 cpm; the critical has therefore been raised only to a higher level in the operating range, i.e., it has been raised from 100 to 102.6 RPM.

At 102.6 RPM the 0.1 mm vertical hull girder vibration measured at 100 RPM would be increased by at least the frequency increase squared. This is assuming a flat frequency response characteristic of the hull girder (not close to a hull girder critical) as well as a non-cavitating propeller. Assuming a frequency squared increase, the vertical hull girder vibration amplitude becomes,

$$X = 0.1\ (102.6/100)^2 = 0.105 \text{ mm}$$

with the exciting moment from (149) increasing to

$$M_e = 4.56 \times 10^5 \text{ N-m}$$

at the new resonant frequency $\omega = \omega_n = (2\pi)\ (513/60) = 53.72$ rad per sec. The house top fore-and-aft vibratory displacement amplitude resulting from the foundation stiffening is changed, from Equation (152), to,

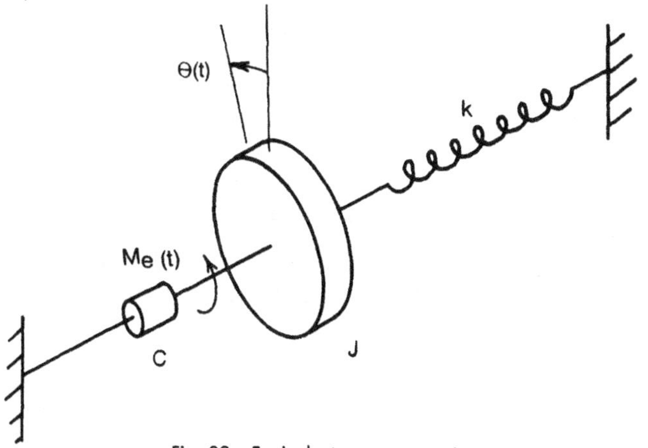

Fig. 39 Equivalent one-mass system

$$U = \frac{15\ (4.56 \times 10^5)/0.866 \times 10^{11}}{2\ (0.05)}$$

$$= 0.79 \text{ mm}$$

This is an increase in vibration of 5 percent over the original 0.75 mm level! The inadequate stiffening has simply raised the critical to a higher point in the operating range where the excitation is more intense. Some care is obviously required here in order to achieve a satisfactory result.

It would be intelligent at this point to evaluate the stiffness increase required in order to achieve a satisfactory vibration level. It is necessary to move the critical above the full power RPM of 105. This establishes the exciting frequency at the full power RPM,

$$f = 525 \text{ cpm} = 8.75 \text{ Hz}$$

$$\omega = 8.75\ (2\pi) = 55 \text{ rad per sec}$$

On consulting Fig. 35, a limiting house-top fore-and-aft vibratory velocity amplitude of 5 mm per sec is selected at this frequency. This corresponds to a house-top displacement amplitude,

$$U = 5/\omega = .09 \text{ mm}$$

The exciting moment, for use in Equation (151), is now, continuing to assume a frequency squared variation in the hull vertical displacement amplitude, from Equation (149),

$$M_e = (55)^5\ (3 \times 10^5)\ (5)\ (.1)\ (105/100)^2/1000$$

$$= 5.0 \times 10^5 \text{ N-m}$$

From (151), for $2\zeta\omega/\omega_n << [\,1 - (\omega/\omega_n)^2]$ for $\omega/\omega_n < 1$,

$$U = \frac{hMe/K}{1 - (\omega/\omega_n)^2} \tag{155}$$

Then with $\omega_n = \sqrt{K/J}$,

$$K = hMe/U + J\omega^2 \tag{156}$$

Substituting the values, the required combined stiffness is,

$$K = 1.74 \times 10^{11} \text{ N-m per rad}$$

This requires more than doubling the as-built combined effective stiffness of 0.882×10^{11} N-m per rad, Equation (147).

Little can normally be done to change the house stiffness; functional requirements of the house usually will not permit the modifications necessary to accomplish any significant increases in house casing section moment of inertia and shear area. Assume that stiffening of the under-deck supporting structure is the only possibly effective structural modification that can be accommodated. The required K_f is, from Equation (147),

$$K_f = \frac{1}{1/K - k_H \bar{r}^2} = 1.02 \times 10^{12} \text{ N-m per rad}$$

Therefore, meeting the vibration limit of 5 mm/sec at the house-top will require increasing the torsional stiffness of the underdeck supporting structure by a factor of,

$$\frac{1.02 \times 10^{13}}{1.35 \times 10^{11}} = 7.5$$

and this would be impossible in any real case. For example, if the two parallel pillars of the sub-section 3.4 example were doubled in number from 2 to 4 and moved 3 meters aft to line up under the house after bulkhead, rather than under the house sides (Fig. 22), K_f would be increased to only,

$$K_f = 1.35 \times 10^{11} + 2(5 \times 10^8)(8)^2$$
$$= 1.99 \times 10^{11} \text{ N-m per rad}$$

which is still a factor of more than 5 below the required value.

At this point, the virtual impossibility of rectifying the problem through structural modifications should be clear, and attention would be turned to ordering a new propeller.

(e) A Propeller Change. Considering an alternative 4-bladed propeller, the critical would be shifted to,

$$100(5/4) = 125 \text{ RPM}$$

which is well beyond the operating range. With the foundation unchanged, the house-top vibration at the full power RPM of 105 would be, from Equation (155),

$$U = \frac{15(5 \times 10^5)/0.822 \times 10^{11}}{1 - (105/125)^2} \times 1000$$
$$= 0.129 \text{ mm}$$

which assumes an unchanged propeller excitation level. The new house-top displacement corresponds to a velocity amplitude of,

$$0.129(55) = 7 \text{ mm/sec}$$

This would probably be acceptable, on the basis of Fig. 35.

Another possibility for the propeller would be to change to 6 blades and lower the critical well below full power to,

$$100(5/6) = 83 \text{ RPM}$$

At full power in this case,

$$U = \frac{15(5 \times 10^5)/0.822 \times 10^{11}}{|1 - (105/83)^2|} \times 1000$$
$$= 0.15 \text{ mm}$$

and the velocity is,

$$0.15(55) = 8.25 \text{ mm/sec}$$

This might also be acceptable, according to Fig. 35. The possible disadvantage to 6 blades is the resonance at 83 RPM. At 83 RPM, the exciting moment, Equation (149), should be down by at least frequency squared (which ignores any reduction at all in the hull girder vibration level):

$$M_e = 5 \times 10^5(83/105)^2 = 3.1 \times 10^5 \text{ N-m}$$

so that the resonant amplitude should be, at most, from Equation (152),

$$U = \frac{15(3.1 \times 10^5)/0.822 \times 10^{11}}{2(0.05)} \times 1000$$
$$= 0.565 \text{ mm}$$

For

$$f = 6(83)/60 = 8.3 \text{ Hz and}$$
$$\omega = 2\pi f = 52 \text{ rad per sec}$$

the vibratory velocity amplitude would be,

$$0.565(52) = 29.4 \text{ mm per sec}$$

While this level is excessive (Fig. 35) it would not necessarily disqualify a 6-bladed propeller, as continuous operation at any particular lower RPM is not usually critical, and 83 RPM could be simply avoided except in passing.

References

Baier, L.A. and Ormondroyd, J. (1952), "Vibration at the Stern of Single Screw Vessels," SNAME Spring Meeting, New Orleans, LA., May.

Boswell, R.J., Kim, K.-H., Jessup, S.D., and Lin, G.-F. (1983), "Practical Methods for Predicting Periodic Propeller Loads," DTNSRDC Report 83/090, October.

Bourceau, G., and Volcy, G.C. (1970), "Forced Vibration Resonators and Free Vibration of the Hull," *Bulletin Technique du Bureau Veritas*, April-July.

Breslin, J.P. (1959), "A Theory for the Vibratory Effects Produced by a Propeller on a Large Plate," *Journal of Ship Research*, December.

Breslin, J.P., and Kowalski, T. (1964), "Experimental Study of Propeller-Induced Vibratory Pressures on Simple Surfaces and Correlation with Theoretical Predictions," *Journal of Ship Research*, Vol. 8, No. 3, December.

Breslin, J.P. (1970), "Theoretical and Experimental Techniques for Practical Estimation of Propeller-Induced Vibratory Forces," SNAME *Transactions*, Vol. 78.

Bureau Veritas (1979), "Recommendations Designed to Limit the Effects of Vibrations on Board Ships," Bureau Veritas *Guidance Note* N1 138-A-RD3, June.

Burrill, L.C. (1934-35), "Ship Vibration; Simple Methods of Estimating Critical Frequencies," *Transactions*, NECI, Vol. 51.

Cox, G.G., and Morgan, W.B. (1972), "The Use of Theory in Propeller Design," SNAME, Chesapeake Section, February.

Cumming, R.A., Morgan, W.B., and Boswell, R.J. (1972), "Highly Skewed Propellers," SNAME *Transactions*, Vol. 80.

DenHartog, J.P., *Mechanical Vibrations* (1956), McGraw-Hill, Fourth Edition.

Denny, S.B. (1967), "Comparisons of Experimentally Determined and Theoretically Predicted Pressures in the Vicinity of a Marine Propeller," NSRDC Report 2349, May.

Dyne, G. (1974), "A Study on the Scale Effect on Wake, Propeller Cavitation, and Vibratory Pressure at Hull of Two Tanker Models," SNAME *Transactions*, Vol. 82.

Hadler, J.B., and Cheng, H.M. (1965), "Analysis of Experimental Wake Data in Way of Propeller Plane of Single and Twin Screw Ship Models," SNAME *Transactions*, Vol. 73.

Hammer, N.O., and McGinn, R.F. (1978), "Highly Skewed Propellers—Full Scale Vibration Test Results and Economic Considerations," Ship Vibration Symposium '78, Washington, D.C., October.

Harrington, R.L. (1971), *Marine Engineering*, SNAME.

Hirowatari, T., and Matsumoto, K. (1969), "On the Fore-and-Aft Vibration of Superstructure Located at Aftship (Second Report)," JSNA *Transactions*, Volume 125, June.

Holden, K.O., Fagerjold, O., and Ragnar, F. (1980), "Early Design-Stage Approach to Reducing Hull Surface Forces Due to Propeller Cavitation," SNAME, *Transactions*, Vol. 88.

Huang, T. and Groves, N. (1981), "Effective Wake: Theory and Experiment," DTNSRDC Report-81/033, April.

Hylarides, S. (1978), "Some Hydrodynamic Considerations of Propeller-Induced Ship Vibrations," Ship Vibration Symposium '78, Washington, D.C., October.

Jacobs, W., Mercier, J., and Tsakonas, S. (1972), "Theory and Measurements of the Propeller-Induced Vibratory Pressure Field," *Journal of Ship Research*, Vol. 16, No. 2, June.

Johannessen, H., and Skaar, K.T. (1980), "Guidelines for Prevention of Excessive Ship Vibration," SNAME *Transactions*, Vol. 88.

Kagawa, K., "A Study on Higher Mode Vibration of Ships (1st Report)" Journal of JSNA, Vol. 143, June, 1978.

Kennard, E.H. (1955), "Forced Vibrations of Beams and the Effect of Sprung Masses," DTMB Report 955, July.

Krylov, A.N. (1936), *Vibration of Ships* (in Russian).

Kumai, T. (1968), "On the Estimation of Natural Frequencies of Vertical Vibration of Ships," *Report of Research Institute for Applied Mechanics*, Vol. 16, No. 54.

Lee, C.-S. (1979), "Prediction of Steady and Unsteady Performance of Marine Propellers With or Without Cavitation by Numerical Lifting Surface Theory," Ph.D. Thesis, MIT, Dept. of Ocean Engineering.

Lewis, F.M. (1929), "The Inertia of Water Surrounding a Vibrating Ship," SNAME *Transactions*, Vol. 37.

Lewis, F.M. (1963), "Propeller-Vibration Forces," SNAME *Transactions*, Vol. 71.

Lewis, F.M. (1973), "Propeller Excited Hull and Rudder Force Measurements," Department of Ocean Engineering, MIT Report 73-10.

McCarthy, J.H. (1961), "On the Calculation of Thrust and Torque Fluctuations of Propellers in Non-Uniform Wake Flow," DTMB Report 1533.

McGoldrick, R.T., and Russo, V.L. (1955), "Hull Vibration Investigation on SS *Gopher Mariner*," SNAME *Transactions*, Vol. 63.

Military Standard (1969), "Mechanical Vibrations of Shipboard Equipment," MIL-STD-167B (Ships).

Noonan, E.F., (1981), "An Approach to the Limitation and Control of Shipboard Vibration," *The Shock and Vibration Bulletin*, Part 1, Naval Research Lab., Washington, D.C., May.

Pien, P.C. (1958), "Five Hole Spherical Pitot Tube," DTMB Report 1229, May.

Pollard, T., and Dudebout, A. (1894), *Theorie du Navire*, Vol. IV, Chapter LXIX, Paris.

Reed, F.E., "The Design of Ships to Avoid Propeller Excited Vibrations," SNAME *Transactions*, Vol. 79, 1971.

Rutherford, R. (1978-79), "Aft End Shaping to Limit Vibration," North East Coast Institution of Engineers and Shipbuilders *Transactions*, Vol. 95.

Sandström, R.E., and Smith, N.P. (1979), "Eigenvalue Analysis as an Approach to the Prediction of Global Vibration of Deckhouse Structures," SNAME Hampton Roads Section Meeting, October.

Sasajima, H. and Tanaka, I. (1966), "On Estimation of Wake of Ships," *Proceedings* of 11th ITTC, Tokyo.

Schlick, O. (1894-1911), Series of Articles on Ship Vibration in *Transactions*, INA (1884, 1893, 1894, 1901, 1911).

Sellers, M.L., and Kline, R.G. (1967), "Some Aspects of Ship Stiffness," *Transactions*, SNAME.

SNAME (1972), "Longitudinal Stiffness of Main Thrust Bearing Foundations," T&R Report R-15.

SNAME (1975), "Code for Shipboard Hull Vibration Measurements," T&R Code C-1, January.

SNAME (1980), "Ship Vibration and Noise Guidelines," Technical and Research Bulletin 2-25, January.

Stern, F., and Vorus, W.S. (1983), "A Nonlinear Method for Predicting Unsteady Sheet Cavitation on Marine Propellers," *Journal of Ship Research*, March.

Stuntz, G.R., Pien, P.C., Hinterthan, W.B., and Ficken, N.L. (1960), "Series 60—The Effects of Variations in Afterbody Shape Upon Resistance, Power, Wake Distribution, and Propeller Excited Vibratory Forces," SNAME, *Transactions*, Vol. 68.

Sulzer Bros. (1977), *RND.M Marine Diesel Engines, Technical Data*, Sulzer Bros. Ltd., Winterthur, Switzerland, October.

Tanibayashi, H. (1980), "Practical Approach to Unsteady Problems of Marine Propellers by Quasi-Steady Method of Calculation," Mitsubishi Technical Bulletin No. 143.

Thomson, W.T. (1973), *Theory of Vibration with Applications*, Prentice-Hall.

Todd, F.H. (1961), *Ship Hull Vibration*, Edward Arnold, Ltd., London.

Todd, F.H. (1935), "Vibration of Ships," Tekniska Samfundets Handlinger No. 5, Göteborg, Sweden.

Troost, L. (1937-1951), "Open Water Test Series With Modern Propeller Forms," Part I, *Transactions*, NECI, 1937-38, Part II, 1939-40, Part III, 1950-51.

Tsakonas, S., Breslin, J., and Miller, M. (1967), "Correlation and Application of an Unsteady Flow Theory for Propeller Forces," SNAME *Transactions*, Vol. 75.

Vorus, W.S. (1971), "An Integrated Approach to the Determination of Propeller Generated Forces Acting on a Ship Hull," Department of Naval Architecture and Marine Engineering, The University of Michigan, Report #072, March.

Vorus, W.S. (1974), "A Method for Analyzing the Propeller Induced Vibratory Forces Acting on the Surface of a Ship Stern," SNAME *Transactions*, Vol. 82.

Vorus, W.S. (1976), "Calculation of Propeller Induced Hull Forces, Force Distributions, and Pressures; Free-Surface Effects," *Journal of Ship Research*, Vol. 20, No. 2, June.

Vorus, W.S., Breslin, J.P., and Tein, Y.S. (1978), "Calculation and Comparison of Propeller Unsteady Pressure Forces on Ships," Ship Vibration Symposium '78, Washington, D.C., October.

Vorus, W.S., and Hylarides, S. (1981), "The Hydrodynamic Mass Matrix of a Vibrating Ship Based on a Distribution of Hull Surface Sources," SNAME *Transactions*, Vol. 89.

Vorus and Associates (1981), "Analysis of Propeller Cavitation and Propeller Induced Vibratory Forces and Pressures on a U.S. Naval Oiler," Second Report VA181-2, September.

von Kármán, T., and Sears, W.R. (1938), "Airfoil Theory for Non-Uniform Motion," *Journal of the Aeronautical Sciences*, Vol. 5, August.

Ward, G. (1982), "The Application of Current Vibration Technology in Routine Ship Design Work," RINA Spring Meeting *Transactions*.

Wilson, M.B. (1981), "Review of Available Criteria for Identifying the Likelihood of Excessive Propeller-Induced Excitation," DTNSRDC Report SPD-1001-01, May.

Volume II
Nomenclature

Resistance, propulsion and vibration

The following symbols apply to Volume II only. The phrase "stands for" is understood between the symbol and its definition.

A	stands for area, generally, m^2
A_D	developed area, propeller blades
A_E	expanded area, propeller blades
A_O	area of propeller disk
A_L	longitudinal projected area
A_r	coefficient of mode shape
A_P	projected area, propeller disk
A_T	transverse projected area
AR	aspect ratio
AP	after perpendicular
a	linear acceleration
B	breadth, molded, m
b	width of a channel; span of a control surface
C	damping coefficient
$C_{i\ \mathrm{etc.}}$	constants
C_A	model-ship correlation allowance
C_B	block coefficient, ship
C_D	drag coefficient
C_F	frictional resistant coefficient
C_L	lift coefficient
C_m	modal damping coefficient, mode n
C_P	prismatic coefficient
C_Q	torque coefficient
C_R	residuary resistance coefficient
C_S	Sears function
C_T	total resistance coefficient; thrust loading coefficient
C_W	wave-making resistance coefficient
C_V	viscous-drag coefficient; Froude No. based on beam
C_X	2-dimensional added mass coefficient at section x
c	chord of a control surface; distributed damping coefficient
D	depth (molded); diameter; drag force
DkW	developed power, propeller
E	Young's modulus of elasticity
EI	flexural rigidity of a beam
EAR	expanded area ratio
EkW	effective power
e	base of naperian logarithms, 2.7183
F	force, force amplitude, kN
Fn	Froude number
Fn_∇	volume Froude number
FP	forward perpendicular
FW	fresh water
$F(x)$	beam exciting force amplitude distribution
F_n	force, modal exciting, mode n
f	force, component of; coefficient of friction; frequency, Hz
$f(x,t)$	beam excitation force per unit length
G	center of gravity of ship; shear modulus of shaft; lift function, N blades
g	acceleration due to gravity; lift of one blade
h	height, depth, head of water, m
I	moment of inertia, generally
i_E	half angle of entrance on LWL
i	degree of freedom; submergence
J	advance ratio (coefficient); Lewis J-factor; mass moment of inertia
K	stiffness, generally; effective torsional spring constant; form factor
K_T	thrust coefficient
K_Q	torque coefficient
K_n	modal stiffness, mode n
k	stiffness coefficient of a section
L	length of ship; lift of a blade
L_E	length of entrance
L_Q	unsteady blade lift

L_R length of run
LCB longitudinal center of buoyancy
LCG longitudinal center of gravity
LWL load waterline
l length of a section
M mass; moment
M_n modal mass, mode n
M subscript denoting "model"
m mass per unit length; components of moment
N number of propeller blades
n revolution per unit time
O origin of coordinates
P power, generally; pitch of propeller
PR pitch ratio
P_B brake power, kW
P_D delivered power, kW
P_I indicated power, kW
P_S shaft power, kW
P_T thrust power, kW
p_A pressure, atmospheric
p_O pressure far removed from a disturbance
p_V vapor pressure of water
Q propeller torque; number of harmonics in wake; Fourier series
Q_O torque in open water
Q_V volume of water passing through a propeller disk
R radius, m; resistance generally, kN
R_A added resistance; R_{AA} in air
R_F frictional resistance
R_T total resistance
R_R residuary resistance
R_W wave-making resistance
R_V viscous resistance
Rn Reynolds number
RMS root mean square
RPM revolutions per minute
T thrust, propeller, kN
t thrust deduction fraction
U velocity of ship; housetop displacement amplitude, fore and aft
u velocity of incident flow; deckhouse vibration displacement, fore and aft
V volume, generally, m^3
V_L velocity, linear, of a ship
V_A axial velocity; speed of advance of a propeller
V_c critical velocity for laminar flow
V_K speed, knots
V_R relative velocity
V_S sustained sea speed
v components of velocity
$v(t)$ vibratory velocity
W weight, generally, kg
WL waterline
$W_n(x)$ complex displacement amplitude distribution
w deckhouse vertical vibratory displacement; Taylor wake fraction in general; weight of an item
W_F Taylor wake fraction based on frictional identity; W_T thrust identity; W_Q torque identity
X displacement amplitude, in general
x axial coordinate of hull from stern; distance from origin 0 along x-axis
$x(t)$ vibratory displacement
y distance from origin 0 along y - axis
Z ship section place; a general complex function
z function in complex plane of unit circle; distance from origin 0 along z-axis

Vibration Symbol Subscripts

n denotes modal order, number, propeller RPS
i, j component number of $\sqrt{-1}$
m harmonic number, deckhouse mass, cylinder no.
V vertical
M mean
R rocking
e natural
x axial axis (length), coordinate
y athwartship axis, coordinate
z vertical axis, coordinate
t tangiential axis, coordinate
a average
G geometric

Greek Symbols

α (alpha) represents angle of wind force off bow; divergent wave pattern angle
α_n modal phase angle, mode n
α_s angle of skew
β (Beta) advance angle of a propeller blade section; deadrise angle
β_I hydrodynamic flow angle of a propeller blade section
β_G geometric pitch angle
β_n modal phase angle of force mode n
Γ (Gamma) "circulation" in circulation theory
γ (gamma) arctan lift/drag
δ (delta) Taylor advance coefficient, $1/J$
ϵ (epsilon) lift/drag ratio
ζ_n (zeta) modal damping factor, mode n
η (eta) efficiency, generally; vertical component of deckhouse

Θ (Theta) blade position angle
θ (theta) angle of apparent wind off bow
K (Kappa) nondimensional displacement variable
Λ (Lambda) tuning factor
λ (lambda) linear scale ratio, ship to model; advance ratio; mean wetted length-beam ratio, planing craft
μ (mu) mass per unit length; coefficient of dynamic viscosity
ν (nu) Poisson's ratio; visco-elastic constant
ξ (xi) longitudinal coordinate, from stern
ρ (rho) density of a fluid, mass per unit volume
σ (sigma) stress; cavitation number
τ (tau) trim angle, planing craft
ϕ (phi) pitch angle of propeller; velocity potential
ψ (psi) mode shape (eigenfunction)
Ω (Omega) propeller, engine angular velocity
Ω_R characteristic rigid-body frequency; Ω_F flexural frequency
ω (omega) circular frequency, rad / sec
ω_n resonant frequency; n^{th} mode natural frequency

Special Symbols

℄ denotes baseline
℄ centerline
⊠ midlength, in general
∇ *vol* volume of displacement

Froude's "Circle" Notation

Ⓒ resistance coefficient, $(R_t / W) \times (1000 / Ⓚ^2) = 2938\, R_T / W^{2/3} V_K^2$
Ⓚ speed coefficient, $\sqrt{4\,/g} \times V_K / \nabla^{1/6} = 0.5834\, V_K / W^{1/6}$
Ⓢ Wetted surface coefficient, $S / \nabla^{2/3}$

In the foregoing R_T is in long tons; ∇ in long tons; and V_K in knots.

Vibration Symbols

$[B]$ beam coefficient matrix
$|C|$ beam solution constant vector
$[D]$ dynamic matrix of discrete model
$\dot{\forall}$ cavitation volume velocity
$\bar{r}$ radius of gyration
|Θ| vibratory amplitude

Mathematical Symbols

∂ is a partial derivative sign
i is $\sqrt{-1}$
$\approx$ approximately equal to
$<$ less than
$>$ greater than
$\dot{x}$ (one dot over a variable) is the first derivative of the variable
$\ddot{x}$ (two dots over a variable) is the second derivative of the variable
$\propto$ proportional to
∞ infinity
Δ (Delta) a finite increment
Σ (Sigma) summation of
π (pi) ratio of circumference of circle to diameter
$\int$ integral of
f some function of
$\rightarrow$ approaches as a limit
$\equiv$ is identical to
$=$ equal to
$\neq$ not equal to
$\geq$ equal to or greater than
$\leq$ equal to or less than
$\oint$ integration around a closed curve

Acronyms Used in References

ABS American Bureau of Shipping
ATMA Association Maritime Technique et Aeronautique, Paris
ASME American Society of Mechanical Engineers
ASNE American Society of Naval Engineers
ASCE American Society of Civil Engineers
ATTC American Towing Tank Conference
BMT British Maritime Technology (formerly BSRA)
DTNSRDC David Taylor Naval Ship Research and Development Center (formerly EMB DTMB) (now DTRC, David Taylor Research Center)
IESS Institute of Engineers and Shipbuilders in Scotland
IME Institute of Marine Engineers
IMO International Maritime Organization (formerly IMCO)
ISSC International Ship Structures Congress
ITTC International Towing Tank Conference
JSR SNAME *Journal of Ship Research*
JSTG *Jahrbuch des Schiffbautechnischen Gesellschaft*
MARIN Maritime Research Institute Netherlands (formerly NSMB)

NASA	U.S. Space Administration, (formerly NACA)
NAVSEA	U.S. Naval Ship Systems Command
NECI	Northeast Coast Institute of Engineers and Shipbuilders
NMI	National Maritime Institute, (formerly NPL), London
ONR	Office of Naval Research, U.S. Navy Dept.
RINA	Royal Institute of Naval Architects, formerly (INA)
RSE	Royal Society of Edinburgh
SSC	Ship Structures Committee
SNAME	Society of Naval Architects and Marine Engineers
SSPA	Swedish State Shipbuilding Experimental Tank
WEBB	Webb Institute of Naval Architecture

International System of Units (Systeme International d'Unites, or SI) Useful quantities for Naval Architecture

Quantity	SI Unit	Definition	Conversions: English to SI	Conversions: SI to English
<u>Base Units</u>				
Length	meter, m		1 ft = 0.305m	1m = 3.28 ft
Mass	kilogram, kg		1 lb = 0.454 kg	1 kg = 2.20 lb
Time	second, s			
<u>Supplementary Units</u>				
Angle, plane	radian, rad	1 rad = 180°/tt		
Density of solids		kg/cm^3 or t/m^3		
of liquids		kg/L		
Distance	nautical mile, knot	1.852 km		1 knot = 6,080 ft
Force	newton, N	1 $kg\text{-}m/s^2$	1 lb (force) = 4.45N	1 N = 0.225 lb
	kilonewton, kN	103 $kg\text{-}m/s^2$		
Frequency	hertz, Hz	cycle/sec, cps		
Mass	metric ton, t	10^3 kg	1 long ton (weight) = 1.016 t	lt = 0.98 long tons
Power	watt, W	1 N-m/s		
	kilowatt, kW	1 kN-m/s	1 hp = 0.746 kW	1kW = 1.34 hp
Pressure	kilopascal, kPa	10^3 N/m^2	1 lb/in^2 = 6.895 kPa	1kPa = 0.15 lb/in^2
Specific vol.	1/density	m^3/t		
Stress	megapascal, MPa	$MN/m^2 = N/mm^2$	1 long ton/in^2 = 15.44 MPa	1MPa = 0.065 long $tons/in^2$
Volume of solids	m^3			
of liquids	liter, L	$10^{-3}m^3 = d^3$		
Velocity	meters/sec	m/s	1ft/sec = 0.305 m/s	1m/s = 3.28 ft/sec
	knot	1 nmi/hr = 1.852 km/kr.		

Index

This page intentionally left blank